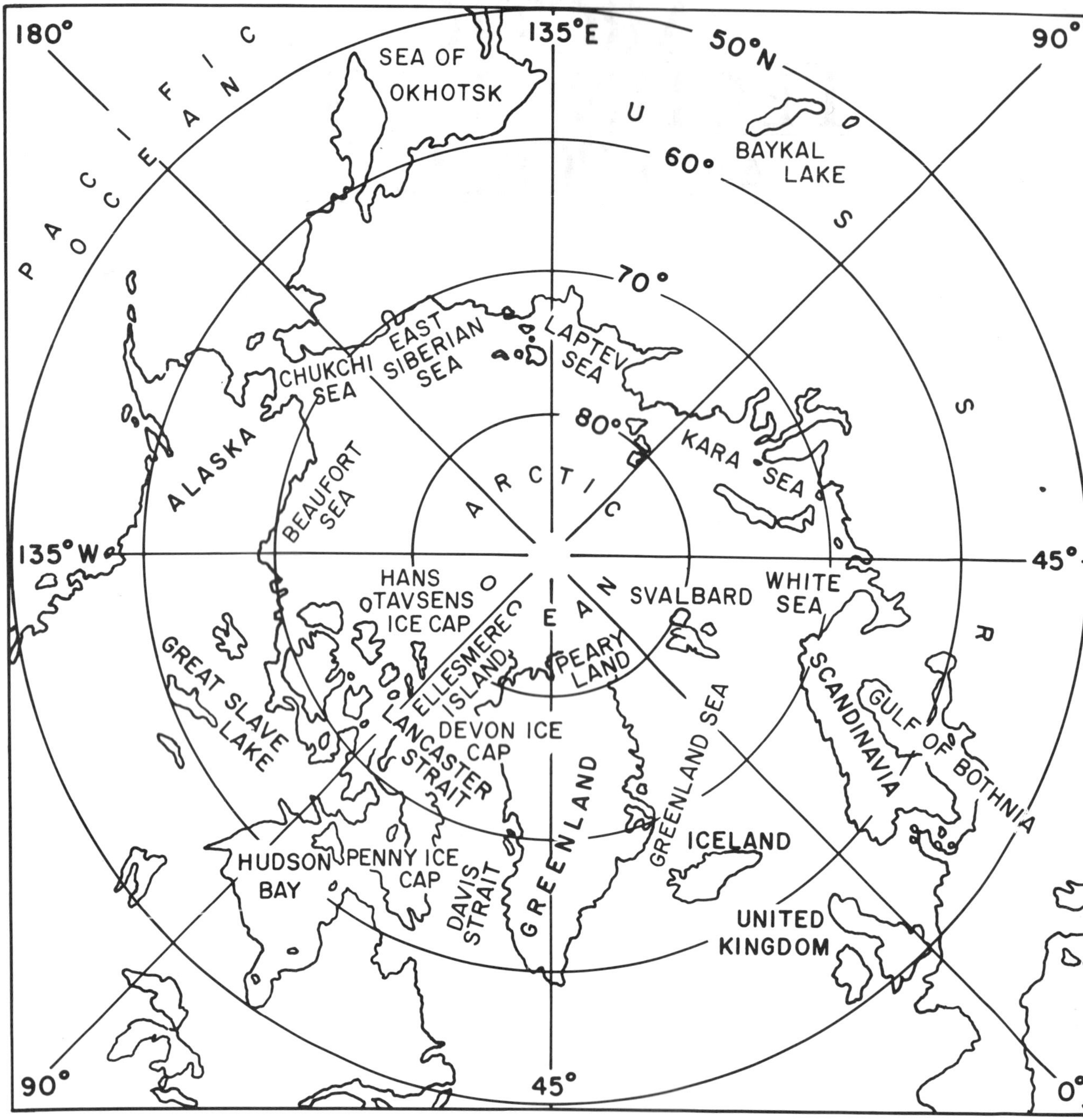

180°
135°E
50°N
90°
PACIFIC OCEAN
SEA OF OKHOTSK
U.S.S.R.
60°
BAYKAL LAKE
70°
EAST SIBERIAN SEA
LAPTEV SEA
CHUKCHI SEA
80°
KARA SEA
ALASKA
BEAUFORT SEA
ARCTIC OCEAN
135°W
45°
HANS TAVSENS ICE CAP
SVALBARD
WHITE SEA
ELLESMERE ISLAND
PEARY LAND
GREAT SLAVE LAKE
LANCASTER STRAIT
DEVON ICE CAP
GREENLAND SEA
SCANDINAVIA
GULF OF BOTHNIA
GREENLAND
ICELAND
HUDSON BAY
PENNY ICE CAP
DAVIS STRAIT
UNITED KINGDOM
90°
45°
0°

ARCTIC TECHNOLOGY AND POLICY

Proceedings of the

Second Annual
MIT Sea Grant College Program
Lecture and Seminar

and the

Third Annual
Robert Bruce Wallace Lecture

Edited by
Ira Dyer
and
Chryssostomos Chryssostomidis
both of
Massachusetts Institute of Technology

HEMISPHERE PUBLISHING CORPORATION
Washington New York London

DISTRIBUTION OUTSIDE THE UNITED STATES
McGRAW-HILL INTERNATIONAL BOOK COMPANY
Auckland Bogotá Guatemala Hamburg Johannesburg Lisbon
London Madrid Mexico Montreal New Delhi Panama Paris
San Juan São Paulo Singapore Sydney Tokyo Toronto

Proceedings of the Second Annual MIT Sea Grant College Program Lecture and Seminar Series and the Third Annual Robert Bruce Wallace Lecture, of the MIT Department of Ocean Engineering held at Massachusetts Institute of Technology, Cambridge, Massachusetts U.S.A., March 2–4, 1983; organized by the MIT Sea Grant College Program and the MIT Department of Ocean Engineering. MIT Sea Grant College Program Report No. MITSG 83-1.

Secretariat:
Elizabeth T. Harding
MIT Sea Grant College Program
Communications Information Office
Massachusetts Institute of Technology
Cambridge, Massachusetts 02139, U.S.A.

Sponsors:

Det norske Veritas
Houston, Texas

MIT Sea Grant College Program
Cambridge, Massachusetts

MIT Department of Ocean Engineering
Cambridge, Massachusetts

National Science Foundation
Division of Polar Programs
Washington, DC

ARCTIC TECHNOLOGY AND POLICY

2 3 4 5 6 7 8 9 0 B C B C 8 9 8 7 6

Library of Congress Cataloging in Publication Data
Main entry under title:

Arctic technology and policy.

(Proceedings of the second annual MIT Sea Grant College Program lecture and seminar series) (Third annual Robert Bruce Wallace lecture)
"The conference was organized by the MIT Sea Grant College Program and the MIT Department of Ocean Engineering"—T.p. verso.
Bibliography: p.
Includes index.
1. Engineering—Arctic regions—Congresses. 2. Arctic regions—Congresses. I. Dyer, Ira. II. Chryssostomidis, Chryssostomos. III. Massachusetts Institute of Technology. Sea Grant College Program. IV. Massachusetts Institute of Technology. Dept. of Ocean Engineering: VI. Series: Proceedings of the MIT Sea Grant College Program lecture and seminar series; 2nd. VI. Series: Robert Bruce Wallace lecture; 3rd.
TA125.A73 1983 551.46′8 83-18403
ISBN 0-89116-361-1

CONTENTS

PREFACE

A glance at a globe indicates the strategic importance of the Arctic Ocean. Adjacent nation states are Canada, Greenland, Norway, the United States, and the Soviet Union. When one includes other states in the upper half of the northern hemisphere, each within easy reach via sea and air routes, one readily concludes that the Arctic is a central element of the world's industrial and military might.

Quite apart from national aspirations, our interest in the Arctic concerns the bounty of its resources. For centuries the Arctic has provided food from its mammalian and fish life and, with appropriate management, could continue to do so forever. Recently, petroleum has been discovered and its exploration begun; most experts believe that the Arctic's resources will one day rival those of the Persian Gulf.

It is no surprise, therefore, that in recent years nations bordering the Arctic have intensified their efforts to resolve national and international policy questions, to develop new technologies, and to understand the characteristics of the ice and other elements of the environment that make working in the region costly and dangerous.

The Second Annual Lecture and Seminar Series, organized by the MIT Sea Grant College Program and the MIT Department of Ocean Engineering, brought together in March 1983 many experts from fields that will be essential to safe and wise use of the Arctic. More than 200 people representing governmental, academic, and industrial institutions from the United States, Canada, and abroad discussed the political, scientific, and technological facets of Arctic development. Two principal questions provided a focus for participants: What knowledge do we have today for working in the Arctic? What are the challenges in science, engineering, and policymaking in the decade ahead?

To answer these questions, the conference was organized into three main sections: (1) *Arctic Policy* looked at the issues of international law, strategic concern, state vs. federal jurisdiction, native rights, and regulatory processes. (2) *Arctic Technology* covered advances in design and construction techniques and materials for offshore platforms and ice transiting ships. (3) *Arctic Science and Engineering* aimed at understanding the geology, geophysics, and oceanography of the Arctic, and the characteristics and properties of the surface ice which dominates the region.

To introduce and put into context the 15 papers presented within these major subject areas, the annual Sea Grant College Program Lecture and the third annual Robert Bruce Wallace Lecture were held as a single event.

The MIT Sea Grant Lecture was established in 1972 with support from the Henry L. and Grace Doherty Foundation to bring issues that affect the development and management of ocean and coastal resources before the public and the marine community. The Robert Bruce Wallace Lecture began in 1980 with support from the Aldemere Foundation to focus on advanced ideas in ocean engineering and give faculty and students an opportunity to meet with eminent figures in the marine field to discuss the potential and problems of technology development for ocean uses.

This year's lectures and the seminars offered participants an excellent opportunity to confer and to build future networks of communication which should further understanding of the Arctic.

Planning the lectures and seminars that made up the Arctic Technology and Policy conference has been a joint effort involving the faculty and staff of the Department of Ocean Engineering, the staff of the Sea Grant College Program at MIT, practicing professionals in the offshore industry, and the Arctic scientific community. We would also like to thank Christine Simonsen, Marilyn Staruch, and Marge Chryssostomidis for their special help. We are indebted to the authors for their outstanding contributions to the proceedings, to Susan Stolz for preparing the index, to the Sea Grant and Ocean Engineering staffs for coordinating the meeting, and to the Division of Polar Programs, National Science Foundation, Washington, D.C., and Det norske Veritas, Houston, Texas, for their partial support of the conference.

Ira Dyer
Chryssostomos Chryssostomidis

MIT SEA GRANT
COLLEGE PROGRAM
LECTURE

ARCTIC POLICY: OPPORTUNITIES AND PERSPECTIVES

R. Tucker Scully

Director, Office of Oceans and Polar Affairs
U.S. Department of State

The polar regions have a fascination for those whose professional or personal experience involves significant contact with them. These beautiful and seemingly remote areas have a powerful attraction. At least this has been the case in my own professional experience--both with regard to the Arctic and Antarctic--though to date the latter has occupied more of my attention.

With all this attraction, however, the Arctic has an elusive quality from the policy perspective. This is not to question the concept of the Arctic, the existence of the Arctic as a region. To paraphrase Justice Potter Stewart in his dissenting opinion in a 1964 Supreme Court case on pornography, "I shall not today attempt to define it ... but I know it when I see it." In many disciplines--in oceanography, meteorology, geology and biology, for example--there is a clear sense of what is meant when one speaks about the Arctic. Likewise, the sociologist, the engineer, and the geographer all have a concrete idea of what is meant by the Arctic. Indeed, if one looks at the distinguished list of participants in this seminar and the panels into which they are organized, it is evident that the Arctic has clear definition from the perspective of many disciplines.

Nonetheless, the picture of the Arctic for the policy-maker can be a cloudy one. Efforts to treat the Arctic as an integrated whole for the purpose of governmental activity--including the definition of objectives and the means to achieve them--have encountered significant difficulties, certainly here in the United States. These difficulties have raised the issue of whether the Arctic is a suitable subject of integrated policy or policies. Put another way, the issue has been raised as to whether the U.S. should have an overall Arctic policy, or perhaps as an overall Arctic science policy. The answer to such questions is not easy. For example, one could legitimately take the view that Arctic matters should be treated on a discipline-by-discipline approach, in which objectives are not defined on a region-wide basis, but on a functional basis. On the international level, Arctic matters could be considered simply as elements in the bilateral relations among the states bordering the Arctic Ocean.

It is this sort of question that most particularly faces the policy-maker in approaching the Arctic. Seeking to provide preliminary answers to this sort of question is the topic of this address. In so doing, I would like to start with geography--or perhaps political geography.

There appear to be a number of functionally adequate definitions of the Arctic for specific purposes. It is far more difficult to come up with a definition which is suitable for the variety of disciplines which must be integrated in a policy sense. When we use the term "Arctic" in a political as well as a physical sense, what are we talking about? What are its boundaries? It is perhaps easiest to take the second question first and attempt to draw a line around the Arctic which will usefully delimit it for policy purposes.

A basic geographic feature of the Arctic, however defined, is that its nucleus is an ocean basin. The Arctic Ocean proper and adjacent seas form a distinct marine area surrounded by islands and continental land masses. (By contrast, Antarctica is a continent surrounded by marine areas.) The primary question in delineating the Arctic as a region lies in how to treat the surrounding islands and land masses. Portions of the land areas adjacent to the Arctic Ocean form an integral part of the Arctic from the point of view of almost any discipline. But how much, and how much viewed from a policy perspective?

The Arctic Circle does not appear to be a useful boundary because it slices arbitrarily through national territories at 66°30' North, and has no close fit with Arctic topography and conditions. The 10°C July isotherm (line bounding the area which has an average temperature of less than 10°C for July) is the boundary for the Arctic preferred by many scientists but, again, arbitrarily cuts across political jurisdictions. Somewhat more useful is the line marking the average southern limit of ice coverage at sea and permafrost on land--although strictly speaking this would include large areas in the interior of China.

As a matter of fact, there is no single physical phenomenon, or combination of phenomena, that satisfactorily bounds the region of our policy interest, and we are thrown back on rather arbitrary limits that reflect the current human use of the territory as well. I would chose a definition of the Arctic designed to create a balance between patterns of human activity and physical characteristics associated with Arctic conditions. This would result in a boundary that varied between 60° and 65° North Latitude.

On this basis and starting with the United States, I would include Alaska, except for the panhandle. Within this definition would also be included the Aleutian Islands and the Bering Sea. In Canada, I would include in the Arctic policy realm the Yukon Territory, all of the Northwest Territories, the northern half of Quebec, and all of Labrador. Further east, the Arctic, for policy purposes, would take in all of Greenland, Iceland, and Svalbard; the northern fringes of Norway, Sweden, Finland and European Russia; and large areas of northern and eastern Siberia.

Having set out a rough geographic framework, it remains to identify, or circumstances which define or affect, the United States view of the Arctic as a policy region. A word first of all on history--at least U.S. history. From this angle, the first fact that strikes one is that the U.S. is a late-comer to the Arctic, having become an Arctic state in the mid-nineteenth century and almost by accident. "Seward's Folly", the purchase of Alaska in 1867, resulted from a series of fortuitous circumstances. The territorial aspirations of Secretary of State Seward were not widely shared by his colleagues at the time and only an unusual set of political and economic circumstances facing the Russian Czar allowed Seward to bring this off.

There were significant economic interests in this country which strongly supported the purchase of Alaska, and, of course, the U.S. has never had cause to regret Seward's initiative. However, though the first explorer to reach the North Pole was an American--Robert Peary--the exploration and development of Arctic territory has played a more minor part in the history of the U.S. than in the development of other countries bordering the Arctic Ocean, most particularly Canada and the Soviet Union. The Arctic has played little role in forging of a national identity in U.S. experience. By contrast it has played a much larger role in the history of the other states included in our geographic region. In fact, it has been comparatively recently that the U.S. has begun to consider itself an Arctic state.

The most significant catalyst for U.S. awareness that it is an Arctic State was provided by the Second World War. On the one hand, scientific and technological innovation stimulated by the war effort accelerated trends in transportation and communication which have laid the groundwork for a revolution in human activity in the Arctic. Radio communications and the airplane, for example, provided for regular and close contact between Arctic areas and their more populous and developed hinterlands. Equally important, the war--and particularly the rapid development of aviation it spawned--demonstrated or perhaps created the strategic significance of the Arctic. This significance lies in the fact that the shortest distances between the major population centers of the Northern Hemisphere lie along great circle routes that traverse the Arctic.

The emphasis upon the strategic and military aspects of the Arctic intensified in the years following the Second World War, with the emergence of the cold war, with its periods of competition and tension between East and West. North America, Western Europe and the Soviet Union all face each other across the Arctic Ocean. And in the era of long-range aircraft and ICBM's, the shortest path of attack and counter-attack between East and West would lie across the Arctic. In fact, one of the most important post-war U.S. activities in the Arctic has been the development of early-warning radar systems in the high Arctic in response to cold war tensions.

The military importance of the Arctic has contributed to the delineation of the region and remains a primary factor in U.S. thinking about the Arctic. Similarily, the strategic importance of the Arctic has influenced the views of other Arctic states and their perspectives on jurisdiction in the Arctic. Soviet and Canadian jurists have elaborated and supported the sector theory according to which states bordering the Arctic Ocean would claim jurisdiction over ice-covered waters north of their land territories all the way to the North Pole. Strategic concerns certainly are an important element in Soviet perspectives on the Northern Sea Route, Canadian concerns regarding the Northwest Passage, as well as attention to Svalbard and the 1920 Treaty on the Status of Spitzbergen.

There is a panel devoted to the subject of jurisdiction in the Arctic which will delve into these and other Arctic jurisdictional issues in detail. From this more general perspective it simply is worth noting that the military/strategic component in perceptions of the Arctic by those bordering states have made them jealous in their approach to jurisdiction in the Arctic, in efforts to secure control over water and air space adjacent to their territory, and in seeking freedom of movement beyond their own areas of jurisdiction. There is some irony in the fact that one of the primary motives for viewing the Arctic as a region from the policy perspective derives from the centrifugal forces generated by military competition and rivalry.

At the same time there are other factors which have emerged, particularly in recent decades, which also contribute to the concept of the Arctic as a region of policy concern. The first of these relates to resource activities in the Arctic.

The lure of resources was one of the primary inspirations to outside exploration and settlement of Arctic areas. Gold, furs, and whales were among the first objectives to attract peoples from the mid-latitudes. Resource activities in the Arctic, therefore, are not new. What is new are the incentives and capacity to undertake large industrial-scale activities in the Arctic. In particular, the search for new energy resources has affected Arctic areas. Large hydrocarbon deposits--oil and gas--have been located in both the Soviet and North American Arctic. There has been a virtual explosion in the technology necessary for operating in these frontier areas in the past decade. Soviet Arctic gas fields are in production, as is the U.S. oil field at Prudhoe Bay. Major investments by Canadian interests have been made both onshore and offshore in the Arctic, and there is the prospect of hydrocarbon activities further and further offshore.

While the specific energy shortages which followed the oil embargo in the mid-70's do not at present exist, the long-term trend toward major hydrocarbon resource development in the Arctic is not likely to reverse itself. Present and anticipated resource activities, and their trans-boundary implications, are changing the face of the Arctic. They constitute a second of the major elements which bring the Arctic into focus as a policy region.

Increased human activity in the Arctic generally, and that attendant upon resource development, have served as catalysts for concern about the environment of the Arctic and the possibility of environmental disturbance in the Arctic extending beyond it to other areas of the globe. There is perhaps a tendency to overstate the sensitivity of the Arctic environment to outside disturbance. Terms such as "unique" and "fragile" are often applied to the ecology of polar regions. At the same time, the conditions under which life exists in the Arctic are extreme and adaptations made by species of a specialized and sensitive character. The margins of survival for plants and animals alike are thin. In addition, the importance of polar regions to world climate has become increasingly clear in recent decades. In the same way that the pattern of circulation within the Arctic Ocean links the coastal areas which border that ocean, the pattern of meteorological circulation links the Arctic with the mid-latitudes. Again, one should resist over-dramatizing such connections but there are legitimate questions about the effects of activities which could modify weather patterns in the Arctic both within the region and beyond.

From the policy perspective, the significance of environmental considerations is not simply the fact that increased human activity can have significant impacts upon the environment. It derives equally from the fact that these impacts are not bounded by jursidctional limits whether on land or at sea. Such effects can be felt on a region-wide basis. The understanding which has developed of the components of the Arctic environment, and its inter-relationships, is a third and very important factor which delineates the Arctic as a policy region.

Rapid change in the nature and scale of human activity in the Arctic have likewise exerted impact upon the indigenous peoples of the Arctic rim, for example, the Eskimo communities in Alaska and other Inuit peoples. These activities can have negative influence upon social and cultural values central to these communities. Efforts to come to terms with rapid change while preserving a sense of cultural and social identity have been elicited from native peoples throughout the area. In fact, these pressures have contributed to a sense of cultural kinship illustrated by the formation of the Inuit Circumpolar Conference. A fourth important factor in delineating the Arctic, therefore, is the concept of region which is emerging--or re-emerging--among its original inhabitants.

The final major feature delineating the Arctic lies in the realm of science. The strongest and most consistent theme in human activity in the Arctic has been the quest to understand the region and its processes and to fathom its influence on the planet as a whole. The Arctic has been an object of coordinated scientific investigation for several centuries, including the scene of cooperative research undertakings during international polar years and the International Geophysical Year of 1957-58.

It is the major advances in Arctic science over the past half century that have made possible the rapid acceleration in the nature and scale of human activities in the Arctic. It is also from scientific study of the area that have emerged those integrative elements which tend toward the concept of the Arctic not only as a field of study but also as an area of collaborative action. Within the sciences there has developed an interdisciplinary approach to study of the Arctic in which inter-relationships between phenomena and processes have taken precedence over individual facts or events. It is from the scientific research activities that the concept of the Arctic as a region has most clearly emerged. At this conference there are specific panels which will illuminate, in detail, contributions which the major scientific disciplines have to made the concept of the Arctic.

The emergence of factors bringing the Arctic into focus as a region in the era following the Second World War has evoked responses in the policy arena. In fact, during World War II itself, U.S. Vice President Henry A. Wallace proposed both to the Department of State and to the Congress that the U.S. should take the initiatives to develop an international treaty for the Arctic which would provide for cooperative efforts to promote transportation and communication in the Arctic and to assist Arctic exploration.

The times, however, were not ripe for an initiative of that sort. The concept of some form of international agreement applying to the Arctic, however, has periodically received policy level attention. During the late 1960's the concept of a "Northlands Compact" emerged. This idea, which appears to have been influenced by the Antarctic Treaty of 1959, looked toward development of an umbrella-type agreement to which Arctic nations could accede for particular purposes. Among the areas identified for cooperative activities were scientific research and the sharing of information and data relating to economic development, environmental protection, and health and medicine. The idea for such a "Northlands Concept" was discussed informally for a number of years, but was shelved in the early 1970's due to lack of sufficient interest among the governments of Arctic nations.

The interest in the Arctic, however, which prompted attention to possible international agreements for the area, also received expression in the form of a White House declaration on Arctic policy. This declaration is contained in National Security Decision Memorandum (NSDM) 144 of December 22, 1971. It provides that the U.S. will support the sound and rational development of the Arctic guided by the principle of minimizing any adverse affects to the environment; will promote mutually beneficial international cooperation in the Arctic; and will at the same time provide for the protection of essential security interests in the Arctic, including preservation of the principle of freedom of the seas and superjacent air space.

In support of these policy objectives the NSDM called for consideration of steps for increasing cooperation with other countries in such fields as exploration, scientific research, resource development and the exchange of scientific and technical data. It also urged improvement in U.S. capability to inhabit and operate in the Arctic and to understand the Arctic environment, as well as the development of a framework for international cooperation (the "Northlands Compact" approach). While, as noted, the "Northlands Compact" did not bear fruit, the general points articulated in NSDM 144 have remained.

The NSDM also provided for establishment of a policy mechanism, the Interagency Arctic Policy Group (IAPG). This group was charged with the responsibility for overseeing the implementation of U.S. Arctic policy and reviewing and coordinating U.S. activities and programs in the Arctic, with the exception of purely domestic matters internal to Alaska.

The general objectives articulated in NSDM 144 have not generated specific policy initiatives. Likewise, the policy group, established in 1971, became essentially inactive with the fading of the "Northlands Concept" idea.

In the past few years, however, interest in Arctic policy has revived. The IAPG has been reactivated and is engaged in seeking to assess the implementation of U.S. interests in the Arctic in the years ahead. In addition, there has been concentrated attention upon the issue of a U.S. Arctic science policy. The Alaska Lands Act of 1981 called for a review and recommendations on the need for redirecting U.S. Arctic research policy. More recently, the Alaskan congressional delegation has introduced legislation in both House and Senate--the Arctic Research and Policy Act--which would create an Arctic science policy council to develop and supervise an integrated Arctic science policy. In addition to such a council, the bill would establish an Arctic research commission to determine research priorities, establish an Arctic data center and provide for research grants. Finally, the National Academy of Sciences, in its <u>A U.S. Commitment to Arctic Research</u>, has recommended:

> 1. United States Government science policy should include a commitment to the support of scientific research in its Arctic territory and in other areas of Arctic interest, as a necessary and integral part of its national policies for economic, technical, and social development, resource development, environmental protection, national security, and international cooperation in the Arctic; and
>
> 2. To assure productive polar research, the United States Government should improve the coordination and effectiveness of Arctic research programs and provide stability and continuity of effort.

In looking at this renewed interest in the Arctic as an area of policy attention, we face a chicken and egg sort of question. Renewed interest reflects appreciation of the changes which have taken place in the character of human activities in the Arctic in recent years--changes which continue at what appears to be an accelerating rate. As I have attempted to show, these changes have been the primary factors in making it possible to define the Arctic as a region for the policy-maker, as well as for others concerned with the area. However, there remain the questions--posed at the outset--of whether it is desirable or feasible to seek an overall Arctic policy or overall Arctic policies.

In returning to these questions, it may be useful to consider, first, whether there are analogies between the Arctic and other regions of the world from a policy standpoint and, second, to look at two aspects of policy: its formulation and its implementation.

First of all, there are parallels between the north and south polar regions. From the point of view of scientific phenomena and processes, the Arctic and the Antarctic both form distinct regions and should be treated as such. The similarities in physical conditions and in the challenges facing human activity between the two areas are obvious. However, from the policy perspective, it is the differences between these areas that are most striking. An obvious, yet often overlooked difference, relates to population. There is no indigenous or even permanent human population in the Antarctic, as is the the case in the Arctic. Second--and perhaps there is a causal connection with the first point--the countries active in Antarctica have established a clear, regional framework for their activities there, the Antarctic Treaty. Within this framework which constitutes a legal and political system, these countries are able to formulate and implement Antarctic policies.

No such regional framework exists in the Arctic. Initiatives over the past several decades to develop regional arrangements for the Arctic have not met with encouraging response. Of those factors which delineate the Arctic as a region, the perception of the area's strategic importance and emergence of large-scale resource activities exert, or tend to exert, centrifugal forces rather than incentives for regional collaboration. In a political and legal sense, these centrifugal forces have an ascendancy. From this angle, the region that most directly resembles the Arctic is not the Antarctic but the Mediterranean. Like the Mediterranean, the Arctic is a marine basin surrounded by countries which exhibit diversity in political systems and diversity in political objectives. This diversity can and frequently does involve conflict among the nations bodering the central marine area. In the 1940's the polar explorer Vilhjalmur Stefansson, referring to the impact of air power upon the strategic importance of the Arctic, noted: "In an air age, the Arctic Mediterranean is the hub of world power." However one evaluates this conclusion, the analogy of the Mediterranean--a middle sea surrounded by major world powers--certainly fits the Arctic.

The analogy may extend further. Even in the Mediterranean, with its political rivalries and sensitivities, there have begun to emerge certain patterns in region-wide collaboration. A good example may be found in the ongoing efforts to develop region-wide mechanisms for dealing with marine pollution in the Mediterranean. The Barcelona Convention and related protocols involve important commitments among the Mediterranean nations to combat marine pollution--a problem which cuts across political divisions and perceptions. The experience in the Mediterranean--with its functionally defined efforts at regional cooperation--may provide a model which is of use in approaching the Arctic.

These comparisons point to a useful distinction between policy formulation and implementation in approaching the Arctic as a policy region. In my view, many of the difficulties which have arisen in the realm of Arctic policy relate to confusion or lack of synchronization between the formulation and implementation aspects of the policy process. On the one hand there has sometimes been a tendency to assume that, since the Arctic can be defined as a region, region-wide policies are required (for example, a "Northlands Compact"). Such assumptions have run aground on the fact that the Arctic does not, like the Antarctic, offer broad scope for implementation of policies on a region-wide basis. Conversely, at other times there has been a tendency--precisely because of the fact that the Arctic at present does not lend itself readily to region-wide policies--to ignore the need for a regional approach in policy formulation. Turning again to the Mediterranean example, it is my contention that perception of the Mediterranean as a region has become a major element in the formulation of policies by Mediterranean states and has lead to the identification of functional areas where region-wide approaches are necessary as well as desirable. This has taken place even though the political consensus necessary for region-wide activities does not exist in many areas.

I believe that there are significant opportunities for policy development and for policy initiatives in the Arctic. To do so requires that definition of U.S. interests, programs and activities relating to the Arctic should take place in a regional context. This does not mean, necessarily, the elaboration of an overall policy, or policies, for the Arctic. What it does mean is that the formulation of interests, objectives, programs and activities in the Arctic include as a basic element consideration of their region-wide implications. It seems to me it is important to establish a region-wide prism as one of the primary means through which we examine the definition of our interests and the conduct of our business in the Arctic. By ensuring that U.S. activities and actions relating to the Arctic can be assessed in the context of the region as a whole, we will best ensure that effective Arctic policies can not only be formulated but also implemented.

To do this requires an inter-disciplinary approach--an understanding of the connections and inter-relationships between the various forms of activities which are taking place, and which may take place in the Arctic. Further, this implies an attention to the mechanisms, both formal and informal, to ensure that the inter-relationships between interests, programs and activities in the Arctic in a geographic sense, in a functional sense, and in a precedential sense, are adequately understood and assimilated. It is for this reason that meetings such as this one, which bring together individuals with a wide variety of backgrounds, disciplines and responsibilities, can play such an important role in identifying what are the real opportunities in the area of Arctic policies. Efforts such as those being undertaken by the Comite Arctique, which seek to provide a forum for intellectual integration in the disciplines relating to the Arctic, are are another example of what I believe to be the necessary development of a context--a mind set--for policy formulation. Finally, of course, we need to replicate this process on the governmental level as well, to create mechanisms which will ensure that the prism (as I have called it) is established and employed--a prism which will permit the identification of legitimate and realistic opportunities for constructing Arctic policies, but which will also avoid elaboration of sweeping but unimplementable policy pronouncements.

I would conclude with the observation that the current revival of interest in the Arctic is on the right track. This revival has been expressed through a variety of channels--through the Interagency Arctic Policy Group, in which I am most directy involved; in the Arctic Research and Policy Act, which received Administration support at the end of the last congressional session; and in the work of the National Academy of Sciences. These activities are not congruent in all respects, but their common features far outweigh any divergences. An important and, to me, encouraging aspect of these trends is the assumption that if the United States is to be able to act as an Arctic nation, it must be able to think as an Arctic nation. In other words, the prerequisite for national policy development in the Arctic is establishing an effective mechanisms to discern where our interests and actions relating to the Arctic do, and where they do not, have region-wide implications.

ROBERT BRUCE WALLACE LECTURE

THE SONG OF SEA ICE AND OTHER ARCTIC OCEAN MELODIES

Ira Dyer

The Robert Bruce Wallace Lecture: 2 March 1983
Massachusetts Institute of Technology

1. INTRODUCTION

My goal is to share some observations of sea ice and other physical features of the Arctic Ocean. These observations are a result of three field programs in the Arctic since 1978 and have been made possible not only by financial and logistic support from the Office of Naval Research but also by the creative help of many colleagues and students.

To this point, our research in the Arctic is largely related to its acoustic properties. Hence the title of my paper. We have in various ways listened to the song created in the ocean by sea ice under stress, and to other melodies of the Arctic environment. For some there is an intrinsic interest in the song, and for others the song is a means of better knowing the singer. While difficult enough, it is somewhat easier to collect a library of songs than to interpret their fundamental messages. Thus I offer my observations on Arctic acoustics more as a systematic collector, perhaps as a speculative interpreter, and possibly also as an evoker of new insights. Such insights, I believe, are relevant to basic engineering questions on the Arctic with potential application to a wide range of problems. For example, noise radiated into the ocean by sea ice relates to its mechanical strength; and its strength, in turn, is of concern in determining loads on petroleum platforms, icebreaker bows, and the like.

2. ARCTIC OCEAN FOCII

Why should we be interested in the Arctic Ocean? The Conference of which this paper is but one part, emphasizes several reasons for concern with the Arctic. Companion papers of the Conference cover in depth, and with useful bibliographies, most of the important technical focii. My immediate purpose is to highlight a few of the more important societal motivations and to assess technological implications of pursuing them.

Many geologists, petroleum producers, and lease-sale managers believe that the Arctic contains oil and gas in amounts which rival some of the richest regions of the world, such as the Persian Gulf. While market enthusiasm for oil has cooled in the current recession, there is little doubt that it will heat up in years to come. Not only will a healthier world economy provide such heat, but also the inexorable reduction of reserves in present oil fields makes essential the development of new ones, such as in the Arctic.

Both onshore and offshore fields will be developed. Either way, transporting exploration and production equipment to the field and then transporting the product to a distribution center is an essential technological challenge. Ice-strengthened or ice-breaking ships will no doubt be directed toward this challenge. A look at a rather early attempt at a solution, the Manhattan (Figure 1), suggests the nature of the challenge. Here the main problem is ice-structure interaction for which we need better engineering knowledge of the mechanical properties of sea ice, including its dynamic strength.

Offshore oil is bound to be important in the Arctic, as it is for other regions. Figure 2 shows that the Arctic Ocean has a rather large continental shelf, some high fraction of which is thought to contain petroleum. While not strictly in the Arctic Ocean proper, I include the Bering and Greenland Seas and other adjacent seas as candidate areas for petroleum exploitation. What is significant is that all these candidates are in a marginal or seasonal sea-ice zone, in that for at least some brief time in the summer they are ice-free. The flip side is that they are covered with sea ice at other times, and this suggests that petroleum platforms would have to be protected against both open water waves and sea ice loads. Offshore engineers already have considerable experience with wave loads in stormy areas, and we can reasonably expect that such experience would be brought to bear successfully on Arctic platform design. The newer and more uncertain element is again ice-structure interaction. Many approaches have been suggested to counter ice loads, most of which take advantage of the limited mechanical strength of sea ice by breaking it at the point of its interaction with the platform.

We do not as yet have a thorough understanding of the ice environment in the Arctic. But in one region of intense interest for offshore oil - the Beaufort Sea north of Alaska and the Canadian archipelago - we do know that sea ice tends to be thicker and rougher than usual. Known as a rubble field, the ice is variously broken up and piled helter-skelter. It is grounded on the shelf and relatively motionless, forming one edge of a high shear zone in which the ice makes a transition to the central Arctic pack ice perhaps 100km away.

It is not my central purpose to discuss Arctic offshore platform design, but the rubble field often encountered leads one to wonder whether surface-piercing platforms will in the end be practical. In temperate zones, platform technology has already advanced to bottom production systems. Use of such systems in the Arctic, at depths of 60m or more, could avoid the ice canopy altogether. Ultimately such deep completions would be considered anyway, as production evolves from shallow water to the edge of the continental shelf and beyond. The Arctic is the kind of place that may well encourage this technological leap to be taken early in the game.

I have seen a draft of an Arctic ice atlas being compiled under sponsorship of the US Maritime Administration (Labelle, et al 1982). In my opinion, this effort is both creative and potentially useful, at least because through it we can readily recognize our shortcomings in understanding the ice field. Knowledge is relatively good on degree of ice cover, including area of coverage and mean floe size, fair on ice drift rates and directions, and not very good on ice thickness and ice roughness. Little is compiled on ice strength. Figures 3 and 4 give some indication of what is meant by floe size and, in this case, by roughness caused by pressure ridges. Remote imaging of large areas by satellite or high-flying aircraft explain why some parameters are better understood. Ice thickness, roughness, and strength, however, are generally obtained by in situ observations and dense point measurements or track lines are required. Thus far, such measurements have not been carried out widely enough.

A quite different focus is formed by military needs. I am not privy to the closed councils of military planners. But I do know from discrete public announcements that the United States intends to increasingly operate in the Arctic Ocean as a counter to presence of Soviet submarines carrying strategic nuclear weapons. U.S. submarines have transitted the Arctic almost from the dawn of nuclear-powered submarines. Now the thrust seems to be long term presence or occupation rather than brief demonstration visits.

The submarine nuclear deterrent in the Arctic implies ice capability. Surfacing through the ice canopy ultimately requires the submarine to break the ice, and we see again the importance of knowledge on ice strength, thickness, and roughness as they affect submarine structure and shape. It seems that the U.S. Navy, as it lays plans for the next generation submarines, will insist on appropriate structure and shape for routine Arctic operations.

Incidentally, the possible advent of Arctic-capable military submarines has technological implications for oil transport by submarine. This transport mode has many advocates among naval architects and ship constructors, and could become a viable candidate as a follow-on to Navy development.

Another influence of the ice canopy pops out in a rather different way. Sonar, which of course uses sound waves, is the principal means of detecting and monitoring intrusive submarines. Evidence exists that transmission of sound in the Arctic Ocean is greatly influenced by ice roughness, as one may see by viewing Figures 5 and 6. Judging from the model wavenumber spectra shown (Mellen and Marsh, 1963), ice roughness is similar to ocean wave roughness. But for a given root-mean-square roughness, sea ice is significantly rougher than ocean waves for large wavenumbers (K larger than about 1/3 m^{-1}, or wavelengths smaller than about 20m). Thus with use of Rayleigh's diffraction grating theory, we can expect significantly more roughness effects in the Arctic when the sound wavelength is less than 40m, or the sound frequency is higher than about 35Hz.

Such different behavior with respect to roughness as shown in Figure 6, where I have estimated the reduction in acoustic energy versus frequency for a transmission range of 500km. We see that ice scattering severely limits transmission in the Arctic. Indeed, ocean wave scattering is so relatively weak that still another process, scattering into internal degrees of freedom of sea water (Urick 1975), is more important. As a practical matter, therefore, we should compare ice scattering in the Arctic with relaxation scattering in the open ocean. We see that to achieve equivalent sonar transmission range, a shift downwards in frequency of about one order-of-magnitude is required.

The implication of lower-frequency sonar for the Arctic is that, for equivalent performance, its linear dimensions must be larger, by as much as one order- of-magnitude. Larger submarine sonar platforms and/ or larger systems fixed on the bottom would be required.

Another way to view Figure 6 is that a sonar capable of detecting and tracking a target in the open ocean at ranges up to 500km would have, without modification, its range capability in the Arctic reduced to about 50km. Areal coverage would be reduced by two orders-of-magnitude. Thus submarines in the Arctic have more stealth, and the mutual strategic deterrent is made more stable. To my mind, this is both a desirable outcome and the most important implication of new-found interest by the military in the Arctic.

While I have confidence that the arguments presented here on sonar performance are at least qualitatively correct, I must make it clear that systematic knowledge of sea-ice roughness and of attendent acoustic scattering, falls below our needs in engineering design. Thus I am pleased that many scientists in the Arctic community, as well as the MIT team, consider research on sea ice as essential.

There is one other societal motivation for interest in the Arctic. Quite apart from petroleum resources or mutual security, many of us care about the Arctic as a relatively unexplored arena for scientific work. When asked about my interest, I often explain first that I have enough of the "little boy" left in me to make the Arctic attractive. Such a lighthearted response hardly satisfies the serious questioner or the introspective self. Frivolity aside, I like Arctic research because it demands fresh approaches to basic questions, awakens interest in physical phenomena rarely met, and rewards innovative use of observational tools. Examples, not necessarily out of my own direct experience, might help make this clear:

- The atmosphere-ice-ocean system is no doubt tightly coupled with a strong need to quantify the exchange elements and the feedback loops.
- Marginal ice zones are oceanographically complicated; vertical fluxes, water mass fronts, and variable ice edges need to be measured and modeled.
- Electromagnetic and acoustic remote sensing equipment need to be developed and validated for wide-area synoptic measurements.
- Primary biological productivity, especially near or in the marginal ice zones, needs to be connected with basic physical processes such as mixing, water transport, and nutrient upwelling.
- The geology, geophysics, and topography of the Arctic basin and its margins have not been explored and modeled as thoroughly as they should.

These scientific needs are intrinsically interesting and worthy of fulfillment. The reader will quickly discern their importance to society at large. Among others they relate: to improved weather forecasting in the northern hemisphere, to safer navigation and piloting near or in the ice, to understanding of the biological chain linked with endangered mamalian species or with commercially important fisheries, and to discovery and exploitation of hard mineral resources. Our interest in the science, perhaps well ahead of societal usage, is an affirmation that the Arctic is important and has a future.

3. THE SONG OF SEA ICE

A hydrophone suspended beneath Arctic ice senses a rich symphony of sounds. The essential quality of these sounds can be represented via spectral analysis, and a composite of such spectra is given in Figure 7. Each spectral sector making up the composite was selected at different sample times. By noting the reasonable agreement between overlapping sectors, one can gain assurance that the composite is at least a rough representative of the family of sounds, as if each family member had been observed at the same time. I call your attention to spectral changes that occur from family to family, for these differences, as well as similarities, can provide clues on the underlying physical mechanisms.

From about 10^{-2} to 5×10^{-1}Hz, the pressure spectral density falls sharply with frequency, about as ω^{-4}, with an rms value of about .5 Pa. My hypothesis is that these fluctuations are caused by quasistatic velocity fluctuations in the flow V past the hydrophone. The change in pressure Δp caused by a fluctuation in flow velocity ΔV is

$$\Delta p = C_D \rho V \Delta V \quad (1)$$

where C_D is the hydrophone drag coefficient and ρ the water density. Thus the rms pressure of .5 Pa would convert to an rms velocity fluctuation of about 3mm/sec for $V \approx 15$cm/sec, and is reasonable when compared to observed fluctuations (McPhee and Smith 1976). Of equal interest is the velocity spectral density which, according to Equation (1), can be obtained by scaling the observed spectrum by $(C_D \rho V)^{-2}$.

Starting at about 5×10^{-1}Hz and extending to about 10Hz, one spectral member exhibits quite narrow peaks. The other one shown in this frequency range does not. These peaks are caused by interaction of current with the hydrophone cable, an effect often quite variable and probably nonlinear. Known as strum, it seems to be caused in our case by second-order length oscillations as a result of transverse cable oscillations induced by vortex shedding. I note in passing that eddies in the Arctic are most energetic in the upper 100m or so (Hunkins, 1981), and cable strum might be a useful indicator of eddy energy. Our hydrophones were suspended on cables down to about 90m.

When strum is absent, it is clear that the Arctic song has a broadly peaked spectrum extending from about 1 to 100Hz and possibly beyond. It peaks at about 10 to 20Hz. (Strum can mask this result below 10Hz but rarely above it.) This broad spectrum is caused by ice action, a mechanism I will speculate on in great detail a few paragraphs from here.

Also suggested by the data is a second broadly peaked spectrum, extending at times from about 150 to 5000Hz. It is not an ever-present feature of Arctic noise, is most intense at times of atmospheric cooling, and was first related by Milne (1972) to thermal cracking. This, too, will be discussed subsequently.

Finally the composite shows a relatively important spectral peak at about 5000Hz which, however, is not always present. It is observed when wind speed is large enough to carry loose snow pellets and impact them on the ice floe. Milne (1967) observed and modeled this phenomenon, but his spectral shape is significantly different. I will not pursue this discrepancy; instead I turn to those mechanisms relevant to ice action.

The spectral peak at 10 to 20Hz invariably has a spectral shape of about ω^2 and ω^{-2} below and above the peak respectively. We have observed this as a regular feature of pack ice noise (Shepard 1979, Chen 1982). I outline here ideas, or more humbly, speculations on its origin. Ice is a crystalline or granular material which, while mostly regular in structure, contains many defects. Under stress, these defects act as locii for structural rearrangements and account for material properties such as plasticity, cracking, and breaking. I believe that defect motions account for the observed noise; and to the extent that this is so, one ought to be able to connect the noise with structural properties.

Evidence that defect motions account for this component of the sea-ice song is given in part in Figures 8 to 10. The first shows a spatial spectrum of the noise centered at 10Hz, with the strongest component arriving at an azimuth of 012°. The time series of the signal arriving from this direction is given in Figure 9; it shows two short-duration bursts separated by about 2 sec. Finally the frequency spectrum of one of these bursts (Figure 10) displays a spectral shape of about ω^4 and ω^{-4} below and above the peak, at first sight suprisingly different than the shape shown in Figure 7. But I delay explanation of this difference in favor of first presenting a physical model for the results of Figures 8 to 10.

Aki and Richards (1980) and Rice (1980) describe elastic waves generated in solids by slip and by tensile opening as a quantitative basis for earthquakes and other damage processes. I picture a burst originating in ice in much the same way. In simple terms, ice under stress accomodates to the stress via slip motions. These, in turn, create elastic waves, the most important of which travel within the ice as a shear wave (Crary 1954, Oliver et al 1954, Ewing et al 1957, Hunkins 1960). The shear wave radiates sound into the water in the vicinity of each hydrophone. By comparing the arrival times among a large number of hydrophones, we have estimated the group speed of these ice waves to be roughly 1360m/sec at about 30Hz and smaller at lower frequencies. The horizontal phase speed of these waves is, of course, expected to be larger than the shear speed, the latter being about 1800m/sec.

Following Aki and Richards (1980), the shear wave displacement radiated from a small slip region in an infinite medium can be written as:

$$u_s = F(r)\, \dot{M}\, (t - r/\beta_1) \quad (2)$$

where F(r) is a function of radial distance r from the slip, of slip orientation, and of material properties (ice density, thickness, and shear phase speed β_1). Also M (t) is the time-dependent slip moment, and $\dot{M}$ its time derivative. M can be equated to $GA\bar{u}(t)$, with G the shear modulus, A the slip face area, and $\bar{u}(t)$ the slip displacement averaged over the area A. As may be seen, the radiated shear wave displacement itself is proportional to the time derivative of this moment, evaluated at a later time r/β_1.

Our ultimate interest is pressure in the water which is governed by its continuity with traction across the water/ice interface. For an unbounded medium traction would be proportional to $\dot{M}$, since the spatial derivative related to strain can be transformed to a time derivate via the functional form, $t - r/\beta_1$. Finally, traction in the ice must be governed by the traction-free ice/air interface which, for small slip regions, brings into play another spatial or time derivative. (The ice/air interface acts as a negative reflecting surface, so in effect, each slip is represented by itself plus an opposite slip above the ice.) Thus the sound pressure p is expected to be:

$$p = H(r)\, \ddot{M}\, (t-r/\beta_1) \qquad (3)$$

where H(r) includes all the parameters in F(r) but in different form, plus additional ones relating to sea water. Although the argument seems complicated, the result is simple: The acoustic pressure is proportional to the third time derivative of the slip displacement $\overline{u}(t)$.

I model $\dot{M}$, i.e. the shear wave displacement, as a Gaussian time function. This results in a sequence of time functions, and corresponding spectral shapes, as shown in Figure 11. The spectral density corresponding to Equation (3) when $\dot{M}$ is Gaussian is

$$S_p = S_1 \left(\frac{\omega}{\omega_o}\right)^4 \exp\left[-2\left(\frac{\omega}{\omega_o}\right)^2\right] \qquad (4)$$

where where S_1 is the peak spectral density corresponding to an arbitrary reference state of the process, where the peak radian frequency is

$$\omega_0 T = 4 \qquad (5)$$

and where T is the duration of the Gaussian pulse at the e^{-1} level (see Figure 11). My effort here is to describe reasonably but simply the characteristics of a single acoustic burst, and the Gaussian model leads to a time pattern and a spectrum fairly representative of the data (compare Figures 9, 10, and 11). To my knowledge, fundamental theories relevant to slip dynamics in ice or ice-like materials are not available, and acoustic sensing of the slip moment as the foregoing shows is possible, may well provide the basic data upon which to build such theories.

The difference in spectral shape between Figures 7 and 10 must now be attended to. When averaging over many bursts, as in Figure 7, it is likely that the duration T is a stochastic function rather than a fixed value. We can estimate T as ℓ/v, where v is the average slip or rupture speed and ℓ the slip displacement. In Figure 11, ℓ is shown and v can be determined from elastodynamic data:

$$v = \beta \frac{(\tau/\tau_o)^n}{1 + (\tau/\tau_o)^n} \qquad (6)$$

where n is some number, τ the resolved slip plane stress, and τ_0 some characteristic (high) stress above which v reaches a plateau speed β. In earthquake seismology, β is taken as the shear speed (Brune 1970); but for crystalline solids, β can be an order-of-magnitude less (Johnston and Gilman 1959, Gutmanas et al 1963, Fitzgerald 1966). Readey and Kingery (1964) have found from laboratory experiments with ice that n is about 2, but Weeks and Mellor (1983) report n to be about 4.

In consequence of Equations (5) and (6), the peak frequency ω_0 varies with stress. To be specific, the variation of slip displacement with stress is needed; Brune (1970) gives the basis for the following:

$$\ell = \ell_1(\tau/\tau_1) \qquad (7)$$

In this equation, as in the definition of S_1, the subscript corresponds to an arbitrary reference state. We see that ω_0 depends upon τ^{n-1}, at least for $\tau \ll \tau_0$.

Energy radiated by a slip is given by (Aki and Richards 1981)

$$E = \eta A \ell\tau \qquad (8)$$

where η is a fixed fraction of total energy which radiates, the remainder being invested in work as the two slip planes grind against each other. Thus the integral of Equation (4), call it $p_1{}^2$, is proportional to $A\ell\tau$. In the simplest case of fixed slip plane area A, p_1^2 is proportional to τ^2 via Equation (7). I am encouraged to think of A as fixed for a given process simply because it is likely to be set or limited by a finite scale of the ice canopy (grain size, columnar spacing, floe thickness, ridge width, etc.).

I do not know how stress varies over the time scales and ice areas typically involved in measurements such as shown in Figure 7. These time-space series are now under study following a path laid down by Hunkins (1975). But if stress values are broadly distributed, it is clear that its distribution will dominate the long-time, large-area noise spectrum. It is, thus, not much of a leap to speculate that the sea ice song can be inverted to give stress distribution. I will sketch such an inversion by starting with an assumed probability density function (pdf) for ω_0

$$P(\omega_o) = \frac{(L-1)^L}{\Gamma(L)\omega_m}\left(\frac{\omega_o}{\omega_m}\right)^{L-1} \exp\left[-(L-1)\omega_o/\omega_m\right] \qquad (9)$$

where L is some positive number and ω_m the value of ω_0 of maximum probability. Equation (9) is the well-known gamma pdf (or chi-square pdf for 2L an integer), and we start with it simply because L can be selected to give a wide range of probability shapes. Next the spectral average over ω_0 is

$$\langle S(\omega)\rangle = \int_0^\infty S_p\,(\omega, \omega_0)\, P(\omega_0)\, d\omega_0 \qquad (10)$$

This integral may be evaluated quickly by noting that S_p is really quite narrow [see Equation (4)], and to first-order can be approximated by

$$S_p\,(\omega, \omega_o) \simeq p_1{}^2 \left(\frac{\tau}{\tau_1}\right)^2 \delta\,(\omega - \omega_o) \qquad (11)$$

in which I have explicitly included the stress dependence (relative to the stress state leading to p_1^2). Thus

$$\langle S\rangle \simeq J\,(\omega/\omega_m)^{\alpha}\,\exp[-(L-1)\omega/\omega_m] \tag{12}$$

where $\alpha = L-1 + 2/(n-1)$,

where $J = \dfrac{p_1^2(L-1)^L}{\omega_m\,\Gamma(L)}\left(\dfrac{\omega_m}{\omega_1}\right)^{2/(n-1)}$,

and where I have concluded that $\tau \ll \tau_0$ since the observed $\langle S\rangle$ shows no abrupt termination at a high value of ω.

One can now select α from the observations. At low frequencies $\alpha = 3$ leads to a good fit, and this can be obtained with L = 3 and n = 2. (Note that L is only weakly dependent upon the choice of n; for n = 4, we would have L = 3.3.) Comparison of Equation (12) with data for L = 3 and n = 2 is made in Figure 12. As arranged by my choice of α, the low frequency fit is satisfactory, but the model falls far short at high frequencies. The difficulty is in Equation (11); the first-order approximation that all energy for a single burst resides at ω_0 is only roughly tenable. One can go to a second-order but still simple approximation to remedy this:

$$S_p \simeq 0.57\, p_1^2 \left(\frac{\tau}{\tau_1}\right)^2 \sum_{k=0}^{3} \left(\frac{k+1}{2}\right)^4 \exp\left[-2\left(\frac{k+1}{2}\right)^2+2\right] \cdot \delta\left[\omega - \left(\frac{k+1}{2}\right)\omega_0\right] \tag{13}$$

In effect this second approximation partions the burst energy at 4 frequencies ($\omega_0/2$, ω_0, $3\omega_0/2$, $2\omega_0$). As a more realistic representative of S_p, it now leads to a much better fit with the data (see Figure 12). Analytical arm-waving aside, it appears that Equation (9) is an adequate pdf for ω_0, with L = 3 and $\tau \ll \tau_0$ as the experimentally derived conditions.

I can now state a tentative but important conclusion. We can transform $P(\omega_0)$ to $P(\tau)$ via $\omega_0 \sim \tau^{n-1}$ to get, for n = 2:

$$P(\tau) = \frac{3\sqrt{3}\,\tau^2}{2\sigma^3}\exp[-\sqrt{3}\,\tau/\sigma] \tag{14}$$

where σ^2 is the stress variance. Thus the stress is chi-square distributed with 6 degrees of freedom (= 2L) and is shown in Figure 13 (see Cramer 1946 for a formal description of this pdf). Whether this result can withstand further experimental and theoretical scrutiny is uncertain. Among changes that may be expected are better choices of n and L, use of other but similar pdfs such as the Erlang or Weibul, and improvements in the description of the slip dynamics. What I do feel comfortable with is Equation (14) as a point of departure, not only in further research, but also in present-day engineering analysis. Equation (14) and its successors are apt to play a role as important in sea-ice engineering as wave-height pfds play in open-ocean engineering.

One final comment seems to be in order and concerns the song identified earlier as thermally related. Its spectrum stands about 20 times higher in frequency. Milne (1972) argues the relevant mechanism here is tensile opening, and Rice (1980) shows how close in character this mechanism is to slip. Given that it might be, one can speculate that its higher frequency means that its rupture speed v is more by an order-of-magnitude or its displacement ℓ is less by an order-of-magnitude. It would be interesting (and valuable) to find out which.

4. MELODY OF THE MID ARCTIC RIDGE

I conclude this paper with brief accounts of two other uses of acoustics in the Arctic Ocean. These relate to sounds from earthquakes along the Mid Arctic Ridge and to discovery of major topographic features from sound echoes, the latter being the topic of the next and final chapter of the paper.

The Mid Arctic Ridge is located between about 83°–87°N and 5°W–120°E, running through this region on a line as may be seen in Figure 2. It is an active rift zone, i.e. a narrow line along which plate spreading and earthquakes occur frequently. While the study of earthquakes from oceanic rifts is well developed (Lilwall 1982), seismicity of the Mid Arctic Ridge is less well known.

Our ice camps in 1980 and 1982 were located within several hundred km of the Mid Arctic Ridge. Consequently we recorded seismic activity from the Ridge while carrying out other acoustic experiments. Keenan (1983) analyzed five such events from the 1980 experiment, and I show data from one of them here. First, however, brief discussion of the physical ideas will enhance our understanding of the observations.

Figure 14 is a cartoon of the physical situation. Slip motion in the crust (earthquakes are like icequakes of the previous chapter!) causes shear (S) and longitudinal (P) waves to be generated. These propagate as head waves along the crust/mantle interface (Aki and Richards 1980), radiate nearly vertically up through the water as sound waves, and are received by our horizontal hydrophone array. Also S and P waves radiate sound waves into the water vertically above the earthquake and are scattered by the rough ice into nearly horizontally directed sound waves, which then transmit to our array. The latter group of sound waves (T) are predominantly waterborne, while the former (S and P) are predominantly earthborne.

A typical seismic event results in an acoustic burst which can last for 50 to 100 sec. One such is contoured in Figure 15* as a frequency/time/spectral density plot. A cut through this plot at the time of maximum spectral density, Figure 16, shows the peak frequency at 5Hz, with an approximate ω^2 and ω^{-4} shape below and above the peak. We are presently modeling this spectral characteristic to uncover, as for icequakes, the slip displacement $\bar{u}(t)$. For earthquakes, the frequency dependence of ice scattering and of S and P wave attenuation in the earth needs to be included.

We can further cut the frequency/time/spectral density plot to distinguish between paths which are predominantly earthborne versus waterborne. Figure 17 partitions the burst into paths with nearly vertical arrival angles (high phase speeds) and nearly horizontal angles (low phase speeds). Early arrivals are thus earthborne (S and P) while the late ones are waterborne (T), the latter carrying about 30 times more energy.

In addition to spatial analysis in the vertical, the array can be used for horizontal analysis (as in Figure 8 for ambient noise). Figure 18 is such a horizontal analysis and shows the T wave arriving at about 099°. When this line of position is drawn to intersect the rift zone, it

*Color figures 15, 17, 18 may be found on pages 25 and 26.

does so on a transform fault 318km from the observation point. The separation in time between the earthborne and waterborne arrivals, as well as the difference in their phase speeds (see Figure 17), can be used independently to check this result. The computed distance by the latter technique is 313km and checks the former within experimental error. Indeed, resolution in earthquake location for the Arctic by other methods is generally considered 10km at best (Lilwall 1982).

Beyond use of the foregoing observations to model earthquake slip dynamics (e.g. seismic moments, slip displacements), we are presently studying:

- total duration of the observed T wave. We observe the duration to be longer than a single slip event. Is its duration caused by a series of slip/stick displacements? Is it caused by swarming or clustering of many individual events? Or does the signal undergo multiple reflections from various layers above and below the earthquake source and so generate a longer T wave?
- shape of the T wave frequency/time/spectral density contours. Can the shape be inverted to yield the fault plane orientation? Can the shape help distinguish between various classes of slip plane motion?
- magnitude of observed earthquakes. The event shown here was too small in magnitude to be observed on the world seismic network (International Seismological Center) but was quite easily observed with our measurement and analysis system. Interest is high on relatively weak earthquakes, since their magnitudes and occurrence rates relate to spreading and other tectonic properties (Lilwall 1982).

5. ECHOES OF DISTANT PLACES

In each of our experiments, we have also observed sound signals backscattered from various Arctic underwater features. These signals are a result of intense impulses created at (or near) our camp by explosion of TNT charges, typically 200kg at 244m depth. Signals backscattered (reflected back toward the camp) are then analyzed by the horizontal array of hydrophones.

A three-dimensional plot of backscattered signals analyzed at 9Hz from one such experiment is shown in Figure 19 (Dyer et al 1982). Significant structure in the backscatter is evident out to the maximum time displayed (3600 sec or 3431/sec after shot instant), which corresponds to a range of about 2500km. In this experiment, the camp was in the Beaufort Sea some 350km north of Alaska. Thus, at a bearing of about 000°, 2500km is the distance from the camp to the European margin. In simple language, we have been able to echo-range across the widest part of the entire Arctic Ocean.

At a bearing and time, say of 180° and 3600 sec, Figure 19 must be showing noise since in that direction the Alaskan coast has intervened long before then. Note, however, that the noise is <u>less</u> than the noise just before shot instant. The underwater TNT shot accounts for this. Shock pressures up to 1 MPa act on the ice in the camp's vicinity, apparently relieving stress and thus decreasing noise for some time. It is not an uncommon effect, but reductions are observed only when noise (and stress) are low to begin with.

With suitable adjustments for energy loss in transmission, and for analysis parameters such as averaging time and array resolution, charts of iso-backscattering strength can be prepared. Figure 20 shows one for the Beaufort Sea. These contours delineate major features, most notably the continental slopes off the Canadian archipelago (along about 125°W) and off the Alaskan coast (along about 72°N). Further west, Figure 20 shows a major submerged feature along about 160°W, known as the Northwind Escarpment. Compare Figure 20 with Figure 2 to gain confidence that the echoes of the Beaufort Sea are indeed meaningful. What is especially interesting to me is strong backscatter from about 73.2^{o}N, 139.0^{o}W. Figure 2 does not show any prominent feature there, nor do other accepted charts, at least not to the degree necessary to account for strong echoes. I am confident that we have indeed discovered a prominent feature, one that we have named in honor of G. Leonard Johnson (Dyer et al 1982).

Other features or differences with known topography have been discovered. Figure 21 shows iso-backscatter contours obtained from 86.4^{o}N, 22.2^{o}W (Williams 1981). The data show clearly the Morris Jessup Rise which, in Figure 2 and other charts, lies in the region of about 84° to 86°N and 15° to 30°W. The acoustic echoes show that it actually extends much further to the north than previously had been thought. At the eastern edge, existing charts also show the Morris Jessup Rise quite separate from the Mid Arctic Ridge. But our data show a definite merger of the two, likely caused by a more westerly extent of the ridge. Indeed our data indicate that the Mid Arctic Ridge is broader and has more relief than shown on accepted charts.

That iso-backscatter charts show new or differently shaped features than those shown on standard charts is only modestly related to resolution limitations of the acoustic technique (Dyer at al 1982, Williams 1981). It is simply the case that bathymetric data in the Arctic Ocean are sparse in many areas, and cartographers must sketch depth contours as best they can. At the same time, I do not suggest replacement of standard bathymetry with acoustic echo-ranging. But as a reconnaissance tool, it can readily highlight areas deserving detailed bathymetric study or it can discover new features in poorly charted areas.

A glance at Figure 6 will show why a low frequency like 9 or 10Hz was chosen for backscatter analysis. Loss due to sea ice scattering is expected to be tolerable at low frequencies, giving an opportunity for sensing major topographic features at long ranges. Stated differently, ocean-wide echo-ranging in the Arctic at say 100Hz would have been as much as 3 orders-of-magnitude more difficult.

6. SUMMARY

Here are the main points. Knwoledge of ice roughness and ice strength is important to societal activities in the Arctic Ocean, including:

- safety of surface-piercing platforms for oil and gas production,
- performance of icebreaking and ice-strengthened ships,
- stealth of (mutually deterrent) submarine nuclear forces.

We are gaining knowledge of ice roughness, though questions remain:

- how general is the K^{-3} roughness asymptote (see Figure 5) over large sample areas and for diverse sample times?

- what physical mechanisms explain K^{-3}, in contrast to planetary topography, which is asymptotic to K^{-2}?

Via inversion of the sea ice song we are on the threshold of better definition of ice strength and ice stress:

- the probability density function for stress (see Figure 13) reflects the combination of variations in imposed stress, and variations of ice dislocations or macroscopic defects. In principle separation of the two can be achieved and is now under study.
- stress probability has profound importance in engineering, since ultimately failure criteria must entail knowledge of a stress extremum in relationship to the mean stress.

Advanced technology can and is being brought to the Arctic. It has its own flavor and often a unique form. Based on progress to date, I know that the next generation will find tasks in the Arctic less an adventure for intrepid pioneers and more a question of application of basic scientific and engineering principles. It is an exciting prospect with many rewards.

REFERENCES

Aki, K. and P.G. Richards, Quantitative Seismology, Freeman and Co. (1980).

Brune, J.N., Tectonic Stress and the Spectra of Seismic Shear Waves from Earthquakes, J. Geophys. Res., 75, 4997-5009 (1970).

Canadian Hydrographic Service, General Bathymetric Chart of the Ocean, Map No. 5-17, Ottawa (1979).

Chen, Y-M., Underwater Acoustic Ambient Noise in the Arctic, MIT thesis (1982).

Cramer, H., Mathematical Methods of Statistics, Princeton University Press (1946).

Crary, A.P., Seismic Studies on Fletchers Ice Island (T3), Trans. Am. Geophys. Union, 35, 293-300 (1954).

Dyer, I., A.B. Baggeroer, J.D. Zittel, and R.J. Williams, Acoustic Backscattering from the Basin and Margins of the Arctic Ocean, J. Geophys. Res., 87, 9477-9488 (1982).

Ewing, W.M., W.S. Jardetzky, and F. Press, Elastic Waves in Layered Media, McGraw Hill Book Co. (1957).

Fitzgerald, E.R., Particle Waves and Deformation in Crystalline Solids, Interscience, Wiley and Sons (1966).

Gutmanas, E.Y., E.M. Nadgorni, and A.V. Stepanov, Dislocation Movement in Sodium Chloride Crystals, Soviet Phys/Solid State, 5, 743-747 (1963).

Hunkins, K., Arctic Ocean Eddies and Baroclinic Instability, CU-2-81 Technical Report No. 2, Lamont-Doherty Geological Observatory (1981).

Hunkins, K., The Oceanic Boundary Layer and Stress Beneath a Drifting Ice Floe, J. Geophys. Res., 81, 3425-3433 (1975).

Hunkins, K., Seismic Studies of Sea Ice, J. Geophys. Res., 65, 3459-3472 (1961).

Johnston, W.G. and J.J. Gilman, Dislocation Velocities, Dislocation Densities, and Plastic Flow in Lithium Fluoride Crystals, J. Appl. Phys., 31, 129-144 (1959).

Keenan, R.E., Seismic Events in the Arctic Ocean, MIT Thesis (1983).

LaBelle, J.B., J.L. Wise, S.C. Atkinson, R. Becker, Jr., M.S. Floyd, J.A. Fredston, P. Marshall, R.P. Voelker, R.H. Schultz, G.M. Wohl, and D. Belloli, Alaska Marine Ice Atlas, University of Alaska and Arctec, Inc, (Draft) (1982).

Lilwall, R.C., Seismicity of the Oceanic Rifts, in Continental and Oceanic Rifts, G. Palmason, ed., Geodynamics Series, 8, 63-81 (1982).

McPhee, G.M. and J.D. Smith, Measurements of the Turbulent Boundary Layer under Pack Ice, J. Phys. Oceanog., 6, 696-711 (1976).

Mellen, R.H. and H.W. Marsh, Underwater Sound Reverberation in the Arctic Ocean, J. Acoust. Soc. Am., 35, 1645-1648 (1963).

Milne, A.R., Thermal Tension Cracking in Sea Ice: A Source of Underice Noise, J. Geophys. Res., 77, 2177-2192 (1972).

Milne, A.R., Sound Propagation and Ambient Noise Under Sea Ice, in Underwater Acoustics, Vol. 2, V.M. Albers, ed., Plenum Press (1967).

Oliver, J., A.P. Crary, and R. Cotell, Elastic Waves in Arctic Pack Ice, Trans. Am. Geophys. Union, 35, 282-292 (1954).

Ready, D.W. and W.D. Kingery, Plastic Deformation of Single Crystal Ice, Acta Metallurg., 12, 171-178 (1964).

Rice, J.R., Elastic Wave Emission from Damage Processes, J. Nondestruct. Eval., 1, 215-224 (1980).

Shepard, G.W., Arctic Ocean Ambient Noise, MIT thesis (1979).

Stephens, R.W.B. and A.A. Pollock, Waveforms and Frequency Spectra of Acoustic Emissions, J. Acoust. Soc. Am., 50, 904-910 (1971).

Williams, R.J., Backscattering of Low Frequency Sound from the Topographic Features of the Arctic Ocean Basin, MIT thesis, (1981).

Urick, R.J., Principles of Underwater Sound, McGraw Hill Book Co. (1975).

Weeks, W.F. and M. Mellor, "Mechanical Properties of Ice in the Arctic Seas," in Arctic Technology and Policy, I. Dyer and C. Chryssostomidis, ed., Hemisphere Publishing Co. (1983).

Figure 1. Voyage of the Manhattan to the Arctic in 1969. The Manhattan was an experimental tank-ship/icebreaker to test the feasibility of transporting Prudhoe Bay oil to eastern markets through Canada's Northwest Passage. It lost out to the pipeline which now takes Prudhoe oil to an open port on Alaska's southern coast.

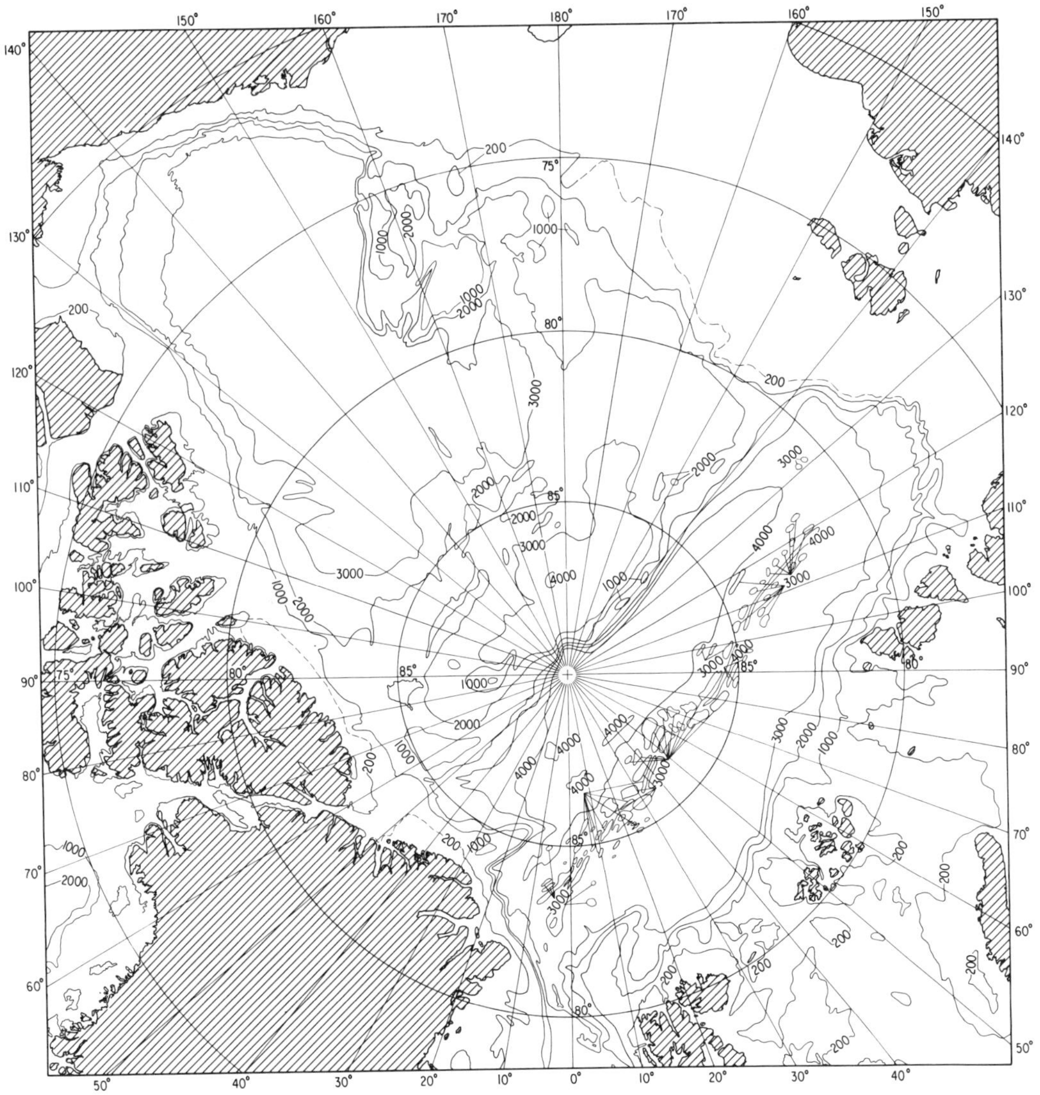

Figure 2. Bathymetric chart of the Arctic Ocean with contour depth in meters. Major features with approximate coordinates are: Alaskan coast (75°N, 140°-160°W), Canadian archipelago (78°-83°N, 60°-125°W), Greenland coast (83°N, 10°-60°W), European and Asian margin (80°N, 10°-160°E). The shelf area (depths less than 200m) is quite large. Redrawn from Canadian Hydrographic Service (1979).

Figure 3(a). General view of pack ice in central Arctic, typical of April. A pressure ridge runs from the lower left to the upper right, and other ridges and refrozen leads (breaks) can be identified.

Figure 3(b). Closer view of refrozen lead, showing smoother and darker ice. Old or weathered pressure ridges can be seen on the older ice floes.

Figure 4(a). A weathered pressure ridge can be quite handsome, but poses dangers to icebreakers or petroleum platforms as well as difficulties in over-the-ice transport.

Figure 4(b). A relatively fresh pressure ridge can also pose difficulties, but the ice blocks are less consolidated and presumably the ridge has less strength. Note the record of layering in the ice.

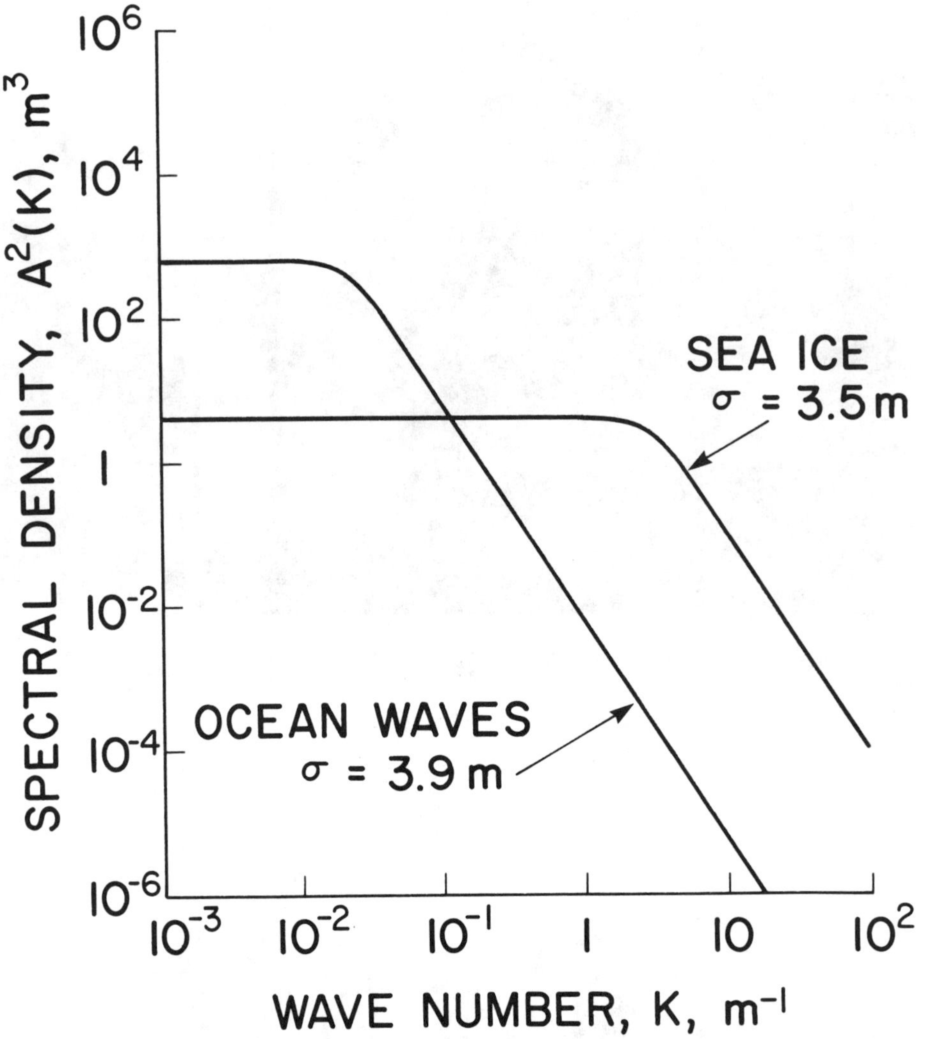

Figure 5. One-dimensional wavenumber spectra of sea ice roughness and ocean wave roughness (Mellen and Marsh 1963). Each has a high wavenumber dependence of K^{-3}.

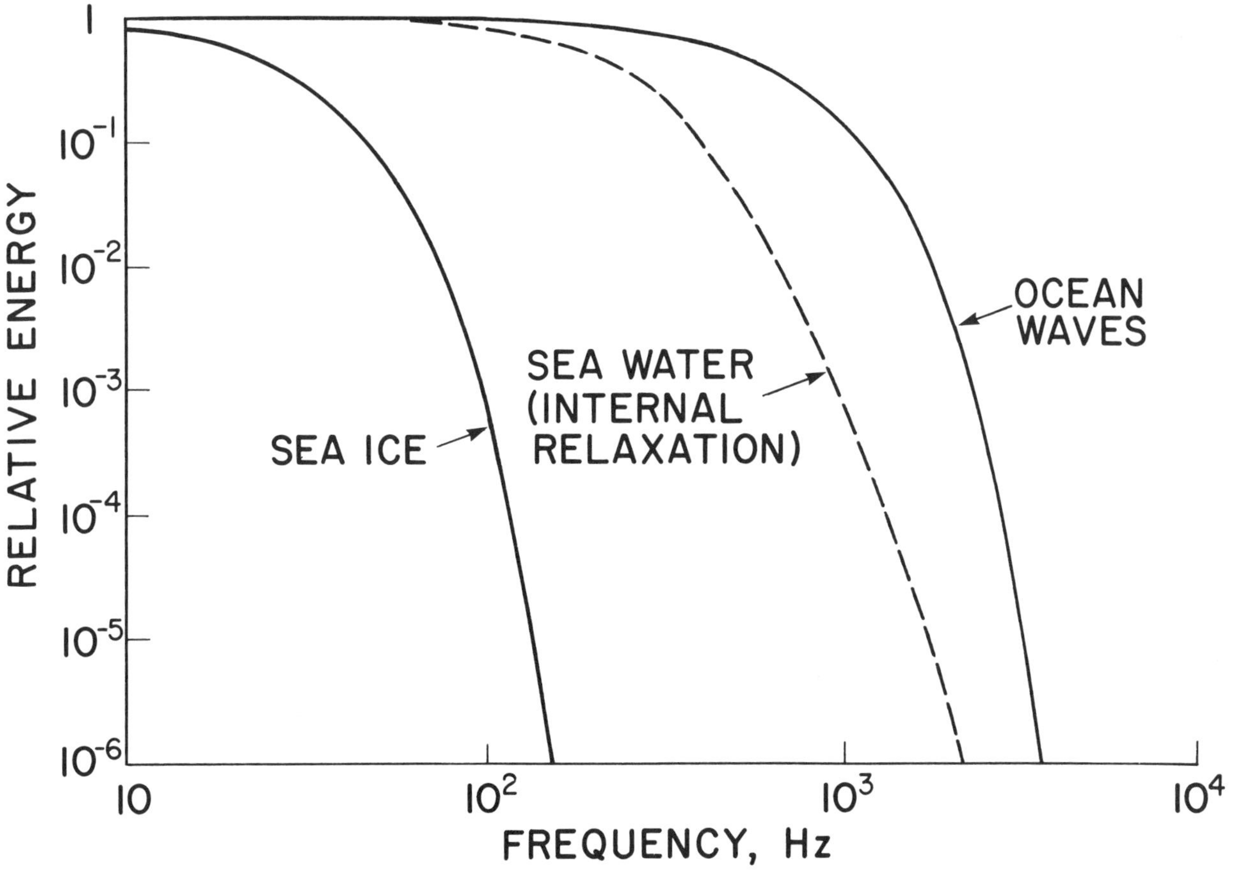

Figure 6. Forward scattering reduction of a sound signal for transmission range of 500km in deep water. Shown also is scattering into the internal relaxation processes of sea water, an effect more important than ocean wave scattering but less important than sea ice scattering.

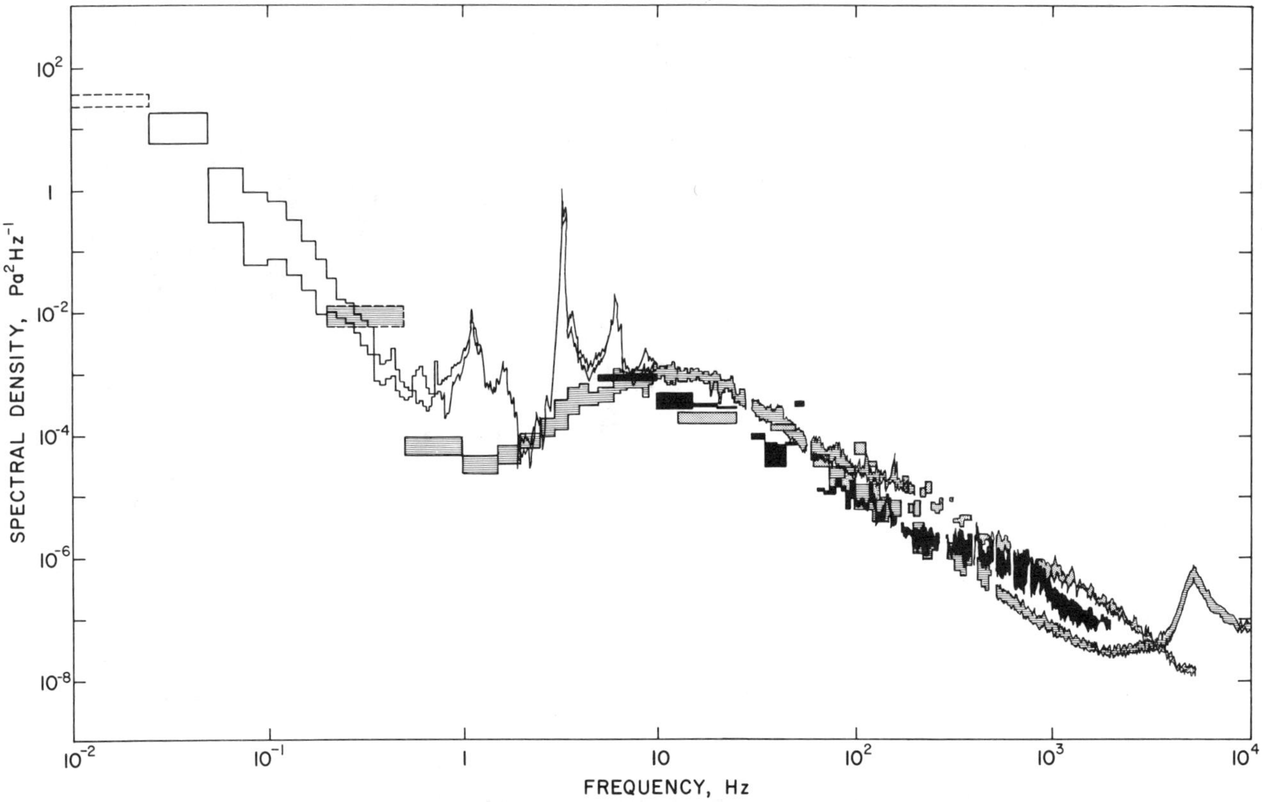

Figure 7. Traces of data analyzed at different times in April 1982 on pack ice at about 83°N, 20°E. Traces selected fo form a compatible composite and to illustrate various noise characteristics. Interruptions in some traces occur where electrical hum at the line frequency and its odd harmonics dominate the analysis bin. A mechanically coupled sub-harmonic was sometimes dominant, and the trace was also interrupted there. (When lowest analysis bin is located at frequencies less than 1 Hz, its spectral dentisy it quite uncertain and is shown in broken lines.)

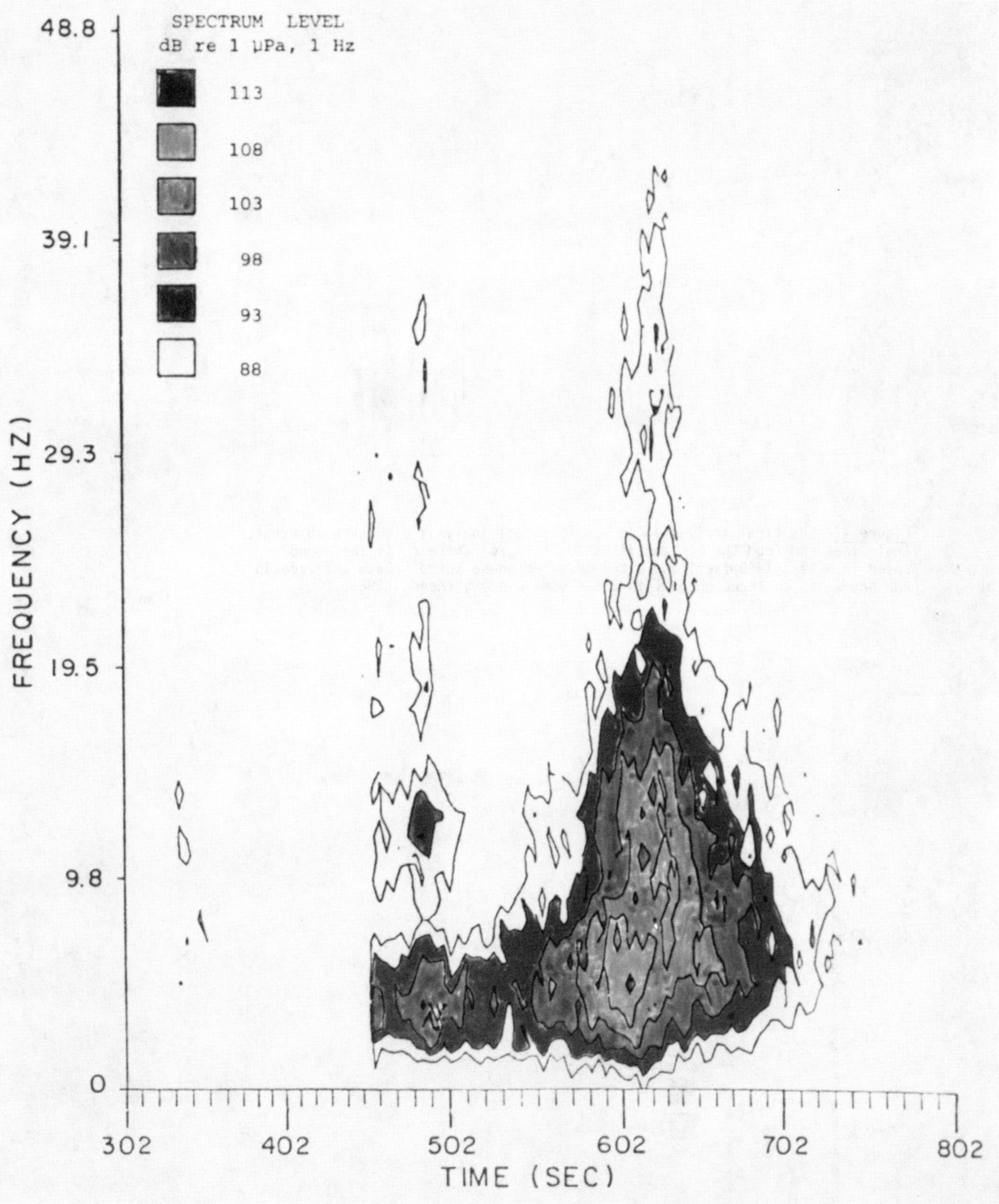

Figure 15. Sonogram of event on 14 April 1980. Each spectral cut is taken in a 2 sec time window. The data are high-pass filtered at 18dB/octave with a roll-off of at 1Hz of 8dB (Keenan 1983).

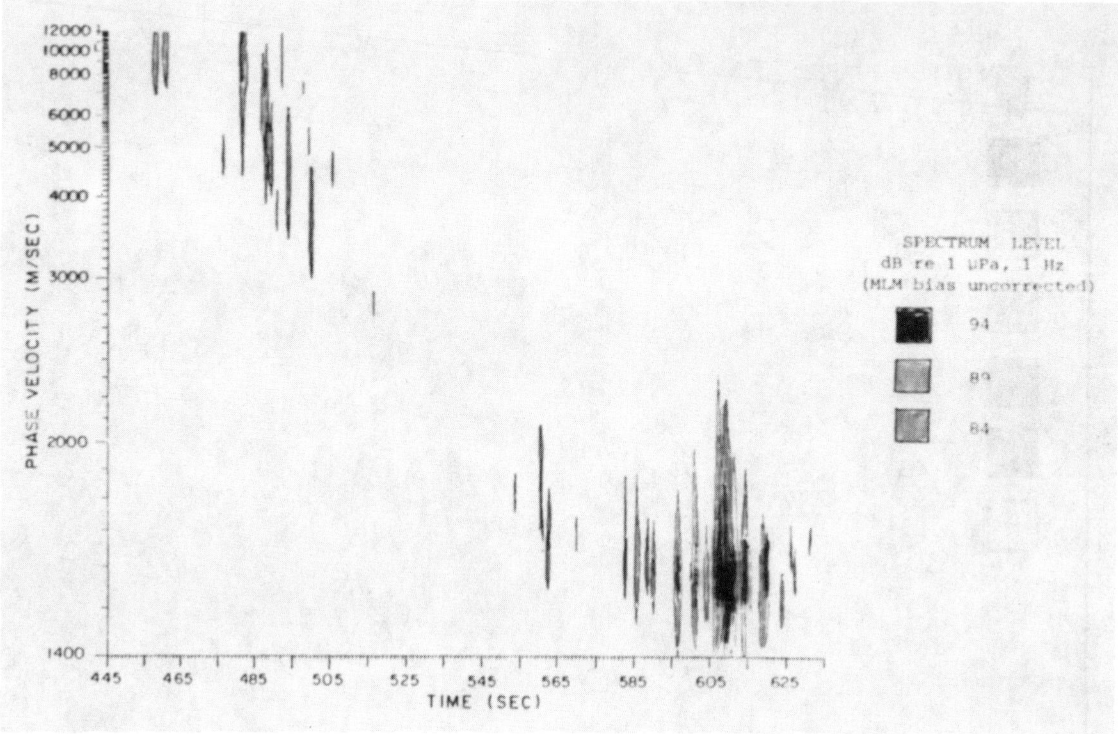

Figure 17. Vertical arrival angle (phase speed) analysis of Figure 15 event. Angle measured from the horizontal is $\cos^{-1}(c_w/c)$ where c_w is the sound speed in water (1440m/sec) and c the observed phase speed. Data anlayzed in a 2Hz bandwidth centered on 5Hz, in 2 sec time windows (Keenan 1983).

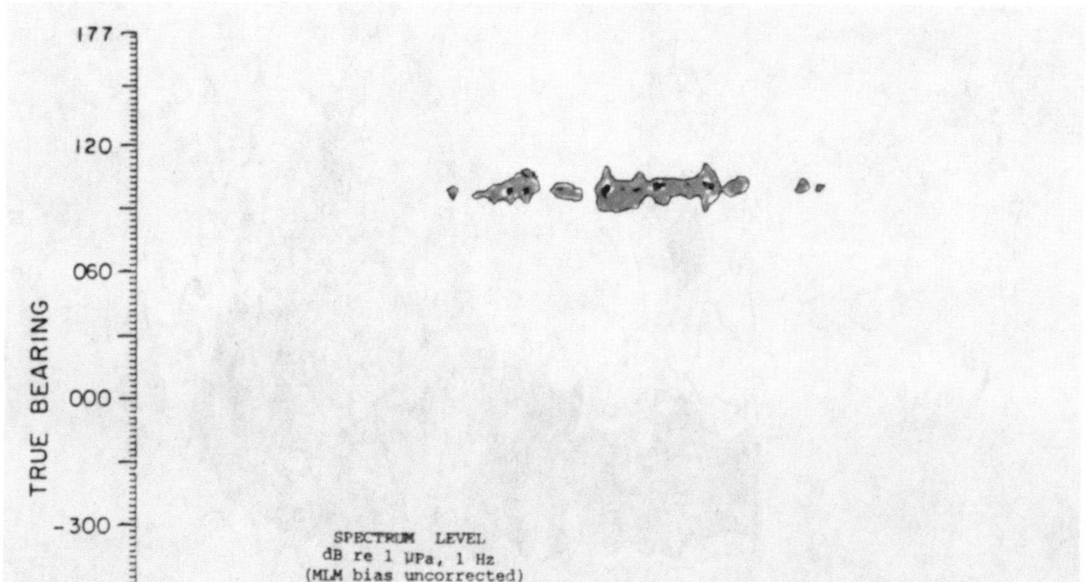

Figure 18. Horizontal arrival angle of Figure 15 event. Data analyzed in 2Hz bandwidth centered on 16Hz in 2 sec time windows (Keenan 1983).

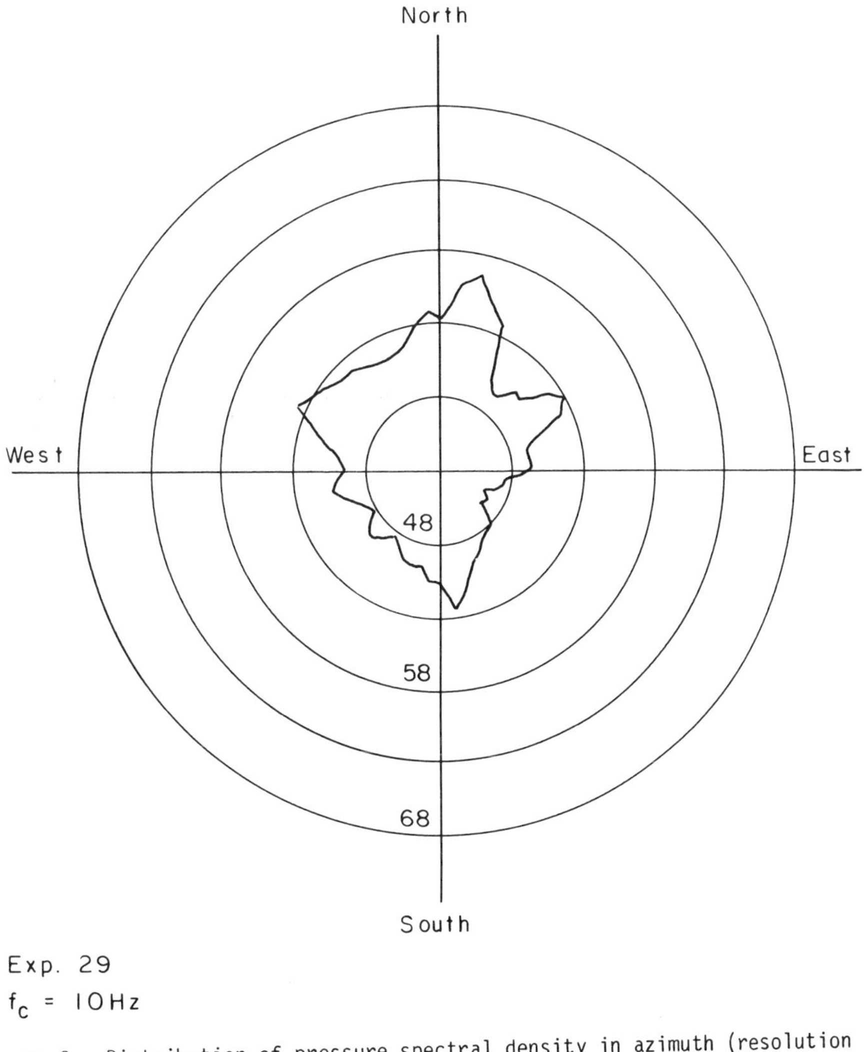

Figure 8. Distribution of pressure spectral density in azimuth (resolution about 8°). Data obtained in April 1978, under relatively quiet conditions, in the Beaufort Sea, over deep water, with fairly solid pack ice. A large horizontal hydrophone array was used to obtain the spatial distribution. Scale is in dB re 1 μPa and 1Hz, i.e. 68dB corresponds to a spectral density of 6×10^{-6} Pa^2/Hz. Frequency bandwidth is 4Hz centered on 10Hz. A spatial maximum occurs at about 012° (Shepard 1979).

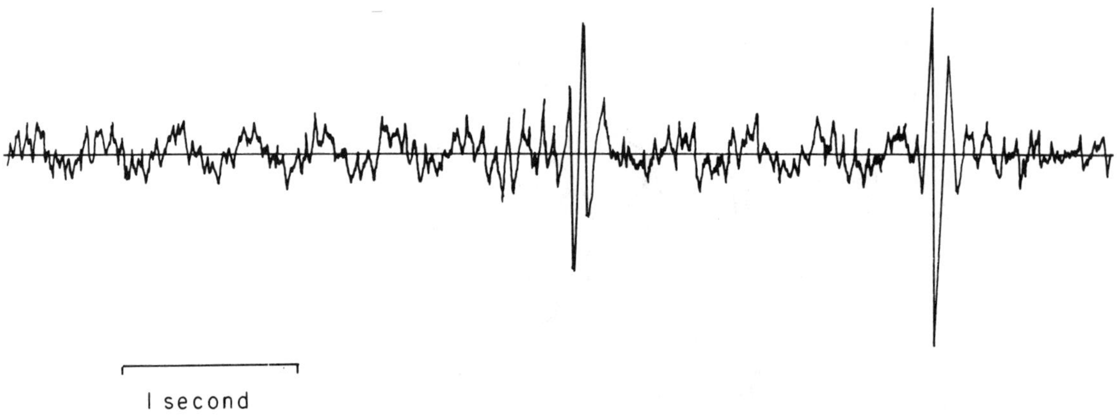

Figure 9. Time series observed in direction of maximum noise of Figure 8, 012°. Frequency bandwidth is 4 to 30Hz. Fluctuations appearing other than the two bursts represent noise of weaker and/or more distant events, noise leaking through the side lobes of the horizontal array, and strum (Shepard 1979).

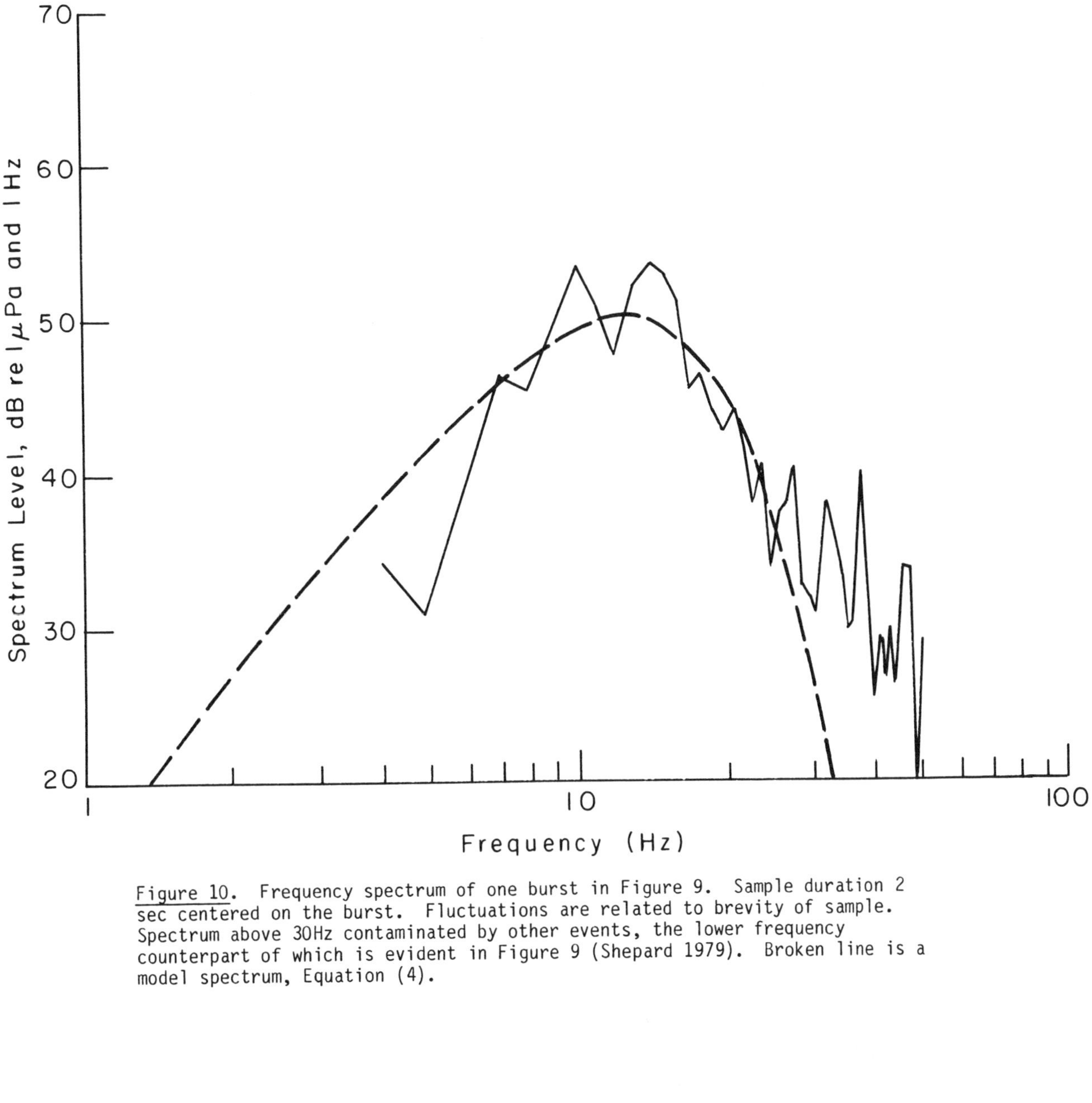

Figure 10. Frequency spectrum of one burst in Figure 9. Sample duration 2 sec centered on the burst. Fluctuations are related to brevity of sample. Spectrum above 30Hz contaminated by other events, the lower frequency counterpart of which is evident in Figure 9 (Shepard 1979). Broken line is a model spectrum, Equation (4).

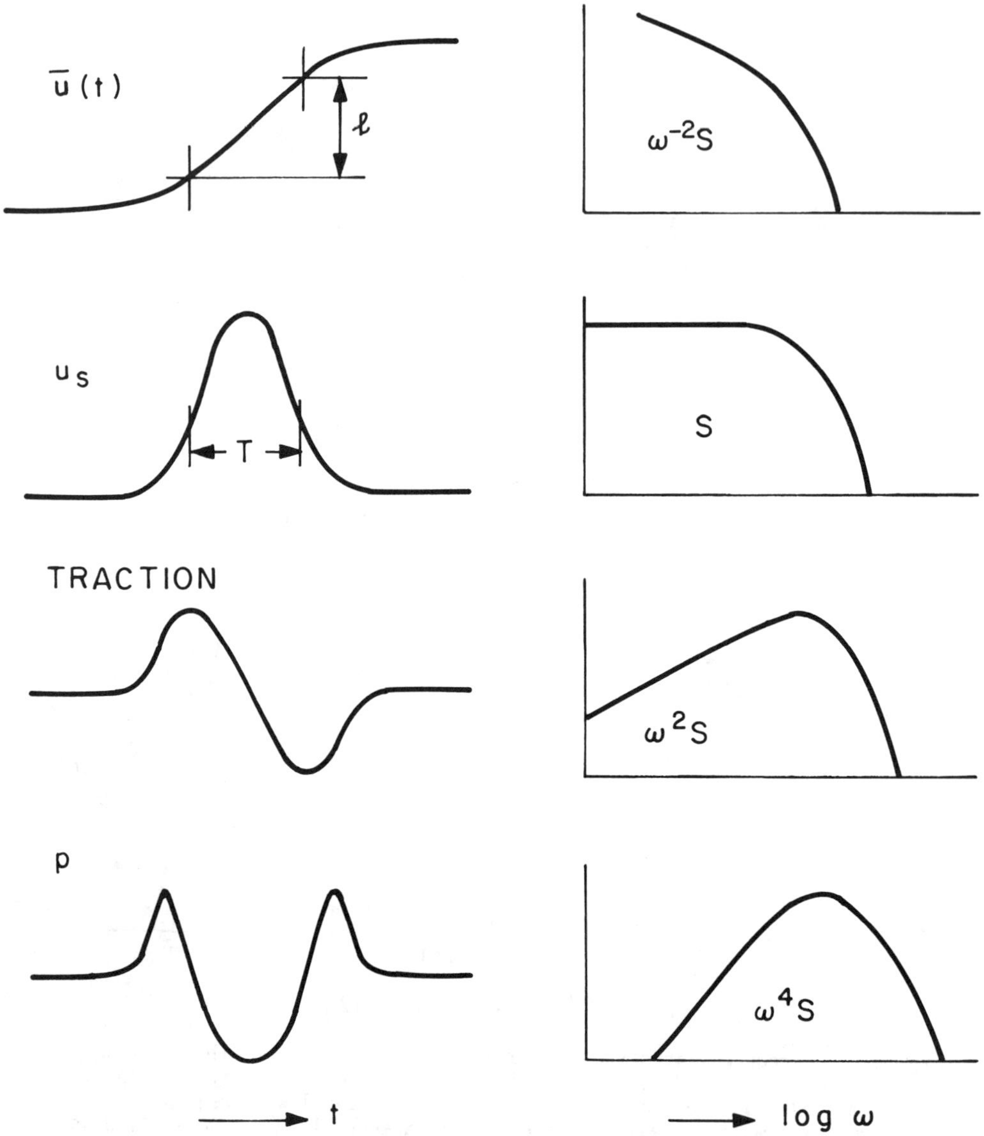

Figure 11. Model time functions and spectra for slip displacement $\overline{u}(t)$, shear wave displacement u_S, traction in an unbounded medium, and sound pressure in the water p. Time functions are linear - linear sketches, spectra log - log. Instead of a Gaussian pulse for u_S, Brune (1970) uses $u_S \simeq t \exp(-at)$ which would lead to a spectral shape for sound pressure the same as shown at low frequencies, but independent of ω at high frequencies. From the data, the Gaussian pulse appears to be a better candidate. Stephans and Pollock (1971) proposed a Gaussian pulse to model very high frequency (10^5Hz) emission spectra in nondestructive materials testing.

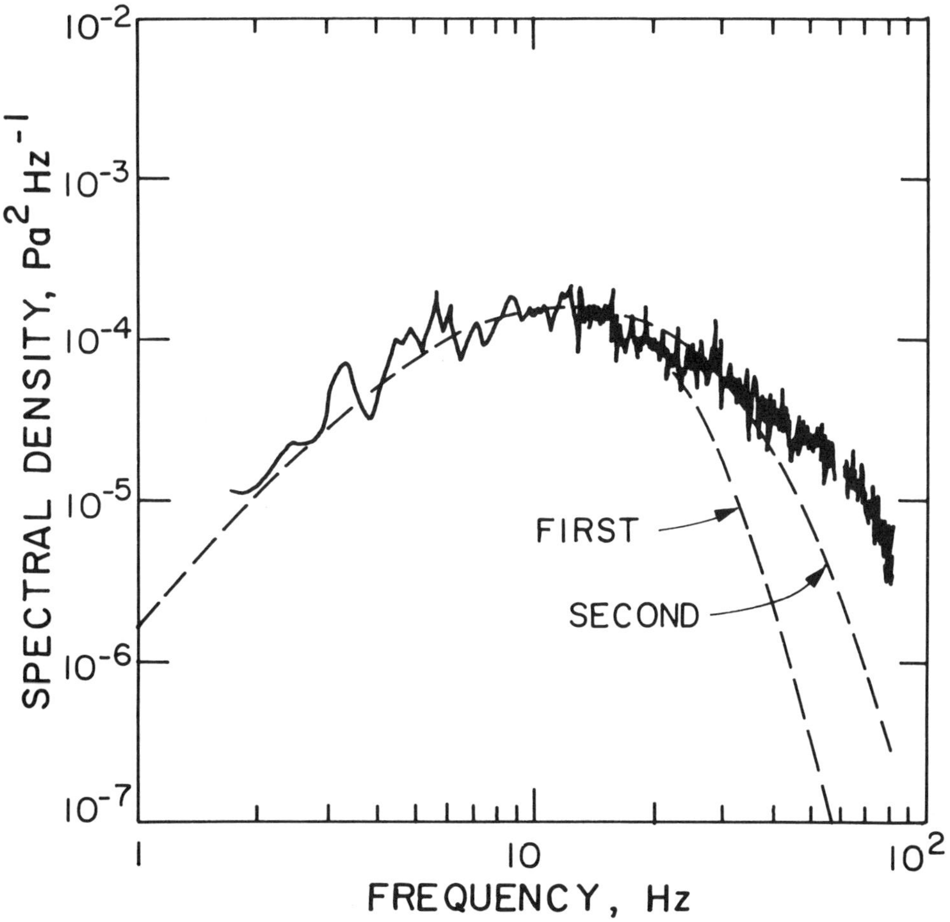

Figure 12. Noise spectral density obtained in April 1980, eastern Arctic Ocean over Pole Abyssal Plain (deep water) generally covered with broken pack ice. Noise is in quiet range of all conditions observed. Averaging time 60 sec, sensing area > 1km^2 (Chen 1982). Noise at 1Hz and 60Hz not shown because of strum and line frequency hum respectively. Broken line is the model spectrum $\langle S \rangle$ based on Equation (11) (first approximation) and on Equation (13) (second approximation). Some of the remaining under-estimation of noise may be due to another ice-action spectral distribution with its peak at about 200Hz.

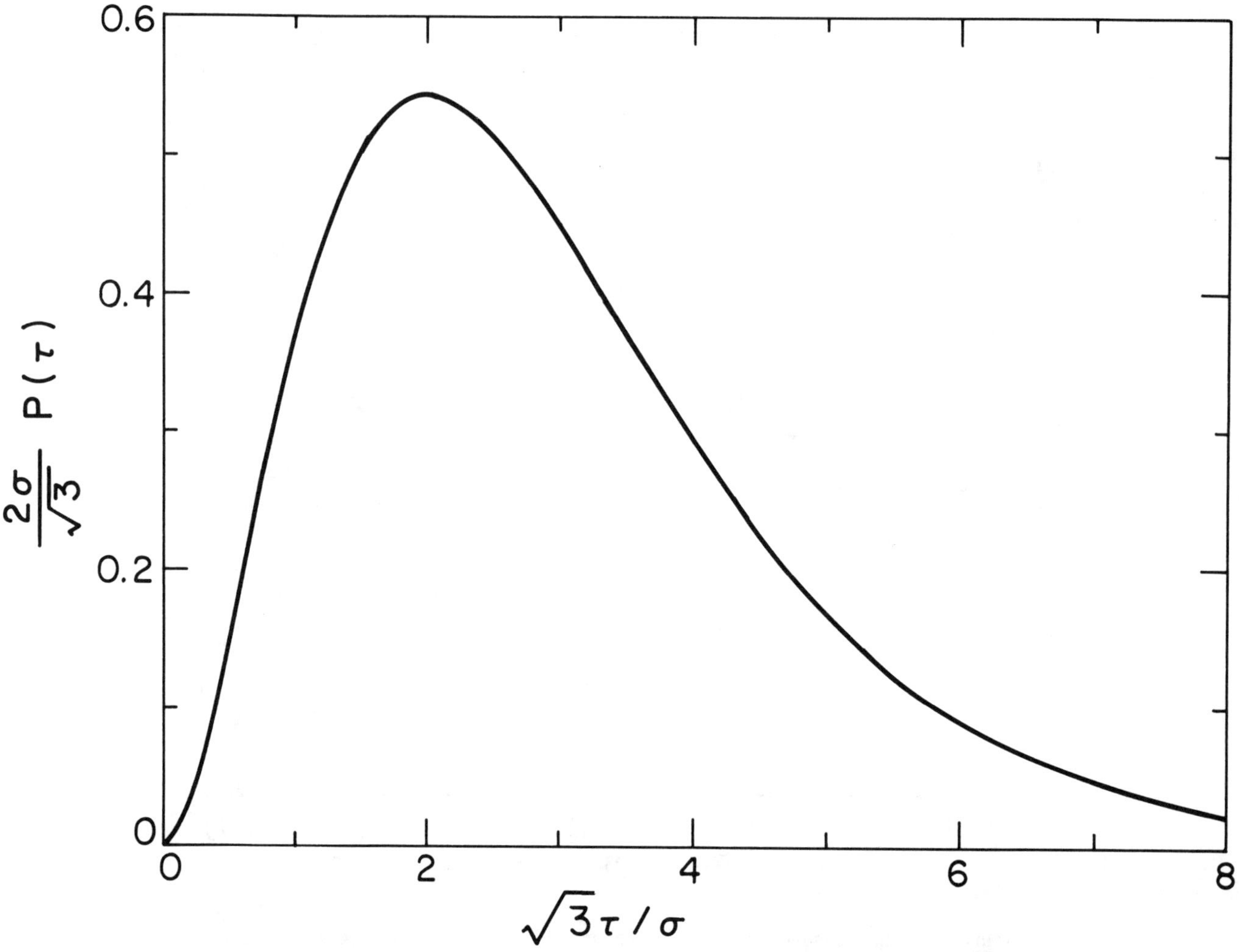

Figure 13. Stress probability density in pack ice derived from acoustic noise. The stress variance is σ^2, the mean $\sqrt{3}\sigma$, and the most probable value $2\sigma/\sqrt{3}$.

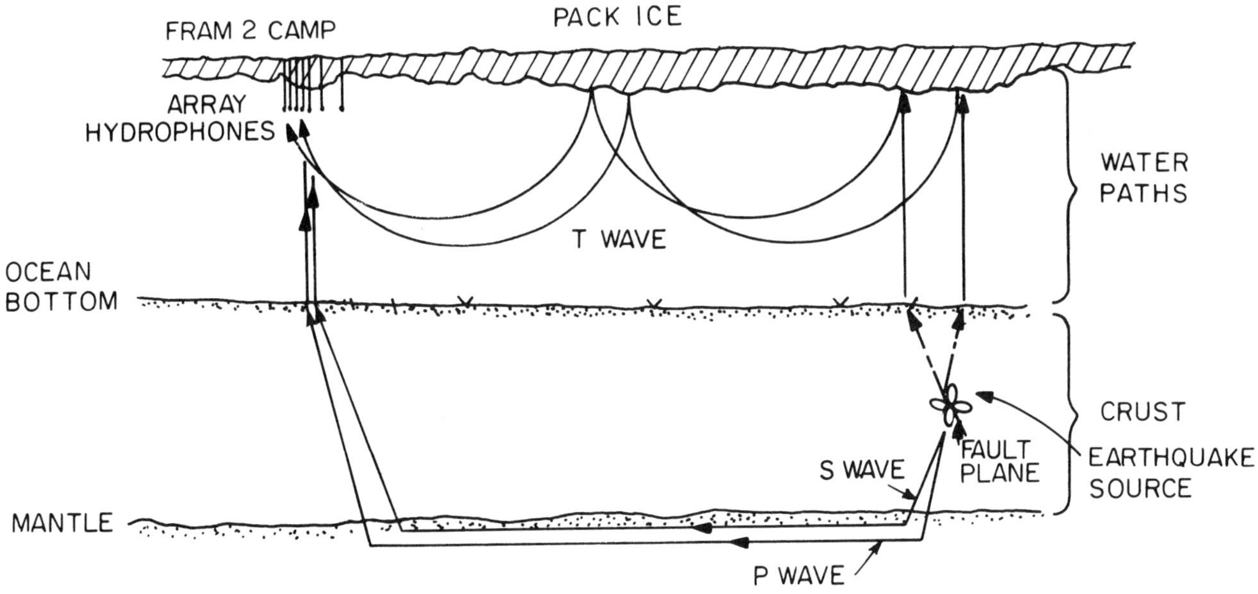

Figure 14. Scheme of seismic and acoustic propagation from an earthquake. Vertical scale grossly exaggerated (ice thickness about 3m, water depth about 4km, crust and sediment thickness about 10km, distance between source and array about 300km).

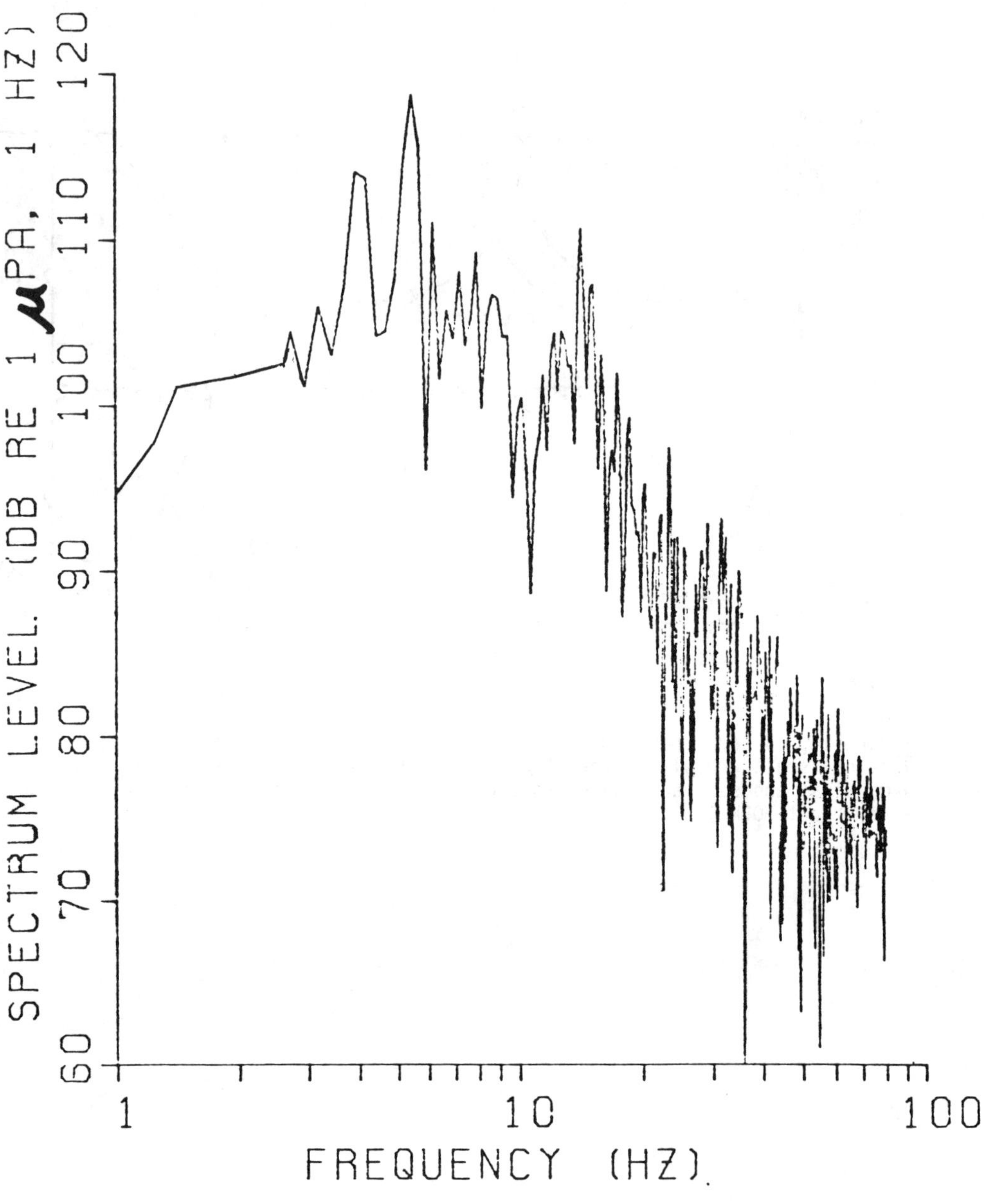

Figure 16. Spectral cut of Figure 15 at time of maximum spectral density (Keenan 1983). Values from 1 to 3Hz corrected for filter roll-off.

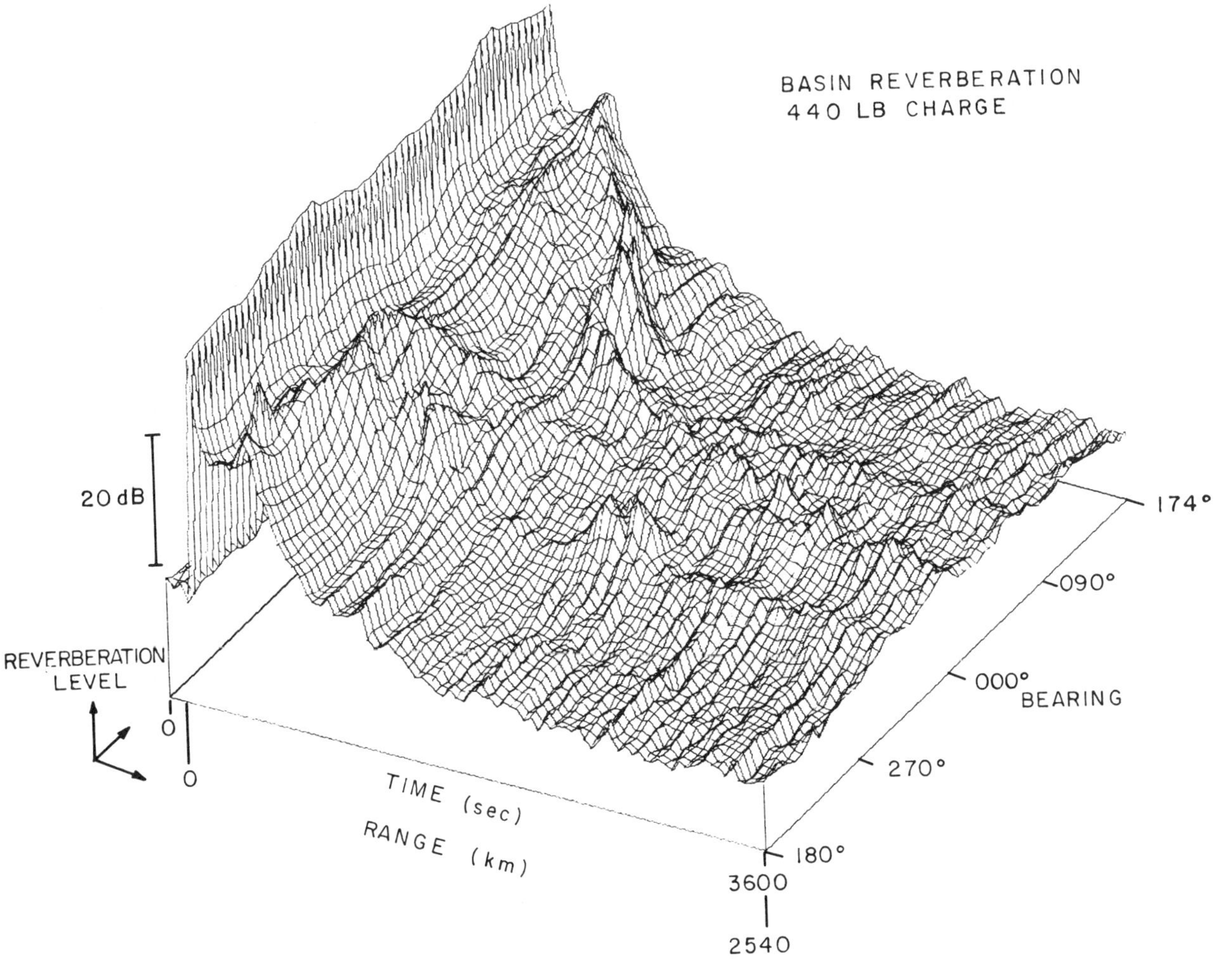

Figure 19. Arctic basin backscatter or reverberation, uncorrected for transmission loss or analysis parameters. Charge mass 200kg and detonation depth 244m. Array depth 61m, center frequency 9Hz. Signal level at the shot time (169 sec) is not physical, representing saturation of the hydrophones which quickly recover. Experiment centered at about 73.5° N, 150.5° W on pack ice in the Beaufort Sea with water depth about 3.8km, and carried out in April 1978.

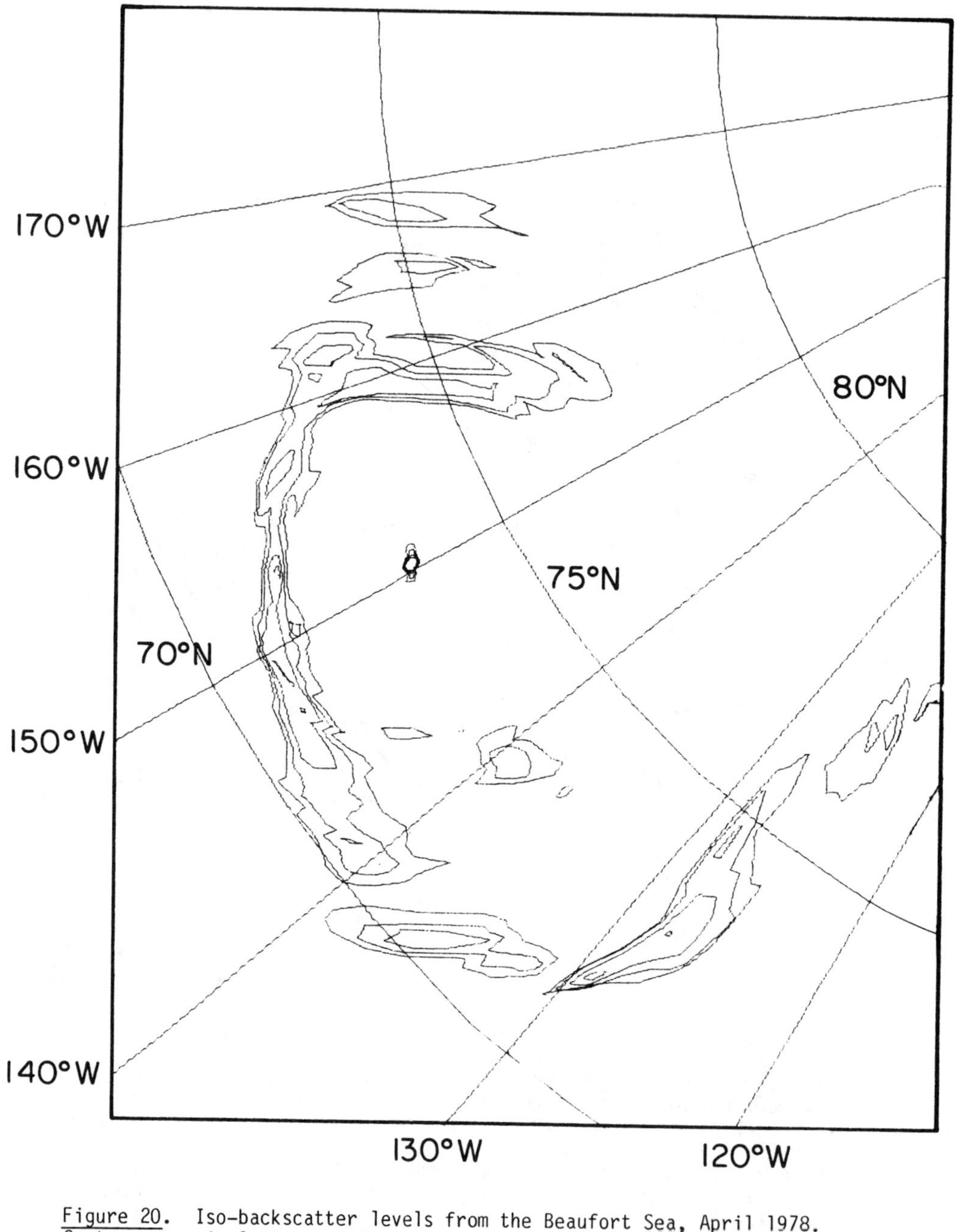

Figure 20. Iso-backscatter levels from the Beaufort Sea, April 1978. Contours are in 3 dB steps from -27 to -36 dB (see Dyer et al(1982) for interpretation of numerical values). Analysis frequency 9 Hz.

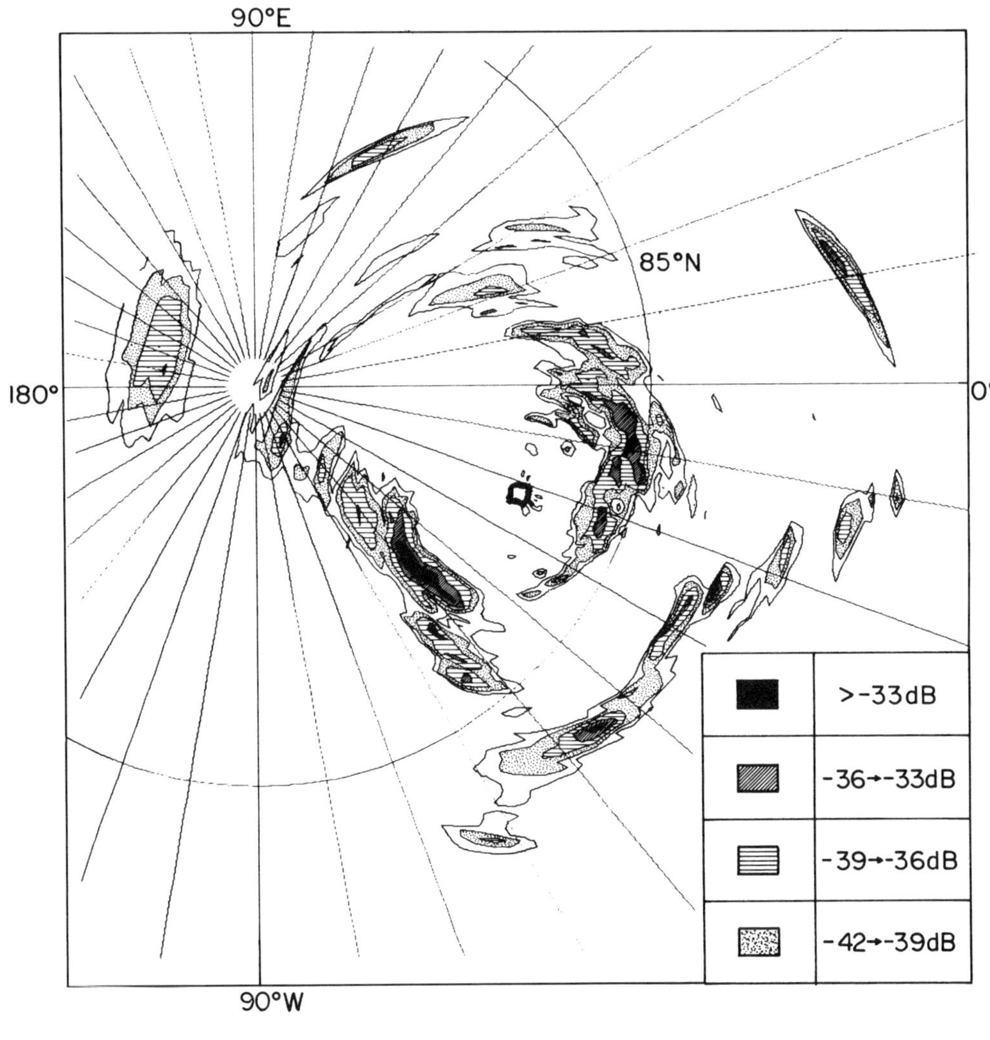

Figure 21. Iso-backscatter levels from the eastern Arctic, April 1980. Analysis frequency 10.5 Hz (Williams 1981).

MIT SEA GRANT COLLEGE PROGRAM SEMINAR

ARCTIC POLICY I

INTRODUCTION

Judith T. Kildow

Department of Ocean Engineering
Massachusetts Institute of Technology
Cambridge, Massachusetts

Not long ago it was considered isolated and remote. But now the Arctic has come of age. Modern society has come to the Arctic and with it the concerns and issues of our times. The Arctic is immense, so much so that changes brought by the buffeting political winds of the past decade are not readily visible, but rather must be seen as changes in perceptions and circumstances.

Issues that affect the Arctic, and through which make it possible to view this continuing process of change, can be identified as follows:

1. National sovereignty and jurisdictional claims.
2. National defense
3. Gas and oil and other mineral developments
4. Transportation and navigation
5. Oceanic and atmospheric research
6. Living resource and environmental management
7. Native interests and self-determinism

Perceptions of shortages in natural resources, particularly petroleum, pushed prospectors to remote Arctic areas. Advances in science and technology followed naturally, as did improvements in transportation, providing easier access to the entire region. This growth took place along with growing awareness of the importance of the region for national defense. At the same time, the Arctic's role as a nursery for world-wide weather became a factor in all northern hemisphere nations.

But despite this growing importance of the region in international affairs, the human systems needed to regulate change in the Arctic have lagged behind.

The five nations bordering the Arctic Ocean (U.S., Canada, Norway, USSR, and Denmark/Greenland) are becoming increasingly aware of the importance of the region in which they have a shared interest. On several occasions in the past, experts have suggested that these circumpolar states might best manage policy issues through cooperative efforts.

However, if the issues are scrutinized closely, this form of management or cooperation might be too broad in some cases and too limited in others. The policy issues, themselves, dictate who would be most interested in which areas. (see Figure 1)

For example, in addition to the five nations that border the Arctic Ocean, there are three more nations, Finland, Sweden and Iceland, which are subject to the same environmental conditions and are considered to be in the Arctic Environmental Region, (north of the July 10° Celcius isotherm). Finally, there are four more nations which have major interests in the Arctic: Great Britain, France, West Germany and Japan.

Some issues, such as national defense, petroleum and other mineral developments, lend themselves to national zonal management. Others, such as jurisdictional disputes, principally involve the nations in dispute. Transportation and navigation issues would involve these nations with commercial or military shipping interests. Oceanic and atmospheric research, particularly relating to climate, but also to ice qualities and other scientific phenomena, are probably of interest to all twelve nations. (see Figure 2)

In view of current Law of the Sea negotiations, other considerations must be noted which will influence modes and levels of cooperation and collaboration. While these are simple examples, they provide an international framework for thinking about the policy and legal matters which relate to the Arctic. Mr. J. Lawrence Hargrove and Ms. Nancy Kellner address this framework in their paper.

As far as U.S. interests and current management arrangements are concerned, many of the same questions might be asked. With more than 14 U.S. agencies having Arctic programs, the question arises, who should participate in which issues and coordinate with which levels of government? (see Figure 3) There is only one state involved, Alaska, but there are more than six state agencies in Alaska involved in Arctic affairs, and numerous boroughs and Native American reservations. At the same time, there are private sector interests including major industries involved in development. Add to these the industrial associations and the public interest groups, which mostly represent environmental and native peoples' interests

The U.S. has several options for management of Arctic issues, utilizing governing arrangements currently in place. Policy making can be carried out within the seven issue areas or across them, according to the level(s) of government directly involved. (see Figure 1)

Given increasing activities in the issue areas, one could conclude that better coordination, planning and cooperation would help avoid conflicts that may occur as well as take advantage of opportunities. Yet, there is an inertia against formal coordination. There is a need to develop sufficient evidence to overcome this affinity for the status quo. Are there sufficient problems now and in the near future to promote a coordinating mechanism?

For example, do the costs of delay in developing OCS resources warrant instituting public intervention mechanisms to speed up development? Or, are uncertainty and ignorance resulting from the lack of coordination of R and D impeding development and effective use of this area? Obviously, benefits result from reviews by more than one agency. Redundancy always provides the checks and balances in the system. But the question that must be addressed, using the case of the TransAlaska Pipeline (TAPS), for example, is: Were the benefits derived from environmental review and redesign of the project greater than the costs resulting from delays and redesign?

There are several guidelines which should influence Federal Government Policy in the Arctic.

- The government must provide a regional systems perspective concerning the range of seven issues previously mentioned.
- The government should coordinate Arctic regulatory activity among the federal agencies.
- The government must assure coordination of Arctic research as well as set priorities and assure that the research will be carried out either by government sponsorship or under government regulation.

The past few years has seen a piecemeal attempt on the part of national and international legal systems to recognize the unique features of the Arctic area. Mr. John Gissberg who has written about the current management regime for Alaska and the United States as he views it from his far northern perspective, mentions some of that legislation.

Some have said that although oil and gas development and military matters have dominated our interests in the Arctic up to now, and that science and environmental affairs have been pushed to the side, development and military interests will ultimatley have to rely on environmental and scientific information, demonstrating a need to push those concerns to the forefront of their priorities. If there is truth in that observation, then the 12 nations with Arctic interests will have to make immense efforts, pulling together to set long term research programs. It will take an especially inspired effort to pull in the most uncooperative member of that group, the Soviet Union, the nation which seems to pay least attention to this observation.

*FIGURE 1

ISSUES AREAS (7)	POLITICAL ARRANGEMENTS (6): INTERNATIONAL	INTERNATIONAL - NATIONAL	NATIONAL	NATIONAL-STATE	STATE (ALASKA)	STATE-LOCAL
• JURISDICTIONAL CLAIMS						
• NATIONAL DEFENSE						N.A.
• OIL, GAS, AND MINERAL DEVELOPMENT						
• TRANSPORTATION AND NAVIGATION						
• RESOURCES AND ENVIRONMENTAL MANAGEMENT						
• OCEANIC AND ATMOSPHERIC RESEARCH						N.A.
• NATIVE INTERESTS AND SELF-DETERMINISM						

*Figures prepared in collaboration with Dr. Jens Sorensen, M.I.T.

FIGURE 2*

NATIONAL INTEREST / ISSUE AREAS	JURISIDICTIONAL CLAIMS	NATIONAL DEFENSE	OIL, GAS & MINERALS	TRANSPORTATION & NAVIGATION	OCEANIC & ATMOSPHERIC RESEARCH	RESOURCES & ENVIRONMENT	NATIVE INTERESTS
USSR	●	●	●	●	●	●	
CANADA	●	●	●	●	●	●	●
GREENLAND DENMARK	●	◐	●	●	●	●	●
USA	●	●	●	●	●	●	●
NORWAY	●	◐	●	●	●	●	◐
ICELAND	◐	◐	◐	●	◐	●	
SWEDEN		◐		◐	◐		◐
FINLAND				◐	◐		◐
GREAT BRITAIN	◐	●	◐	◐	●	◐	
FRANCE		◐			◐		
WEST GERMANY		◐		◐	◐	◐	
JAPAN		◐	●	◐	◐	●	

ARCTIC OCEAN NATIONS: USSR, CANADA, GREENLAND DENMARK, USA, NORWAY

ARCTIC ENVIRONMENT NATIONS: USSR through FINLAND

ARCTIC INTEREST NATIONS: USSR through JAPAN

HIGH INTEREST = ● / LOW INTEREST = ◐ / NO INTEREST = BLANK

*This chart represents a first cut approximation at identifying national interests in the Arctic, from a study underway at M.I.T.

FIGURE 3

NATIONAL PROGRAMS

NATIONAL SCIENCE FOUNDATION

- POLAR PROGRAMS

DEPARTMENT OF DEFENSE

- ARMY (COLD REGIONS ENGINEERING)
- AIRFORCE (EARLY WARNING)
- NAVY RESEARCH
- NAVY OPERATIONS

DEPARTMENT OF ENERGY

DEPARTMENT OF INTERIOR

- MARINE MINERALS SERVICE
- FISH AND WILDLIFE SERVICE
- BUREAU OF INDIAN AFFAIRS
- GEOLOGICAL SURVEY

DEPARTMENT OF COMMERCE

- FISHERIES SERVICE
- WEATHER SERVICE
- COAST AND GEODETIC SURVEY

DEPARTMENT OF TRANSPORATION

- COAST GUARD

THE INTERNATIONAL LEGAL REGIME OF THE ARCTIC

John Lawrence Hargrove and Nancy J. Kellner
American Society of International Law

1. INTRODUCTION

The Arctic has long been popularly regarded as too harshly forbidding an environment to allow protracted periods of most kinds of human activity except on the rigorously constricted basis of what Arnold Toynbee called, in referring to the hunter-gatherer Inuit cultures, "arrested civilization". Especially in the last decade or so, however, it has come to be widely acknowledged that the Arctic is becoming at once less of a mystery, and a more important technical challenge. This widened awareness was stimulated primarily by discoveries of large oil and gas reserves, which made mastery of that environment at least for purposes of hydrocarbon exploitation an economic and strategic imperative for some of the five Arctic nations, and raised the prospect of extensive use of Arctic land and ocean areas for purposes of transport of products and equipment. But it was contributed to also by the development of submarine technologies which made possible protracted navigation under the polar ice, and, in the decade following World War II, strategic planning on the part of the two superpowers for the use of the Great Circle route for the delivery of airborne nuclear weapons against each other.

The practical concerns of states with the Arctic--military, economic, scientific, environmental, or political--have thus been increasing. This paper sets out, in summary fashion, some of the features of international law (the law governing among sovereign nation-states) that for various of these practical reasons seem to be of special interest in their application to the Arctic, and explains why this is so. It also examines possible needs for clarifying the law or developing new law or institutions for this region in the future.

It would be convenient but a bit misleading to say that we are concerned with "the international law of the Arctic", since, for the most part, international law takes no notice of the Arctic (or indeed either polar region): with narrow exception, the same general law applies in these areas as elsewhere. That this is so has become clear, in the development of the law, only within the last few decades, and it is instructive to note that the law could well have been arranged otherwise. The outcome of developments during this period might conceivably have been a general law which treated one or both polar regions--both land and water--as distinct areas for various purposes. In that event the likeliest analogues in actual international law would have been the concept of the "high seas", recognized for several centuries as an area in which the basic rule is a broad *laissez faire* principle denying any state exclusive prerogatives and ensuring equal access for all states, and the much later but quite similar concept of "outer space". Or the law could have developed in the opposite direction, granting special exclusive prerogatives to states most intimately connected geographically to the areas--a development in fact encouraged, with some practical if not legal result, by certain Artic states in their advocacy of the "sector theory" as the legal basis for determining sovereignty over Arctic land and (it was sometimes suggested) ocean areas [1]. [See also discussion below under "National Claims" and "Jurisdictional Uncertainties."]

In fact, neither of these directions was followed, as we shall see in the following pages, and there is no special international law of the Arctic or Antarctic as there is an international law of outer space. Nor, for that matter, is there even an "international law of the sea of the Arctic" in this sense. What is special, rather, is a set of perplexities or legal uncertainties resulting when states seek to apply general international law in the peculiar polar environments. In the Antarctic, these difficulties have been met to some extent by a special international contractual--i.e. treaty--arrangement among a few states with special scientific, economic, environmental or political interests in the area: the Antarctic Treaty [2]. This treaty does not change general international law in respect of the area, nor could it, being of very limited participation. Rather, it creates rights and obligations among a limited group of parties designed to facilitate the pursuit of at least some of these interests, while preserving others against serious injury. The same is true of a new treaty on Antarctic living resource exploitation adopted in 1980 [3], and will also be true of any arrangement resulting from the current discussions of a regime for mineral resource exploitation.

With the exception of the highly specialized treaty for the protection of the polar bear, there are no such special treaty arrangements for the Arctic region [4]. The question of whether there should be, and if so, what kinds of interests they should seek to deal with and among what group of states, has been fairly frequently raised in recent years [5], and we will recur to it. Our own conclusion is that the further development of international law respecting the Arctic, in the decades immediately ahead, is unlikely to take place in the form of any major revisions of the fundamental law so as to take special notice of the Arctic (a course for which, aside from the advocates of the sector theory, there seems never to have been an unequivocally committed political constituency). Instead, it will probably consist in clarification of that law in its application to the Arctic particularly as regards ocean spaces, coupled with an overlay of treaty arrangements among a limited group of states dealing with specific practical interests.

In any event, the only prudent assumption as a matter of policy is that some substantial legal development should and probably will take place. The Arctic falls clearly into that familiar class of environments once relatively inaccessible and uninteresting, which technology, economics and politics have combined--often, in this century, with great abruptness--to render otherwise. In such situtations it is common wisdom that an established legal framework acceptable at least to the principal parties in interest becomes more desirable in rough proportion to the increasing intensity of human activities and importance of the interests at stake. That principle is now especially relevant to the Arctic, where some of the activities in prospect require large capital investment and involve high political stakes. Clarification of existing law or elaboration upon it may well be necessary to reduce enterpreneurial risks to economically acceptable levels and to keep the risks of politicl conflict or environmental injury to tolerable minimums.

2. THE INTERNATIONAL LAW OF SPACES

2.1 Land Territory

National Claims. There have been stirrings of concern since the mid-19th century regarding international law in the Arctic in the form of disputed claims of national sovereignty over discovered land masses. With the exception of Alaska, these land masses are islands lying wholly or partly north of the Arctic Circle. The terms of the purchase of Alaska in 1867 [6], which seem almost frivolous by any contemporary measure of the value of that vast and only partly Arctic territory, presumably provide some indication of the level of importance attached to the earlier disputes. In any event most actual disputes over land territory were settled, on an *ad hoc* and bilateral basis, by the mid-1920's although the dispute between Denmark and Norway regarding portions of Eastern Greenland was decided by the Permanent Court of International Justice only in 1933, and in 1932, Norway had still not recognized Soviet claims to sovereignty over the Franz Josef Land archipelago [7].

There seems, in any case, never to have been any impetus to establish the kind of regional regime "freezing" disputed claims and establishing common rights that was set up in the Antarctic Treaty. As to Arctic islands, this is probably in part because there were fewer such claims, and because the relative proximity of the continental land territories of claimant countries to disputable islands made these claims at once more plausible *prima facie* and more readily reassertible on the basis of research expeditions or other national activities [8]. It is probably also due in part to the fact that, unlike the southern polar ocean, the Arctic sea lanes had long had a perceived potential importance. As to the continental territories, it is doubtful that a "common rights" approach could ever have been politically conceivable. Geopolitically, in short, the Arctic and the Antarctic are radically different.

An exception to the foregoing is the settlement regarding the Svalbard Archipelago concluded in 1920 in the Svalbard Treaty among Norway, the United States, Canada, Denmark and a number of additional European countries [9]. The treaty resolved disputes as to the territorial status of Svalbard by recognizing the sovereignty of Norway subject to the equal rights of all parties to enjoy access and carry on economic activities. While the Soviet Union was not among the original parties (it became a party later), the treaty provided that Russian nationals were to have the same rights as the contracting parties (Article 10). The Soviet Union in turn, unilaterally recognized Norwegian sovereignty over the archipelago and drew the Soviet "sector line" around Svalbard to reflect this position [10], though there is now considerable controversy over the Soviet contention that the economic rights provided for on land also extend to the continental shelf areas around the islands. [See discussion below, "Ocean Areas: Jurisdictional Uncertainties."]

As already indicated, there has been some effort on the part of Canada and the Soviet Union to assert sovereignty at least over Arctic land masses on the basis of a "sector" principle, according to which sovereignty would be divided among the Arctic states by lines drawn from the East-West extremities of the land territory of states abutting the Arctic ocean, northward to the Pole. The principle has not been accepted by other states and has been explicitly contested by the United States and Norway, among others. At the same time, it appears that one of the practical objects of its assertion--uncontested sovereignty over the islands of the Canadian archipelago and islands lying north of the Soviet Union--has been accomplished through the prolonged acquiescence of other states, despite the limited character of the actual occupation of the territories in question, and--except for Wrangel Island and Franz Josef Land--with relatively few actual disputes [11]. It has been pointed out, moreover, that the lenient criteria respecting occupation as a basis for establishing sovereignty that were applied by the Permanent Court of International Justice in the Eastern Greenland case make any actual contest of Soviet or Canadian sovereignty extremely unlikely [12].

Nevertheless, in view of the unacceptability of the sector approach to other nations, Canada has apparently been careful to support claims to land territory with other arguments, particularly in recent years. The Soviet Union, for its part, formalized in 1926 its claim to sovereignty over "all lands and islands already discovered or discovered in the future" within their "region of attraction" (bounded by the meridians 32°04'35"E. Long. and 168°49'30"W. Long.) which were not at that time recognized by the Soviet government as territories of any foreign state [13].

As will be seen, moreover, the sector theory, though moot as to sovereignty over land territory, is still at issue in some potentially important disputes over boundaries of the various ocean zones between adjacent Arctic states.

Issues Relating to Native Claims. While dealing with the international law question of national sovereignty over land areas, it should be borne in mind that, in addition to national states, groups of indigenous peoples may have certain legal rights with respect to Arctic lands. In fact, many of these groups have actively claimed such rights, and as commercial development of the region becomes more widespread and intensive, it is likely that native peoples will continue to seek primary control over the disposition of lands and resources which they have traditionally seen as their birthright, and upon which they continue to rely for sustenance and shelter.

It is important to note, however, that these claims have by and large not been claims of the right to exercise sovereignty over specified areas--the basic prerogative, in international law, of independent states-- but rather have been claims of rights under national law of ownership of particular areas or other special entitlements (such as rights of political or cultural autonomy, easements of access for hunting or fishing or other purposes, and others) [14]. They do not, therefore, raise issues about the international status of land areas, at least in respect of underlying sovereignty over those areas. This is true even though these claims may be based in part on agreements with the national government that, arguably, enjoy the status of international agreements--i.e. treaties--or upon newly emergent principles of international human rights law, in particular those which establish certain minimum standards relating to economic, social and cultural conditions of life.

International law and legal institutions do, of course, make provision in some detail for claims by a people that its international law right of "self-determination" has not been satisfied, and for the exercise of that right through--in the usual case--the achievement of national independence and the sovereignty over specified territory which being an independent national state implies. The typical "people" to which these provisions have been applied, however, during the period of decolonization since the Second World War, is that of a recognized colony, rather than indigenous peoples or other groups, within the population of a national state, that for whatever reasons may have secessionist aspirations.

As we shall note below, the historic position of indigenous peoples, upon which their claims may be based, may have a specific bearing on certain international law issues relating to ocean areas, since the traditional usages of these areas may be invoked by national governments in seeking to establish a status as national rather than international waters of certain ocean areas. [See discussion under 'Jurisdictional Uncertainties as to Ocean Areas', below.]

More importantly from the point of view of public policy, these indigenous groups constitute the main or only historically settled population in some areas that may be the areas most directly affected by resource-management and environmental decisions and the resulting stream of activities. Their right to have their interests taken into account in the working out of such decisions is recognized in various ways at the international level [15] as well as the national, and their ability to insist on this right with practical effect seems to be growing, though the specific terms of settlement may not be identical to the original claims and, further, the extent to which these peoples are granted hearing at all depends heavily on the country involved.

In the United States and Canada, for example, the federal governments have established native land claims settlement programs, which provide generally for settlement packages combining land grants and financial compensation with varying degrees of political autonomy [16]. Also recognized is the obligation of the national governments to consult native groups in the event of federally orchestrated or regulated future development of the area, e.g., as in the case of pipeline construction which might tend to disrupt the migratory patterns of caribou or other species. This is true in Greenland and Norway, as well, though in these two countries there has been a much higher degree of integration of native and non-native peoples particularly geographically, but to a certain extent culturally as well. In the case of Greenland, which was only recently granted home rule by Denmark, there is the additional question of national political and economic independence. In the Soviet Union, there appears to have been little action of this kind in respect of native peoples.

On the transnational plane, the first Inuit Circumpolar Conference was held in 1977 in Greenland, bringing together representatives from all the Arctic states except the USSR. There have been in this connection some efforts by several native groups to establish a kind of 'panarctic' policy with respect to rights over mineral and other natural resources, as well as to attain the status of recognized nongovernmental organizations (NGO's) within the United Nations system. The UN system has, in recent years, been more directly concerned with the rights of indigenous peoples, if only in a small way thus far. It would seem likely that these efforts will continue, and that Arctic native peoples will become more vocal in U.N. fora, especially as commercial activity increases.

2.2. Ocean Spaces

The International Law of the Sea. The main bulk of the earth's surface north of the Arctic Circle is salt water, and for this reason if no other, those portions of general international law that determine the rights and duties of states with respect to ocean areas or regulate various ocean activities could be expected to loom large in the international legal regime of the Arctic. As already indicated, this law, as with the law of land territory, almost without exception is the same for the Arctic as for other parts of the earth's surface.

This is not to say that, as a practical matter, the application of what is called generally "the law of the sea" to the Arctic will always be satisfactory, since for much of the year nearly 90% of the ocean surface is frozen. The ice pack and ice islands (the general permanence of which may vary considerably) are for the most part not specially provided for in existing law, and because they permit some activities (hunting, scientific research from semi-permanent stations, and hydrocarbon exploration, among others) that are normally carried out on land or from ocean vessels may present special problems. Nonetheless, a few remarks characterizing in barest outline the main body of the law of the sea, especially insofar as this law bears on the Arctic, are in order here.

It should be noted first that the law of the sea at the present moment is in a state of some uncertainty. It has been the object of efforts carried on intermittently for perhaps fifty years, and continuously for the last decade and a half within the United Nations, to draft a modern and comprehensive body of treaty law supplementing and accomodating to contemporary needs rules developed over several centuries. These efforts produced the massive United Nations Convention on the Law of the Sea adopted in December of 1982 by the Third United Nations Conference on the Law of the Sea, but with a significant though small minority of states (including the United States) declining to vote for or sign the treaty [17]. The treaty itself is not yet in force, and will not be until the required 60 states have ratified it--a process likely to consume several years at the least. Nevertheless its text is already an important element in determining what the law is, for two reasons: First, it carries forward many features from earlier law of the sea treaties, also produced by the United Nations in the course of the 1950's, which are in force and have wide if not universal acceptance. Secondly, most of its provisions were widely supported during the treaty negotiations and almost universally at the conclusion of those negotiations, and to some extent are already reflected in the actual practice of states. Thus there is some basis for arguing that they constitute "customary international law" even before the Treaty's entry into force and more strongly so after entry into force.

Such argument has already been contested, however, by some signers of the treaty, to the extent that it might be put forward by states like the United States that will wish to claim the legal force of portions of the treaty acceptable to them while avoiding (through refraining from becoming parties to the treaty) portions that they find unacceptable. The principal objection of states refusing to sign the treaty was directed to the deep seabed mining provisions [18]. Even if there were no such controversy, it is in the nature of customary international law that establishing both the existence and the content of its rules is likely to be subject to greater uncertainties than is the case with the written law contained in treaties in force.

Presumably the greatest uncertainty as to the state of the law of the sea will be as to those portions of the new treaty which consist most nearly completely of new law, neither carried forward from previous treaties nor reflecting pre-existing customary law. The most important such new portions of the treaty are those establishing for the coastal state a broad coastal band of exclusive jurisdiction for economic purposes (the "Exclusive Economic Zone", or EEZ), and those creating an international deep seabed area beyond national jurisdiction, and establishing arrangements for mining of seabed minerals in that area.

The main function of the law of the sea is the drawing of geographically definable ocean zones within which states have different jurisdictional prerogatives. Like the law of land territory, it mainly lays down rules governing which states legally may do what, or must refrain from what, in what areas.

Only secondarily--but still to a much greater extent than in the case of land territory--does it establish rules of conduct prescribing how those prerogatives are to be exercised, and to what ends.

For example, there are some highly general principles of conduct such as those enjoining cooperation among states or making states legally responsible for the harmful consequences of their actions. These are found in general international law and are reflected somewhat more specifically, for example, in the new Convention on the Law of the Sea, discussed below. Additionally, that Convention (as the pre-existing law of the sea does to a lesser extent), contains a good deal of material which is concerned, sometimes in considerable detail, with accommodating possible overlaps or conflicts in the exercise of the various prerogatives which it lays down. An important example is the provisions of the Convention concerned with reconciling the rights of free navigation of the ocean with the rights of coastal states to regulate activities for economic, environmental, or various other purposes in zones of national jurisdiction. Moreover, the new treaty sets out separate, fairly detailed sections relating to the conduct of scientific research (particularly in the international seabed area); the development and transfer of marine technology; and protection of the marine environment.

The New UN Convention on the Law of the Sea. The jurisdictional map drawn by the law of the sea is that of a variety of sometimes overlapping zones in all of which (with the exception of "internal waters") there is a mix between the prerogatives of the coastal state and those of all other states. The new treaty redefines in important respects criteria for the determination of both lateral boundaries and seaward limits which could well hold implications for current disputes in the Arctic. Of perhaps even greater interest for the Arctic is the Treaty's attempt, just mentioned, to balance in each of several coastal zones the interest of the coastal state in protection against pollution from vessels with that of other maritime states in unhampered navigation. It does so by various requirements that coastal state environmental rules for vessels in effect be no more stringent than accepted international rules--which all maritime states may have had a say in formulating--but makes an important exception for the Arctic.

The territorial sea, that which is closest to the land territory of the coastal state and therefore the most heavily national zone, is set in the Treaty at a maximum of 12 nautical miles (Article 3), an extension of the traditional three-mile limit which the United States, notably, has not recognized but which has over the years been adopted by many states, including the Soviet Union. The treaty contains specific instructions for the drawing of baselines from which the territorial sea is to be measured, but where the coastlines of two states are opposite or adjacent to each other, the median line prevails, except where "historic title or other special circumstances" require variance from these provisions (Article 15).

The concept of the territorial sea is that of a narrow belt of ocean waters adjacent to land in which the sovereignty of the coastal state--including its power to regulate specific activities such as transportation--is like that which it enjoys on land territory, subject only to narrowly circumscribed, if quite important, rights of passage by foreign vessels. Within that belt, the right of "innocent passage" of all foreign vessels is subject to regulation by legislation of the coastal state in respect of--among other things--environmental protection, conservation of living resources, and scientific research (Article 21 (1)), but these laws and regulations cannot apply to the "design, construction, manning or equipment" of these vessels, "unless they are giving effect to" already established international standards (Article 21, (2)).

By contrast, in the regimes established for straits used for international navigation, which are or could be enclosed by a 12-mile territorial sea, and for the new 200-mile exclusive economic zone, the balance between the prerogatives of coastal states and those of other states interested in navigation is shifted significantly in the direction of the latter. States bordering straits used for international navigation may adopt laws and regulations relating to "transit passage" (Part III, Section 2) through the strait, for the "prevention, reduction and control of pollution" by vessels, but they may do so only "by giving effect to applicable international regulations regarding the discharge of oil, oily wastes and other noxious substances" in the strait (Article 42(1)(b), emphasis added). Moreover, unlike that of the territorial sea, the straits regime does not make any distinction between warships and other vessels, nor does it require submarines to surface while in transit. This new right of "transit passage" applies as well to aircraft.

In the exclusive economic zone, as with straits, the basic rule as to coastal state environmental regulation is that laws and regulations for the prevention, reduction and control of pollution from vessels are enforceable by the coastal state, but must give effect and conform to generally accepted international standards and rules (Article 211(5)).

However, the treaty makes an exception to this rule which, though in theory potentially applicable as well in the Antarctic, was drafted with specific regard for the Arctic, and is the first of its kind. Article 234, often called the "ice-covered exception", applies to areas "where particularly severe climatic conditions and the presence of ice covering such areas for most of the year create obstructions or exceptional hazards to navigation, and pollution of the marine environment could cause major harm to or irreversible disturbance of the ecological balance." It gives the coastal state the right to "adopt and enforce non-discriminatory laws and regulations for the prevention, reduction and control of marine pollution from vessels in ice-covered areas within the limits of the exclusive economic zone", and therefore exempts the coastal state from the requirement to keep its own regulations within the limits of stringency set by international regulations.

As to pollution in connection with sea-bed activities, artificial islands, structures or installations in the EEZ, laws and regulations adopted and enforceable by the coastal state must be "no less effective than international rules, standards and recommended practices and procedures." (Article 208(3), emphasis added). States have the further obligation to try to harmonize such laws and regulations at the "appropriate regional level" (Article 208(4)).

The EEZ, a new concept in the law of the sea, constitutes the most important nationalization of formerly common resources ever effected in international law. The zone is an area adjacent to the territorial sea, which extends beyond the coastal baselines to a maximum of 200 nautical (Articles 55, 57). As on the high seas, freedoms of overflight, navigation, and the laying of submarine cables and pipelines are preserved for all states in the EEZ (Article 58). However, in its EEZ, the coastal state has the sovereign right to explore, exploit, conserve and manage the living or non-living resources of the waters, seabed or subsoil (Article 56(1)(a)).

In addition to these basic economic prerogatives, the coastal state has important rights in the EEZ with respect to marine scientific research, with potentially restrictive effects on research activities by foreign vessels. Research is to be conducted with the consent of the coastal state, though consent is not to be "unreasonably" delayed or denied, in keeping with the express obligation on the part of all states to promote and facilitate marine scientific research for peaceful purposes (Articles 239, 246). Research must conform to certain "principles" which as a practical matter would give an unsympathic coastal state considerable ability to impede research activities (Article 240). The coastal state may, moreover, reasonably restrict or prevent such research if the proposed activities have direct significance for the exploration and exploitation of living or non-living natural resources or involve the construction or operation of artificial islands or other structures (Article 246). In any event, the coastal state has jurisdiction over these installations or research activities (Article 56).

On the continental shelf, the coastal state has the same basic rights and obligations with respect to marine scientific research and environmental protection as in the exclusive economic zone, including the right to pass laws and regulations with respect to artificial islands, installations and seabed activities. However, the coastal state may not withold consent for marine scientific research in such continental shelf areas as may extend beyond 200 miles from the territorial sea baseline (see below), except for special, publicly designated areas in which "detailed exploratory operations" or exploitation are or will be taking place (Article 246(6)). In addition, the coastal state may regulate and determine the course for the laying of submarine cables or pipelines, although they may not prevent other states from using the continental shelf for these purposes (Article 79).

The legal continental shelf may extend to a distance of 200 nautical miles from the territorial sea baseline or to the outer limits of the submerged natural prolongation of its land territory (the continental margin), whichever is greatest, with a maximum outer limit of 350 nautical miles from the coastal baseline (Part VI, Article 76). It is likely that this would effectively place most of the seabed hydrocarbons within the exclusive jurisdiction of coastal states, which have the sovereign right to authorize and regulate any drilling to be done on the shelf (Article 81). Moreover, once any outer limits beyond 200 miles from the coast have been reviewed and, as necessary, revised by a special Commission on the Limits of the Continental Shelf (Annex II), those limits are considered "final and binding", thus reducing in large measure the risk of future disputes relating to, for example, the division between national and international mineral resources (Article 76(8)).

However, there is no specific criterion for delimiting the continental shelf between states with opposite or adjacent states (a problem of particular relevance in the Arctic. [See "Jurisdictional Uncertainties", below]. In the case of a dispute on this point, the new treaty stipulates merely that delimitation shall be by agreement "on the basis of international law . . . in order to achieve an equitable solution" (Article 83), whereas in the 1958 Convention on the Continental Shelf, the line of equidistance was to be considered the boundary unless another line were justified because of special circumstances. Failing agreement, states are required to resort to binding dispute procedures set out in a separate part of the treaty (Part XV).

One wholly new aspect of the Treaty is that any exploitation in the area beyond 200 miles of non-living resources be, in effect, taxed. These "payments and contributions" are to go into an international fund managed by the new Seabed Authority [see discussion of deep seabed, below], and will eventually be distributed to states parties on the basis of "equitable sharing criteria", emphasizing the needs of developing countries (Article 82).

The high seas constitute the least national zone, where all states have equal rights and obligations, and are defined only as "all parts of the sea" not included in the EEZ, territorial sea, internal or, as appropriate, archipelagic waters of a state, which would include waters superjacent to the continental shelf (Part VII, Article 86). Rules of the high seas apply as well to the exclusive economic zone, providing they are not incompatible with the regime of the EEZ (Article 58). The basic defining characteristic of the high seas is that no part of it may be validly subjected to the sovereignty of any nation (Article 89), a provision which greatly weakens any Arctic sector claim to what would under the treaty be high seas. [See discussion of "Jurisdictional Uncertainties", below]

On the high seas, all states may exercise the freedoms of navigation, overflight, scientific research (subject to other relevant provisions), and fishing (subject to conservation and management conditions in the Treaty), as well as the freedom to lay submarine cables and pipelines, and to construct artificial islands and other installations permitted under international law (Article 87). Various more specific duties with respect to operations on the high seas are also laid down [19].

If the exclusive economic zone can be considered the most effective nationalization of what were formerly common resources, the new regime of the deep seabed can be said to do roughly the opposite: Under pre-existing law of the sea, it can be argued that all states would have freedom of access to exploit the resources of the deep seabed, analogous to the freedom to fish on the high seas. Under the new treaty, however, the "Area" (defined as the deep seabed, ocean floor and subsoil thereof beyond the limits of national jurisdiction) and its "resources" (all minerals including polymetallic nodules) are designated as the "common heritage of mankind" (Article 136). Mineral exploitation is controlled and regulated by the International Seabed Authority (in conjunction with its various consultative organs), whose activities are supported through fees levied on mineral production and/or profits. The "Enterprise", an organ of the Authority, is an intergovernmental mining company which is concerned with the economic development of the Area, much like any other company (Annex IV). However, there are special provisions to ensure the availability of the necessary technology to the Enterprise (Annex III, Article 5(5)), and individual applications to the Authority for mining privileges must propose for development two "sufficiently large" areas of "sufficient and equal estimated commercial value". The Authority will designate one to be awarded to the applicant, and the other to the Enterprise or to developing countries. (Annex III, Article 8).

As noted above, the concept of the international seabed area is that of a "common heritage of mankind", in which activities are subject, in considerable degree, to limitation by duties owed to the international community as a whole and possibly by regulation emanating from an agency representative of that community. (The conflict between this concept and that of the freedom of the high seas was accomodated in detailed provisions regarding the international area, but not to the satisfaction of the United States and other states that have declined to sign the Treaty because of objections to the seabed provisions.) Accordingly the Treaty's provisions on conduct of seabed research in the international zone contain a number of restrictive if vaguely formulated obligations. For example, scientific research in the area "shall be carried out exclusively. . . for the benefit of mankind as a whole"; states are required to "promote international co-operation in marine scientific research in the area by participating in international programmes and encouraging co-operation in marine scientific research by personnel of different countries and of the [seabed] Authority" and by engaging in other co-operative programs (Article 143).

Jurisdictional Uncertainties as to Ocean Areas. Even if it is accepted that questions of sovereignty over land territory are well settled, and that their settlement has entailed no general recognition of a sector principle or other special international law (for land or sea) in the Arctic [*], there remain several unsettled issues about national rights in ocean areas, and a number of these stem from efforts by Canada and the USSR to introduce special considerations of law or fact into the application of general international law in the region.

For one, the sector principle survives in the advocacy of these two states of boundaries separating their coastal zones from those of adjacent or opposite states. When this method of demarcation -- rather than, for example, the "equidistance" principle which is embodied in the 1958 Convention on the Continental Shelf, is applied to boundaries involving Norway and the the United States (in case of the USSR) or the United States (in the case of Canada) it produces some quite significant disputed areas.

Another set of disputes arises from differences as to whether earlier legal agreements on the disposition of land areas also govern, on the basis of general international law principles, the division or the status of the appurtenant ocean zones (most importantly, the continenal shelf and, potentially, the EEZ). In the case of the US-USSR and the US-Canada shelf boundaries, the Soviets and Canadians have used this argument in addition to or in place of the sector approach, with at best, dubious justification [20].

The same kind of question has been raised concerning the delimitation of the continental shelf between Norway (Svalbard) and the Soviet Union (Franz Josef Land) in the Barents Sea, and the status of the Svalbard shelf. The Soviets have invoked the 'equal access' language of the Svalbard Treaty in defense of their claim to rights of equal access to the continental shelf and ocean areas around the islands (for which there is little basis, as the treaty does not mention the sea or shelf zones), while the Norwegians reject this view and support the "equidistance principle" which is set out in the 1958 Convention on the Continental Shelf [21].

A further set of disputes, also

---

* Indeed it would appear quite doubtful (Footnote continued)

*(continued) that either power would in the end have wished to press the sector theory through to its most extreme logical conclusion -- namely, that there is no part of the Arctic Ocean in which states enjoy high seas rights. It is hardly likely that the Soviet Union, for example, would have any interest in calling into question its high seas right to move freely in and above the open waters offshore of other Arctic states, or exit freely into the Bering or Norwegian seas, in return for the only dubiously enforceable right to exclude the United States and others from navigation in a Soviet "sector" of what is now the Arctic high seas. In addition, it has been pointed out that the long traditions by all the Arctic states of sending research teams all over the Arctic Ocean indicates that they subscribe in practice to high seas freedoms and so implicitly, sector theories notwithstanding, recognize the existence of Arctic high seas (D. Pharand, "The Legal Status of the Arctic Regions", op. cit., at 84.)

relating not to boundaries but to the status of the maritime zones themselves, grows out of the efforts of Canada and the Soviet Union to invoke the special historic, geographic, climatic or physical facts of life in the Arctic not in the application of general international law but to produce outright exceptions to it, applicable in special circumstances. Particularly as pressed by the Canadian government, this strain of policy produced in the new Treaty what could be described as the only genuine polar-region exception to general international law, the "ice-covered exception" stated in Article 234. Of a similar sort are the claims advanced by Soviet jurists, if not unambiguously asserted by the Soviet government, that the Kara, Laptev, East Siberian and Chukchi Seas constitute internal waters of the Soviet Union [22].

Finally, there remains the question of the jurisdictional status of the waterways of the Northwest Passage, through the Canadian archipelago, and of the Northeast Passage along the northern coast of the Soviet Union. The former was the issue which underlay the dispute over the 1969 voyage of the U.S. icebreaker Manhattan through the Northwest Passage and the 1970 Canadian legislation establishing a 100-mile environmental protection zone (the Artic Waters Pollution Prevention Act), and was instrumental in producing the "ice-covered exception" in the new treaty, just mentioned. In light of the text of the new treaty, the issue is whether the right of "transit passage" through straits used for international navigation will apply to these waterways, to permit commercial or other vessels to navigate one or another of the routes constituting the Northwest Passage. It may well be that Canada does not intend, as a practical matter, to impede passage even of foreign oil tankers. But it is safe to presume that it will continue to insist, on whatever ground, on the right to subject such passage to environmental criteria, and would prefer to be able to exclude military vessels if it wishes. There appear to be several legal grounds, none completely uncontroversial, on which Canada might base these positions. (1) Because of the historic usage of the waterways by native peoples, it can be argued that they have the status of internal waters, excluding any international rights of passage. A difficulty with such an argument is that the concept of historic rights in the law of the sea is of dubious validity except where specifically mentioned (in other connection--see, for example, Article 15 of the new UN Convention) in treaty language. (2) Because there has never been commercial international navigation through these waterways, it may be argued that they cannot be international straits but are at most territorial sea, in which Canada has the rights with respect to environmental regulation and passage of military vessels that that status implies. The concept of a "strait used for international navigation" is not defined in the new treaty, however, and the outcome of a judicial determination of this issue is uncertain. (3) In any event, Canada can argue that its right to regulate for environmental purposes is augmented by the "ice-covered exception" in the new treaty, although the status of that provision, particularly in a dispute between Canada and the United States, might itself be uncertain [23].

The situation with respect to the waterways constituting the Northeast Passage has been described as "roughly the same" as that of the Northwest [24]. Soviet practice with respect to the control of foreign vessel traffic through the Vil'kitsky Straits and the coastal seas has apparently not been entirely consistent [25]. A 1971 Soviet law establishing the Administration of the Northern Sea Route did give that organ broad authority to control navigation on environmental or other grounds, although it did not explicitly apply these rights to foreign vessels [26].

There is one further area which thus far has not been treated specifically in international law, and which is peculiar to the polar environments--that is, the jurisdictional status of drifting ice islands, or, perhaps, human activities on the ice generally. The ice has been used by scientific teams (mostly Russian and American) for a wide range of research activities for over a century. The problem of the ambiguous status of ice islands in international law can be thought of in terms of the legal characterization of the islands (i.e., should they be regarded as 'islands', 'installations', 'vessels', or something else?), and the closely related matter of jurisdiction over the islands and the activities conducted on them. Probably the most satisfactory application of the existing law would consider the drifting islands as ships, which would give the "flag state" (however defined) jurisdiction over the activities aboard the ice island on the high seas, and would give the coastal state varying degrees of jurisdiction, depending on the zone in which the island is drifting [27]. This approach is, however, somewhat ill-fitting at least in that the course of the island is often unpredictable, the means of determining a "flag state" is not clear, and the original "vessel" may be in fact shorter-lived than the activities taking place on it. Especially in view of the recent changes in the law of the sea proposed in the new Treaty regarding zones of national jurisdiction (coupled with what are likely to be higher levels of scientific activity), it would seem that there is a need for a specific regime for ice islands.

Specific regulation of marine pollution by global treaties and other arrangements. One important body of regulatory law, superimposed on the basic, mainly jurisdictional law discussed in the foregoing paragraphs, must be noted because of its relevance to the Arctic. That is the collection of treaties of global application that address problems of marine pollution from vessels and from ocean dumping.

International measures on vessel pollution have been developed mainly within the International Maritime Organization (IMO, until recently IMCO--the Intergovernmental Maritime Consultative Organization), a specialized agency of the United Nations concerned with regulation of shipping for safety and environmental purposes. Beginning in 1954 with a convention limiting deliberate discharge of oil by vessels [28], a series of treaties was produced authorizing intervention on the high seas by states to deal with "grave and imminent danger to their coastline or related interests" by oil pollution from an accidental oil spill (1969) [29]; providing compensation for oil spill damage and placing an upper limit on liability for each incident (1969) [30]; increasing that limit (and providing for further increases), and establishing a compensation fund (1971)[31]. Some coverage was provided at the same time as these public compensation arrangements by voluntary arrangements among oil companies: TOVALOP (Tanker Owners Voluntary Agreement Concerning Liability for Oil Pollution) (1969) [32] and CRISTAL (Contract Regarding an Interim Supplement to Tanker Liability for Oil Pollution) (1971) [33]. In 1973, the 1954 oil pollution convention was extended to cover other hazardous substances whether intentionally or accidentally discharged. This treaty, the International Convention for the Prevention of Pollution from Ships (known as the "MARPOL" Convention) [34], is not yet in force. Among other things it set certain tanker design standards and has since been supplemented by more stringent standards [35]. During the same period a new convention on Safety of Life at Sea was concluded, setting a number of equipment and operational standards designed to reduce risks of accidental pollution [36]. It entered into force in 1980.

Meanwhile, a global treaty on ocean dumping -- the deliberate disposal in the ocean of waste substances transported to the site for that purpose -- was concluded in 1973 and entered into force in 1975 [37]. It applies a blacklist/greylist approach to the prohibition of dumping of an expandable list of hazardous substances.

The foregoing treaties and other measures constitute an impressive body of international legislation, applicable in the Arctic as elsewhere in the world ocean, but it is not without certain weaknesses, some of which are especially relevant in regard to the Arctic. First, they have on the whole been very slow to receive the necessary ratifications for entry into force. Secondly, arrangements for compensation for damage from oil spills are at present of doubtful adequacy to meet the readily ascertainable costs of damage (e.g. costs of cleanup operations), and it is questionable whether general environmental damage and loss of amenities--which might be specially great in the Arctic--will qualify for compensation at all, because of difficulties in measurement, establishing causation, and determination of victim [38]. To the extent that this is the case, not only the compensation, but also the deterrent, function of the compensation schemes -- which constitute a kind of public insurance arrangement -- will be badly flawed in respect of the risk of major environmental injury from oil transport casualties in Arctic waters. (Moreover, design, equipment and operational standards have in general not been devised with a special view to operation in the Arctic environment.)

3. INTERNATIONAL LAW IN THE ARCTIC'S FUTURE

When one examines the present legal regime of the Arctic, as we have seen, a number of law-related problems emerge which may be of considerable practical significance for some of the array of important human interests in the area--notably extraction and transport of hydrocarbon and other mineral resources, commercial navigation generally, environmental protection, and scientific (particularly marine) research--not to mention a general interest in the minimization of political conflict in any area of new and increasingly intensive activity. These problems may usefully be grouped as follows:

(1) _Problems of jurisdiction: i.e., problems growing out of uncertainties or actual disputes as to the jurisdictional status of certain areas,_ including disputes over the drawing of national boundaries separating adjacent or opposite national maritime zones, the status of the allegedly closed Soviet seas, the Svalbard continental shelf or economic zone, or the waterways constituting the Northwest Passage; or uncertainties over the status of ice islands. While a number of these problems are straightforwardly bilateral, not only are some of them multilateral, involving the Arctic states, but some engage the interests of non-Arctic states as well (e.g., Japan and other non-arctic maritime powers particularly in northern Europe, and the non-Arctic parties to the Svalbard treaty).

(2) _Problems of management and cooperation_. _Even assuming resolution of some of these uncertainties in the basic, mainly jurisdictional law, there remain problems growing out of the probable inadequacy of that law--alone and unaided by any overlay of more specific treaty arrangements--to provide a sound basis for pursuit of the practical interests just mentioned._

For example, the new law of the sea Treaty would sharply restrict the legal freedom of foreign research activities in Arctic marine areas newly embraced by the exclusisve economic zone of coastal states, and suggests the need for stable and permanent arrangements of some sort to ensure that that freedom is not curtailed in fact.

Environmental management problems, to mention one other area, are not unrelated to potential restrictions on scientific research, and in their own right are more complex and perplexing. The EEZ itself, with its accompanying extension of coastal state environmental prerogatives, was assumed to augur for more effective marine environmental protection by enlarging the powers of the state most directly concerned with each segment of coastal ocean. But it could have an opposite effect, through its enlargement of the coastal state's development prerogatives in the coastal ocean, and removal of the coastal ocean from the area in which all states have an equal interest and voice. And leaving the EEZ aside, if there really are quite special environmental vulnerabilities in the Arctic (of the sort that were invoked in the 1970 Canadian protective legislation, for example, and led in due course to the "ice-covered exception"), it is clear that the poorly developed general law of international responsibility for environmental injury is not an adequate basis for accomodating environmental and development interests. It has not been so regarded in other, somewhat analogous ocean areas which lack those special vulnerabilities such as the Mediterranean, North Sea, and Baltic Sea, even when complemented by the growing body of treaty law on accidental and deliberate pollution from vessels discussed earlier. In any event, as we saw, that body of law is in considerable measure not yet in force, and of questionable adequacy to compensate for, not to mention deter, the form of catastrophic oil pollution damage which it covers. More importantly, it does not cover generalized damage from extractive activities in the ocean, such as an uncontrolled oil flow. Nor does it cover the marine pollution from land-based activities that are understood to constitute the bulk of marine pollution in other parts of the world ocean [39]. Land-based pollution must prudently be presumed to be a significant present and future factor in Arctic environmental management, particularly as the size of populations near the Arctic rim engaged in the full range of activities associated with modern industrial societies--like that of the substantial Soviet population on the Kola Peninsula--increases.

It is noteworthy that the new law of the sea Treaty enjoins states to develop international standards respecting land-based pollution and pollution from seabed activites (such as drilling) within the EEZ or continental shelf, and requires that the coastal state's regulations on the former take the international ones "into account" and on the latter be "no less effective" than the international rules (Articles 207, 208). The same articles also enjoin regional harmonization of policies on these matters. But until international standards are in fact developed--a protracted process at best--the coastal state is left a wide discretion in balancing its development and environment interests.

Indeed, on the important principle of the development of global standards respecting land-based pollution, the relevant language (Article 207) is--perhaps necessarily--so weak that it may be doubted that coastal states will ever be much restrained by such standards. That being the case, regional cooperation particularly in semi-enclosed seas is all the more important.

What are the possibilities for addressing the two classes of problems mentioned above, by clarifying the existing law in general or in its application to particular controversial situations; by changing that law; or by grafting onto it new rules or institutional arrangements? And should these steps be undertaken, if at all, among the Arctic states as a regional group, or in a wider multilateral forum?

To turn to the latter question first, our own view is that the Arctic powers should begin promptly to explore the possibility of regional treaty arrangements. The question properly stated is not whether there are sufficient shared interests among the basin states to warrant regional legal and institutional arrangements among them (perhaps other specially interested states). The proper question is, rather: how strong a political base for such arrangements could the Arctic states reasonably expect to develop and sustain, and consequently how big a cluster of shared interests might such arrangements encompass, at what levels of complexity and burdensomeness?

To be sure, those of the jurisdictional problems that are simply bilateral should be addressed as such, with such speed as bilateral relations call for and can sustain. Some that are legally of a broadly multilateral character, such as the status and use of the Northwest Passage, may in fact be mainly bilateral for the time being. Disputes over the Svalbard coastal zones could, presumably, be addressed in isolation by the parties to the Svalbard treaty. But in planning for dealing even with these more readily isolatable Arctic jurisdictional problems, governments would be well advised to consider, as to each, its possible legal, political or other practical connections with other Arctic matters of more general concern before determining how they should be addressed.

It is also true that one important component of a strategy for dealing in a reasonably timely way with Arctic jurisdictional problems--and to a significant extent law-related problems of management and cooperation--has nothing to do with the Arctic states as such. That is prompt resolution of uncertainties about the law of the sea, growing out of the position of the United States and other non-signers of the new Convention. A careful survey of these Arctic problems, with this aspect in mind, might well reveal that the United States' interest in the Arctic alone constitutes sufficient reason for it to reconsider its position, and for all Arctic states to seek ways to bring the U.S. and other non-signers into the treaty. But even if it now appeared likely that the treaty would enter into force fairly promptly for all states interested in the Arctic, substantial law-related problems, some of a clearly regional character, would remain.

As to the political base for negotiating and sustaining regional arrangements for the Arctic even in times of special political stress, some instruction may be gained by examining prior experience with the negotiation and operation of regional treaty arrangements for the Antarctic and certain regional seas, particularly the Baltic and Mediterranean [40]. Every Arctic power save Canada, and four non-Arctic states that would be most likely to be interested in Arctic regional arrangements (United Kingdom, France, Federal Republic of Germany, and Japan), have participated in at least one of these arrangements, and several in two (or more, if the negotiation of the 1980 Antarctic Living Resources Convention, and the current negotiations on Antarctic mineral resources, are taken into separate account). The obvious and important geopolitical differences among all of these regions should not be underestimated in this connection, but neither should they be given more than their due weight. By exact contrast to Antarctica, the Arctic region comprises the heart of the great land masses of the northern hemisphere, which contain the world's major power centers; and it embraces some of the continental land territory of each Arctic state, as to which there will be no question of any arrangements for common use as in the claimed areas of the Antarctic Continent. But it is worth noting that the Baltic and Mediterranean are ocean areas already under intensive use for a wider variety of purposes than the Arctic, and each is of great military and commercial importance to the littoral states, which have no inclination to curtail in any very fundamental way their individual prerogatives as to either land or ocean areas.

It should also be noted that, in Arctic regional negotiations, there may be a certain analogy between the land territorial claims of the Antarctic claimant states and the claims for exclusive Arctic maritime rights of Canada and the Soviet Union. The original negotiation of the Antarctic Treaty offered a convenient way to set aside for the time being the political burden of the claims. In later years, their continued latent existence, with the possibility that they might be aggressively reasserted, and thereupon opposed by fresh competing claims by presently non-claimant states or vocal opposition from non-parties to the Antarctic Treaty, has constituted one of the inducements to the Antarctic states to negotiate promptly, among themselves as a select group, the potentially important resource arrangements that have been concluded or are now under consideration. It is not inconceivable that the Canadian and Soviet maritime claims, which will have to be disposed of in some way if they are not eventually to become impediments to increased maritime transport in the area, could end up with a similar history in Arctic regional negotiations.

As to the scope, content, and form of possible regional arrangements, experience with the three regional arrangements just mentioned, among others, provides an ample stock of features from which to fashion arrangements appropriate to the Arctic. The interests in scientific research and environmental protection are presumably the clearest candidates for inclusion. They would seem to call for a treaty which, in addition to laying down general principles of cooperation, would establish some form of consultative mechanism with the ability, _inter alia_, to produce agreed recommendations to governments or specialized treaties addressing some of the specific problems just enumerated, and others such as species preservation, or scientific or environmental reserves. In addition, there would be no reason in principle why a regional forum might not be used to address problems of living-resource management, in the case of fisheries problems of a more than bilateral character; the 1980 Antarctic Living Resources Treaty, with its greater than usual emphasis on monitoring and broad ecosystem management, would be an instructive model. More problematical, but certainly worth careful consideration, is the possibility of addressing region-wide jurisdictional uncertainties--perhaps, in the case of controversial claims by individual states, through practical arrangements on navigation that would leave issues of jurisdictional principle to one side. (The problem of the jurisdictional status of ice islands, at least as used for scientific research, would seem susceptible to straightforward negotiation of specific rules, probably in treaty form.)

By contrast, and in distinction to the arrangements now being discussed for the Antarctic, rights to hydrocarbon expoitation would not be embraced by regional arrangements, such rights being of an indisputable national character in all or most areas with hydrocarbon resource potential [41]. But as already indicated, providing an economically and environmentally sound basis for hydrocarbon extraction and transport in the peculiar environment of the Arctic is likely to call increasingly for international and perhaps region-wide cooperation.

FOOTNOTES

1. For a discussion of Soviet official policy and scholarly views, see note, "Canadian and Soviet Arctic Policy: An Icy Reception for the Law of the Sea?", 16 Virginia Journal of International Law 610 (1976).
2. Antarctic Treaty, 1 Dec. 1959, 12 U.S.T. 194, T.I.A.S. No. 4780, 402 U.N.T.S. 71.
3. Convention on the Conservation of Antarctic Marine Living Resources. _Done_ 20 May 1980, T.I.A.S.No. 10240, _reprinted in_ 19 INT'L LEGAL MATERIALS 841 (1980), entered into force 7 Apr. 1982).
4. Agreement on the Conservation of Polar Bears. _Done_ 15 Nov. 1973, T.I.A.S. No. 8409, _reprinted in_ 13 INT'L LEGAL MATERIALS 13 (1974), entered into force 26 May 1976.
5. _See_, for example, Bloomfield, L., "The Arctic: Last Unmanaged Frontier," _Foreign Affairs_, vol. 60, no. 1, pp. 87-105, Fall 1981; Feder, B., "A Legal Regime for the Arctic," _Ecology Law Quarterly_, vol. 6, no. 4, pp. 785-829, 1978; and Johnston, D., Ed. _Arctic Ocean Issues in the 1980's, Proceedings_, Law of the Sea Institute and Dalhousie Ocean Studies Programme, Honolulu, 1982.
6. The United States agreed to pay the USSR the sum of seven million two hundred thousand dollars in gold ten months after the exchange of ratifications of the cession convention, Art. IV, Cession of Alaska,

15 Stat. 539; Treaty Series 301.

7. Pharand, D. "The Legal Status of the Arctic Regions," Recueil des Cours, Hague Academy of International Law (1979-II), Sijthoff & Noordhoff, The Netherlands, 1980, p. 64. For a detailed and exhaustive discussion of international legal issues in the Arctic, see Pharand, D. The Law of the Sea of the Arctic (with Special Reference to Canada), Univ. of Ottawa Press, Canada, 1973.

8. Note, however, the dispute between Norway and the Soviet Union concerning sovereignty over the Franz Josef Land archipelago. Whereas Norway had reportedly sent seven times the number of Soviet research expeditions, the Soviet claim prevailed. (Pharand, D. supra note 7, p. 63.)

9. Status of Spitsbergen (Svalbard), 43 Stat. 1892, Treaty Series 686. High Contracting Parties in addition to those mentioned: United Kingdom (Ireland, India, New Zealand, Australia, South Africa), France, Italy, Japan, the Netherlands, and Sweden.

10. W. Lakhtine, "Rights Over the Arctic" 24 American Journal of International Law 709 (1930).

11. For a historical survey of Arctic claims, see D. Pharand, "The Legal Status of the Arctic Regions", supra, note 7, pp. 62-79.

12. Sollie, F. "The New Development in the Polar Regions", in F. Sollie et. al., The Challenge of New Territories, Norway, 1974, p. 30. At the time, Charles Cheney Hyde observed: "In its judgement of April 5, 1933 concerning the Legal Status of Eastern Greenland, the Permanent Court of International Justice . . . concluded that Denmark satisfied the two requisites for the creation of a right of sovereignty to be derived from continued display of authority (as distinct from a right derived by way of transfer, as by cession), namely, 'the intention and will to act as sovereign, and some actual exercise or display of such authority'. . ." [Publications, Permanent Court of International Justice, Series A/B, No. 53, p. 46] Editorial Comment, 27 American Journal of International Law 732-738 (1933).

13. Sobraniye Zakanov SSSR, 1926, No. 32, p. 203, as quoted in D.E. Bierman, "Soviet Territorial Claims in the Arctic and their Economic and Political Implications", 19 Soviet Geography 492 (1978).

14. There appears to be at least one exception, the Muskapi-Montagnais Band Association (Canada), which claims, in addition to the "right as (an) aboriginal nation to maintain ownership of their national territory", the "right to exist as self-determining people". Moreover, they seek "recognition, self-determination and all rights and priviledges (sic) due other sovereign nations". (Native Claims In Canada: A Summary On an Element-by-Element Basis of Comprehensive Native Claim Proposals That Have Been Presented To The Federal Government and Bureau of Native Claim Settlements; Bureau Of Indian and Northern Affairs, Office Of Native Claims, April 1980) Many of the native peoples included in the 'Summary' cited use the term self-determination when asserting a right to political autonomy, or control over local or regional administration. Even the Muskapi-Montagnais seek a native "Government within (the Canadian Confederation) with jurisdiction over a geographical area and over subject matters now within jurisdiction of Government of Canada or Government of Newfoundland." It appears unlikely, however, that the Canadian government will grant provincial status to the Northwest Territories (or the Yukon) because of their potential mineral wealth and the desire to minimize the likelihood of future jurisdictional conflicts over matters having to do with commercial development or national security.

15. For example, Article 3 of the Polar Bear Convention, supra note 4, stipulates that "any Contracting Party may allow the taking of polar bears by local people using traditional methods in the exercise of their traditional rights and inaccordance with the laws of that Party" or where the bears may have been "subject to taking by traditional means by its nationals."

16. For the U.S., see Pub. L. 92-203 (Alaska Native Claims Settlement Act, 1971), as amended Pub. L. 94-204 (1976). In Canada, the Office of Native Claims was established in July, 1974 within the Deptartment of Indian and Northern Affairs, and the following year, a Joint Sub-Committee on Indian Rights and Claims was created. There are two Indian Rights Commissioners, and since 1977, working groups to revise the Indian Act and Native Rights and Claims Processes.

17. Third United Nations Conference on the Law of the Sea, A/CONF.62/122, 21 International Legal Materials at 1261.

18. Non-signing states included Belgium, the Federal Republic of Germany, Italy, Luxembourg, and the United Kingdom, in addition to the United States.

19. Ships must meet minimum requirements with respect to identification, seaworthiness, use of signals -- particularly for the prevention of collisions, and the competence of its crew according to established international standards for the prevention and control of marine pollution, radio communications, and the safety of life at sea (Article 94). All ships have the duty to render assistance to persons or vessels in distress (Article 98). There are additional provisions enjoining states from committing acts of piracy, illicit drug traffic, slave transport or unauthorized broadcasting on the high seas, and establishing the rights of visit and hot pursuit of vessels where

there is "reasonable ground for suspecting" that they are engaged in one or more of these activities. Lastly, states are to implement the provisions which define wilfull or negligent breakage of or injury to pipelines or cables beneath the high seas as a punishable offense, unless all precautions necessary to avoid such injury are taken. Persons responsible for the break are to bear the cost of repairs.

20. It has been pointed out that the terms of the territorial treaties of 1825 (USSR-UK) and 1867 (USSR-US) do not, in fact, support this position, if for no other reason than that the existence of the continental shelf and the concept of the EEZ were not known, and therefore the intentions behind the agreements could not have been to include them. (D. Pharand, "The Implications of Changes in the Law of the Sea for the 'North American' Arctic Ocean" in Gamble, supra note 5, p.183-184. For more detailed discussion of the uncertain ocean boundaries mentioned in this and the preceding paragraphs in the text, _see_ Professor Pharand's chapter just cited, and Ostreng, W., "The Continental Shelf -- Issues in the "Eastern" Arctic Ocean, ibid. pp. 165-182.

21. Ostreng, W. in Gamble, _ibid.,_ p. 172.

22. _See_ W. Butler, The Soviet Union and the Law of the Sea, 112-114 (1971); Note, "Canadian and Soviet Arctic Policy: An Icy Reception for the Law of the Sea?", 16 _Virginia Journal Of International Law_ 609 (1976).

23. For discussions of this issue, and of the Canadian protective legislation, _see_ Pharand, D., "The Implications of Changes in the Laws of the Sea for the 'North American' Arctic Ocean", in Gamble, supra note 20; and Note, "Canadian and Soviet Arctic Policy: An Icy Reception for the Law of the Sea?", _supra_ note 22.

24. Pharand, D., _ibid._ at p.189.

25. _See_ Note, "Canadian and Soviet Arctic Policy: An Icy Reception for the Law of the Sea?", supra note 22.

26. Statute of the Administration of the Northern Sea Route attached to the Ministry of the Maritime Fleet, September 16, 1971, 11 International Legal Materials 172.

27. _See_ Pharand, D. "The Legal Status of the Arctic Regions", _supra_ note 7, pp.87-105. The only practical experience that might be referred to in this area is found in a case involving criminal jurisdiction for an act involving American nationals on a research ice station which had drifted to within approximately 200 miles of the Canadian coast. Canada's unsolicited waiver of jurisdiction in favor of the United States avoided a dispute over jurisdiction in this instance, but brought up the issue of Canada's reservation of the right to exercise such jurisdiction in future cases. _U.S. v. Escamilla,_ 467 Fed. 2nd. 341 (1972).

28. 1954 Convention for the Prevention of Pollution of the Seas by Oil. _Done_ 12 May 1954 [1961] 3 U.S.T. 2989, T.I.A.S. No. 4900, 327 U.N.T.S.3; with amendments _adopted_ 11 April 1962, [1966] 2 U.S.T. 1523, T.I.A.S. No. 6109,, 600 U.N.T.S. 332; 21 Oct. 1969, [1977] 1 U.S.T. 1207, T.I.A.S. No. 8505; 15 Oct. 1971, _reprinted in_ 11 INT'L LEGAL MATERIALS 267 (1972). See, also, generally: Schneider, J. "Prevention of Pollution from Vessels" in Charney, J.I. Ed., _The New Nationalism and the Use of Common Spaces: Issues in Marine Pollution and the Exploitation of Antarctica,_ Allanheld, Osmun and Co., Inc., Totowa, 1982.

29. International Convention on Civil Liability for Oil Pollution Damage. _Done_ 29 Nov. 1969, _reprinted in_ 9 INT'L LEGAL MATERIALS 45 (1970). _See also_ Protocol to the International Convention on Civil Liability for Oil Pollution Damage, 29 Nov. 1976, in _id.,_ vol. 16, at 621 (1977).

30. International Convention on Civil Liability for Oil Pollution Damage.

31. International Convention on the Establishment of an International Fund for Compensation for Oil Pollution Damage. _Done_ 18 Dec. 1971, _reprinted in_ 11 INT'L LEGAL MATERIALS 284, art. 3 (1972). _See also_ Protocol to the International Convention on the Establishment of an International Fund for Compensation for Oil Pollution Damage, 19 Nov. 1976, in _id._, vol. 16, at 621 (1977). This treaty entered into force only in 1978.

32. 7 Jan. 1969, _reprinted in_ 8 INT'L LEGAL MATERIALS 497 (1969).

33. 14 Jan. 1971, _in_ 10 INT'L LEGAL MATERIALS 137 (1971).

34. _Done_ 2 Nov. 1973, _reprinted in_ 12 INT'L LEGAL MATERIALS 1319 (1973). _See also_ Protocol to International Convention for the Prevention of Pollution from Ships, _done_ 16 Feb. 1978, _reprinted in_ 17 INT'L LEGAL MATERIALS 546 (1978), scheduled to enter into force 2 Oct. 1983).

35. Protocol of 1978 relating to the International Convention for the Prevention of Pollution from Ships, 1973, 16 Feb. 1978, _reprinted in_ 17 INT'L LEGAL MATERIALS 546 (1978).

36. International Convention for the Safety of Life at Sea. _Done_ 1 Nov. 1974, _reprinted in_ 14 INT'L LEGAL MATERIALS 959 (1975). SOLAS was itself supplemented in 1978 by a protocol extending or adding tanker design and equipment standards. _See also_ Protocol of 1978 relating to the International Convention for the Safety of Life at Sea, 1974, _done_ 16 Feb. 1978, _reprinted in_ 17 INT'L LEGAL MATERIALS 579 (1978), entered into force 1 May 1981.

37. Convention on the Prevention of Marine Pollution by Dumping of Wastes and Other Matter. _Done_ 29 Dec. 1972, [1975] 2 U.S.T. 2403, T.I.A.S. No. 8165.

38. _See_ Smets, H., "The Costs of Accidental Oil Spills and their Compensation", in _Essays on International Law,_

Asian-African Legal Consultative Committee, New Delhi (1981).

39. *See* Hargrove, J.L., ed., *Who Protects the Ocean?*, West Publishing Co., St. Paul, 1975, and Whipple, Jr., W., "Land-Based Sources of Marine Pollution and National Controls", in J.I. Charney, Ed., *supra* note 28.

40. *See:* Antarctic Treaty of 1959, *supra* note 2; Convention on the Protection of the Marine Environment of the Baltic Sea Area, *done* 22 March 1974, *reprinted in* 13 INT'L LEGAL MATERIALS 546 (1974), entered into force 3 May 1980; Convention for the Protection of the Mediterranean Sea Against Pollution, *done* 16 Feb. 1976, *reprinted in* 15 INT'L LEGAL MATERIALS 290 (1976), entered into force 12 Feb. 1978 with two Protocols: Protocol on Prevention of Pollution of the Mediterranean Sea by Dumping from Ships and Aircraft, *done* 16 Feb. 1976, *reprinted in* INT'L LEGAL MATERIALS 300 (1976), Protocol Concerning Cooperation in Combatting Pollution of the Mediterranean Sea by Oil and Other Harmful Substances in Cases of Emergency, *done* 16 Feb. 1976, *reprinted in* 15 INT'L LEGAL MATERIALS 306 (1976); see also Protocol for Protection of the Mediterranean Sea Against Pollution from Land-Based Sources, *done* 17 May 1980, *reprinted in* 19 INT'L LEGAL MATERIALS 869 (1980).

41. One possible exception is the Lomonosov Ridge, an elongated undersea mountain across the Arctic Basin. The Ridge has been regarded by many scientific authors as probably a continental fragment (*see* Lawver, Grantz and Meinke, "The Tectonics of the Arctic Ocean", *infra* in this volume). To the extent that this geologic status might indicate the presence of hydrocarbon deposits, the possibility arises of hydrocarbon resources falling within the area beyond national jurisdiction and being subject to management by the international authority established by the Law of the Sea Convention. *See* Pharand, D., *op. cit. supra,* note 20. Alternatively, it is perhaps conceivable that the presence of oil and gas resources would tempt one or more of the Arctic coastal states to claim this itinerant fragment as a "submarine elevation" that is a "natural component of [its] continental margin," and thus includable in its legal continental shelf under the terms of Article 76(6) of the new Convention.

ALASKA'S ROLE IN ARCTIC DEVELOPMENT: LEGAL AND POLITICAL CONSIDERATIONS

John G. Gissberg

Assistant Attorney General
Alaska Department of Law
Pouch K, Juneau, Alaska 99811

No directed or comprehensive federal or state legislation addresses human activities in the arctic regions of the United States. Nevertheless, a number of federal and state statutes govern man's activities in the arctic. This array of independent statutory controls has dampened the development of an effective, coordinated policy-making process that could promote the rational use of arctic resources to the benefit of local residents and the nation.

Relationships between the State of Alaska and the U.S. Government are increasingly characterized by contentiousness; new initiatives and commitments are necessary to establish cooperative arctic policies that benefit the citizens of the nation and the residents of the state.

1. INTRODUCTION

Historically, nearly all development in Alaska has been controlled by outside interests and little has been done to provide long-term benefits to permanent residents of the territory. Indeed, the primary impetus for statehood was dissatisfaction over the control of local fisheries harvests by outside managers in Washington, D.C., and the fact that the primary beneficiaries of this exploitation were nonresident seafood processors and fishermen. Statehood was designed to help Alaska provide for itself by localizing resource management decisions and maximizing the in-state benefits of harvest activities conducted in Alaska. Yet, in many areas, outside control persists and continues to frustrate local input into the decision-making that affects Alaskans. Such barriers to local responsibility frequently polarize state-federal issues and frustrate attempts to establish the coordinated policy programs so badly needed in critical areas such as the arctic. It also necessitates political activism by local communities and Native groups who may perceive their interests are not being represented by either state or federal policy positions.

The proposed Arctic Research and Policy Act of 1981 (S. 1562), introduced on July 31, 1981, by Alaska Senators Frank Murkowski and Ted Stevens and Washington Senator Henry Jackson, may be a first step in supporting an arctic commitment that reflects the local, state, national, and international values in the area.

2. LEGAL PROTECTION OF SUBSISTENCE USES IN THE ARCTIC

2.1. Introduction

A key component of a successful arctic policy is a close working relationship with the permanent residents in the local communities. This working relationship cannot succeed unless the importance of traditional way of life and cultures is understood and appreciated. In particular, drastic changes in economic status and awareness resulting from oil development and the Alaska Native Claims Settlement Act of 1971 (43 U.S.C.† 1601-24) have led to increased concern for the local subsistence economies and the way of life associated with it.

The most commonly publicized subsistence problem in arctic Alaska features Eskimo reliance upon an annual bowhead whale harvest. A harvest taken under strictly regulated quotas, it presently allows a limit of 17 bowhead whales taken in any one year during the three-year period 1981-83. The overall total may be 45 bowhead whales taken or 65 struck. In view of increasing restrictions imposed upon commercial whaling harvests in Japan and a recent vote by the International Whaling Commission to place a moratorium on all commercial whaling for three years [1], Japanese delegations to the International Whaling Commission have pointed out the significance of cultural and protein contributions made by the harvests of whales to the Japanese [2]. However, unlike the Japanese harvests which are sold and con-

†Legislation: U.S.C., United States Code; CFR, Code of Federal Regulations; AAC, Alaska Administrative Code; SLA, Session Laws of Alaska.

sumed nationwide, the Eskimo hunt satisfies only local subsistence uses and, in addition, represents a critical cultural activity for the residents.

The importance of the Alaska harvest is not only difficult for other nations to appreciate, it is also the subject of occasional misunderstanding by American citizens. For example, it would not be correct to assert that supermarket foods can satisfy rural residents' subsistence needs. In the first place, the regular air transportation service to regional centers such as Bethel, Nome, Kotzebue, and Barrow is not provided to more than 200 villages in outlying bush areas. In the second place, air freight costs would be prohibitive. Further, the health of Eskimos suffers when they shift from locally available diets and substitute commercial products in their daily meals [3].

This might be especially so because commercial advertising propaganda rarely focuses on nutritious vegetables, fruits, or other staples, instead celebrating the alleged benefits of fast-cook, ready-prepared meals. In addition, no reliable cash economy exists to pay for the food necessities that are adequate substitutes for traditional diets.

Animal harvests are also important in arctic Alaska because the taking of large mammals requires special community cooperation and help from all village residents. In particular, the huge bowhead whale, a marine mammal as big as a semitrailer, is subject to successful harvest through cumulative assistance of all villagers. Whale hunting has been found to be "perhaps the most important single element in the culture and society of north Alaska whale hunting communities" [4]. The hunt occurs in springtime when whales migrate north and along the Arctic Ocean coast during annual feeding migrations. The enormity of undertaking to harvest and prepare a bowhead whale limits harvests and neutralizes any suggestion of a superficial "macho" motivation in hunting such superb creature.

When hauled onto the ice near a village or town, everyone -- men, women, and children -- joins in the exhausting task of cutting the meat and salvaging all parts for use in clothing, utensils, and other domestic needs. Final sharing in the product by all residents provides a unity to the villagers' lives and is cause for a community-wide potlatch of joy and appreciation.

2.2. Statutory Protections

Several recent federal and state laws have acknowledged the subsistence interests of Alaska communities. Much of this increased legislative interest in the north corresponded to the discovery of oil at Prudhoe Bay in 1968. For example, the Alaska Native Claims Settlement Act (ANCSA) of 1971 (43 U.S.C. 1601-24) was enacted to alleviate conflicts in federal, state, and Native claims to land ownership in Alaska. It established 12 regional and over 200 local Native corporations that received title to approximately 40 million acres of land in Alaska. This was supplemented by provisions for a monetary settlement of $1 billion. ANCSA primarily resolved past claims and did not directly provide for subsistence. In fact, it has even been suggested that Native leaders may have to forego subsistence activities in order to attend to the corporate responsibilities required in the law [5].

Nevertheless, section 17(d)(2) of ANCSA directed the Secretary of the Interior to set aside up to 80 million acres of federal land for parks, national forests, refuges, and wild and scenic rivers. This was accomplished in a subsequent act, the Alaska National Interest Lands Conservation Act of 1980 (16 U.S.C. 3101-3233), which has important subsistence provisions.

Three other important statutes of the last decade have also included specific consideration for subsistence harvests: (1) the Marine Mammal Protection Act of 1972 (16 U.S.C. 1361-1407), (2) the Endangered Species Act of 1973 (16 U.S.C. 1531-1543), and (3) the Migratory Bird Treaty Act of 1972 (16 U.S.C. 703-711).

The first two ban or strictly prohibit harvests of certain species. The Endangered Species Act prohibits the taking of creatures designated as endangered species (section 1538 (a)(1)), but provides an exception for any "Alaska Native who resides in Alaska or any non-Native resident of an Alaska Native village if such taking is primarily for subsistence purposes" (section 1539(e)(1)). The Marine Mammal Protection Act (MMPA) bans the taking of marine mammals in order to achieve a special management goal of an "optimum sustainable population" of marine mammals (section 1361(2)). Like the Endangered Species Act, the MMPA recognizes the reliance of Alaska Natives and their culture upon subsistence harvests of marine mammals, so it provides a qualified exception to the general prohibition. Thus, section 1371(b) of the MMPA allows harvests of marine mammals by

> an Indian, Aleut, or Eskimo who resides in Alaska and dwells on the coast of the north Pacific Ocean or the Arctic Ocean if such taking (1) is for subsistence purposes or (2) is done for purposes of creating and selling authentic Native articles of handicrafts and clothing . . . and (3) in each case is not accomplished in a wasteful manner.

The Migratory Bird Treaty Act, in section 712(1), allows the Secretary of the Interior to promulgate regulations

> to assure that the taking of migratory birds and collection of their eggs, by the indigenous inhabitants of the State of Alaska . . . shall be permitted for their own nutrition and their essential needs . . . during seasons established so as to provide for the preservation and maintenance of stocks of migratory birds.

The most comprehensive federal statutory statement on subsistence is found in the Alaska National Interest Lands Conservation Act (ANILCA) of 1980 (16 U.S.C. 3101-3233). It provides for the subsistence use of fish and wildlife by rural Alaska residents on federal public lands as a priority use when the harvestable surplus cannot satisfy all commercial, sport and subsistence uses. Four cities --Ketchikan, Juneau, Anchorage and Fairbanks-- are suggested to be urban centers where subsistence uses do not occur [6]. Although this legislation does not confer special rights on Natives, it does distinguish between urban and rural residents. Its three, key operative sections state a definition of subsistence similar to the Marine Mammal Protection Act reference; establish the priority subsistence use in times of resource shortages; and creates a system of public participation to assure local input into the regulatory process that determines seasons, areas, and other conditions of each harvest (sections 3114-3116).

ANILCA protects subsistence uses in approximately 227 million acres of federal land in Alaska (section 3112(1)). However, the state may continue existing management prerogatives on federal land if state laws of general applicability provide for the required public participation system of regional councils and a rural definition of subsistence uses that receive a priority in times of resource shortages (section 3115(d)).

Prior to passage of ANILCA, the state had already adopted its own subsistence statute (ch. 151, SLA 1978); implementing regulations in 1982 clarified the scope of the state's subsistence priority (5 ACC 99.010-.020, 1982). The Secretary of Interior affirmed compliance in a letter dated May 14, 1982.

The implementing regulations that had been adopted by the joint Alaska Boards of Fisheries and Game identified subsistence uses in Alaska as customary and traditional uses by residents in rural areas. However, many urban residents opposed the developing rural focus because the state statute did not include a geographical restriction on subsistence uses. Nevertheless, practical experience with application of the law since 1978 seemed to have evidenced the survival of customary and traditional uses for personal and family consumption only in rural areas. Thus, the boards' regulatory actions would have indicated that subsistence uses may not exist in urban areas of Alaska.

Extreme opposition to differential treatment between urban and rural residents and also dissatisfaction for preferential treatment to subsistence uses over sport and commercial harvests led to a November 1982 initiative to repeal the state's statutory subsistence priority. If successful, it would have prevented enforcement of a mandatory subsistence priority on private and state lands in the state. However, the proposition failed by a vote of 79,679 to 111,770.

After general elections in November 1982, the new governor discussed creating a task force on subsistence to review existing laws and practices. In late February 1983, the Deparment of Fish and Game and the two boards of fisheries and game were given this charge.

Neither the federal nor the state subsistence priority requires exploration and exploitation of nonliving or plant resources to be automatically subordinated to subsistence considerations. However, continental and ocean development for oil, gas, mineral, timber, and other resources may not proceed in disregard of subsistence considerations. For example, ANILCA, in section 3120, requires the head of a federal agency responsible for issuing land use permits on public lands in Alaska to determine that:

> (A) such a significant restriction of subsistence uses is necessary, consistent with sound management principles for the utilization of the public lands,
>
> (B) the proposed activity will involve the minimal amount of public lands necessary to accomplish the purposes of such use, occupancy, or other disposition, and
>
> (C) reasonable steps will be taken to minimize adverse impacts upon subsistence uses and resources resulting from such actions.

A lawsuit filed in the United States District in Juneau, Alaska on January 13, 1983 (City of Angoon et al. v. Marsh et al., No. J83-0001 Civ.) challenges an Army Corps of Engineers permit for a log transfer facility near the village that allegedly would interfere with local subsistence harvest of marine and shore animals.

Other subsistence protections survive through the National Environmental Policy Act (NEPA) of 1969 (42 U.S.C. 4321-4370) and the Coastal Zone Management Act (CZMA) of 1972 (16 U.S.C. 1451-1464). NEPA requires the well-known Environmental Impact Statement (EIS) for all proposed "major Federal actions significantly affecting the quality of the human environment" (43 U.S.C. 4322(2)(c)). The EIS is the basis for discretionary decisions whether to authorize a proposed project or whether to modify development plans to mitigate environmental damage. EIS requirements result in environmental evaluations before oil and gas sales and again prior to development activities [7].

The CZMA provides more substantive protection for subsistence because it requires federal actions that will affect a state's coastal zone to be "consistent" with the state's approved management program (16 U.S.C. 1456(c)(1)-(3)(B)). Alaska's approved program does acknowledge subsistence interests and provides, in 6 AAC 80.120(a), that "state agencies shall recognize and assure opportunities for subsistence usage of coastal areas and resources." More specifically, subsection (d) requires that potentially conflicting activities cannot be authorized without conducting a study of "the possible adverse impacts of the proposed potentially conflicting use or activity upon subsistence usage" and "appropriate safeguards to assure subsistence usage." Effective since 1978, these safeguards have not been the subject of any formal complaint on noncompliance.

3. BOUNDARY DISPUTES

Though state lawmakers have sought to harmonize subsistence sections of state law with congressional enactments and administration officials have worked to develop a degree of consistency in the application of such statutes, territorial disputes have not been as amenable to cooperative resolution.

3.1. Beaufort Sea

The hundreds of millions of dollars of oil and gas revenues that are frequently at stake in boundary disputes nearly always mask the consequences of oil and gas development that will affect local life long after the nonrenewable resources have been exhausted. Boundary disputes have traveled up the Pacific coast from Dixon Entrance in Southeast Alaska to Cook Inlet [8], to Bristol Bay, and now to the Beaufort Sea where rights to oil and gas leases are before the U.S. Supreme Court.

The continental shelf of the Beaufort Sea

is the oceanic extension of the oil-rich North Slope adjacent to Prudhoe Bay in the Arctic Ocean. When offshore tracts were auctioned in December 1979, ownership to many of the parcels was unclear. In anticipation of the problem that would be caused by title uncertainties, the United States and the State of Alaska entered into a cooperative agreement to conduct the sale and resolve legal title questions in a court action. Filed in May 1979, the case was lodged as a case of original jurisdiction with the U.S. Supreme Court [9].

Under the Submerged Lands Act of 1953 (43 U.S.C. 1301-1315), submerged lands within the three-mile limit of the territorial sea vest with the states (sections 1301(a), 1311(a)). The baseline from which this three-mile territorial sea is measured is defined in section 1301(c) as "the line of ordinary low water along that portion of the coast which is in direct contact with the open sea and the line marking the seaward limit of inland waters [emphasis added]." Submerged lands outside the territorial sea were designated as the "outer continental shelf," an area which subsequent provisions in the Outer Continental Shelf Lands Act of 1953 (43 U.S.C. 1331-1343) reserved to the federal government.

Because the Department of State has elected to measure the limits of the territorial sea using an arcs and circles procedure that establishes a three-mile limit line approximating the actual coastal configuration instead of a straight baseline concept that follows the general sinuosities of the coastline [10], several misshapen discontinuities come to describe the outer margin of the territorial sea (Figure 1).

Alaska is convinced that barrier islands along Alaska's arctic coast from 146° to 148° W. represent the "direct contact" referred to in the Submerged Lands Act. Three of the main island groups are within six miles of the coast and the others are just slightly beyond the distance that would otherwise provide continuous state territorial jurisdiction from the Alaska mainland to the seaward limits of the islands. However, because of the federal government's reliance upon arcs and circles as the basis for drawing baselines, federal areas of "outer continental shelf" extend into inshore areas landward of the McClure Islands, the disputed island of Dinkum Sands, and the Midway Islands.

These intrusions result in illogical jurisdiction responsibilities that find state subsoil rights extending three miles offshore where they are interrupted briefly by a narrow band of federal authority before finding state claims reestablished by the immediate proximity of the islands. Only after exceeding the seaward three-mile limits of the islands is the final, uninterrupted, "outer continental shelf" established.

This anomaly is not only contrary to rational effective planning programs but it seems to violate the purposes of the Submerged Lands Act. The Act was passed in 1953 as a response to the California case in 1947 that invalidated state claims to ocean areas and set the stage for corrective legislation [11]. As a compromise measure designed to eliminate boundary controversies between the federal and state governments, it should not be interpreted in a manner that generates further confusion and controversy between the parties. Thus, the phrase "in direct contact with the open sea" could effectively avert unnecessary disputes if it is not read to include inshore submerged lands that are isolated from the continental shelf and do not extend directly into the open ocean area.

Further, in the second California case, United States v. California, 381 U.S. 139, 168 (1965), the court indicated it would not countenance failure to ignore the internationally

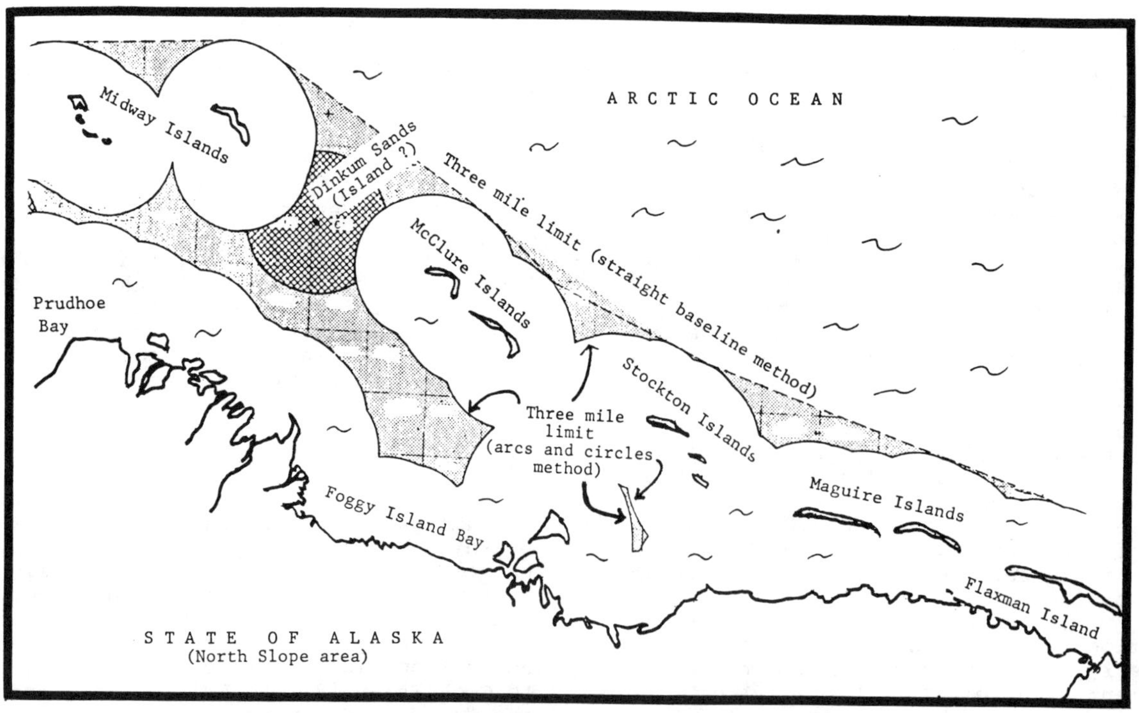

Figure 1 Baseline measurements in arctic Alaska.

approved straight baseline concept if it appeared the more restrictive arcs and circles procedure had been selected because it benefitted the United States in a federal-state dispute as opposed to valid foreign affairs reasons.

In the suit, the state attempted to impose the internationally recognized straight baseline method upon the federal government so as to enclose Santa Barbara Channel in the territorial sea adjacent to California. The court found that Catalina Island, at 11-21 miles offshore, was not so proximate as to justify the territorial extension. However, the Alaska barrier islands are nearby and, in addition, no obvious or significant international question is posed in the Beaufort Sea case and the measurement method seems primarily designed to improve the extent of lands under federal oil and gas lease authority.

Even though the isolated additional pockets and enclaves of federal jurisdiction that are created within state waters by such a reading of the Submerged Lands Act indisputably produce increased oil and gas revenues, the countervailing costs are significant.

In the first place, the cost of original litigation in the Supreme Court is considerable. All motions must be filed in the nation's highest court or with a special master who lives in California thousands of miles from the disputed area. Even though the case was filed nearly four years ago, it still lingers, awaiting completion of necessary discovery procedures regarding the government's selection of the baseline measurement methods for the Beaufort Sea and Arctic Ocean.

In the second place, the uncertainties arising from extended litigation have been further compounded because the local Inupiat Eskimo community has registered its own claims to the offshore seabed out to 60 miles from the baseline. Their claims are based on the same aboriginal rights that were extinguished in the Alaska Native Claims Settlement Act for the state. However, ANCSA did not address Native rights to outer continental shelf regions. Therefore, as a result of this omission, local residents have initiated their own lawsuit to resolve Eskimo claims to the seabed and water column above the outer continental shelf. Consequently, while the state seeks to minimize federal offshore lands, local residents might profit by a contrary conclusion maximizing the federal area.

The Inupiat case was filed in the U.S. District Court for Alaska in Anchorage (<u>Inupiat Community of the Arctic Slope v. United States</u>, Civ. No. 81-019), where an <u>October 1, 1982</u> ruling dismissed the action on the grounds that outer continental shelf claims are within the exclusive purview of the federal government. An appeal has been filed with the Ninth Circuit Court of Appeals.

Thus, while arctic litigation represents a primary concern between federal and state authorities, efforts designed to promote cooperative research and policy are subject to postponement and inaction while the property claims are resolved.

3.2. United States-Canada Boundary Claims

Boundary disputes between the United States and Canada are well known, as they have arisen at Canada's southern boundaries at Maine and Washington. However, a related dispute at the northern United States-Canada border, at Alaska, receives less publicity and certainly less attention from Department of State officials. Nevertheless, like other disputes, it causes uncertainties and endless excuse for deferring arctic policy questions.

The dispute addresses whether the United States-Canada border in the Arctic Ocean follows a northerly extension of the 141° W. longitude line or, instead, makes a sharp northeasterly turn at the coastline to correspond to a right angle drawn to the beach tangent at 141°W.

Little progress is being made to resolve this issue in spite of the fact that an absence of other contentious issues or resource enticements in the area would seem to facilitate timely resolution of the boundary problem before it becomes embroiled in other pressures.

4. FISHERIES POLICIES

4.1 Introduction

The Bering Sea yields 1,300,000 metric tons of fisheries products annually [12]. Most of this harvest is attributed to Japanese and Soviet factory ships which dominate the catch with Alaska pollock and yellowfin sole. American fishermen seek out halibut, king crab, and Tanner crab in the vast expanse of the shallow continental shelf of the Bering Sea and inshore fishermen target on five species of Pacific salmon and Pacific herring ripe with valuable roe. In addition, in January 1983, witnesses at hearings regarding processing operations by foreign fishing vessels within the state's internal waters under section 1856(c) of the Magnuson Fishery Conservation and Management Act suggested capelin, lamprey, and sea snails as candidates for future interest.

The influence of arctic conditions on these fisheries is enormous. The winter ice pack has advanced well into the commercially productive area of the Bering Sea where it has ripped away the surface buoys marking locations of large, seven-foot by seven-foot crab pots that will never again be retrieved. Damage done in January and February 1980 resulted in compensatory payment of $3,018,945.57 to 57 fishermen for "acts of God" related to sea ice. The Fishermen's Protective Act of 1967 (22 U.S.C. 177 et seq.) had been amended in 1978 to provide for such compensation but the provisions were subsequently repealed in 1980 [13].

4.2. Fisheries Conflicts

The Magnuson Fishery Conservation and Management Act (MFCMA) of 1977 (16 U.S.C. 1801 et seq.) has spawned a number of state-federal fisheries conflicts. The MFCMA is applauded for its controls over foreign fishing in the Fishery Conservation Zone from 3 to 200 miles seaward of the territorial sea baselines, but the Act's primary role in the state-federal arena is based on its establishment of a national fisheries management program (section 1851). The national program with its system of regional fishery management councils is most applicable in coordinating and supervising the management of fish species subject to the jurisdictions of several states rather than operating as a comprehensive subordination of traditional state fisheries authority to an overriding federal scheme. In particular, Alaskan regulation of fisheries harvests has been acknowledged judicially to extend beyond the three-mile limit of the territorial sea when the fishery evidences an appropriate nexus to the state's territorial resources and activities [14].

Thus, in Alaska, where fishing grounds are insulated from other states by the Pacific coast of Canada, the MFCMA has been gaining a reputation for challenging the local management benefits achieved by statehood. Nevertheless, cooperative resolutions of conflicts in several fisheries have been able to preserve the integrity of traditional state management authority while complying with national standards set forth in section 1851.

King crab. The Bering Sea king crab fishery, where approximately one-third of the fleet represents nonresident fishermen [15], was one of the first domestic fisheries to be subjected to extensive federal review. Draft fishery management plans for king crab have been under review since March 1979 even though federal and state courts have acknowledged the validity of extraterritorial state management programs beyond the three-mile limit in the absence of conflicting federal regulations [16].

This intense interest seems to have arisen from the original MFCMA requirement in section 1852(h) that "each council . . . prepare and submit to the Secretary a fishery management plan with respect to each fishery within its geographical area of authority." Despite indications that strict compliance has not been necessary [17], the North Pacific Fishery Management Council has followed the requirement to adopt national plans for the Bering Sea.

At the same time, there has been an admission that continued state management responsibilities are essential for the conservation and optimum yield of the resource. Because of this reluctance to scrap the proven state system and start again with new federal managers, more recent council proposals have favored the formal federal adoption of existing state regulations or the development of a framework plan with accompanying notices in the Federal Register to supplement state administrative procedures regarding public notice and extensive public hearings [18].

These cumbersome maneuvers apparently will not be necessary because new amendments to section 1852(h) embodied in Public Law 97-453 (1983) have limited the council's responsibilities to preparing fishery management plans only for fisheries "that require conservation and management." This important reemphasis of the Act upon conservation considerations should help assure the vitality of the king crab resources by minimizing the unnecessary waste and duplication that have been caused by imposing federal management plans on established domestic fisheries and diverting management discussions to jurisdictional questions.

Tanner crab. Federal-state give and take has also been occurring in the Tanner crab fishery. Initially, federal regulations, modeled after existing state rules, presented no conflict. The same regulatory program applied to fishermen in the state's territorial sea and the outlying Fishery Conservation Zone. However, annual public hearings by the Alaska Board of Fisheries have led to frequent regulatory adjustments in response to changing stock conditions and fishing operations. Unfortunately, federal procedures require that all regulatory changes be subjected to a lengthy internal review process in Washington, D.C., at least 6,000 miles from the fishing areas. As as result, accompanying federal approval has sometimes lagged so far behind new state regulations that fishermen have had to conduct their operations under conflicting rules. For example, season openings have varied (5 AAC 35.510(2) vs. 50 CFR 671.26), as has the permissible configuration of the crab pots (5 AAC 35.125, top-loading pots, vs. 50 CFR 671.26, side-loading pots).

The MFCMA seems to resolve such conflicts by indicating that states can regulate all vessels "registered under the laws of the state" (16 U.S.C. 1856(a)) but the Alaska supreme court, in dicta, recently warned, "to the extent that there may be a conflict between state fisheries regulations and federal regulations . . . Alaska's authority to regulate fisheries . . . has been superseded" [19]. This section of the MFCMA has not been subjected to direct judicial review, so, as might be expected, both federal and state officials have been posturing in support of their respective authorities. In the meantime, fishermen attempting to harvest the resource in a manner consistent with a sustained yield are tempted to select the most liberal provision so they will not be penalized at the marketplace by competition from less conservative operators.

Fortunately, the January 1983 MFCMA amendments will reduce the potential for such conflicts in the future.

The first and most promising amendment greatly abbreviates the extended federal review period that has frequently delayed timely approval by the Secretary of Commerce of regulations proposed in fishery management plans submitted by the regional councils. Under the new provision, section 1854 has been rewritten to require "immediate" review of a fishery management plan. In addition, "immediate" publication in the Federal Register begins a 75-day public review opportunity as soon as possible. After completion of the public review period, the secretary has 20 days to reject or approve the plan; failure to act signifies approval.

The other important amendment to the MFCMA regards the scope of emergency regulations provided by section 1855(3). Formerly, emergency actions were limited to a specific resource emergency. The new language addresses emergencies existing in "any fishery" as that term is defined in the MFCMA to include the act of fishing as well as the stocks themselves as identified on the basis of geographical, scientific, technical, recreational, and economic characteristics in section 1802(7).

The conflicting Tanner crab regulations were the first to be harmonized under the secretary's expanded emergency powers. As a result, the 1983 Tanner crab season benefitted from an orderly start on February 10 and similar gear was required in both the territorial sea and the Fishery Conservation Zone.

Bering Sea herring. A third arctic fisheries dispute was averted without the help of statutory amendments in January 1983 when the Alaska Board of Fisheries agreed to approve a regulation providing for consideration of an offshore trawl fishery for herring in the Bering Sea. The approval followed widespread local opposition to offshore operations that exploited the same stocks sought for valuable herring roe when they migrated to coastal spawning bays and beaches each spring.

Initially, compromise did not appear likely. Prior to the effective date of the MFCMA in March 1977, foreign fleets had taken up to 145,000 metric tons of Pacific herring in the Bering Sea [20]. Accordingly, in light of a minimal domestic effort the North Pacific Fishery Management Council's proposed herring management plan in 1979 acknowledged the role of foreign harvesters in being available to utilize the surplus not taken by U.S. fishermen. Under the MFCMA, section 1821(d)(2), foreign fishing may be conducted on "that portion of the optimum yield of the fishery that will not be harvested by vessels of the United States."

However, since the local coastal fishery was rapidly expanding, residents expressed fear that the offshore fleets could decimate one or more of the many local herring stocks that dispersed to the spawning grounds each spring. The enormous size of the offshore feeding populations -- in the hundreds of thousands of metric tons -- made it difficult to show that foreign fishing at past levels of effort could result in the "overfishing" that the first national standard of the MFCMA sought to avoid (section 1851(a)(1)). Nevertheless, the traditional subsistence harvest in the coastal waters and the growing commercial importance of a catch that leaped from 2,500 metric tons in 1977 to 25,000 metric tons in 1982 combined to prove the potential of a fully utilized inshore fishery that did not require a supplemental offshore fishing effort either by foreign boats or by joint ventures using American fishing vessels and foreign processing ships.

A redrafted fishery management plan provided for the local needs and helped spur an increase in the domestic fleet of seine and gillnet vessels from 65 to 335 boats between 1977 and 1982 [21].

The final federal management plan protected a substantial spawning population of some 115,000-200,000 metric tons of herring during the time the various local stocks were intermingled for feeding purposes in the offshore areas of the Bering Sea. Further, a lid of 2,000-10,000 metric tons was put on any offshore fishing that might be necessary in the future to achieve full utilization of the stocks.

Although the state originally assumed a very protective role over the stocks during their extraterritorial migrations, the new management plan favored the inshore fishery and allowed the Alaska Board of Fisheries to provide for the possibility of an offshore effort if inshore harvests did not fully utilize available stocks [22].

5. CONCLUSION

Both state and federal governments are involved in the ultimate resolution of arctic policy issues. Unfortunately, state-federal conflicts are most frequently decided by time-consuming litigation in which the parties assume adversarial roles. Because of the high stakes, both real and perceived, final judgments do not always erase the emotions of lengthy litigation. On the other hand, more cooperative approaches would allow both parties to focus their undivided attention on developing arctic policies and programs that will promote comprehensive arctic planning that reflects the true potential of the American arctic, instead of the present uncoordinated efforts so inadequate in comparison to the unified programs in other countries [23].

REFERENCES

1. See, e.g., Bonker, Hon. Don, Whale Protection Will Be Enhanced Through the Governing International Fishery Agreement With Japan, Congressional Record, E5045, December 9, 1982.
2. Japan Whaling Association, The Whaling Issue: Time for Clean Thinking, Tokyo, 1982.
3. International Whaling Commission, Report of the Nutrition Panel for the Aboriginal/Subsistence Whaling Panel Meetings 5-6, pp. 26-27, 1979.
4. Council on Environmental Quality, Eleventh Annual Report, pp. 22-24, 1980.
5. Price, Monroe E., A Moment in History: The Alaska Native Claims Settlement Act, U.C.L.A. -- Alaska Law Review, vol. 8, pp. 89-102, 1979.
6. Senate Report 96-413, Committee on Energy and Natural Resources, p. 233, November 14, 1979.
7. Note, The Outer Continental Shelf: Bridging the Informational Gap Through Regulations, Ariz. St. L. J., vol. 1978, pp. 635-636, 1978.
8. United States v. Alaska, 422 U.S. 184, 1975, is an example of hotly contested litigation over oil and gas rights to submerged lands.
9. U.S. Constitution, art. III, sec. 2; United States v. Alaska, Original Jurisdiction Case No. 84, 1979.
10. Convention on the Territorial Sea and Contiguous Zone, Done at Geneva, on April 29, 1958 (T.I.A.S. No. 5639), arts. 4-7.
11. United States v. California, 332 U.S. 19 (1947).
12. U.S. Department of Commerce, Fisheries of the United States, 1981, April 1982.
13. Personal Communication, Ray Montgomery, Financial Services Section, National Marine Fisheries Service, Washington, D.C.; Pub. L. 95-376 (1978), Pub. L. 96-289 (1980), Pub. L. 96-561 (1980).
14. Hjelle v. Brooks, 377 F. Supp. 430 (D.C. Alaska 1974); State v. Bundrant, 546 P. 2d 530 (Alaska 1976), appeal dismissed sub nom. Uri v. Alaska, 429 U.S. 806 (1976); F/V American Eagle v. State, 620 P. 2d 657 (Alaska 1980), appeal dismissed, 454 U.S. 1130 (1982).
15. Testimony of North Pacific Fishing Vessel Owners Association to North Pacific Fishery Management Council, December 6, 1980.
16. F/V American Eagle v. State, 620 P. 2d 622, footnote 10, supra note 14.
17. Under 16 U.S.C. 1854(c), the Secretary of Commerce adopts plans only "if such fishery requires conservation and management."
18. Alaska Administrative Procedure Act,

Alaska Statutes, Title 44, Chapter 62.

19. *F/V American Eagle v. State* at footnote 10, supra note 14.
20. North Pacific Fishery Management Council, *Bering-Chukchi Sea Herring Fishery Management Plan and Final Environmental Impact Statement*, p. 31, Anchorage, Alaska, June 1981.
21. Fried, Stephen, et al., *Pacific Herring Stocks and Fisheries in the Eastern Bering Sea, Alaska, 1982*, Report to the Alaska Board of Fisheries, Alaska Department of Fish and Game, Juneau, Alaska, December 1982.
22. 5 AAC 27.730, .885, .930 (1983).
23. Senate Report No. 97-660, Arctic Research and Policy Act of 1981, October 18, 1982.

ARCTIC POLICY II

INTRODUCTION

Donn K. Haglund

Department of Geography
University of Wisconsin-Milwaukee
Milwaukee, Wisconsin

Good morning! My name is Donn Haglund and I am chairman of the Department of Geography at the University of Wisconsin-Milwaukee. My specialty is cultural-economic geography of the Arctic and Sub-Arctic and on December 31st I completed a three year stint as Chairman of the Board of Governors of the Arctic Institute of North America. For the past eleven years I have offered what is believed to be the only University-based Arctic Winter Field Course in the western world. My presence have today however is in order to substitute for Professor Louis Rey, President of Comite' Arctique International, the world organization, multi-disciplinary in character, of Arctic specialists. Louis has very recently left his home in Switzerland to serve for several weeks as Visiting Professor at the University of Alaska and was unable to attend this conference due to the prior commitments. He has asked me, in my capacity as Vice-President of Comite' Arctique International, to fill in for him.

I am sorry that Professor Rey could not be with you. All of us would have benefitted from his participation in this conference. But then, if he were here I would not have this opportunity. Thank you, Louis, for not being able to fulfill this assignment. Your loss is my gain.

On his behalf and my own, we bring you greetings from Comite' Arctique International. We hope to see all of you at one or more of the annual Arctic oriented international conferences that are conducted or co-sponsored by our organization. Should you have any questions concerning Comite Arctique International and its activities, please contact me some time during this conference.

Now to our speakers for this session.

Our first paper, "Regulatory and Environmental Issues Associated with Arctic Marine Transportation" was prepared by D. Bruchet and M. Robertson who are associated with the Arctic Pilot Project of Petro-Canada, headquartered in Calgary, Alberta. The paper will be delivered by Mr. P. Douglas Bruchet.

For those who may not be familiar with the Arctic Pilot Project it is "designed to produce and liquify 7.1 million cubic meters of natural gas per day in the Arctic and move it to Eastern Canadian markets in ice-breaking ships. Arctic Pilot Project proposes to test the feasibility of producing natural gas from wells in the Arctic Islands, transporting the gas by a 160 kilometer buried pipeline, transforming the gas into liquified natural gas (LNG) and shipping the LNG by ice-breaking carrier to a regasification plant in southern Canada - all on a year-round basis." (Quotations from a 1980 summary published by APP).

The proposal has not met with universal acceptance in Canada and neighboring Greenland. Environmentalists, Native (Inuit) leaders in both countries, a number of Canadian governmental regulatory agencies, legislative bodies and task forces have all had a diversity of "says" on the Project and its proposals.

We are fortunate to have a distinguished spokesman for APP with us to describe the geographical, historical and sociological setting of the Proposal, the range of regulatory processes impacting upon the Proposal and the environmental issues that apply (the effects on whales, ringed seals, birds, ice, the possible problems associated with crossing ice disturbed by ship tracks, the controversial issue of underwater sound and others which will undoubtedly surface eventually) to the implementation of the project. "That year-round transportation in Arctic waters is both technically feasible and environmental acceptable" is a conclusion to which our speaker will now direct us.

As the author of the study, "Maritime Transportation to Support Polar Resource Development" recently published by the U.S. National Academy of Sciences I eagerly anticipate Mr. Bruchet's concurrence in many of the conclusions we asserted in that publication from this side of the forty-ninth parallel.

The second of our papers this morning is by Ian J. Jordaan, Head, Research and Development, Det norske Veritas (Canada) Ltd. After receiving his baccalaureate and master's degrees in engineering at the University of the Witwaterstrand, South Africa, Dr. Jordaan completed his Ph.D. at King's College of the University of London in 1969. For the next thirteen years Dr. Jordan was a member of the Department of Civil Engineering at the University of Calgary, in Alberta. Last year he left the university to accept his present position with Det norske Veritas at their Canadian corporate headquarters, also in Calgary. Dr. Jordaan is the author of over forty publications in his chosen field.

Today he speaks to us on the subject, "Risk and Safety Assessment for Arctic Offshore Projects". After defining risk and applying it to concepts of safety and decision-making Dr. Jordaan will acquaint us with risk analysis methodology and its application, through examples, to Arctic offshore projects, suggesting a number of scenaries. Environmental factors are considered and probability theory applied to risks in Arctic operations. Questions of risk aversion and of proness to risk in utility theory are developed as our speaker draws our attention to practical safety objectives before suggesting conclusions.

REGULATORY AND ENVIRONMENTAL ISSUES ASSOCIATED WITH ARCTIC MARINE TRANSPORTATION

D. Bruchet, M. Robertson

Arctic Pilot Project, Petro-Canada
Calgary, Alberta Canada
January, 1983

1. INTRODUCTION

The relationships between resource development, traditional land use and conservation, and the course of development - how much, how fast and where, at times has set the stage for the rather emotional debate on Arctic Development. The Canadian North is composed of the land and water areas encompassed by the Yukon and Northwest Territories and their associated offshore areas, and constitutes almost 50% of Canada's land mass. The area has many regional differences, as shown in the Northern Land Use Planning Policy,[1] and is one of several circumpolar regions bordering the Arctic Ocean. Accordingly, certain international issues are also discussed in this paper. The introductory sections of this paper present a brief background on: the Canadian North, federal policies for northern development, and the current regulatory controls over arctic marine transportation in Canada. Specific references are made to the Arctic Pilot Project, a unique Canadian Arctic venture, which is later described in more detail.

Although most land in Canada's North lies above the 60th parallel, activities in southern Canada and the rest of the southern world can have significant effects on the North. Early encounters between native people and Europeans occurred mainly due to whaling, the fur trade, and exploration of the Northwest Passage. The nomadic lifestyles of the northern natives began to change with the appearance of fur trading posts, whaling stations and missions. The establishment of government health and education services in communities during the 1950s and 1960s resulted in nearly all native people adopting permanent residences in the communities. Native people continue their traditional harvest but now with the aid of high-powered rifles, motorboats and skidoos. 1981 statistics indicate that the majority of the Canadian non-native northern populations resides in the capitals of Yellowknife and Whitehorse. The majority of the native population lives in 60 small communities scattered throughout the North. The languages spoken are either Inuktitut or Athapaskan, together with English. The northern wage economy now is based on the export of minerals, limited quantities of fur, fish and native arts and craft products, and the existence of government services. Large quantities of oil and gas may be produced in the future and will add significantly to the North's economic base.

The evolution of Canadian political boundaries, the transfers of Rupert's Land in 1870 and the Arctic Archipelago in 1880 from Great Britain are important historic events. Of recent importance is the report by C.M. Drury on "Constitutional Development in the Northwest Territories, 1980"[2] and the plebiscite held by the Government of the Northwest Territories where the majority of voters favoured division of the Territories into two parts. At present, both the Yukon and Northwest territorial governments have a Commissioner and a fully elected legislative body. In the case of the Northwest Territories, the legislature consists of 22 members.

Because of the native northerners' aboriginal occupancy of the land, many native people are opposed to extensive mineral exploration, and other non-renewable resource-related activities. This opposition has led to the formation of strong native political organizations with a central demand for the settlement of their land claims. Land claims, until resolved, will continue to be a major issue to be faced by any resource development project in the Northwest Territories.

Particularly in the last thirty years, the issue of Canadian sovereignty has been an important consideration. In the 1950's Canada moved Inuit into the Arctic Islands to confirm assertions of sovereignty over that area. Further, the Manhattan voyages of 1969 and 1970 served as a catalyst in the development and enactment of specific legislation to protect Canada's sovereignty over the Arctic Islands, and to initiate environmental protection over Canadian waters.

Land claims and environmental impact-related matters have also resulted in the formation of the Inuit Circumpolar Conference, an association of Inuit political groups from Alaska, Canada and Greenland. Among other things, this group is concerned with the Right of Innocent Passage through Baffin Bay and Davis Strait, and the control and routing of vessels carrying hazardous cargoes. The response from the Greenland Home Rule and the Inuit Circumpolar Conference has been to oppose not only Canada's Arctic Pilot Project, but also other projects using the route between Canada and Greenland. There are two levels of negotiations on ship routing currently underway: one is the Arctic Pilot Project Working Group, comprised of Danish, Canadian, Greenlandic and Petro-Canada membership, and the second is the Marine Environmental Cooperation Agreement (MECA) between the Governments of Canada and Denmark. There are also various studies on arctic marine transportation now underway which address national, as well as international implications of Arctic marine transportation.

2. REGULATORY PROCESSES

The last two decades have been a period of significant social and technological change in the Canadian Arctic. Exploration for minerals and hydrocarbons has been very active, resulting in several discoveries of major deposits. As well, modern resupply programs for northern communities have been developed. Two mines in the High Arctic, Polaris and Nanisivik, are now in production, shipping lead-zinc concentrate on a seasonal basis by marine mode. The summer resupply of northern communities is by ship, supported by Canadian Coast Guard icebreakers. Much of the resupply of the Arctic Islands drilling ventures of Panarctic Oils Ltd. also takes place by ship. In 1982, approximately 39 500 tons of resupply materials were delivered by ship into the High Arctic. In addition, the 1982 resupply of the Beaufort Sea saw the first large hydrocarbon carrier move into that area to support Dome Petroleum's fleet of drillships and icebreaking support vessels.

Much of the present marine transport takes place in the open water period, or with icebreaker support. However, some ships, such as the M.V. Arctic, an ice class bulk carrier, and the M.V. Kigoriak, an icebreaker used in the Beaufort Sea, have significantly extended the traditional arctic shipping season.

The requirement for more reliable year-round transportation modes arose in the early seventies with the delineation of major oil and gas deposits in Canada's Arctic Islands.

Subsequently, major hydrocarbon finds in the Beaufort Sea have resulted in a need for marine transportation from that area. For example, since 1961, industry (mainly through Panarctic Oils Ltd.) has spent approximately $947.0 million Canadian dollars in energy exploration of the Canadian Arctic Islands. Discoveries to date amount to 18 trillion cubic feet of natural gas, and three-quarters of a billion barrels of crude oil. These reserves alone add 25% to Canada's gas reserves, and 15% to Canada's conventional oil reserves.[3] In the Beaufort Sea, the Geological Survey of Canada estimates oil reserves at 6.9 to 9.4 billion barrels.

The use of ships rather than pipeline to move arctic oil and gas is advantageous because:

1. smaller scale reservoirs may be exploited
2. there is flexibility in terms of delivery point
3. the major investment needed for pipeline systems is not required, and
4. reserves can be developed and marketed in a shorter time frame.

The government of Canada recognized the need to move Canada's arctic resources to market at an early stage. In 1970, the Honourable Jean Chretien, then Minister of Northern Development, discussed the Arctic Waters Pollution Prevention Act Bill (AWPPA) in the House of Commons, and identified Canada's northern interests as: Canadian security, northern economic development and the preservation of the ecological balance of the area. In passing the AWPPA, the government expressed its desire to develop northern technology in a controlled manner, and to encourage transportation through the Northwest Passage. In his 1970 speech, Mr. Chretien stated that the Canadian Government wanted Arctic waters to be opened up to commercial shipping.[4] By passing the AWPPA, the Government established shipping zones and mandatory ship standards in order to ensure that vessels operating in the Canadian Arctic are sufficiently strong to withstand expected ice conditions in their zones of operation. At the same time, it was recognized that Canada, one of the largest arctic countries in the world, was falling severely behind both in the technology of arctic transportation and in the exploitation of arctic resources. In 1971, The Science Council of Canada[5] recommended that Canada turn its attentions immediately to the problem of navigation in the Arctic in order to fully realize the development and utilization of northern resources.

Again, in 1975 and 1979, the Science Council of Canada identified the need for Canadian development of arctic transportation, particularly through smaller scale pilot projects in the North, with meaningful evaluation of such projects to obtain full value from the knowledge gained.[6,7]

In 1977, a seminar on "Natural Gas from the Arctic by Marine Mode" was sponsored by the Atlantic Provinces Economic Council and the Science Council of Canada[8]. This conference noted that although the federal government had stated in 1973 its intention to move into areas of northern development and to achieve world-recognized excellence in operating on ice-covered waters, by 1978, in fact very little had been done and rapid steps needed to be taken to enhance Canadian technological, scientific and industrial capability in the north.[9]

In spite of the Arctic Waters Pollution Prevention Act, Science Council studies, stated government policy, and a recognized need for Canada's northern resources, little positive action has been taken so far in the area of hydrocarbon transportation.

The first specific proposal to use the marine mode for Arctic hydrocarbon extraction was made in 1977 by a consortium led by Petro-Canada known as the Arctic Pilot Project (APP). The Project involves drawing gas from the large natural gas reserves at Drake Point on Melville Island, overland transportation of the gas by pipeline 167 km southwards to a natural harbour at Bridport Inlet, liquefaction at that site, and removal of the gas to markets using two Arctic Class 7 LNG carriers. The process undergone by the Arctic Pilot Project since its inception well illustrates the dilemma facing northern developers trying to operate within the Canadian regulatory framework. Critical to any investment decision, of course, is the degree of risk involved, and the question of how much a company must spend before receiving approval to proceed. The APP has spent about 60 million dollars in developing and defining the Project over the last six years. It is no longer possible for companies to spend such sums before an approval in principle is given. The front-end costs and time frame must be reduced to meet the requirements of the highly competitive world in which we live. It should also be remembered that Canada's competitors in the area of gas exports are countries such as Algeria, Australia, Indonesia and the Soviet Union. In a globally competitive business, any hindrance or risk factor unique to a single competitor puts that competitor at a disadvantage. In Canada, duplicative, lengthy public hearings have become a competitive disadvantage.

The Canadian Northwest Territories is an area of shared jurisdiction between the federal and the territorial governments. While the territorial government has responsibility through an elected council for such areas as social welfare, health, community management, wildlife management, etc., the federal government, largely through the Department of Indian Affairs and Northern Development, retains control over most non-renewable resource management, including oil and gas exploration and production. Since 1972, no firm northern development policy has been enunciated. All projects are required to proceed in a relatively ad hoc way, depending upon the areas of jurisdiction within which they fall. In the case of the Arctic Pilot Project, the Project was required to undergo a public environmental and socio-economic review under the Federal Environment Assessment and Review Process, although neither of the two heavy metal mines in the same area had done so, nor had the approximately one billion dollars of exploration activity in the Arctic Islands. The Beaufort Sea exploration activities did not undergo a public review process, however, the development proposals for that region currently are in the early stages of such a review. The Environmental Assessment and Review Process is required by cabinet policy rather than by any piece of legislation and results in a review panel being formed, the issuance of a series of guidelines for an environmental impact statement, the production of an impact statement by the proponent, a series of public hearings to review the Project and the publication of a report by the Panel. In the case of the APP, this process took approximately three years from start to finish. In October 1980, the Environmental Assessment and Review Panel on the APP recommended that the Project was environmentally acceptable provided certain conditions were met[10].

Notwithstanding such a wide review, the Department of Indian Affairs and Northern Development, which retains general authority over northern development, also took the initiative for a regulatory review and referred the Project to the National Energy Board which recommends to the federal Minister of Energy, Mines and Resources. Traditionally, the NEB has dealt with issues such as interprovincial pipeline licenses, export to the United States, rate base matters, etc. and has not been involved in a major way with any High Arctic proposals requiring marine transportation. However, in the APP's case, in addition to dealing with the issue of gas export from Canada, the Board was requested to hold a general inquiry into the Project. This hearing began in February 1982 and was adjourned by the Board in August 1982 due to opposition to the fact that APP was exploring the possibility of new markets.

An encouraging development during 1982 was the establishment of a Ministry of Transport Control Authority to regulate all shipping in the Canadian Arctic. This Authority is now advised by an environmental committee, which includes membership from industry and government to allow for the development of environmentally-compatible shipping regulations for ships using Canadian waters. In 1981, the federal government also established the Canada Oil and Gas Lands Administration as part of the proposed Canada Oil and Gas Act. This new body, referred to as COGLA, is responsible for administering oil and gas activity on Canadian lands, including exploration and production, and to coordinate the development of related Canada benefits plans and the resolution of environmental concerns.

Concurrent with these other activities, the Government of Canada undertook a study known as the "Lancaster Sound Green Paper". This exercise was designed to identify land use options for the Lancaster Sound area, which serves as the entry and exit point of the Canadian Northwest Passage. The Green Paper began in 1979 and included a series of public workshops to discuss and review the present resource base, present land use and future options. A "draft" Green Paper for public discussion was produced in December, 1980, and a final "Green Paper" was released on July 30, 1982.[11] The final paper suggested two options for a regional planning process, and the government then started planning a further series of public reviews to determine which land use option was most needed for the area. The paper was unclear regarding the significance of the early 1970s policies, of the Arctic Waters Pollution Prevention Act, or of the Environmental Assessment and Review Process report on the Arctic Pilot Project which had indicated that the Project could proceed. At this time, the 1980 Environmental Assessment and Review Process report was being considered by the National Energy Board as another input to their process, and all the issues aired at the EARP hearings were being raised and discussed at length in the National Energy Board hearing. The advice to the proponents of the APP was that the Department of Indian Affairs and Northern Development would receive the reports of the National Energy Board and the Environmental Assessment and Review Process and would recommend to Cabinet a position on approval or disapproval of the Project. To date the APP has spent approximately five years in the regulatory process.

In October, 1982, the Department of Indian Affairs and Northern Development issued a new policy on "Land use Planning in Northern Canada." This policy was "... not intended to replace any existing planning mechanisms; rather, it is designed to complement them."[1] This plan identifies a process of establishing Regional Planning Areas, Northern Land use Planning Committees, Northern Land use Planning Commissions and Area Planning teams to produce land use plans for the six large regions and fifty sub-regions identified. In this draft, it is interesting to note that the Lancaster Sound/Parry Channel region is divided down the centre of that channel, and falls into three regions and eight sub-regions. Naturally, proponents wonder how this new process will affect their interactions with the regulatory process in the Canadian North.

The above clearly shows the convoluted and overlapping structure of the Canadian Regulatory Process. The Environmental Assessment and Review Process, National Energy Board, the Department of Indian Affairs and Northern Development, the Canada Oil and Gas Lands Administration, the Department of Fisheries and Oceans, and the Department of Environment all identify environmental issues under their mandates. Therefore, companies must constantly determine who is handling what to ensure that steps in this process are not missed, and that further delays do not occur.

This dilemma is exacerbated by the perceived need of regulatory bodies to review projects within a static situation. It is typical of large projects to evolve and change from concept to final design during the whole development phase. Options must be kept open to allow the best engineering, economic and technological parameters to be used when the project is constructed. Lengthy reviews therefore, must be willing to accept that designs cannot be final, and that there may be changes both during the review period and before construction commences. It is unrealistic to expect companies to move large projects to a final design stage before substantial assurances have been given that the project will proceed.

This paper will now discuss the types of environmental issues that have arisen as a result of the reviews held to date on arctic marine transportation. Specific references will be made to the APP.

3. ENVIRONMENTAL ISSUES

It should be kept in mind that many ships currently travel arctic waters, mainly in the June to November period. These include oil tankers, general freighters of several sizes, Coast Guard icebreakers, and an ice class bulk carrying freighter. None of this shipping has undergone any review other than being required to meet the conditions of the Arctic Waters Pollution Prevention Regulations. However, hydrocarbon development proposals are being subjected to major public hearings and regulatory review.

The Arctic Pilot Project ships will carry liquefied natural gas, a non-polluting cargo which rapidly vapourizes when placed in contact with water. Consequently, there are none of the risks inherent in oil transportation where cargo loss could lead to a major environmental problem.

The majority of the concerns raised in environmental reviews of the APP deal with possible interactions between ships and marine mammals, although the issues of ice regime changes and ship track crossings also have been widely discussed. The seven major issues can be described as follows:

3.1 Entrapment of Whales in Icebreaker Tracks

Whales, particularly narwhal and beluga, tend to mill around the ice edge in Lancaster Sound in the spring, waiting for break-up to allow a westerly migration to the summering areas. Early arriving animals follow open leads into the fast ice. Concerns were raised that the whales may follow the icebreaker tracks into Parry Channel at this time, and become trapped when the tracks refreeze. Observers were placed onboard Canadian Coast Guard icebreakers in two successive years to determine whether this phenomenon occurred as the ships entered the fast ice of Lancaster Sound in the spring. Observations from the ships and helicopter surveys of the icebreakers' tracks revealed that no whales were present in the track.[12] The whales, as one might expect, seem to know the difference between an open lead and a ship track filled with ice rubble.

3.2 Physical Interference with Ringed Seals

Studies by the Department of Fisheries and Oceans and the Canadian Wildlife Service[13] identify Barrow Strait as an important area for ringed seal. The ringed seal is a major hunting item for the Inuit, and a major prey species for polar bears. The potential for year-round icebreaking to interfere with ringed seal denning and pupping in Barrow Strait was thus investigated. Initial calculations made for the Environmental Assessment Review Process in 1980, estimated that 1.0% of the annual recruitment could be effected assuming a worst case scenario i.e. with the ships taking a different course on each passage. Subsequent considerations have indicated that there may be fewer passages during the pupping period than originally predicted, that the ship will be able to reuse the same track, and that the pressure ridges, where most dens are located, will generally be crossed at a 90° angle. These developments lower the estimate of annual recruitment potentially effected to 0.2%, a portion considered biologically insignificant compared to the 40% mortality which occurs naturally during the first year of life.

Ringed seals are also being studied in Baffin Bay. Some people have suggested that the ringed seals which are harvested inshore by the Greenland Inuit may overwinter in the offshore areas of Baffin Bay. Studies of appropriate ice habitat conditions were undertaken using remote sensing.[14] The results of these studies were, in turn, used to develop a model of the ringed seal population dynamics in the Baffin Bay region.[15] Investigations of the age structure of the inshore population by aging of seal teeth, and of possible migration routes by tagging, were also undertaken. The results are not yet sufficiently clear to provide a decisive answer, but it does appear that there is some population interchange between the two areas.

3.3 Bird Colonies

The APP is not expected to effect the marine-associated birds which summer in the High Arctic[16,10]. APP ships will, of course, carry liquefied natural gas which is a non-polluting cargo even in the unlikely event of a spill. If a spill occurred, the gas would very rapidly dissipate into the atmosphere. The operational corridor of the proposed carriers is located in the center of the passage over 25 km from the coasts. No impacts are anticipated from passage of the ships. The ships will carry a reserve of diesel fuel which, if spilled, could threaten any seabirds in the immediate vicinity. However, considering the double hull of the carrier, the location of the storage tank high above the waterline, and the seasonal presence of the birds, the probability of such a spill is very low.

3.4 Ice Edge Effects

During the EARP hearings, attention was drawn to the possibility that passage of ships through the ice in Parry Channel might alter freeze-up and break-up patterns. The APP conducted research into the recurring ice crack patterns, ice crack morphology and behavioural processes in Resolute Passage and Barrow Strait[17]. As testified before the National Energy Board in 1982, evaluation of this material indicates that the forces involved with the ships' passage are far too small, compared with those naturally-occurring forces, to influence the freeze-up and break-up of the Lancaster Sound area.[18] This conclusion is confirmed by review of Landsat photographic images of a Canadian Coast Guard icebreaker penetrating the Lancaster Sound ice edge with no detectable alterations to the fast ice[19].

3.5 Ship Track Crossing

Hunters from Resolute Bay cross Barrow Strait by skidoo to hunt mainly caribou on Somerset and Prince of Wales islands to the south, and polar bear in Barrow Strait itself. The potential for the icebreaker's track to interfere with these travel patterns has thus been studied. An initial theoretical analysis predicted that refreezing rates would allow safe crossing of the track within a few hours during the winter period[20]. To determine where and when hunting trips across Barrow Strait occur, a resource harvest study was initiated in January 1981 in conjunction with the Hunters and Trappers Association of Resolute. Three actual ship track crossing trials have been completed, using an icebreaker stationed at McKinley Bay in the western Arctic. The tests were carried out in November, March and June. In all three tests, the ship track was safely crossed by a skidoo and loaded komatik within 2.5 hours of the ship's passage. In an attempt to summarize and interpret the many investigations of the ship track crossing question, the Arctic Pilot Project prepared and submitted to the National Energy Board a report entitled "Ship Track Crossing"[21]. This study concluded that for the large majority of the winter season, the track will be filled with ice rubble that will refreeze very quickly and will not affect hunting activities. For a very short period at spring break-up, there may be areas of open water in the ship track that will not refreeze. Such occurrences will be similar to the natural situation encountered when trying to cross leads and cracks in the ice at this time of year.

3.6 Integrated Ship Routing

One condition attached to the Environmental Assessment and Review Process Panel's approval of the APP was "the integration of physical factors and biological factors so as to minimize adverse impacts on wildlife. "An Integrated Route Analysis" (IRA) has been developed, covering the 5200 km shipping route, to delineate an operational corridor that ensures public and ship safety with minimal environmental impact, while maintaining economic feasibility. The document is an evolving routing guide, intended to incorporate new information as it becomes available, which will result in further refinements of the operational corridor. The IRA was first published in April 1981, and was then revised in December 1981.

3.7 Underwater Sound

The impact, or lack of impact of ship-generated underwater sound on marine mammals has been a prominent area of investigation for the Arctic Pilot Project. The issue was raised at the public Environmental Assessment and Review hearings in 1980, and a recommendation was then made that further work should be done to better understand the possible interactions between marine mammals and shipping[10]. This matter is not exclusive to the Arctic Pilot Project as all ships generate some noise and even considerably smaller ships now operating in the area produce comparable levels of noise to that expected from the APP carriers. Present shipping also takes place during the time of year when most marine mammals are in the area. Notwithstanding these facts, little was known of the interactions between marine mammals and ships.

To develop a scientific, rather than a speculative context for evaluating the issue, the APP convened a workshop of recognized international experts to define the parameters of the underwater sound equation relative to the APP[22]. Progressing from this workshop, subsequent studies were undertaken to identify critical areas for wintering marine mammals[23], and to determine noise levels produced by APP ships[24]. A computerized sound transmission model was developed to predict the zone of influence of various source levels. The Project also is preparing a noise research plan to guide research on underwater acoustics during the pre-operational and operational periods.

This issue has been focused on by certain groups who predict large-scale environmental problems due to ship noise in arctic waters. Recent studies done by the APP and others do not support this view.

4. CONCLUSIONS

The main issues discussed above are not unique to any particular shipping project, but should be addressed by any Arctic shipping operator. Nevertheless, the onus seems to have been placed exclusively on year-round operators to address these issues, even though arctic shipping has been a reality for many years. One reason given as possible justification for this approach is the fact that year-round shipping does not yet exist. However, most marine mammals use the Lancaster Sound and Parry Channel during the open water period, when most shipping now takes place.

At this point it should be mentioned that the Arctic Pilot Project has been required to address environmental issues and concerns south of 60° latitude as well. The Project appeared before joint federal/provincial environmental reviews in the provinces of Quebec and Nova Scotia, both of which were being considered as possible regasification terminal sites. In addition, a review of the southern shipping route and terminal operations was conducted by the federal Ministry of Transport. Along with environmental concerns, significant emphasis was placed on economic and industrial benefits during these reviews. In all three review cases, the APP received conditional approval.

Despite the time, money and effort devoted by industry to the study of arctic issues, there are, of course, areas in which some interests feel information is lacking. The question to be addressed, however, is not how much we do not know, but whether we know enough to allow year-round transportation to commence. Apart from the basic legal issue of whether or not ships meeting the Arctic Waters Pollution Prevention Act requirements can in any way have their right of innocent passage regulated by Canada, there is a clear concensus emerging that Canada does have the information necessary to make the informed decision that year-round transportation in arctic waters is both technically feasible and environmentally acceptable. Recognizing that all the answers to detailed environmental questions raised by frontier developments cannot be available before a project begins to operate, the APP included, as part of its management plan, a $220 million Research and Development Program to study northern environmental, socio-economic, technical and operational issues. These funds will be spent under the direction of a comprehensive advisory group, including northerners, academics, technical and government people, and will work under a policy of "northern studies done by northern people in the north". This concept, and the inclusion of such a large Research and Development fund to be dedicated to northern studies, is a major commitment by industry to Canada's northern research efforts.

For many years environmental interests were excluded from project planning, now they are an integral part of such planning in all northern development activities. However, the cost of being in the mainstream of decision-making is being willing to make useful, constructive decisions based on the information available. Indecision, or a constant search for further information, will ultimately relegate environmental issues to the back rows of both government and industry. Such a move would be extremely destructive, but can only be prevented by the generation of high levels of accountability and responsibility by academic, government and industrial environmental advisors.

The matters raised in this discussion of northern marine transportation, from a regulatory and environmental perspective, are not necessarily unique to the marine mode. Several project review procedures have been used in the past, and the process continues to evolve, for example, with the recent creation of the Canada Oil and Gas Lands Administration. The process is, however, still so lengthy, slow and poorly defined that it is considered a major risk factor in project development.

The point has been made that when a project is the first to go through the regulatory process, it will understandably suffer from the learning curve that any new procedure requires. However, it can be shown that because the process is so frequently adjusted, every new project proposed has had to enter new regulatory waters. The Mackenzie Valley Pipeline, the Arctic Pilot Project and the Beaufort Sea proposals are examples which support this. The search for effective, yet more efficient methods of regulatory review is encouraged and must be actively pursued to produce a timely, yet comprehensive review and approval process.

The environmental issues associated with northern transportation are well-defined. It is now time for a concerted industry-government program to be developed to expedite northern transportation projects by concentrating on the few areas still in question to fine tune the knowledge base. This will give industry the necessary guidelines for any operational constraints within the lead time which can be accommodated in project design. It is obvious to all that industry has now generated the majority of information in northern environmental study over the last ten years, and that we can be expected to behave in a responsible manner. What is needed is less public review and supervision and a greater degree of synthesis and cooperation in working toward common goals.

I would like to conclude with a quote that comes to mind by a gentleman named Geoffrey Vickers: "We seem to be in an infinite regress, leading sooner or later to a choice made by applying a process which cannot yet be specified to data which remain obscure."

REFERENCES

1. Department of Indian Affairs and Northern Development, Northern Land Use Planning Policy, (draft), Ottawa, October 14, 1982.

2. Drury, C.M., Constitutional Development in the Northwest Territories - Report by the Special Representative, pp1-194, Canadian Government Publishers, Hull, Quebec KIA OS9, 1979.

3. Senate of Canada: Proceedings of the Special Committee of the Senate on the Northern Pipeline Wednesday, June 9, 1982 Issue No. 28.

4. Department of Indian Affairs and Northern Development, Address by the Honorable Jean Chretien, Minister of Indian Affairs and Northern Development to the House of Commons on Second Reading of the Arctic Waters Pollution Prevention Bill, April 16, 1970.

5. Stewart, R.W., for the Science Council of Canada, Special Study No. 16, Ad Mare: Canada Looks to the Sea: A Study on Marine Science & Technology in Canada, pp1-173, Ottawa, 1971.

6. Science Council of Canada, Report No. 23, Canada's Energy Opportunities, Ottawa, March 1975.

7. Science Council of Canada, Roads to Energy Self Reliance - The Necessary National Demonstrations, Ottawa, 1979.

8. Science Council of Canada, and the Atlantic Provinces Economic Council, Seminar on Natural Gas from the Arctic by Marine Mode, Ottawa 1977.

9. Science Council of Canada, Report No. 26, Northward Looking,A Strategy and a Science Policy for Northern Development, Ottawa August, 1977

10. Federal Environmental Assessment Review Office, Arctic Pilot Project (Northern component) Report of the Environmental Assessment Panel, pp1-108, Ottawa, October 1980.

11. Dirschl, H.J., for Department of Indian Affairs and Northern Development, Green Paper, The Lancaster Sound Region: 1980 - 2000 Issues and options on the Use and Management of the Region, pp1-101, Ottawa, 1982.

12. Hatfield Consultants Ltd., Observations of Marine Mammals and Sea Bird Interactions with Icebreaking Activities in the High Arctic, for the Arctic Pilot Project, Vancouver, December 1979, December 1980.

13. Kingsley M.C.S., Stirling I., Calvert W.. High Arctic Seal Distribution, Canadian Wildlife Service, Department of Environment, Ottawa, February, 1981 and February 1982.

14. Marko, J.R., Satellite Imagery Studies of Ringed Seal Habitats in Baffin Bay, unpublished report by Arctic Sciences Ltd., Sidney, B.C. for LGL Limited, Toronto, 1981.

15. Miller G.W., Davis R.A., Finley K.J., Ringed Seals in the Baffin Bay Region, Habitat Use, Population Dynamics and Harvest Levels, LGL Limited, Toronto, 1982.

16. McLaren, P.L., Distribution of Sea-Associated Birds in Eastern Lancaster Sound and Western Baffin Bay, unpublished report by LGL Limited for Petro-Canada Exploration, May-July, 1979.

17. Acres Consulting Services Ltd., Ice Crack Morphology in Barrow Strait, filed as Exhibit 199 at National Energy Board Hearing into the Arctic Pilot Project, Ottawa, July 1981.

18. Steffan, K., Prepared Testimony submitted to the National Energy Board Hearing (Marine Special Interest Panel) for the Arctic Pilot Project, Exhibit 638, Ottawa, June 1982.

19. Arctic Pilot Project, submitted to the National Energy Board Hearings, as Exhibits 644 A, B, C, Photographic Enlargements of Entry Point of the M.V. Louis S. St. Laurent Into Landfast Ice Edge in Lancaster Sound area, Ottawa, June, 1982.

20. Arctec Canada, Study of Bridging a Shiptrack in Landfast Ice, submitted to the National Energy Board Hearings for the Arctic Pilot Project as Exhibit 75, Ottawa, February, 1982.

21. Arctic Pilot Project, Ship Track Crossing Report, filed as Exhibit 685 at National Energy Board Hearings into the Arctic Pilot Project, Ottawa, July, 1982.

22. Arctic Pilot Project, The Question of Sound from Icebreaker Operations: Proceedings of a Workshop, pp1-350, Toronto, February 23, 24, 1981.

23. McLaren P.L., Davis R.A., Marine Mammal Distribution off West Greenland and in Southern Baffin Bay and Northern Davis Strait, LGL Limited, Toronto, Ontario, 1981, 1982.

24. Arctic Pilot Project, Revision to Section 3.2.3.1 and Section 3.2.3.2 of the Integrated Route Analysis, Volume 2 filed as Exhibit 168 at National Energy Board Hearings into the Arctic Pilot Project, 1982.

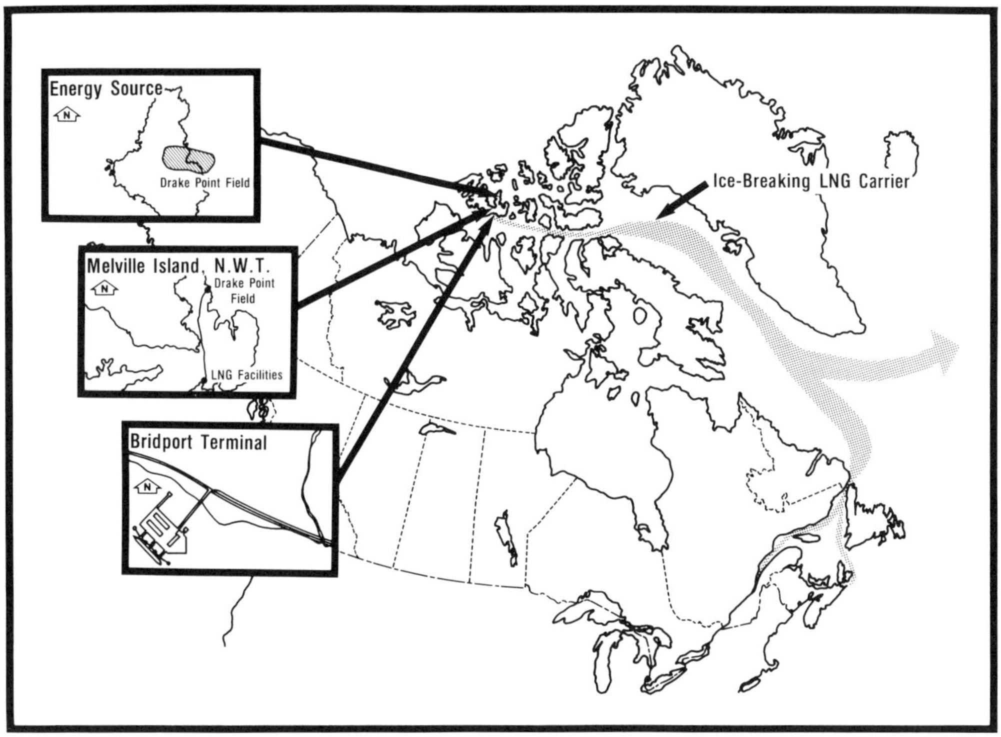

Arctic Pilot Project Components

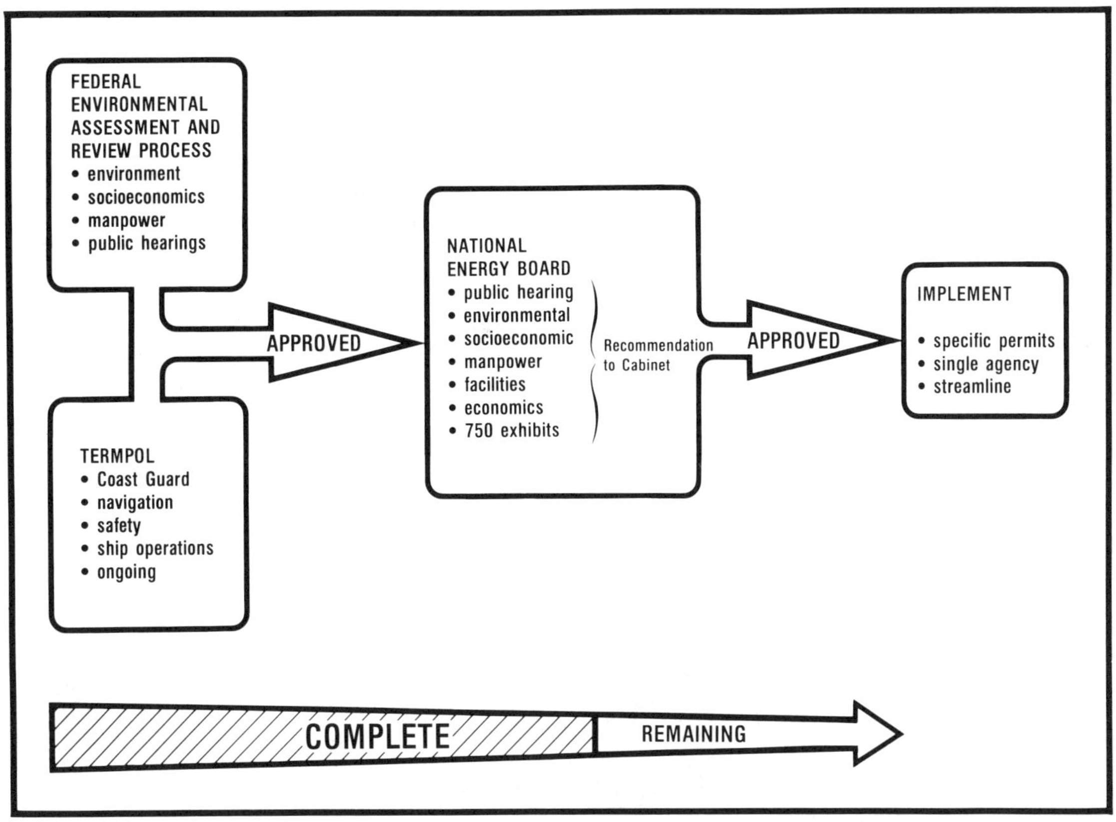

APP Regulatory Process

RISK AND SAFETY ASSESSMENT FOR ARCTIC OFFSHORE PROJECTS

Ian J. Jordaan

Det norske Veritas (Canada) Ltd.
Calgary, Alberta, Canada

Risk is defined in terms of both the probability, generally small, and the undesirability of an event. Risk analysis is a systematic procedure requiring estimation of probabilities and an analysis of the consequences of the event. It can be seen as part of the decision-making procedure and is aimed at determining an optimal decision from a set of possible actions, in addition to delineating levels of risk. The methodology is most suitable for problems in frontier areas of technology such as are involved in Arctic projects. Some typical problems are discussed in a general way, such as the choice between fixed and mobile production systems in iceberg-infested waters. Environmental loading and the probabilistic analysis of pressure ridges as a Poisson process are discussed. An example of application to the risk of scour of a pipeline system is given. The use of utility theory for combining attributes is suggested as a fruitful area for further development. Practical guides for dealing with problems of risk are discussed.

1. INTRODUCTION

It is the task of the engineer to plan and conduct offshore projects with due regard for the protection of human life and the environment. The questions that arise with respect to such protection may be placed under the single umbrella term "safety."

The development of Arctic areas poses challenges of a high order. Apart from the more obvious climatic difficulties, such as the effect of low temperatures on metals, there are uncertainties with respect to the environmental data which add a dimension to the problem. The river engineer may well have many years of records of floods; the offshore engineer, likewise, will in all likelihood be able to base his designs on records of some length regarding wave heights. When considering the frequency and size of multi-year pressure ridges in the Arctic, there is no equivalent data bank.

To ensure a certain level of safety requires, inevitably, that the parameters in the analysis of offshore projects be quantified. Thus, numerical criteria for design, for operational facilities, and so on, need to be set. The motivation in the present paper is to suggest that an overall methodology exists for approaching such problems and for developing policy. The methodology may be adapted to the different areas of decision-making and it will therefore be introduced in a general way.

2. WHAT IS RISK?

The term "risk" is defined in Webster's as "the chance of injury, damage or loss" or "a dangerous chance." This involves two aspects: first, probabilities; and second, the consequences which will include possible fatalities, injuries, environmental damage, loss of production, the installation, plant and equipment. Risk is therefore related to *undesirable events* as well as the chance aspect. Occasionally risk is defined as the probability of failure alone, but I consider that some measure of the consequences should be included. There is more risk if 1000 people might die with a certain probability than if 10 might die with the same probability. As emphasized by Tveit [1] the important aspects are *probability* and *undesirability*.

3. RISK, SAFETY AND DECISION-MAKING

There are several levels upon which decisions are made or influenced. For instance, governments and regulatory authorities will issue and interpret regulations aimed at controlling the level of safety. Such regulations may contain a "zero-risk" type of recommendation [2], e.g. to the effect that structures should be capable of resisting any "foreseeable" environmental loading, etc. Such a regulation does not explicitly account for risk. It is impossible (and quite uneconomical to attempt) to obtain a completely "safe" structure or system. A more realistic approach is to delegate responsibility for judging whether the risk is excessive or not to a regulatory agency. As noted by Starr and Whipple, [2], the mandate might be to provide against "unreasonable" risks.

The question of influencing decisions regarding acceptability is a question of the interaction of the public and the regulatory authorities. Controversial areas can lead to extremely strong and motivated pressure groups, such as in the question of nuclear energy programmes. Such issues will not be considered further here, but the attainment of a public consensus is an important consideration. The whole area of levels of risk is a sensitive one deserving much attention.

The remaining part of the present work will address the analytical tools that an engineer might use in evaluating risk and in making decisions. One tool commonly mentioned in the literature is cost-benefit analysis. The intention is to weigh costs against the possible benefits arising from implementation of a project/proposal. However, the formal application of this method involves some rather arbitrary assumptions, for example, the following:

(1) All consequences are converted to costs; this raises some awkward questions for intangibles, value of human life, etc.

(2) The usual formulation of the method does not permit the inclusion of probabilistic aspects, i.e. all costs are deterministic.

(3) Cost-benefit ratios are not the best indicators for making optimal decisions. A more comprehensive and flexible analytical tool is based on decision theory. This will be outlined in the next section.

4. RISK ANALYSIS METHODOLOGY

The decision-making process can be easily understood and visualized by means of a decision tree [3,4]. A typical simple example of a decision tree is illustrated below in Figure 1. The branches emanating from the decision fork (□) represent decisions whereas those from the fork denoted (o) represent events involving chance. The tree illustrated is a simple example and the tree decided on for any particular study would consist of a series of decision (□) and chance (o) forks. The number of decisions or outcomes is, in principle, unrestricted and continuous parameter spaces can also be used. An example of the latter would be "volume of oil spill". Practical analysis requires that a reasonable set be decided on.

To relate the tree of Figure 1 to the previous discussion, one should make the following associations.

(1) The chance events should relate to the key possibilities in the risk/safety assessment. For instance, one might consider various possibilities such as blowouts, pipeline failure, tanker accidents or collisions between ships and fixed or mobile platforms. Other undesirable events are fires, explosions and accidents in general. Some scenarios typical of the Arctic are illustrated in Figure 2. Decision theory is not necessarily linked to undesirable events - any random quantities can be considered.

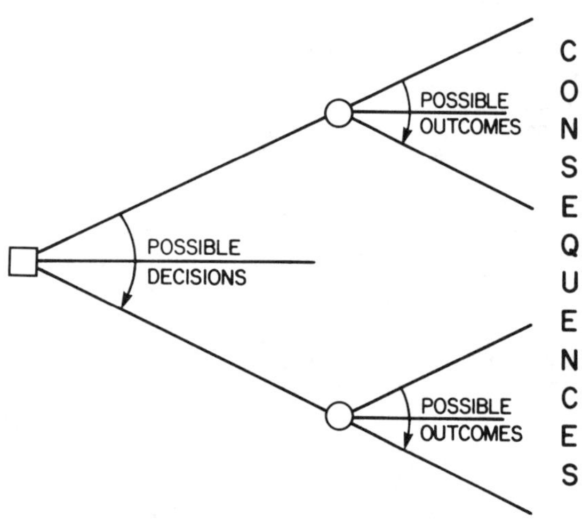

Figure 1. Elementary Decision Tree

Here, we emphasize the point that risk analysis is associated with undesirable happenings, and is therefore a part of decision theory.

(2) The "consequences" shown in Figure 1 would involve quantities "at risk." These should be quantifiable, and examples such as the following will illustrate the kind of quantity (attribute) to be measured.
(i) Cost. As far as risk analysis is concerned, this relates to economic risk and includes allowance for possible loss of production, etc.
(ii) Fatalities.
(iii) Injuries.
(iv) Measures of pollution, such as volume of oil spill, area affected, etc.

If there is a single attribute, it is possible to calculate the expected value of the attribute i.e., the sum of the attribute values weighted by their probabilities, at the conclusion of the analysis. For the various decisions under consideration, one can obtain that decision which gives the smallest expected cost, expected deaths and so on. However, the use of expected values can lead to difficulties if certain limitations of this approach are not appreciated. These will be discussed in section 6.

5. APPLICATION TO ARCTIC OFFSHORE PROJECTS

5.1 Overview

The methodology introduced above is quite general and can, in principle, be applied to any project. A few examples will be outlined so as to clarify the approach in practical problems.

A typical problem of decision regarding production systems in iceberg-infested waters relates to the choice between a gravity platform and a collection system with mobile drilling and loading systems (Figure 3). In the latter case, the system will disconnect in the presence of the larger ice-features but might be expected to have some limited capability to operate in pack ice and to resist small icebergs [5]. A complete risk analysis of the choice between the two systems would require considerable effort and should include consideration of attributes such as initial cost, disconnection costs, repair, delay time, possible fatalities, injuries and pollution. It would be necessary to develop probabilistic models of loading, using the rate of arrivals of icebergs, and possibly statistical descriptions of their kinetic energies, such as has been developed by Blenkarn and Knapp [6]. Other scenarios could also be incorporated, such as a mobile gravity system [7] or a system with protective berms to fender the larger icebergs.

A myriad of subsidiary risk/decision problems arise in consideration of the problem outlined above, for example strategies for iceberg avoidance; whether iceberg towing is an economical procedure for the mobile gravity systems mentioned above; how pipeline systems should be protected against possible iceberg scour; what loading to use in the design of fixed or floating systems; and so on.

Another area that might be mentioned as an introductory example is Arctic shipping. Here the probability of encountering hazardous ice is the essential aspect. It is important to be able, in the first instance, to define what is meant by "hazardous ice". This could be multiyear ridges with a keel greater than a certain depth, or an iceberg/bergy bit of a certain mass. Given an encounter with such a feature, the change of various kinds of damage is estimated. The decision might relate to the installation of a detection system (Figure 4). Having dealt with some

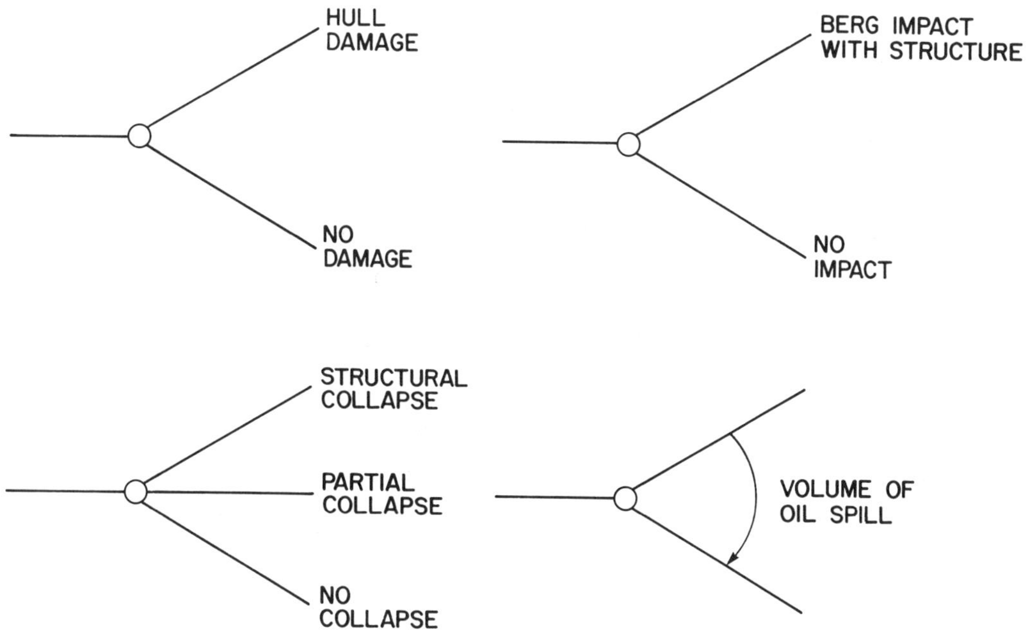

Figure 2. Some Scenarios for Events in the Arctic Involving Risk

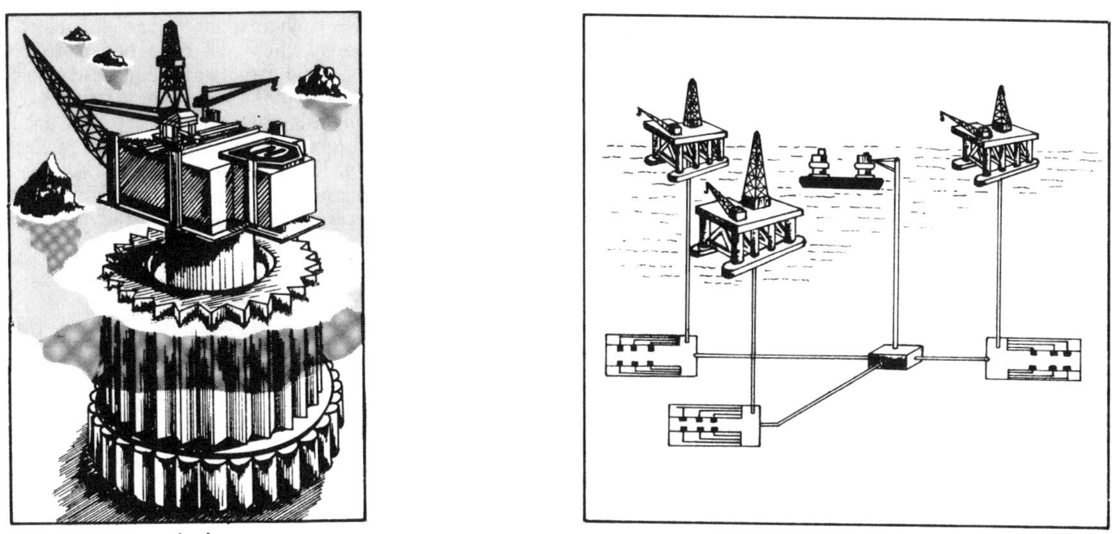

(a) (b)

Figure 3. Possible Fixed and Mobile Production System for Iceberg-Infested Waters. The Sketch of Fixed System is based on a Concept by Norwegian Contractors; the Mobile System is discussed in Reference [5].

problems in a general way, more specific discussion followed by an example problem will be introduced.

5.2 Environmental Loading of Structures and Ships

In the introduction, the fact that certain loadings on structures (wind, water) have long histories of study was mentioned. Quite sophisticated probabilistic models have been developed and implemented in design. The situation with regard to the Arctic environment is not quite as advanced; in many ways the appropriate models have yet to be tailored to a specific end-purpose. In the present work, the modelling of ice features will be addressed but it should be noted that there are many other environmental aspects, such as the effect of temperature and other climatic variables.

The question of end-purpose may be highlighted by contrasting the requirements for ships and structures. A structure is essentially fixed and has to resist whatever forces are applied (more precisely, the force should be related to some risk level). However, a ship has the ability to detect and avoid certain features, and, in addition, the ship does not have to resist the kinetic energy of a moving piece of ice. One may summarise this briefly as follows:

(1) A structure has to resist either the forces imposed due to movements of landfast ice, or the forces transmitted by moving ice masses. In the latter case, the kinetic energy of the ice mass is of interest (together with the ice properties - not addressed in detail here). The kinetic energy is dissipated by crushing/plastic flow of the ice. Fracture of an ice sheet into pieces is also possible.

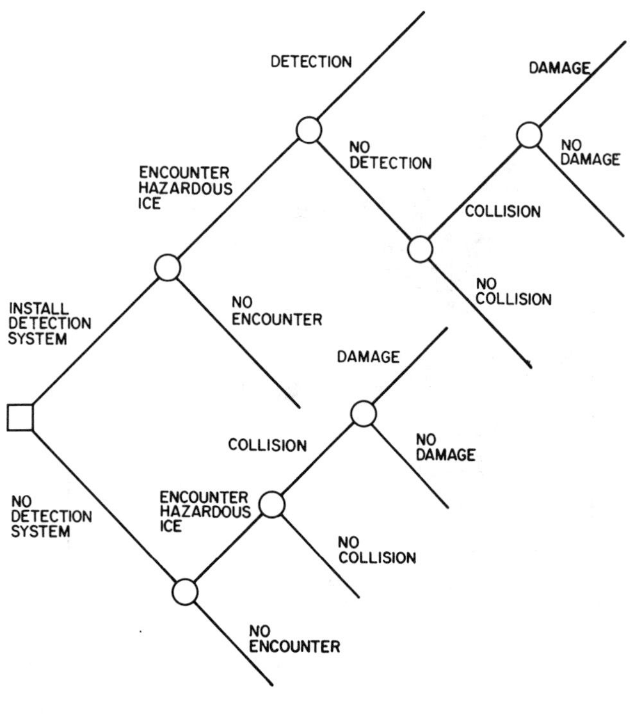

Figure 4. Decision Tree regarding Installation of Detection System in Arctic Shipping

(2) In the case of shipping, it is the shape, topology and properties of the ice masses that is of importance.

A typical statistic that might be used in design of a fixed installation is the kinetic energy of icebergs. Figure 5 shows a typical histogram developed at DnV (Calgary), based on statistics for the Davis Strait [8]. Arrivals might be treated as a Poisson process.

In other parts of the Arctic, icebergs are comparatively rare. Pressure ridges, and in particular multiyear pressure ridges are of interest. A striking analysis of pressure ridges has been conducted by Hibler and his associates [9, 10 for example]. The analysis was based on both theoretical and empirical considerations. The probability distribution of depth of a ridge (below a certain fixed distance) was derived by finding the most probable distribution. Actually the method is essentially the same as the maximum-entropy method which has received much attention in recent literature; see Tribus [11], for example. The resulting probability distribution, found on the basis of Hibler's formulation and constraints, is the truncated Gaussian distribution.

Hibler also obtained the probability distribution of n ridges along a line segment of length L, n=0,1,2 ...., which was the Poisson distribution. The conclusion is that the occurrence of ridges may be modelled as a Poisson process. One may conclude, also, that the probability density distribution of the distance z to the kth pressure ridge is given by the Erlang distribution. The probability distribution of n is

$$p(n) = \frac{e^{-\lambda L}(\lambda L)^n}{n!}, \quad n = 0,1,2.... \tag{1}$$

where λ = rate of process, i.e. number of ridges per kilometer. A special case of the Erlang distribution is the negative exponential distribution which gives the probability density for the distance to the next ridge. Excellent correlation with measurements was found by Hibler and his colleagues.

The Poisson process has a long history in applications of probability theory, early examples of which are the study of yeast particles in beer, the number of men killed by the kick of a horse in Prussian cavalry units, and radioactive disintegration. The rate λ of the process for ridges varies with geographical location [9, 10], and in [10] an analysis is presented of optimal routing, such that the largest ridges are avoided. The analysis required information on ridge lengths. There is considerable potential to be gained from extending this kind of approach.

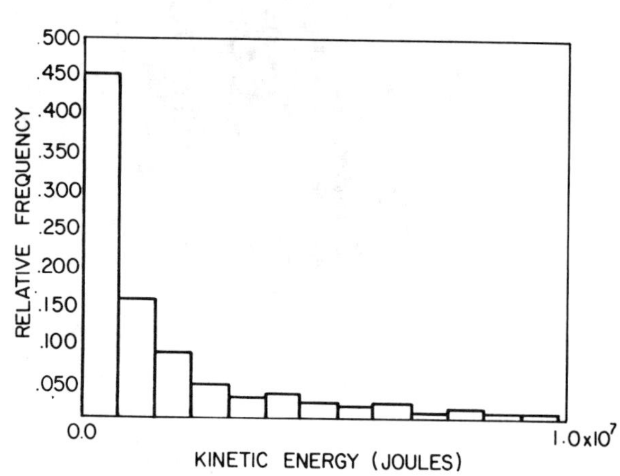

Figure 5. Histogram of Kinetic Energy of Icebergs in the Davis Strait

From the analysis above, it is a simple matter to obtain new Poisson processes with different rates, e.g. in considering ridges greater than 20m the rate would be λq where q is the proportion of ridges greater than 20m. However, this approach cannot be used in every case. In some Arctic zones, multiyear ridges occur in floes and a modified model is necessary.

The models above should be most useful in studies of ships in the Arctic; for fixed structures the kinetic energy of floes might be of interest. Indeed, there are seasonal variations that must be considered. The area of interest may be in the shear zone during summer and in pack ice during winter. An interesting analysis of these effects has been presented by Wheeler [12].

5.3 Example: Undersea Pipelines

Undersea installations, including pipelines and cables, can be damaged by floating masses of ice, in particular by the deeper keels of pressure ridges and by icebergs. The particular problem to be addressed here is that of protection of undersea pipelines by placing them in trenches as described by Mellor [13]. The ploughing of the seabed by pressure ridges will be studied for the continental shelf of the Beaufort and Chukchi Seas. Mellor shows a convincing set of data which gives an exponential probability distribution for the gouge depth s, i.e.

$$f(s) = \mu e^{-\mu s}, \quad s > o \tag{2}$$

It can easily be deduced that the probability of a gouge being greater than any specified value s_1 is e^{-us_1}.

It is worth noting that the distribution above for the gouge depth is for significant gouges defined as those with depths greater than 0.3m. Also, the distribution is somewhat different in form from the truncated Gaussian distribution found for the depth of pressure ridges; one might expect the distributions to be similar, but the empirical evidence presented by Mellor is accepted for the present work.

The formulation of the problem is shown in Figure 6. The question to be decided is the optimal depth for trenching; if too shallow, there is a high risk of the pipeline being damaged whereas the trenching itself is an expensive operation. In the present analysis, the number of gouges is taken as random, and will be modelled as a Poisson process using equation (1). The rate of "arrivals" is taken at 7 per km-yr based on the data in the paper by Mellor, considered to be appropriate for water depths of 25m. The pipeline is taken as 100km long with a lifetime of 25 years. This gives an expected number of gouges as 17,500; however, this has to be multiplied by the factor $e^{-\mu s_1}$ where s_1 is the depth to the top of pipeline. μ was taken as $3m^{-1}$, based on data by Mellor.

The cost associated with the burial was taken as $\$20,000\ (s_1 + d)^2$ per km where d is the diameter of the pipe in metres; this was based on reference [14]. The costs associated with a gouge are the sum of (1) loss of production, (2) repair costs and (3) costs associated with possible pollution, cleanup, including "intangibles." Estimates were made of these quantities and two extreme allowances were made for the intangibles - either $15 million and $900 million. The depths considered were 4.2, 4.4, ... and so on (in metres). The optimal depth (i.e. giving minimum expected cost) was 4.4 and 4.8 respectively, for the two extreme values regarding the intangibles.

Other strategies could be considered - constructing two pipelines (Figure 7) or optimising the spacing of block valves. Some of these have been considered during the present study.

6. COMBINING ATTRIBUTES: UTILITY THEORY

As noted previously, it is not entirely satisfactory to use money alone as an attribute. Expected values are useful indicators for decision-making but they must be used with care. The main difficulties relate to the following aspects:

(1) These might be two different situations with equal expected values, the first with a high value (the value could be cost, or some other attribute) and low probability, the second with a lower cost and a higher probability. Although the expected values are equal, we might prefer the second option. This is essentially a question of risk aversion [3, 4].

(2) There is a dilemma when one is confronted with several attributes; how does one compare one "mixture" of expected values, say of cost, fatalities and pollution damage, with another "mixture"? Clearly the

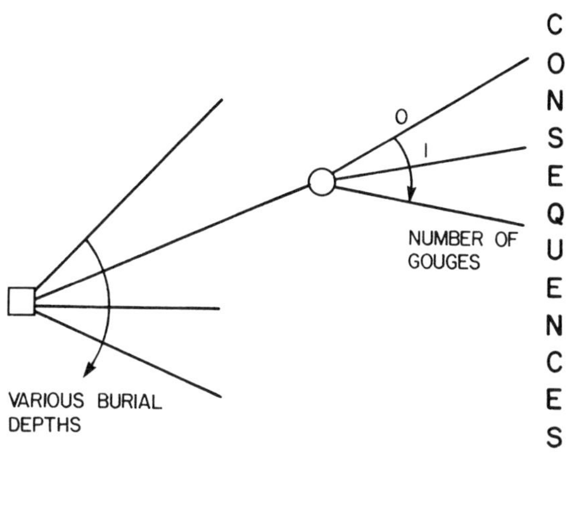

Figure 6. How Deep Should the Trench Be?

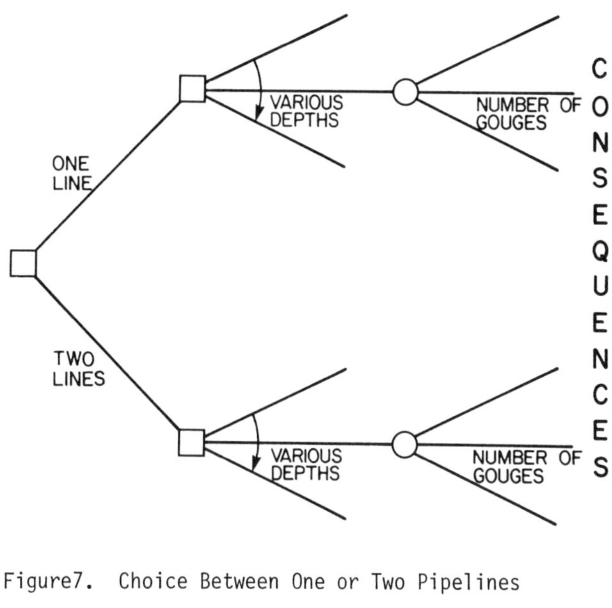

Figure7. Choice Between One or Two Pipelines

different values have to be "weighted" in some manner but it is not at all obvious how to do so.

On what basis can one resolve these difficulties? The most promising area with potential for future development appears to be utility theory [4]. This provides a detailed and logical methodology for dealing with both of the difficulties noted above. Some of the attractive and useful aspects of the theory will be highlighted.

First, the question of risk aversion is dealt with by having a non-linear relationship between the attribute and utility. This has been dealt with extensively in the literature especially with respect to money, and does not need further elucidation. However, the question of risk aversion in the case where "number of fatalities" is the attribute does merit some discussion. It is common experience that large accidents involving many casualties cause more public reaction and attract more media attention than do small accidents. Whether this implies risk aversion in society as a whole is doubtful (see also reference [15]). Some authors have even suggested risk-prone or mixed prone and averse curves.

Whether one is risk-averse can be determined by one's attitude to a lottery such as that shown in Figure 8(a). This shows a situation with risk (lottery) involving a 50-50 chance at 0 and 200 units respectively. The "units" could be any attribute and there is no reason why fatalities or injuries should not be considered. What number of the units (for sure) would one exchange for the lottery? Clearly, the expected value of the lottery is 100. If we would exchange 105 units for the lottery, this indicates some risk-aversion since one would "sacrifice" an additional 5 units (over the expected value) to avoid the chance of obtaining 200 units. Similarly, the exchange of 95 units for the lottery indicates risk-proneness. In considering the lottery of Figure 8(a) one should think of the <u>maximum</u> number of units that one would exchange for <u>the lottery</u>. In this way one obtains an "equivalence" to the lottery.

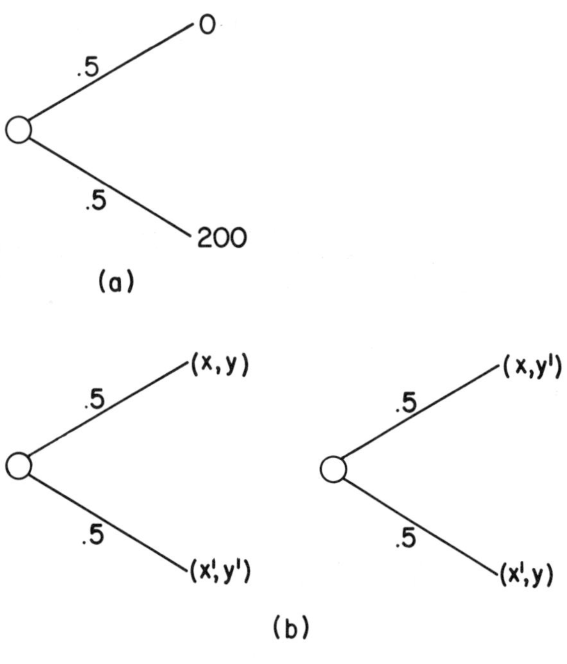

Figure 8. Lotteries for Utility Theory

Proneness to risk, if used in a decision-making as a policy, will bias decisions towards those resulting in larger accidents (even if other factors, such as expected fatalities are kept constant). I consider that a slightly risk averse function to be most reasonable; in addition, if one accepts the notion of <u>constant</u> risk aversion [4], then a utility function of <u>the form</u>

$$u(x) = b(1-e^{ax}) \tag{3}$$

results. If a is small, as one would expect for the case of slight risk-aversion, then $e^{ax} \cong 1 + ax$ and

$$u(x) \cong - (\text{const.})\, x, \tag{4}$$

i.e. a simple linear relationship results and the number of fatalities is a measure of disutility.

A further illustration of the use of utility theory relates to multi-attribute functions. Many theorems have been developed [4] to aid in the logical treatment of the case where there are several attributes to be taken into account. Consider, for example, the two attributes x, the number of fatalities and y, cost. If we consider two sets of values (x, y) and (x', y'), and if we are indifferent between the two 50-50 lotteries shown in Figure 4(b), for all arbitrarily chosen (x', y') then we have additive independence [4]. The indifference noted seems to be reasonable in the present case and consequently we can write the utility function as

$$u(x,y) = k_1\, u_1(x) + k_2\, u_2(y) \tag{5}$$

where u represents utility and k_1 and k_2 are constants. Using equation (4), we can write

$$u(x,y) = -k_1\, x + k_2\, u_2(y) \tag{6}$$

The importance of this equation is that the criterion of maximum expected utility can be shown to constitute a logically satisfactory basis on which to choose between possible decisions.

7. COMMENTS ON PRACTICAL SAFETY OBJECTIVES

The first issue to be settled is a question of aversion to large accidents. Following the arguments of the preceding section, it is suggested that the media attention that follows large accidents is more a result of the news "industry" than an expression of risk aversion in society as a whole.

A "target" value of probability of failure p_f has been suggested by some authors; for instance Flint and Baker [15] suggest

$$p_f = 10^{-4} \frac{K_s}{N_r} N_d \tag{7}$$

where N_d = design life in years, N_r = number of people at risk and K_s = social criterion factor, varying from 5 for hazardous activities to .05 for structures providing sanctuary. One sees that the socio-economic factor has been addressed. The probability of failure per year averaged over N_d years is then

$$p_f = 10^{-4} \frac{K_s}{N_r} \tag{8}$$

The use of the factor N_r to express risk aversion might be questioned on the basis of the comments above. A value of p_f that is independent of N_r would better express the risk neutrality (or very slight aversion) advocated herein. Socio-economic factors do, however, have to be taken into account.

In order to obtain a better appreciation of possible target values, the following tables have been compiled; the first on the basis of statistics Canada 1978 figures and the second based on reference [16]. A target cut-off probability of 10^{-4} per year, for the total probability of situations which should be disregarded in precertification reviews, has been suggested by Fjeld [17]. This appears to be very reasonable in light of the values in Tables 1 and 2.

8. CONCLUDING REMARKS

The overall methodology outlined constitutes a flexible tool for decision-making and should play a major role in evaluating Arctic offshore projects. It does, however, have to be tailored to any specific end-use, whether precertification of a project or a comparison of competing systems. The techniques for most applications need to be further developed, and potentially offer a practical tool for rational decision-making in many novel areas.

Table 1. Risks for the Population as a Whole

Cause or Activity	Annual Risk /10,000 Persons	Hours of Exposure per Person	Risk per 10^8 Person-Hours
All Causes	71.6	8760	81.7
All Causes -			
age 20-24	11.2	8760	12.8
age 40-44	24.2	8760	27.6
age 60-64	156.9	8760	179.1
Heart Disease	24.5	8760	28.0
Cancer	15.8	8760	18.0
Lung Cancer	3.5	8760	4.0
All accidents	6.85	8760	7.8
Motor vehicle accidents	2.20	300*	73.3
Accidental falls	.77	-	-
Accidental drowning	.28	-	-
Accidents due to fire	.31	-	-

*Estimate

Table 2. Risks in Particular Activities or Occupations

Activity	Risk per 10^8 Person-Hours*	Annual Hours of Exposure	Annual Risk /10,000 Participants
Mountaineering (International)	2700	100	27
Air Travel (International)	120	100 - passenger 1000 - crew	1.2 12
Deep Water Trawling	39	2900	17
Coal Mining	21**	1600	3.3
Construction sites	7.7**	2200	1.7
Structural failures	.002**	5500	.001

*for persons exposed
** U.K. data

ACKNOWLEDGEMENTS

I wish to thank my colleagues John Appleby, Dale Babala, and Touraj Nasseri for their advice and assistance in preparing this paper.

REFERENCES

1. Tveit, O., Risk Analysis. An Aid in Decision-making, Seminar on Industrial Safety. Riyadh, Saudi Arabia, March, 1980.

2. Starr, C. and Whipple, C., Risks of Risk Decisions, Science, Vol. 208, pp. 1114-1119, 1980.

3. Raiffa, H., Decision Analysis: Introductory Lectures on Choices under Uncertainty. Addison-Wesley, 1968.

4. Keeney, R.L., and Raiffa, H., Decisions with Multiple Objectives: Preferences and Value Tradeoffs. John Wiley and Sons, 1976.

5. Vandall, P.E. Jr., Ice and Offshore Production on the East Coast of Canada, Conference on Canadian Drilling and Downhole Technology, Edmonton, Alberta, September, 1981.

6. Blenkarn, K.A., and Knapp, A.E., Ice Conditions on the Grand Banks. Canadian Institute of Mining and Metallurgy, Special Vol. 10, pp. 61-72, 1969.

7. Di Tella, V. and Sebastiani, G., Production System in Arctic Waters by Using a Fully Integrated TSG Platform, Proc. Sixth Int. Conf. on Port and Ocean Engng. under Arctic Conditions, Vol 1, pp. 39-48, 1981.

8. Nielsen, H.A., Dietrich, J., and Fabricius, J., Environmental Conditions on the Continental Shelf off West Greenland, Offshore Tech. Conf., pp. 533-546, 1977.

9. Hibler, W.D., Weeks, W.F., and Mock, S.J., Statistical Aspects of Sea-Ice Ridge Distributions, Journ. Geoph. Res., Vol. 77, pp. 5954-5970, 1972.

10. Hibler, W.D., Statistical Variations in Arctic Sea Ice Ridging and Deformation Rates, Proc. Ice Tech Symp., Montreal, Canada, Society of Naval Architects and Marine Engineers, 1975.

11. Tribus, M., Rational Descriptions, Decisions and Designs, Pergamon, New York, 1969.

12. Wheeler, J.D., Probability Distributions for Structure Loading by Multiyear Ice Floes, Proc. Sixth Int. Conf. on Port and Ocean Engng. under Arctic Conditions, Vol. II, pp. 643-652, 1981.

13. Mellor, M., Subsea Trenching in the Arctic, CRREL Report 81-17, September 1981.

14. Pilkington, G.R. and Marcellus, R.W., Determination of Pipeline Trench Depths in the Beaufort Sea, Conference on Canadian Offshore Drilling and Downhole Technology, Edmonton, Alberta, 1981.

15. Flint, A.R. and Baker, M.J., Risk Analysis for Offshore Structures - the Aims and Methods, Proc. Conf. on Design and Constr. of Offshore Structs., Inst. Civ. Engrs., London, pp. 87-90, 1977.

16. Rationalisation of Safety and Serviceability Factors in Structural Codes, CIRIA Report 63, 1977.

17. Fjeld, S., Safety Evaluation of Field Development Concepts, Hibernia Symposium, 1981.

ARCTIC TECHNOLOGY I

INTRODUCTION

F.P. Dunn

Shell Oil Company
Houston, Texas

The Arctic environment presents as formidable a challenge to our engineers and scientists as does deep water. Moreover, it is a totally new environment for most of us. We are accustomed to dealing with the problems confronting us in over 1000 feet of water--wave forces, wave heights, currents, earthquakes, dynamic response of structures, etc. In the Arctic we encounter most of these, but in addition we are faced with the awesome forces generated by moving ice--forces sometimes more than an order of magnitude greater than wave forces imposed on a large Gulf of Mexico structure. The Cognac structure, a permanent drilling and production platform installed in 1025 feet of water in the Gulf of Mexico, was designed for a total lateral load of 14,000 kips, including dynamic amplification. A structure whose sole purpose is to drill one or two exploratory wells in less than 100 feet of water in the Beaufort Sea must be designed to resist as much as 200,000 kips--quite an increase.

The first offshore structures designed for ice loads were those installed in Alaska's Cook Inlet, nearly 20 years ago. These were designed to resist four-foot thick sheet ice. The design ice pressure used for most of them was 300 psi. Fourteen structures were installed, and they are all still there.

More recently, our Canadian neighbors have led the way in Arctic development, starting with gravel islands ten years ago in the MacKenzie River Delta area. As they moved into deeper water, the haul distances and the volumes of dredged fill required made islands prohibitively expensive. Economics thus dictated steel and concrete structures set on prepared subsea berms. Both types have already been built. The Canadians have also pioneered in the use of drill ships for exploratory wells, using ice breakers for support. They are now building a conical floating drilling vessel, to be held in place with 12 anchors, supported by ice breakers.

On the American side of the Beaufort, we are following the Canadian lead, using gravel islands out to about 50-60 feet, then probably switching to steel or concrete structures, set either on prepared berms or designed to be set on the sea bottom like the North Sea structures. Ten years from now, when several of such structures have been installed (hopefully also some production platforms), we will still be arguing the merits of steel vs. concrete for the Arctic.

These structures must be designed to resist multi-year pressure ridges, which can generate local forces as high as 1000 psi, depending on the area and shape considered, and global forces of 500 kips per horizontal foot of structure exposed to the ridges. Several different concepts exist: some have vertical faces, others are of conical shape; some are segmented, others monolithic; some rely on piles for lateral resistance, others rely solely on bottom friction. The next few years will provide the test period for such concepts.

I might mention in passing that the exploration structures we are building and will build in the next few years present an excellent opportunity to learn much about the Arctic environment, so that we can be better prepared to design the larger production platforms and structures in even deeper water--to 200 feet at least. We will learn more about ice forces and movement from sensors imbedded in the structures and in the ice surrounding the structures. We will learn to quantify the benefits of purposely built rubble piles surrounding the structures. We will learn a lot about the merits of different types of slope protection on our gravel islands. Instrumentation of such prototype structures over a period of years will prove far more beneficial to the industry than building instrumented test structures and then waiting. We may wait for years before we learn something of consequence, simply because the statistical probability of getting anything close to the design ice load on any one structure per year is something less than 10%.

Let me touch upon the other areas of the Arctic where we'll be operating in the next few years.

In the Bering Sea, we will not be faced with the large ice forces we'll encounter in the Beaufort Sea. As a matter of fact, the Southern Bering Sea will present no more structural challenges than we have already faced and solved in the North Sea. As we move further north into the Navarin Basin, we will have to take ice floes into account, but these should present no insurmountable problems. The major problem in this area is going to be one of transportation economics--how much it is going to cost to transport the gas and oil. Norton Sound will not be nearly as formidable as the Beaufort Sea, though we do have significant amounts of first year ice to contend with. The big problem here, again, is oil and gas transportation. The Chukchi Sea will present the same problems as the Beaufort Sea, with the additional complication of extremely costly transportation.

In summary, we face a number of significant technical problems in the Arctic, and we certainly do not have all the answers yet. I am confident, however, that as we have shown before in other areas time after time, our engineers will solve problems, given the opportunity. They won't solve them overnight, and they will make some wrong turns; but the problems will be solved, and, if there's enough oil and gas up there to warrant development, then we'll find ways to develop the reserves safely and, hopefully, economically.

STEEL STRUCTURES IN ICE COVERED ARCTIC WATERS

Robert E. Smith
Director, Civil Engineering
ARCO Oil and Gas Company
Dallas, Texas 75221

1. INTRODUCTION

The title of this conference is "Arctic Technology and Policy: An Assessment and Review for the Next Decade". Two major questions are to be addressed: What knowledge do we have today for working in the Arctic? What are the challenges in science, engineering and policy making during the decade ahead?

The majority of papers and lectures deal with offshore arctic issues. Therefore; this is essentially a marine conference dealing with arctic science and technology and with arctic policy of local, national and international scope.

The offshore arctic resource which we will be exploring for and hopefully producing in the next decade is petroleum. Other resources may come into the picture, but for the next decade, petroleum will be the focal point.

The organizers asked that discussion from the participants be encouraged. The presentation by Mr. Maxwell immediately following these remarks deals basically with the same subject. Therefore, this presentation will be brief. The status and potential of steel structures for petroleum activities in offshore Arctic will be reviewed. Several critical environmental work areas will be suggested.

A reliable prediction of steel structures in the Arctic in the next decade is not possible. Steel structures will be competing against concrete, frozen earth, ice and hybrid structures. In addition, the timing of construction of exploratory and production structures depend on many things including timing of lease sales, developing economic and safe designs, satisfying regulations and the concerns of private citizen groups and the price of oil. Last, but surely not least, construction of structures will depend greatly on what we discover in terms of petroleum reserves. Nevertheless, we will speculate on what is in store for steel structures offshore Arctic in the next decade or so.

During preparation of this discussion a story attributed to the late Professor J. Frank Dobie, historian and humorist of the southwest United States, came to mind. Prof. Dobie remarked, during a graduation address in a half serious half humorous tone, that "most dissertations were simply the digging up of dry bones from several graves and burying them into a deeper grave". Granted, there has been repetition in the literature; but there also has been some appropriate burying of bones. Wise development of the Arctic requires much debate and airing of critical issues. The critical issues, important data needs and viable approaches are being brought to the attention of all concerned groups. An objective of this conference is to assist in this focusing.

2. ASSESSMENT OF ARCTIC OFFSHORE TECHNOLOGY

Ice conditions offshore Alaska north of the Arctic Circle are a principal environmental constraint. Arctic ice makes these areas truly frontier for petroleum exploration and development. Frontier in the sense that the industry had not operated in these extreme ice conditions prior to the 1970's. There still remain many "firsts" in terms of structures and operations. An illustration of this is the fact that as of 1982 there is not a permanent petroleum platform in the Arctic offshore North America, Soviet Union or Scandinavia.

The National Petroleum Council in their report, U.S. Arctic Oil and Gas, December 1981, advised the Secretary of Energy that...."the basic technology is available to safely explore for, produce, and transport oil and gas in most of the U.S. Arctic". This report correctly places some qualifications on this conclusion. Regarding petroleum development offshore U.S. Arctic, the qualifiers basically are four:

Economics. Key requirements are large reserves and sufficient market price for oil and gas. Exploration, production and transportation of petroleum in the arctic offshore are very expensive because of both remoteness and environment.

Time. Primary lease terms should be at least 10 years with an automatic "suspension of production provision" for marginal reserves. Considerable time is needed after leasing to explore and define reserves, to design, permit and construct production and transportation systems and to drill development wells. The time between leasing and delivering oil to a transportation system could be 9 years for shallow water locations close to the existing oil fields and pipelines on the North Slope of Alaska. For deeper water and remote locations in the Chukchi and Beaufort Seas, this time could be 15 years.

Regulations and Policy. Regulatory systems need some redesign. Agencies need to be staffed adequately. Local, state and federal regulatory systems are complex, overlap and need more coordination. Federal energy policy is not clear. Leasing needs to be kept on schedule to allow efficient planning and timely exploration of these virgin areas.

Research and Development. Establishing safe and efficient facilities and operations will require considerable effort by industry, concerned private groups and the government. This effort should be continuous and coordinated as much as practical.

The "can do" statement of the National Petroleum Council is based in part on the industry's experience on the North Slope, the Trans Alaska Pipeline, Cook Inlet, Gulf of Alaska, Beaufort Sea and North Sea. In particular, the experience gained from exploration activities offshore Alaska and Canada contributes to this "can do" opinion. This experience, technology development and future study needs are summarized in several status reports and proceedings of technical conferences and workshops. [1,2,3,4,5] Mr. Maxwell's paper will illustrate the knowledge and experience gained thus far in the Canadian Beaufort.

The author has been involved in most of these frontier activities and shares this optimism, but with perhaps a slightly more guarded posture. Even though industry has been studying and operating offshore Arctic for ten years, we very probably have not yet experienced an "extreme ice event" or the more severe "maximum credible ice event". Yet, this exploration phase affords us an opportunity to "test the water" before more costly permanent structures are built. Industry are taking advantage of this opportunity. We simply must not become overly confident and think the Arctic will be like she has been during our seasonal visits. Reliably predicting the extremes of the Arctic at a specific site is one of our most difficult problems.

As a general guideline permanent structures will need to survive a maximum credible ice event of a recurrence interval of possibly 500 years or more with predictable but severe damage. A permanent structure should function reliably for twenty plus years with only seasonal maintenance and minimum down time for an ice event of return intervals of 100 years or less.

It appears that economic structures for these environmental criteria are feasible. Designs will require considerable scientific and engineering studies of both environmental and structural systems. Risk taking is unavoidable regardless of the quality and amount of study.

3. APPLICABILITY OF STEEL STRUCTURES IN ARCTIC WATERS

3.1. Exploration Structures

Chief competition for steel structure in the shorefast ice zone and shallow waters (60 ft or less) will be temporary earth islands, earth filled caissons, mobile concrete drilling structures and open water floating drilling vessels with adequate ice forecasting and ice management support vessels. Some candidate steel structures for these conditions are cones, vertical sided gravity structures and monopod jackups. Figures 1,2 and 3 are example sketches of these steel structures. Steel structures will be hard pressed to win out over earth islands in the shallow water depths and even in deeper waters given an economic borrow source and a suitable marine dredge.

Chief competition for steel structures in the shear zone and beyond in deeper waters (60 to 200 ft.) will be mobile concrete conical drilling structures, earth filled caissons placed on below water berms and floating drilling vessels during the ice free season. This list is basically the same as for the shorefast ice zone except that earth islands probably will not be competitive. A candidate for year round operations is the conical shape structure shown in Figure 1. For less severe ice exposure or during a restricted drilling season, the steel monopod, steel piled barge (Figure 4) or monopod jackup are candidates.

The ice free season in the southern part of the Chukchi Sea is significantly longer than in the northern part. With reliable ice forecasting, conventional exploratory drilling vessels or jackups are considered possible in the southern Chukchi Sea.

For areas beyond 200 ft water depth it appears that conical shaped structures resting on earthen berms are a possibility. Experience gained in shallow waters in the next 10 to 15 years will aid us in designing exploratory structures for these depths. Design ice events for 200 ft water and deeper probably will not be any more severe than for 150 feet depth. The cost differential will be the berm or a submerged steel raft on the sea floor. Another possibility would be to use vertical sides on a monolithic steel or concrete structure below the depth of significant ice strength. This depth may be less than 150 ft. Ice studies are needed to confirm this possibility.

3.2 Production Structures

The operational life of a production structure at one site is 20 to 30 years.

The size, complexity and cost of a permanent drilling and producing structure are several times that of a exploration system. The cost of a production facility could exceed 1 billion dollars. A production structure in the Arctic may support 20 to 60 producing wells, production facilities, two drill rigs and crews totalling over 100 personnel. Therefore, for the same location, environmental design criteria for production structures will be more severe than an exploratory structure.

For shallow water (0 to 50 feet) and locations well within the normal shorefast ice, earth filled islands with substantial slope protection probably will be chosen over steel or concrete structures. Production and drilling facilities will be constructed on barges or in modular form and transported to these islands.

For 50 to 200 feet water depth and beyond the shorefast ice, conical structures are the probable structural approach. Steel will compete with concrete. Hybrid structures such as an earth berm and caisson or combination steel and concrete may be a better approach than a pure steel or concrete structure.

Designs of production structures will evolve from exploration structures. The knowledge gained during exploration will influence the design of permanent structures.

4. SELECTED STUDY ITEMS FOR THE 1980-90's

While much research and engineering studies have been made to develop structural designs for offshore arctic structures, there remains a myriad of tasks for the next decade or two. These tasks range from collecting environmental data, to monitoring performance of first generation structures, to developing design criteria, to building fabrication yards, to final design.

The literature contains many papers and reports from which one can obtain listings of research and development needs and necessary design studies. Two topics are suggested here as critical to the design of permanent structures in the Arctic. These topics have not received enough emphasis to date.

4.1. Influence of Water Depth on Ice Wasting and Force

Past modelling efforts of ice action against and around structures have not accounted for the effects of shallow water.

Similarly, present ice force prediction formulae do not account for the effects of shallow water. The basic question is "What are the effects of bottom contact and resistance on ice wasting around a structure and the ice force on the structure"?

Some predicted extreme and maximum credible ice events involve ice thicknesses approaching or equalling the water depth. This being the case, the seafloor will interact with the moving ice as it breaks against or moves up on the structure. This interaction between the ice and the seafloor could cause the resultant ice force on the structure to be somewhat larger and the rideup to be higher than for the case of no seafloor interaction. Figure 5 is a simplified illustration of this problem.

Even if we find that ice rideup and ice force are not influenced greatly by the ratio of water depth to ice thickness, surely the amount of ice rubble formed around a structure will be more if this ratio of water depth to ice thickness is near one then when it is two or more. In a situation of large and continuous ice movement and a water depth to ice thickness ratio of one, considerable rubble will form which may cause serious logistic-access problems. Also, this large rubble dam may cause high bottom scour from currents and thus expose pipelines and structural foundations.

4.2. Extreme and Maximum Credible Ice Events and Design Approach for These Events

Permanent production and drilling platforms traditionally have been designed to withstand extreme environmental events with considerable safety margin. In the Gulf of Mexico and North Sea a steel platform will resist the "100 year storm" and wave loading within the elastic stress range. Present design of platforms for earthquakes are similar. Platforms are design to resist a "100 year earthquake" within the elastic or linear range. These are called "extreme events".

In the past few years design considerations for more severe, less frequent events have been made. The best example is in earthquake design. Platforms are designed to survive a "maximum credible event" with substantial damage. Under a maximum credible earthquake the platform would suffer plastic yielding perhaps, visible distortions and complete loss of some structural members, but the structure would have a high probability of surviving with no loss of life and no loss of wells fluids. The structure may be rendered useless by the maximum credible event, yet the probable loss would be only investment and income.

In the design of permanent arctic structures, a similar design approach will be required. Clearly, we will need to predict these extreme and maximum credible events and then to design and construct structures to withstand these events within predictable, acceptable damage limits. Figure 6 illustrates this design problem.

Acceptable "happenings" under maximum credible events could be severe local rupturing of the structure, a rubble pile of extra ordinary extent or even a movement of the structure off location with breakage of wells above their safety values. These possibilities may first appear unacceptable, but this type design approach is considered prudent. We now have the computer capacity to model progressive failure of complex structural systems. The stakes are high in the Arctic. The environment is obviously very severe. It is difficult to see any other design approach.

5. ACKNOWLEDGEMENTS

The offshore Arctic is truly a frontier area. Our best technology will be required to evaluate its potential and wisely develop its resource. Social and environmental impacts are unavoidable. To minimize these impacts and to make them acceptable will require much thought and open debate between all concerned groups.

During the past decade or more the ability and willingness of all concerned groups to constructively discuss and to reach reasonable conclusions have improved. The author wises to acknowledge this progressive move and to encourage its continuance. This conference is an illustration of this willingness to discuss these issues upfront and with candor. We should continue our efforts to reach informed, wise decisions.

The structures illustrated in this brief discussion are examples of many proposed structures for the offshore Arctic. The conical structure shown in Figure 1 is the Arctic Cone Exploration Structure-ACES. The ACES and the steel pile barge, Figure 4, are designs of Brian Watt and Associates. While the ACES structure is concrete, a similar design could be made for steel.

Figures 2 and 3 are structures proposed by Global Marine Development Company. The vertical sided structure in Figure 2 is a concrete structure and is named Concrete Island Drilling System or CIDS. A similar steel structure design is possible.

6. REFERENCES

1. "U.S. Arctic Oil and Gas", National Petroleum Council Submittal to Secretary of Energy, Library of Congress Catalog Card No. 81-86072, December, 1981.

2. Proceedings, Outer Continental Shelf Frontier Technology Symposium, U.S. National Research Council, Washington, D.C.

3. "Report of Oil Industry Capability in the Deepwater Beaufort and Chukchi Seas," Alaska Oil and Gas Association, Lease Planning and Research Committee, 1983.

4. Proceedings, Annual Meeting, Canadian Shipbuilding and Ship Repairing Association, Montreal, February 1981.

5. Proceedings, Fourth, Fifth and Sixth International Conferences on Port and Ocean Engineering Under Arctic Conditions, 1977, 1979, 1981.

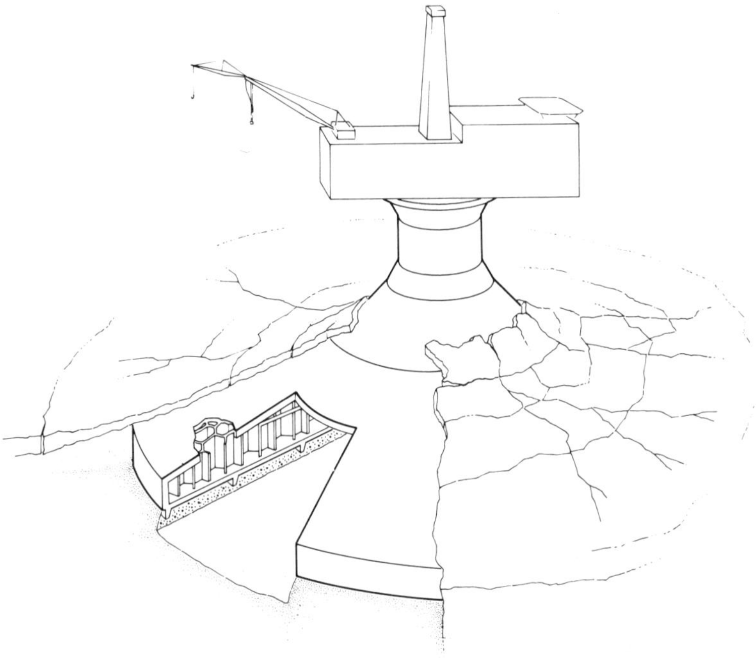

FIGURE 1. CONICAL SHAPED EXPLORATION STRUCTURE (CONCRETE OR STEEL).

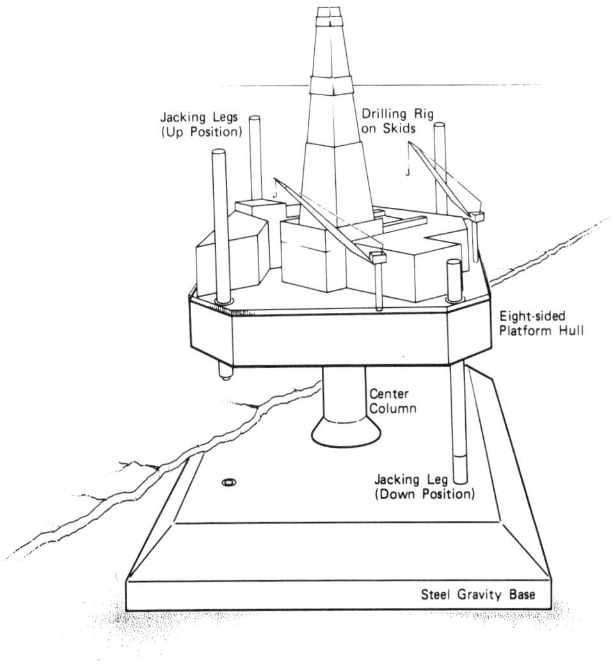

FIGURE 3. MONOPOD JACKUP EXPLORATION STRUCTURE (STEEL).

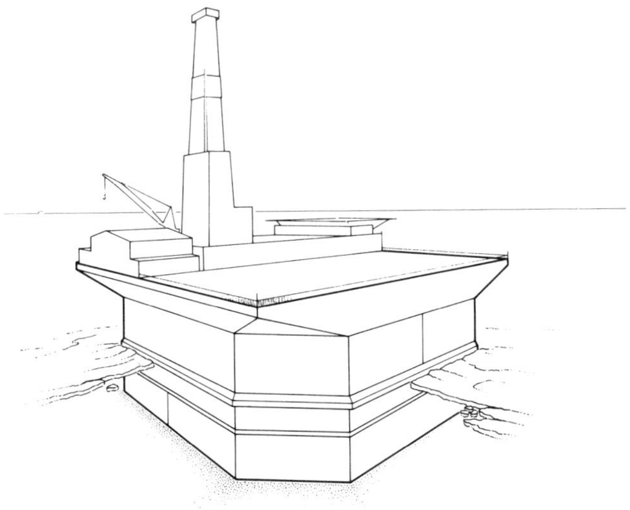

FIGURE 2. VERTICAL SIDED EXPLORATION STRUCTURE (CONCRETE OR STEEL).

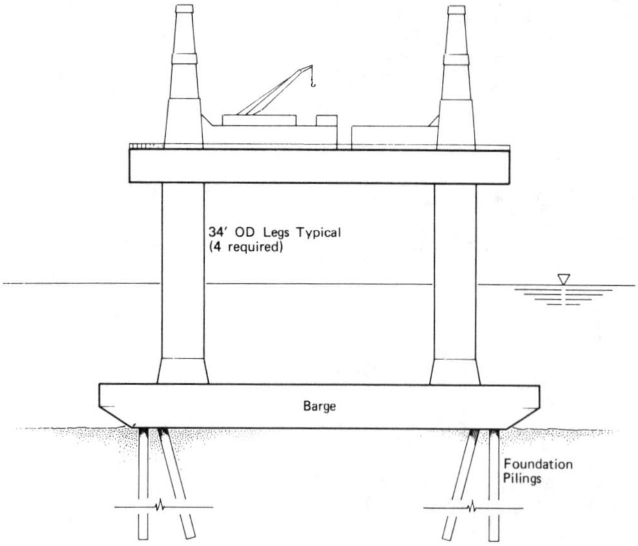

FIGURE 4. STEEL PILED BARGE EXPLORATION STRUCTURE.

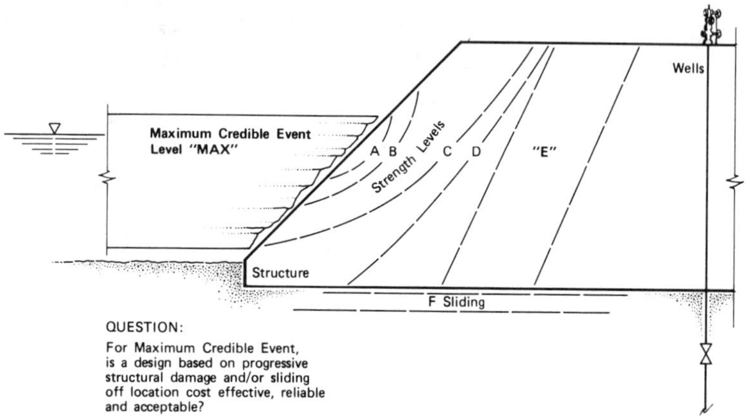

FIGURE 6. DESIGN APPROACH FOR MAXIMUM CREDIBLE ICE EVENT.

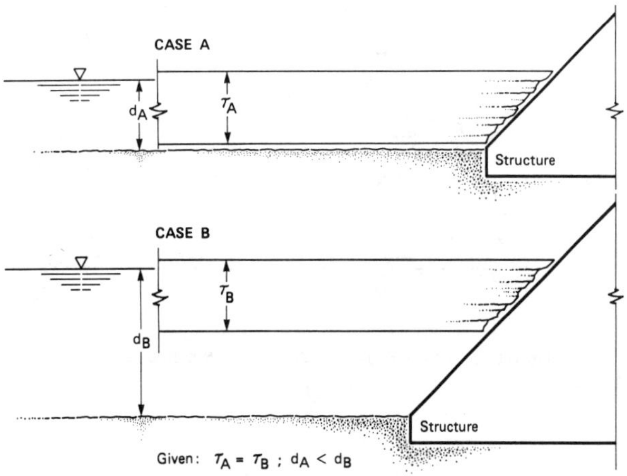

Given: $\tau_A = \tau_B$; $d_A < d_B$

QUESTIONS:

1. Will Ice Wasting be the same for both cases?
2. Will Ice Force be the same for both cases?
3. Is **Case** A a "real" design problem?

FIGURE 5. INFLUENCE OF WATER DEPTH ON ICE WASTING AND ICE FORCE.

ARTIFICIAL ISLANDS AND STEEL STRUCTURES IN THE BEAUFORT SEA

S.V. Maxwell, D. McGonigal, and C. Graham

Gulf Canada Resources Inc.
Calgary, Canada

This paper presents a brief history of the evolution of artificial islands and caisson structures for oil exploration in the Beaufort Sea. Following a brief summary of the oceanographic environment and seafloor geology of the Beaufort, the factors influencing design and construction of structures are discussed with examples from oil industry experience in the north. Next, several case histories of islands and caissons are described. The current planning philosophy for the building of production platforms is also discussed, and finally the areas where new developments are needed are outlined.

1. INTRODUCTION

The search for new sources of hydrocarbon energy has carried the petroleum industry into increasingly hostile offshore environments. Recent exploration has confirmed the presence of major hydrocarbon accumulations on the continental shelf of the Canadian Beaufort Sea which extends some 120km from shore to water depths of approximately 180m as shown in Figure 1. Potential recoverable reserves for the area have been estimated at 1.3 million cu/m (8.0 billion bbl.) of oil and 1.5 trillion cu/m (55 trillion cu/ft) of gas.

As a southern extension of the Arctic Ocean, this area is characterized by a seasonal ice cover which presents formidable challenges to the design and operation of exploration and production systems. Since ice covers the sea for at least nine months of the year most offshore construction activity, with a few exceptions, has taken place in the short open water season, while drilling from islands has been done in winter and summer.

In the shallow waters of this area, exploratory drilling from artificial islands has been carried out since 1973. The deepest island to date, Issungnak, was built in 1978 and 1979 in a water depth of 19m. Since 1976, exploration has extended into the deeper waters (up to 70m) using ice reinforced drill-ships. In the summer of 1981 the first concrete caisson retained island was constructed at the Tarsiut location (23m). A steel caisson structure built from a supertanker was placed on the Uviluk location (30m) in the summer of 1982.

Plans for the next several years call for an escalation of exploration activity, and a concurrent detailed evaluation of production structures and systems to bring Beaufort Sea oil economically to market. The exploration effort will involve the building of additional artificial islands in shallow water (up to 10m) and the increasing use of innovative concrete and steel structures in the intermediate water depths (10-40m). Demonstrational production structures may be seen in the Beaufort Sea by the late 1980's. These structures must safely withstand the harsh environment, and at the same time must be cost-effective. Since the costs are so high, this is an area where considerable R&D effort will be focused in the next few years.

2. ENVIRONMENTAL CONDITIONS

The Beaufort Sea environment has been described as one of the most hostile in the world with operational constraints ranging from the extremely low temperatures and darkness of winter to subsea permafrost (Hnatiuk, 1976). However, sea ice is the most significant constraint in this area.

The southeastern Beaufort Sea is usually ice covered nine months of the year. Figure 2 illustrates the natural division in winter into 3 zones i.e. a stable landfast zone and the moving polar pack and transition zones. With the onset of winter in early October, freeze-up commences and progresses seaward with the growth of landfast ice. This ice may be subdivided into two zones, a smooth area extending 25km offshore to about the 9m water depth and beyond that a rough pressure-ridged area extending to the 18-33m water depth range where the land fast ice stabilizes in early January. The undisturbed winter ice within the landfast zone reaches a thickness of 2m by May. Multi-year ice floes and small glacial ice islands may occasionally be incorporated into the landfast ice some years. A recurring open water lead exists along the outer edge of the landfast ice which opens and closes under the influence of offshore and onshore winds respectively. At its outer edge, the landfast ice is characterized by dense pressure ridging which is partially grounded on the sea-bed. The ridges have an average thickness in the order of 8m. Figure 3 shows a view of the edge of the landfast ice.

Beyond the landfast ice is a transition or seasonal pack ice zone. Here, the ice canopy is in sporadic motion which creates changes in ice type, floe size and ice thickness distributions. This area is also characterized by first year pressure ridges which are produced by the motion and resulting deformation of this sea ice. Most pressure ridges in this area have sails in the 1-2m range, however, occasionally they may exceed 8m above sea level and have keels exceeding 30m. Multi-year ice floes, 4-8m in thickness, multiyear pressure ridges, and glacial ice islands up to 30m thick may also be incorporated in to the ice cover. These features which sometimes drift into the zone of operations pose the main hazards to structures (Figures 4 and 5).

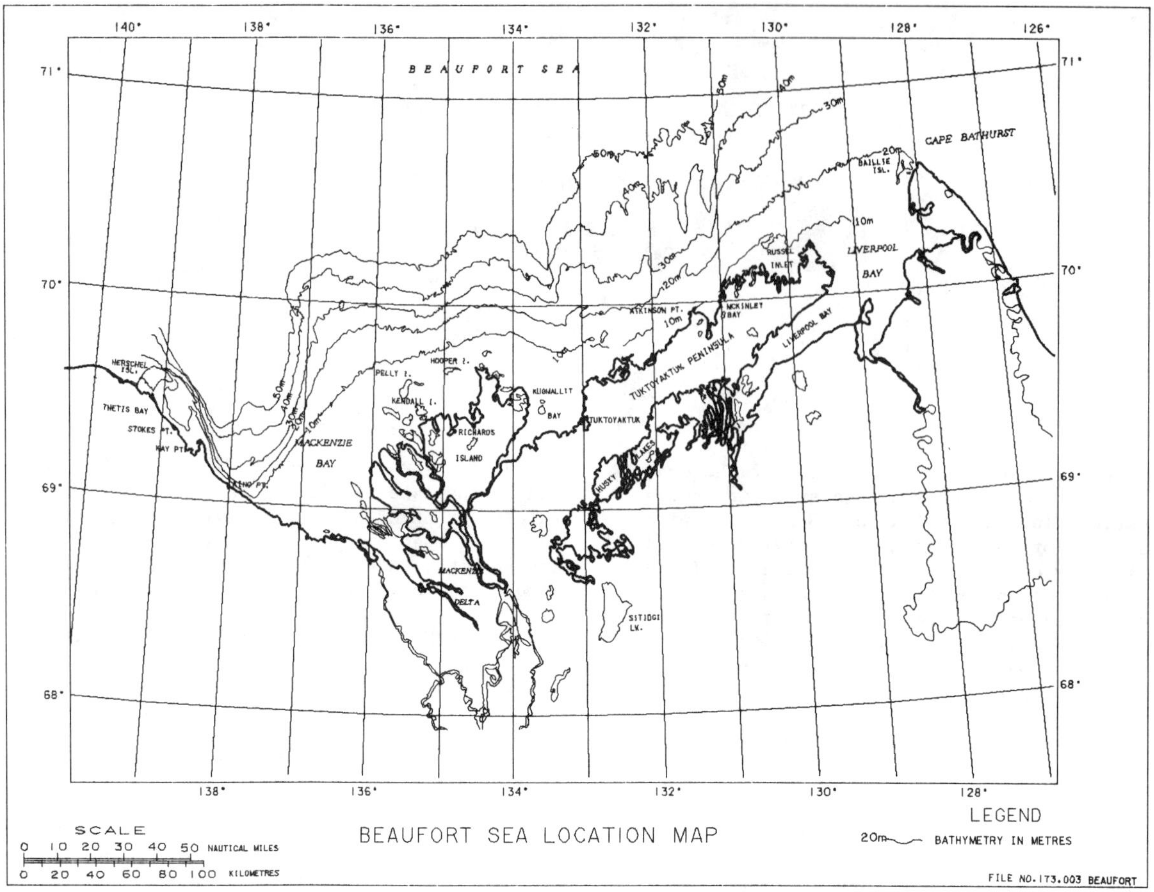

FIGURE 1 BEAUFORT SEA AREA MAP

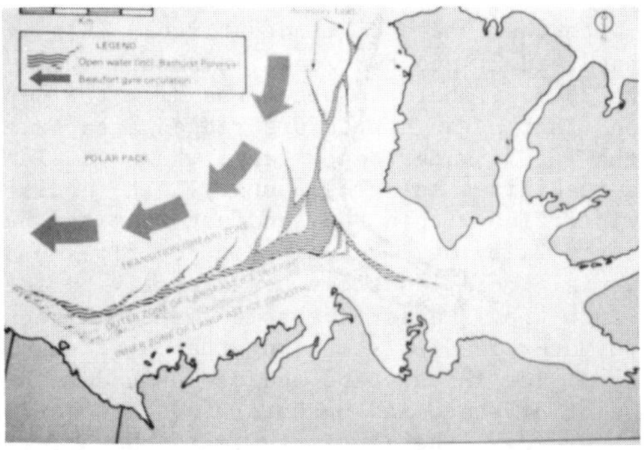

FIGURE 2 SCHEMATIC OF SEA ICE ZONES

FIGURE 3 VIEW OF LANDFAST ICE EDGE

FIGURE 4 PRESSURE RIDGES IN THE SHEAR ZONE

FIGURE 5 AN ICE ISLAND FRAGMENT

Beyond the seasonal pack ice zone exists the polar pack which is comprised primarily of old multi-year ice. The overall movement of the ice in the seasonal and polar pack ice zones is in a clockwise rotation in the Beaufort Sea. Yearly ice movements on the outer edge of this gyre average under 3km per day but in spring can reach 24km per day.

Breakup occurs in late June or early July. Summer ice conditions are variable depending on the direction of the prevailing winds. Figure 6 shows the number of days of open water by year at a typical drill site in the Beaufort Sea in (30m) of water. Some years no open water existed, while other years there have been 120 days. This graph does not include those periods of thin ice during which drilling vessels can also operate with ice-breaker support.

Waves are generally minor because of reduced fetches and the presence of sea ice. Significant waves however can reach 5m. Currents are generally less than 50cm/sec. Total darkness in winter, extended periods of extreme cold, drifting snow and fog hamper operations in the Beaufort Sea.

3. SEAFLOOR CONDITIONS

A simplified stratigraphy of the region is that the eastern Beaufort Shelf consists of thick sand beds overlain in part by a veneer of soft silty clay and that the western portion consists of very stiff clays and silts overlain by a veneer of very soft to firm clays. To understand the nature and variability of the foundation soils, the geologic process contributing those soils is very briefly described. The soils within 100m of the present seabed are of Pleistocene age. During that period there were several cycles of glacially related changes in the sealevel and sediment load in the ancestral Mackenzie River which profoundly affected the nature, distribution and characteristics of the surficial sediments and permafrost.

During an early Wisconsin glacial retreat, when sealevel may have been 60m below present, the predecessor of the Mackenzie River flowed into the eastern portion of the Beaufort Sea. A large deltaic plain consisting of sands developed on the eastern Beaufort Shelf, while pro-delta muds were deposited in the west. Subsequently, the sea level was depressed to about 100m below present, exposing a broad mature deltaic plain with meandering channels. Most of the Beaufort Shelf was exposed to subaerial erosion and sub-freezing temperatures. Below the surface, ground ice bonding developed permafrost to a depth of several hundreds of meters and the near surface sediments were extensively reworked and irregularly modified.

When the late Wisconsin glacier receded about 10,000 years ago, the Mackenzie River carried away the meltwater of a third of the continent. Several deep channels were scoured into the deltaic sands (eastern Beaufort) and pro-deltaic muds, to depths of 50m to 80m below present sea level. These deep channels have been infilled and generally hidden by more recent deltaic sediments consisting of soft silts and clays. As the sea level slowly recovered to its present level, the shoreline moved across the region, reworking the near seabed sediments. Sediments deposited during the recent period, consisting of pro-delta silts and clays subsequently covered the surface. These recent sediments are thickest in the western Beaufort and thin to the east where they are carried by littoral drift. These soft silts are being actively remoulded to depths ranging from 1m to 5m by ice scour. The present seabed topography does not give much evidence of the older seabed surface and hence the complex foundation conditions.

As the sealevel rose the permafrost which had developed in exposed areas would initially have been melted in the shallow warmer fresh water area. As the water recovered further, the colder more saline water retarded and in many areas stopped the thermal decay of the permaforst. Beneath much of the Beaufort area, continuous permafrost exists from 50-100m below the seabed to depths as great as 500m. In the upper 100m of soil the permafrost is discontinuous, occurring in isolated islands, and is generally associated with sandier soils and/or decreased salinity. Most of the permafrost samples recovered to date have negligible excess ice contents. Ground temperature ranges between $0^{o}C$ and $-3.5^{o}C$ and will be very sensitive to the long-term thermal disturbances which will result from production of warm oil.

In summary, the western area consists of 2m-10m of very soft to firm silty clays overlying stiff to hard silts and clays, the eastern regions consists of 0m-6m of soft silty clays overlying dense fine grained sand, and the central area consists of 2m-10m of soft silty clays overlying dense sand. Several deeply scoured meltwater channels have been infilled with 10m to 30m of soft to firm silty clays. This, however, is a generalization, because most stratigraphic contacts are by nature of the described deposition, irregular.

The Beaufort Sea is classed as a Seismic Zone 3 under the National Building Code. This is the most severe rating and will influence design philosophies. The majority of offshore events in the last 50 years have occurred in what is known as the Beaufort Sea Seismicity Cluster, in deeper waters (approx. 200m) and to the north of $70^{o}30'$.

The Beaufort continental shelf is covered by long shallow troughlike features caused by grounding of large ice formations such as pressure ridges. In the 15-30m depth range, the scour intensity is quite high with mean frequencies of 6-12 scours per km and mean depths of roughly 1-2m. Scours of up to 4-5m have been encountered in areas with thick soft seabed sediments. While scour is a concern for subsea pipelines and well completions, it will not affect the operational safety of a structure.

4. DESIGN CONSIDERATIONS

In general, for islands and caisson designs the following interrelated parameters govern the overall effectiveness and costs of a design:

- Water Depth
- Space Requirements
- Ice Effects
- Wave Effects
- Foundation Conditions
- Geotechnical Stability
- Structure Response
- Availability of Construction Materials
- Availability of Suitable Construction Equipment and Techniques
- Length of Construction Season for Various Construction Operations

For steel or concrete structures built in the south and transported into the Beaufort the draft and stability in the floating mode also becomes important, as does the set-down operation. Superstructure icing is another significant consideration.

4.1. Water Depth

Artificial islands appear to be economical for

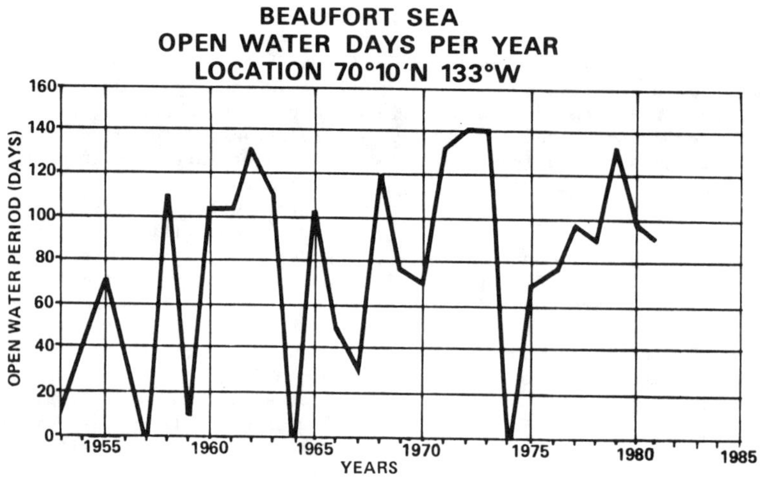

FIGURE 6 VARIATION IN OPEN WATER SEASON

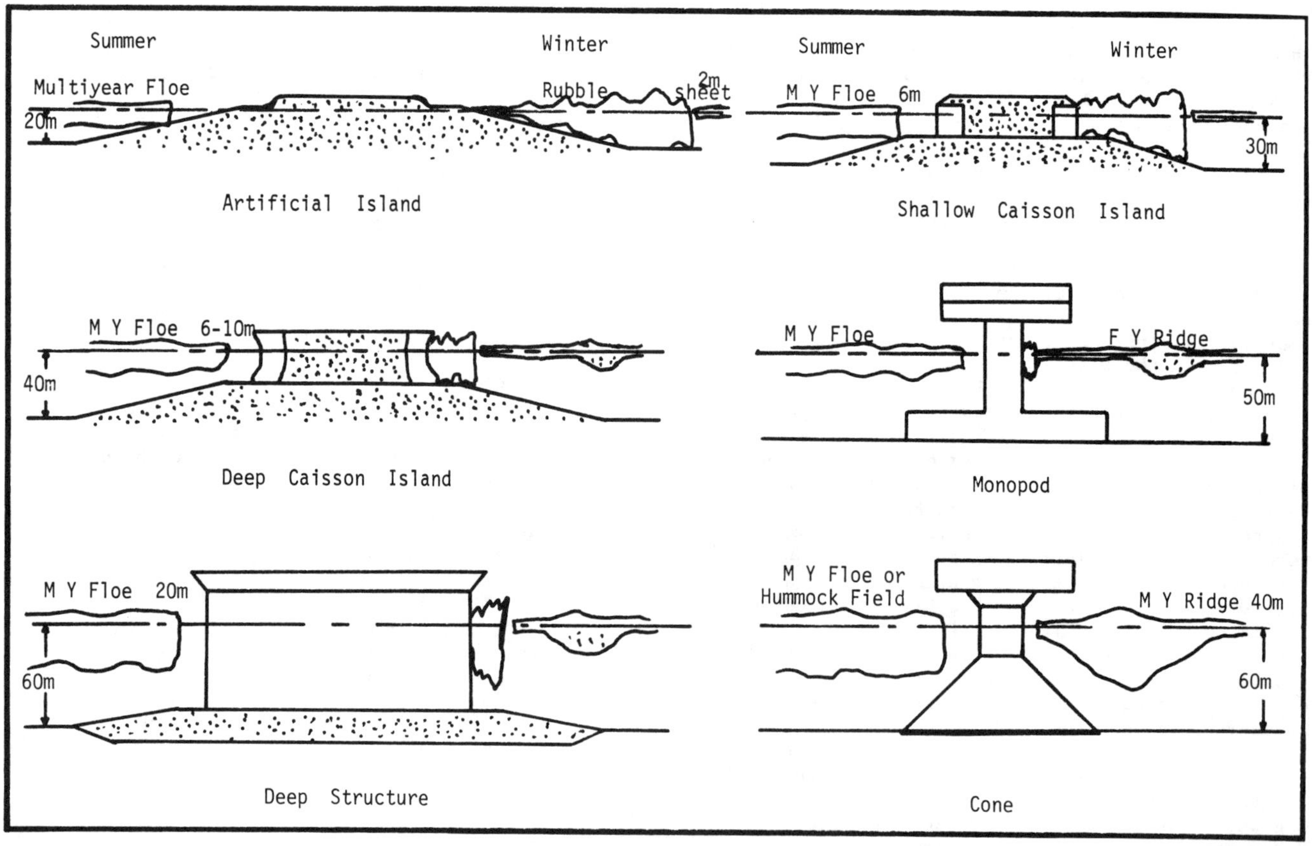

FIGURE 7

TYPICAL ICE CONDITIONS AT STRUCTURES IN DIFFERENT WATER DEPTHS

exploration out to about the 15m water depth. Beyond that, the volumes of fill required, the distance of the site to granular borrow, and the multiple season construction schedules make islands prohibitively expensive. For a production scenario, however, this may change depending on oil reserves and other factors. The mid range water depths (15-40m) provided the challenge which resulted in the caisson-berm types of structure being built. Here a subsea berm of moderate volume can provide a base for a water-penetrating caisson. The deeper waters (40-200m) are the exploration territory for drillships at this time; however, a lucrative reservoir in this water depth may be produced from either a huge island, a caisson-berm, or a gravity type of structure. Figure 7 compares structures which have been used or proposed in different water depths to illustrate this factor. Another important consequence of increasing water depth is the increased risk of impact by thick multiyear ice floes and thus the potential for increased ice forces. This is discussed more in the section on Ice Effects below.

4.2. Space Requirements

The normal exploration well from an artificial island requires a 100m diameter working surface area. This can be augmented in a steel caisson by building in extra storage space below deck. For a production island or structure the space requirments will be considerably greater, corresponding to the area of a 250m diameter island. If an island is the chosen platform then the volumes of fill required become very large and costly. For some applications, therefore, it may be cheaper to consider stacking the construction facilities atop a steel or concrete structure.

4.3. Ice Effects

The major effect of ice which influence design and operation of structures are global ice forces (horizontal and vertical), high local ice pressures, rubble formation, and ice pile-up and ride-up.

For artifical islands and caissons one of the chief concerns is the lateral force exerted by a moving ice sheet or multi-year floe. Design ice forces vary with the geometry of the structure, the mode of ice/structure interaction, and with water depth, since thicker ice is able to contact the structure in deeper water. Detailed discussions of force calculation methods appear in Kry, 1980, and Croasdale and Marcellus, 1981. It is beyond the scope of this paper to discuss the theory. Simply expressed, the current philosophy is that loads should be calculated by comparing the results of the "Limit Stress" and "Limit Force" calculations, and choosing the lesser of these forces. The "limit stress" approach has traditionally been used at artificial islands in shallow water up to 20m and landfast ice. Here, the design load (F) is generated by a 2m thick fully developed ice sheet crushing with a stress, Σ, against the entire width (D_r) of the island plus its surrounding rubble field. Thus:

$$F = \sigma H D_r \tag{1}$$

This condition is illustrated in Figure 8.

In the past, the action of thick ice and a "limit-force" approach has not been applicable since the island slopes are so extensive that large floes would ground out and expend their energy at some distance from shore. The more recent designs for caissons and deeper water structures can achieve less conservatism and more economy if the "limit-force" approach applies. This approach, illustrated in Figure 9, assumes that a large, thick feature impacts the structure and is brought to rest (Stage 1). Subsequently, the combined force of the pack ice in a ridge-building mode pushing against the floe plus the wind and current drag force over the floe is transferred to the structure (Stage 2). Finally (Stage 3), the situation develops such that the force of pack ice drag is added to the Stage 2 force. Thus:

Stage 2

$$F_{LF} = C_{10} P_a V_w{}^2 L^2 + (0.5) C_c P_w V_c{}^2 L^2 + WL \tag{2}$$

where L is the length and width of the ice feature; P_a is the air density: P_w is the water density; C_{10} and C_c are drag co-efficients; V_w is the wind speed; V_c is the current speed and W is the average ridge building force over width L.

Stage 3

$$F_{LF} = L^2(1 + 0.25 \tan B)\,(C_{10} P_a V_w{}^2 + 0.5\, C_c P_w V_c{}^2) + L_w\,(1 + u \tan B) \tag{3}$$

where B is the angle of a rubble field behind the ice feature and u is the friction co-efficient operating at the rubble boundary.

An integral part of the limit force approach has been the consideration of the initial impact during which the large feature is brought to rest. This happens either by indentation of the structure into the ice floe, or, if the ice is thick enough, by arresting action of a soft berm. The submerged slopes of islands and subsea berms which support caissons have served this purpose. Using one of a number of methods, it is possible to calculate the size of berm necessary to absorb this initial impact. It is a function of the energy of the impacting floe (speed and mass), the ice thickness, the allowable load that can be transmitted to the caissons and the characteristics of energy absorption by plowing within the berm. For a specific analysis such as the Gulf Mobile Arctic Caisson, many combinations of ice loading scenarios are considered including different ice thicknesses, floe sizes, rubble formation, and angles of attack. Results showed that beyond a moderate ice thickness the limit force method would apply, while for average and thinner ice floes, the limit stress method would govern.

The field of ice-structure interaction and ice forces is a developing one. There are other methods which are less conservative and by which even greater economies could be realized, however, these require further research and verification before they would be employed.

Local ice pressures up to 20,009 kPa (several thousand psi) must be designed for caisson structures to avoid the prospect of severe puncturing or buckling of the plates and ribs of a steel structure, or spalling of a concrete structure. The estimation of these local loads is one of the most difficult problems facing the designer. While ice breaker hull pressures can be used as a guide, they cannot be compared directly to fixed structures. This is an area where ongoing research may enable greater economies in terms of steel weight to be realized.

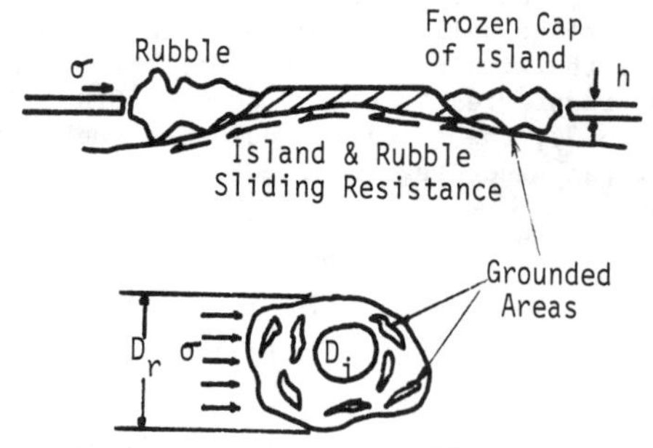

FIGURE 8 LIMIT STRESS INTERACTION

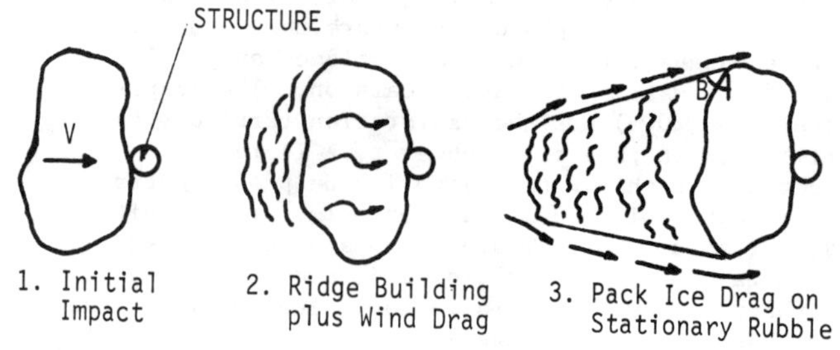

FIGURE 9 LIMIT FORCE INTERACTION

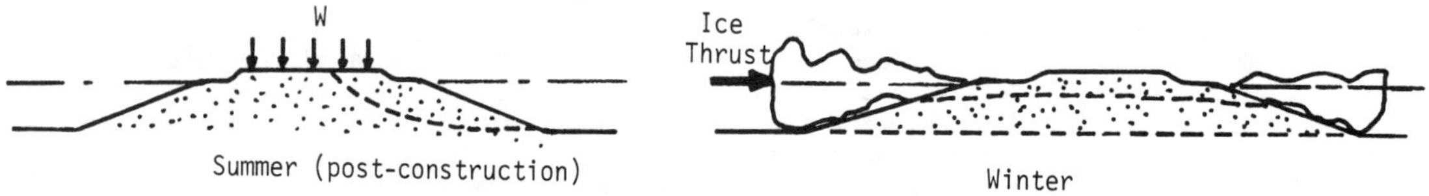

FIGURE 10 POTENTIAL FAILURE MECHANISMS AT AN ARTIFICIAL ISLAND

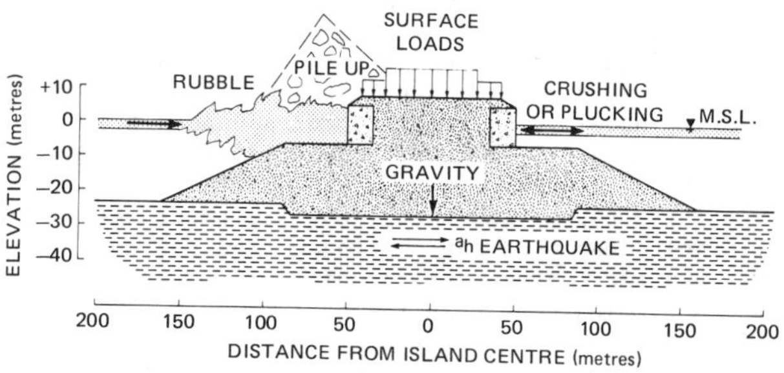

FIGURE 11 LOADING CASES AT A CAISSON ISLAND

NOTE: VERTICAL EXAGGERATION 2:1

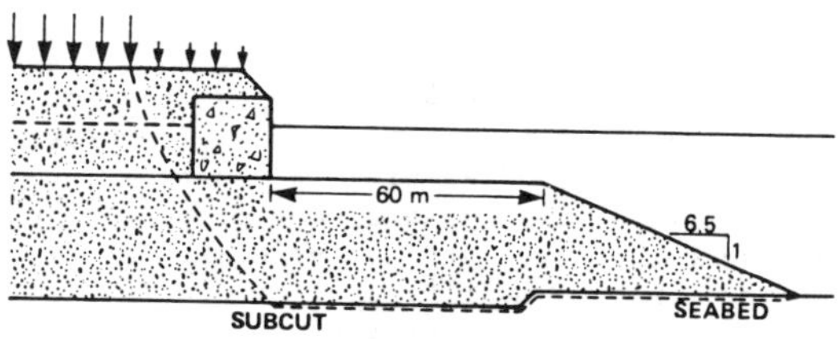

FIGURE 12 POTENTIAL FAILURE MODE UNDER GRAVITY LOAD AT A CAISSON ISLAND

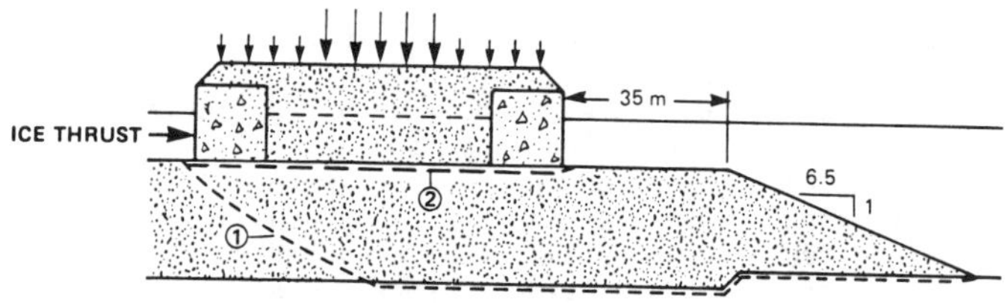

FIGURE 13 POTENTIAL FAILURE MODE UNDER HORIZONTAL ICE LOAD AT A CAISSON ISLAND

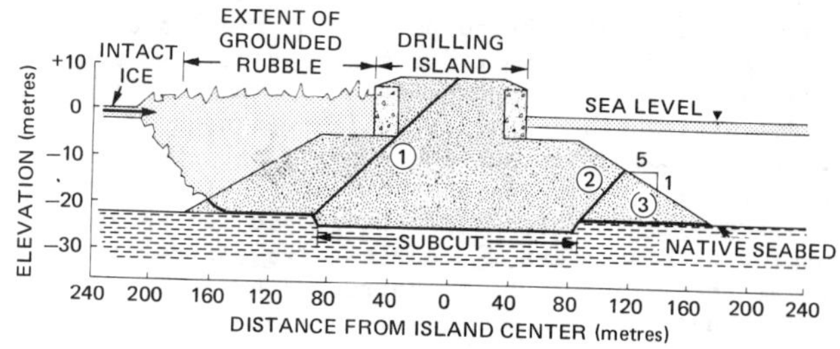

FAILURE MECHANISMS

1. Passive failure to island surface
2. Seabed and berm failure, entire structure
3. Seabed failure, entire structure

FIGURE 14 POTENTIAL FAILURE MECHANISMS FOR ICE LOAD TRANSFERRED BY GROUNDED RUBBLE

Rubble formation around structures has the multiple effects of increasing the area over which force is applied, as discussed in the section on ice forces, while at the same time offering increased resistance to sliding, and also some protection/energy absorption capability against thick ice. Here is another area where continuing research may lead to a less conservative approach to the design of islands. Around production structures, rubble could prove to be a problem of ice management for year-round shipping.

Ice ride-up and pile-up and subsequent inundation of working surfaces is a potential problem at artificial islands and caisson-berm structures. Island beaches are usually designed with a change in slope to initiate pile-up at a safe distance from the facilities (Figure 10). Caissons must be designed with sufficient freeboard to prevent overtopping in ice pile-up (Figure 11).

Other potential ice effects at steel and concrete structures are adfreeze which can result in vertical forces and augmented horizontal forces, and ice scouring of berms which is a maintenance rather than a design problem. At a narrow structure such as a cone (Figure 7), the clearing forces can be of the same order of magnitude as the initial breaking forces.

4.4. Wave Effects

Wave forces, wave erosion, spray and wave overtopping are the key effects, particularly during the late-summer storms which typically generate 3-4m waves.

Wave forces, while low compared to ice forces, could prove to be a significant design consideration for caisson structures because of cyclical loading and the impact effects. Wave forces are not a concern in the design of artificial islands because the waves break on the beaches.

Wave erosion can be a serious problem during the construction of artificial islands in the topping-off stage of construction when large waves during a late-season storm can sometimes prevent attempts to bring the island surface above the level of the sea. This may cause construction to be prolonged into thin ice conditions. Once the island has been completed, fill can also be wave-eroded from the beaches. To combat this, Esso has used either a sacrifical beach when fill material is plentiful, or a slope-protection system such as sand-filled bags, soil-cement, or concrete mats laid in the wave zone. Caisson-berm structures could potentially suffer quite damaging erosion which would undermine the base of the caisson. At Tarsiut, a protective rock layer (riprap) was placed to prevent this problem.

Wave spray has not been a concern at artificial islands due to the breaking of the waves at a safe distance from drilling activity. However, significant wave spray occurred at the two vertical walled caisson structures, Tarsiut and Uviluk, when storm waves slammed directly against the caisson and their energy was deflected upwards. Wave spray 50m in the air directly above Tarsiut restricted helicopter traffic on some occasions. Wave spray can also result in superstructure icing. Future designs will take precautions to lessen this effect by the use of increased freeboard and wave deflection barriers (which also serve as ice rubble deflectors).

Wave interactions at shallow submerged berms can result in an overtopping tendancy if the freeboard is too low. Artificial islands have typically been designed with a 5m freeboard. Both the Gulf Mobile Arctic Caisson and the Esso Caisson have over 7m of freeboard to the working surface and wave deflectors which add another 4-5m of freeboard. This reflects a more thorough understanding of the extreme wave height expectancy and the wave-interaction problems at caisson-berm structures. The 100-year design wave is in the order of 8m.

4.5 Foundation Conditions

Foundation conditions are important in determining stability against conventional gravity loading, stability with respect to various modes of ice pressures, determination of short and long-term consolidation settlements, and thaw settlements where frozen soil exists. The soil conditions have been described previously in this paper under the heading Seafloor Conditions. Because the soil conditions are so variable, it is essential to carry out detailed site investigations with representative insitu and laboratory tests at each site.

The required foundation conditions are a function of the type of structure and the loading it transfers to the soil. The following general comments on foundation conditions and design/construction treatments are most pertinent to the island and berm-caisson structures employed to date.

The area east of 133.5° longitude has 0-6m of soft silty clays overlying dense fine-grained sand. Where possible, locations with thin overburden and minimal foundation preparation would be selected. For sites with deeper overburden stripping would be necessary; however, sand fill borrow material is within a short haul distance.

The area west of 135° longitude has 2-10m of very soft to firm silty clays overlying stiff to hard silts and clays. For the design of Gulf's Moble Arctic Caisson (MAC), it is considered necessary to strip soft soils down to a depth where the undrained shear strength exceeds 70kPa. Within the Tarsiut block, site investigations show this to vary from 2m to 5m below the seabed. The Tarsiut N-44 site, where a caisson retained island was built, required stripping of 2m to 3m. The thickness of soft soils increases further west and are 5m to 10m nearer the Mackenzie Trough. In excess of 10m of soft soils are encountered in some locations between 135° and 136° longitude, where scoured glacial meltwater channels have been infilled with recent soft sediments. Although no bottom-founded structures have been used in areas with an excess of 5m of soft soils, stripping and replacement with imported sand fill is considered the most viable foundation treatment. Multi-season stage loading and vertical which drains to accelerate consolidation (and increase soil shear strength) are several of the possible alternatives where deep stripping or long borrow material haul distances are too costly or impractical with available construction equipment.

The central area, between 134° and 135° longitude, consists of dense sand with 5m-10m of soft silty clay overburden. Although sites with thinner overburden are available, they may not be situated near the desired drilling location. Granular borrow materials are less than 25km distant; therefore, stripping of soft soils and replacement with sand is probably the most viable option. Infilling of a wide glacial meltwater scoured channel, recently named the Kugmallet Channel, has resulted in 10m-20m of very soft to firm silty clays over dense sand in the 133.5°-134° longitude area. Excellent sand borrow sources exist less than 15km eastward. Site specific stratigraphy, type of drilling structure, available lead time, economics and the capability of available construction equipment would be several of many factors in the decision process of whether to strip and replace with sand, or to stage load and/or install vertical wick drains in the existing soft soil to accelerate consolidation.

4.6 Geotechnical Stability

Loading Mechanisms: Forces exerted on or by the structure must be resisted by the shear strength of the soil in order to prevent island failure. Conventional gravity forces include the weight of the sand fill, caissons and operational equipment. Large horizontal forces are exerted by the ice, which may act on the island/structure through ice sheet contact, ice rubble contact, grounded ice rubble contact or impact of large multi-year ice floes. Cyclic forces can result from earthquakes. Horizontal acceleration of less than 10 percent gravity is used in design, depending on the location and design life. Figure 11 shows potential loading cases for a shallow caisson in winter. Pulsating loads from waves and monotonically repetitive ice loads have also been analyzed in the design of caisson structures. Probabilistic studies are carried out to determine what percentage of each design loading mechanism may act simultaneously during the life of the structure.

Effect of Foundation Soils: Caisson structures are generally designed to be founded on a subsea sand berm, partially to improve bearing conditions in clay areas and to allow a fixed caisson height to be matched with various water depths. The berm takes the shape of a subsea island. Berms and islands have slide slopes between 6H:1V and 20H:1V; therefore, the length of potential shear failure planes increase significantly with depth in the fill. In regions with sand foundations, the berm fill is generally weaker than the dense foundation sand. Consequently, the critical shear failure planes are in the upper portions of the berm or island fill, where the failure surfaces are short and offer least resistance.

Previous sections of this paper noted that the soft weak clays were removed in locations with clay foundations. Soft materials provide a preferential shear failure surface below the berm/clay contact. Even when soft materials are removed, critical shear failure surfaces are often in the firm to stiff clay along the base of the sand berm or island.

Effects of Fill Material: Most of the islands and berms are constructed of hydraulically placed, uniform, fine grained sand (0.35 mm $>$ D50 $>$ 0.15mm). Fill densities have been measured as 50% to 60% average Relative Density average over the fill depth; however, zones as loose as 20% to 40% Relative Density occur. The soils of low density exhibit lower shear strength, require higher deformation to mobilize their strength and have a higher potential for liquification and cyclic mobility under earthquake loading. Experimentation with placement methods and extensive insitu testing are carried out in order to determine the lateral and vertical extent of looser zones, and to improve design and construction practices. Caution must be exercised to reject silty materials or clay lumps from being included in the berm, as these can result in lower strength, poor drainage and detrimental differential settlements. Clay lumps have been a problem where borrow areas are thin or contain clay layers within the sand.

Artifical Island Stability: Although islands are more difficult to construct through the wave zone, and are susceptible to wave erosion damage during operation, they are generally safer than caisson structures with respect to general geotechnical failure. The larger surface area and flat side slopes of islands result in less concentrated gravity and ice loading and longer shear failure surfaces. Potential failure mechanisms are shown in Figure 10.

Slope failure under gravity loading is possible but unlikely, as a result of the very flat hydraulic fill slope. Soft weak clay foundation soils would be necessary for the failure mechanism. Failure would be most critical during and immediately after construction, when the soils have not had time to consolidate and gain strength. Design concepts using gravel or rock bunds to steepen the side slopes and reduce fill quantities have been considered. Slopes of 3H: 1V are envisaged for this design. Slope failures for these concepts would become more critical and stiff foundation clays would be essential. Local slope failures within the island fill itself could occur if erosion undercutting significantly steepened the slopes.

Sideslopes cause the attacking ice sheet to break in flexure. Rubble fields form around the island early. They remain grounded on the slopes and tend to act as a protective buffer. Decapitation under the horizontal ice loads is more critical than passive wedge failures. The shear failure surface tries to form as high as possible, where the failure surface is shortest and the frictional resistance the least. After the rubble consolidates and acts more as a solid block, it drives potential failure surfaces deeper into the island where they are less critical.

As discussed earlier in this paper, large multi-year ice floes may impact islands during the summer. They have considerable kinetic energy. The energy is dissipated as the ice plows into the island slopes. Studies show that the ice could be arrested before doing damage to the island crest working surface. Although some maintenance would be required, the slopes would generally be sufficiently self-healing once the ice feature thaws or moves away.

Settlement as a result of consolidation of seabed soils has not been a significant problem for exploration islands; however, islands have not been constructed in those areas having considerable thickness of soft compressible foundation soil.

Settlements resulting from thaw consolidation of underlying permafrost are not a problem for exploration islands. These settlements must be considered in the design of production islands where warm flow through multiple production strings for long durations could thaw the entire depth of permafrost beneath the island. Permafrost will be more critical to drilling and completions than structure settlement.

Caisson Structure Stability: Caisson structures provide a water penetration system and reduce the required fill volume. They are generally placed on a granular berm, the thickness of which is a function of caisson draft and water depth. Berm crest depths (below water) for systems employed or under construction vary from 5m for Tarsiut to 22m for Gulf's Mobile Arctic Caisson. Deeper caissons apply more concentrated gravity load to the seabed and accept more horizontal load from non grounded ice, and are therefore more critical with respect to geotechnical stability.

Failure mechanisms for gravity loading, horizontal loading from free ice sheets or non-grounded rubble and horizontal loading from grounded rubble are shown in Figures 12 to 14 respectively. As previously noted, areas with dense sand foundations have critical failure shear surfaces within the berm only.

Failure under gravity loading would be a conventional bearing capacity failure. Preferential failure shear surfaces would be in the weaker clay just below the fill; however, deep-seated failures are possible. Depending upon fill properties, the factors of safety could decrease during earthquake loading.

The horizontal force from an ice sheet or non-grounded rubble could cause passive failure within the caisson fill, decapitation just below the caissons where the shear surface is shortest, or sliding just below the weak sand fill/clay foundation contact, as shown in Figure 13. In addition, potential failure modes which must be analyzed are rotation at the base and overturning for tall structures such as the MAC and downstream ice plucking or pulling of caissons for Tarsiut where caissons are not connected.

Figure 14 demonstrates a situation where rubble is grounded on the berm slopes. Should the berm crest be narrow, a passive failure of the berm and caisson could occur. Another potential failure mechanism is the entire berm and caisson structure sliding along a weak clay seabed contact. Where shallow water depths permit, the grounded rubble may accumulate sufficiently to ground into the seabed beyond the berm. This can serve to reduce the horizontal loads on the berm and caissons. The grounded rubble protects the structure.

The softer, weaker, more compressible soils are removed in order to improve sliding resistance against ice pressures. This effectively reduces the consolidation settlement. Uniform stripped foundation conditions and placement densities of the hydraulic fill are necessary to avoid detrimental differential settlement below the caisson structure. Settlements from the thawing of permafrost have been discussed previously in this section.

4.7 Structure Response to Loadings

This is important for operational procedures and setting safety limits for shut-downs or evacuations as well as for design. Therefore, force and deflection measuring devices have to date been incorporated in the construction phase and/or the operational phase. These measurement systems become more and more complex for the new designs.

Artificial islands, for example, have been instrumented in winter with ice-force panels frozen into the surrounding ice and rubble, geotechnical inclinometers through the island sand, thermistor strings in the ice rubble to monitor consolidation freezing, and wireline ice motion devices and survey markers to measure lateral motion of the ice relative to the seabed. The instruments have been monitored continuously with telemetering capability between the Beaufort and Calgary. The system has been used as a warning of impending high ice forces so that various ice defense strategies may be implemented.

In the case of the vertical walled caissons which have been built, similar installations have been or will be placed in the ice surrounding the structure, i.e. pressure panels, thermistors, movement stations. With caissons, there is the advantage of a vertical wall for mounting large force panels at the water/ice level and internal members for strainmeters. Geotechnical pressure sensors in the berms and foundations, and inclinometers through the caissons and sand core into the berm have been emplaced. As with the artificial islands, these instruments have been continuously monitored for operational safety indications and all data telemetered to a base and back to Calgary. It is salient to note that in the case of the Tarsiut Island winter monitoring project, the measured force levels never approached the design levels, nor did the deflections and soil pressures exceed the design predictions.

Future generations of structure will presumably have similar systems with perhaps more parameters being measured if an unusual shape warrants it.

4.8 Available Construction Materials

Islands and berms are constructed by hydraulic placement of sand fill. The sand is generally uniform fine grained and has less than 8 percent fines. The geological history described under the heading "Seafloor Conditions" demonstrates that there are large sand deposits east of 135° longitude. The most favorable borrow sources are east of 133.5° longitude, where areas with negligible overburden are common. Localized areas with thin overburden, such as a borrow pit at Isserk, do occur infrequently between 134° and 135° longitude. West of 135° the foundation soils are generally clay. Small deposits of granular material, probably beach deposits left during the sea level rise do exist, but are difficult to find. Sand and gravel were taken from such a location, named Issigak, to construct part of the Tarsiut berm.

Sand deposits are reasonably close to exploration sites in the eastern Beaufort. In order to construct islands and berms in the western areas, sand must be barged up to 100km. In addition, more borrow material is required to replace the layer of soft foundation soil which is removed. The cost of constructing a berm in the west may be 3 to 5 times the cost near the borrow sources in the east. The use of caisson structures to reduce borrow quantities is attractive.

Another method of reducing borrow volumes is to steepen the side slopes. Sand has been placed hydraulically with sideslopes between 5H:1V and 20H:1V, whereas gravel could be placed at 3H:1V. Subsea gravel borrow is very rare. Sources are thought to exist near Herschel Island, further west, however, quantities have not been substantiated. Fill from some of the Esso Resources earlier islands consisted of gravel trucked over the winter ice from inland borrow pits.

Riprap materials have consisted of quarry rock from inland pits near Inuvik. Processing, transportation and placement of this material is very expensive. Alternatively, Esso has used large sand filled bags to provide erosion protection for exploration structures of short life. Because of the scarcity of gravel and riprap, concept designs for longer life production structures have minimized their usage with caissons, or considered man made alternatives such as cement blocks, sand asphalt, or reinforced sand.

4.9 Availability of Equipment and Techniques

An obvious but significant fact about the North American Arctic is that there are no facilities for the building of steel or concrete structures in the area. These types of components must be built in the south. Construction of artificial islands, berms, and foundations is, however, something that is done on-site and with local materials. Considerable cost savings may be realized in the future through the development of high powered dredging fleets and through techniques to minimize material requirements.

In shallow water it is possible to construct an island in one season with available equipment. As construction extends to deeper water it will be necessary to use dredges of increased capacity which have the reach to operate in deeper water. Clearly, if construction time of a berm can be reduced to one season from two seasons, the cost savings can be enormous and schedules more reliable. Another way to save time in an area where local materials are not available would be to use large capacity dumping barges with a fairly high speed, to minimize travel times from borrow sources. These items of equipment do not yet exist in the Beaufort; however, they are contemplated for the near future.

New techniques of construction could yield similar savings of cost and time. Steepening of slideslopes of hydraulically placed sand berms to 5H:1V from the more usual 15H:1V have been achieved by controlled placment of dredged material. This cuts the required volume significantly. Greater side-slope steepening may be achieved by bund-construction which is illustrated in Figure 18. Gulf Canada has pioneered the development of slope strengthening by a polymer cell construction technique which could steepen side slopes to 2H:1V.

5. ARTIFICIAL ISLANDS

The first offshore drilling in the Beaufort Sea was undertaken in 1973 from a dredged gravel artificial island constructed in 3m of water over a period of two summers. To date, 24 wells have been drilled from 21 artificial islands, three of the wells having been drilled directionally. The artificial islands have been constructed by various techniques, with water depths increasing with time, and construction techniques becoming more efficient. The island building techniques have included:

1) Dredging through floating pipe and pipe mounted on wooden piles.
2) A sandbag dyke, later filled with silt and gravel.
3) Gravel trucked from land over the ice.
4) A barge-mounted drilling rig sunk within a dyke.
5) Sand hauled by dump scows from a dredge.
6) Stationary suction dredge and floating line.
7) Stationary cutter suction dredge.
8) Mobile suction hopper dredge.

Capability has increased from 3m water depth at Immerk, the first island, to 19m at the Issungnak oil discovery. All islands were built in landfast ice. Most of the islands used sandbag protection as shown in Figure 15 to prevent summer wave erosion. Some earlier ones were built only to provide a platform for winter exploratory drilling and were then allowed to self-destruct. The side slopes of dredged artificial islands have been in the order of 15H:1V. At islands for which sand had to be hauled long distances, there was a great economic incentive to reduce the island volumes through steeper slopes. Steeper slopes have now been achieved through "controlled" placement of the sand through stingers or dredge arms. Another reason for the high cost of artificial islands has been difficulty in completing the above-water portion without the crown being washed away by waves and swell. Sometimes this crown or cap could not be achieved until commencement of freezeup, leaving little time to move drilling equipment onto the site. The use of sand-filled caissons on subsea berms both reduces the volume of sand required and solves the problem of the crown washing away.

6. CAISSON ISLANDS

6.1 Tarsiut-Caisson Retained Island

The first caisson island in the Beaufort Sea was built for Gulf Canada and its partners as a drilling platform for the Gulf et al Tarsiut N-44 delineation well. The site was excavated to a depth of 2-3m, and construction commenced in the fall of 1980. Carefully controlled placement of the sand enabled berm slopes of 5H:1V to be achieved compared to the 15H:1V common at previous artificial islands. The reasons for choosing the caisson over the artificial island were to reduce the amount of fill which had to be hauled some 100km, and to eliminate the problems encountered in placing the final cap on artificial islands. The underwater berm was completed in the summer of 1981, gravelled and carefully levelled. It was built in 22m of water and reached to within 5m of the water surface.

The concrete caissons were built on the west coast of Canada and moved to the Beaufort Sea on a specially built barge. This submersible barge was sunk, the concrete caissons floated onto the barge, and then the barge was raised and towed to the Beaufort Sea. Each of the four concrete caissons shown in Figure 16 was 11m high, 80m long and 15m wide, and weighed 5,300 tons. The floatable caissons were compartmentalized so that cells could be filled with water for ballasting and later filled with sand on site. The caissons were not fastened together when placed, but the gaps at the corners were closed with steel doors and filter cloth to prevent the sand from leaking out of the core. The weight of the sand-filled caissons and the sand core prevents the structure from sliding when subjected to severe ice loads. The core was built up 2.5m above the caissons, levelled to receive a specially built drilling rig, and the slopes were protected with sandbags. This provided a freeboard of 7.5m. It became evident during the storms that occurred near the end of the open water season that wave overtopping and spray would be a problem the following summer.

In winter, as planned, the sea ice grounded on the underwater berm and formed an annular rubble pile around the caisson protecting it from direct impact and extreme ice forces. In the case of Tarsiut, the regulatory agencies required relief well capability before penetrating hydrocarbon bearing zones. This required flooding of the ice rubble pile to build a grounded ice island 150m from the well and 100m in diameter, large enough to accept a helicopter-transportable drilling rig and supplies. This pad was built with an 8m freeboard to prevent sliding under lateral ice loading. Figure 17 shows the island in winter with surrounding rubble and the artificial ice island in the background.

The concrete caissons, the sand fill and the surrounding ice were extensively instrumented to monitor pressures and movements. This monitoring served to assure the safety of the operation as well as to collect valuable information for application to future drilling and production platforms.

The well was drilled from late December 1981 until mid-April 1982, after which extensive flow testing was carried out. In preparation for the open-water season, upgrading of the surface of the island was undertaken, commencing in April 1982. One thousand two-ton gabions or rock-filled wire baskets were airlifted to the island by helicopters and set on the caissons in place of the sandbags. At the corners, sheet piling and pipe piling were driven to prevent waves from washing sand through the doors between caissons. This work was completed by May 31, on schedule. As soon as the ice had broken up and cleared, another thousand gabions were

FIGURE 15 ARTIFICIAL ISLAND ISSUNGNAK

FIGURE 17 TARSIUT CAISSON-BERM STRUCTURE IN WINTER

FIGURE 16 TARSIUT CAISSONS DURING INSTALLATION

FIGURE 18 SINGLE STEEL DRILLING CAISSON SSDC

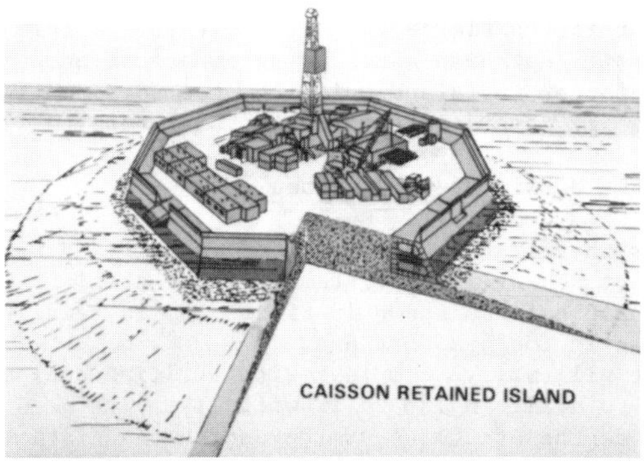

FIGURE 19 ESSO RESOURCES SHALLOW STEEL CAISSON

barged to the site and placed to increase the freeboard by another three meters for a total freeboard of ten meters. Also, additional pipe piles were driven, and rock up to 1.5m in diameter was placed in the water around the toe of the caisson. The caisson has been hit by severe storms and has withstood the 3.5m significant waves, although the island was, on two occasions, evacuated to ensure the safety of personnel.

6.2 Uviluk Steel Drilling Caisson

Dome Petroleum is operating for Gulf and other partners a caisson called the Single Steel Drilling Caisson (SSDC) which was placed on an underwater berm in a 30m water depth in the summer of 1982 at Uviluk site. The steel caisson in this case is half of a supertanker as shown in Figure 18. It has been reinforced to resist ice pressure. It has a concrete-filled double hull. It is ballasted with water and not filled with sand as are the other caissons. It, too, is re-usable at other sites; in fact a second berm was started for it in summer 1982 at Nerlerk. The Uviluk berm required only about 500,000cu/m of sand. The caisson is about 160m long, 50m wide and is about 25m high. It is set on a berm 9m below water surface. The high freeboard protects against wave overtopping, spray and ice ride-up. Should a relief well be required, a helicopter-transportable drilling rig could be placed at the opposite end of the SSDC. This caisson will be used in that portion of the Beaufort Sea beyond landfast ice as were the concrete Tarsiut caissons. In this region, pack ice is in motion most of the winter.

6.3 Esso Caisson-Retained Island

The Esso steel caisson arrived in the Beaufort Sea during the summer of 1982, although it will not be placed in position until summer 1983. It consists of eight steel segments each weighing 1,000 tons and measuring 43m long. The eight segments will be held together by 16 steel cables, 76mm thick, to form a large doughnut 100m in diameter. The steel caisson has a height of 12.2m and walls 13m thick at the base and 8m thick at the top. The $55 million caisson will be assembled in calm waters and then towed to the Kadluk site in 14m of water and ballasted down onto a subsea berm as shown in Figure 19. Some 80,000cu/m of dredged sand will be placed into the interior of the caisson to provide the sliding resistance against the ice forces which may develop. This caisson can be refloated and used at other locations. Other benefits of the system are the reduced volume required for the island and the ability to pierce the water with a platform which will not wash away during construction. Esso, as a pioneer in island building, has advanced the technology considerably over the past ten years.

6.4 Gulf Mobile Arctic Caisson

As the first major component of Gulf's new Beaufort Sea Drilling System, a large steel floatable caisson shown in Figure 20, is being built for use in the Beaufort Sea.

It will arrive complete with drilling equipment ready to drill in 1984. It will likely be used in water depths of 20m to 40m beyond the landfast ice. A berm will first be built to within 22m of sea level using dredged sand. After careful levelling of the berm, the caisson will be ballasted down onto the berm surface and the central core of the caisson filled with 115,000cu/m of dredged sand. It is primarily the weight of sand within the caisson which will prevent it from sliding under the extreme forces exerted by large pressure ridges and multi-year ice. The caisson itself weighs 33,000 tons. The hull will be insulated to prevent the sand core from freezing during the long cold winters. The shape of the caisson was chosen based on model tests to prevent ice or waves from reaching the deck and to minimize spray from waves. The diameters of the caisson are 111m at the bottom and 87m at the top. The height is 29m and the freeboard, excluding the wave deflector, 7m as shown in the cross-section in Figure 21.

The Mobile Arctic Caisson (MAC) will only be moved during the summer. It should drill and fully test one well per year and if the well were a discovery, a second directional delineation well could be drilled without re-supplying the caisson with drilling supplies. The MAC can be repositioned to another site quickly because the drilling equipment need not be moved off and berm volumes will not be large.

Use of the MAC will further extend the technology and will provide valuable data for the production phase.

7. PRODUCTION STRUCTURES

The ultimate choice of a structure for a specific site will be based primarily on safety and cost-effectiveness. These in turn are dependent on the design requirements whose influences were described earlier, as well as fabrication, construction, installation, operation, and maintenance costs. It is also evident that the size and properties of the oil reservoir will partially dictate the characteristics of a structure or group of structures. The selection of a structure must strike an acceptable balance between the cost-effectiveness and the human, environmental and financial risks.

Many different types of production structures have been proposed including artificial islands, steel gravity cones, and massive concrete caissons. Figures 22-25 show examples of some concepts. One is a method of converting a caisson island such as Tarsiut to a production island by erecting a boundary of caissons and filling it with dredged sand. Another is the massive Arctic Production and Loading Atol (APLA) which is a pair of dredged islands in deep water with a tanker harbor in between. There is the Arctic Monocone suitable for deep water and able to withstand deep ice features. There is a shallow-water gravity platform which is less robust because of reduced hazards. These and many other concepts are under consideration for use in the production phase.

8. AREAS FOR RESEARCH AND DEVELOPMENT

In the previous sections on design, the influence of multiple factors was discussed. The following summarizes the areas where additional research and development could have significant benefits and cost savings.

1. Ice/Structure Interaction and Ice/Berm Interaction
   - less conservatism in ice failure pressures would lower requirements and costs
   - understanding of absorption of ice forces in winter by rubble formation
   - more data on summer impacts of multi-year ice
   - berm scour research to optimize berm design
   - local ice pressures

FIGURE 20 GULF CANADA RESOURCES' MOBILE ARCTIC CAISSON

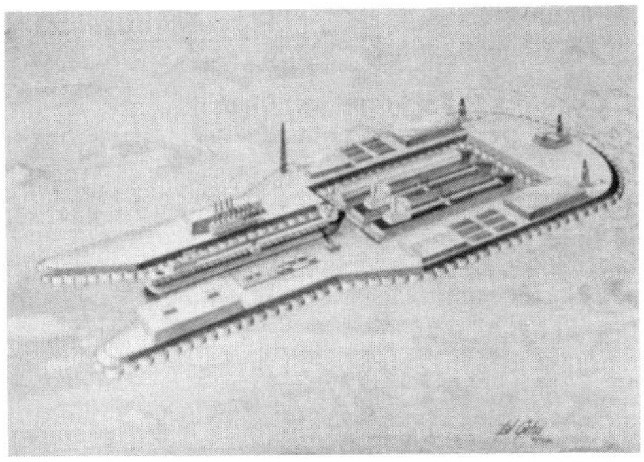

FIGURE 22 CAISSON ISLAND PRODUCTION CONCEPT

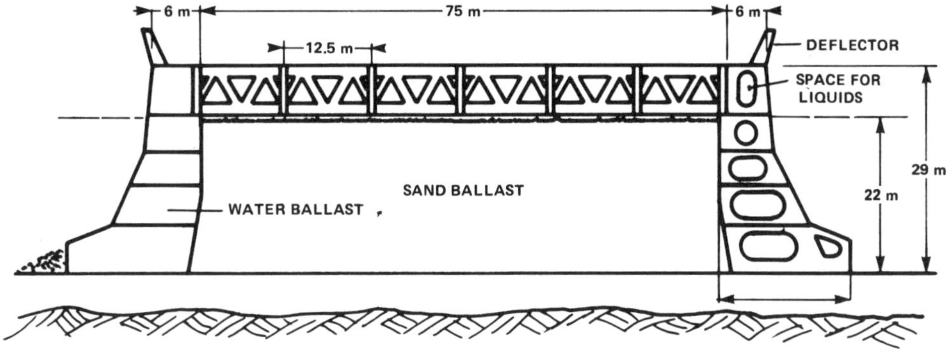

FIGURE 21 GULF MOBILE ARCTIC CAISSON CROSS-SECTION

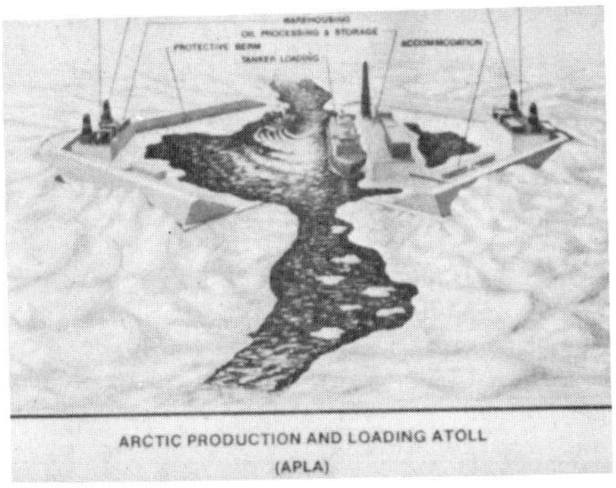

FIGURE 23 ARCTIC PRODUCTION LOADING ATOL

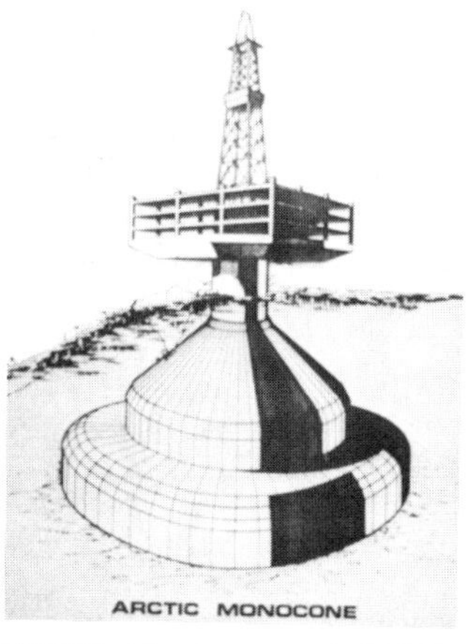

FIGURE 24 ARCTIC MONOCONE

FIGURE 25 MONOPOD

2. Structure Response to Ice Loading
   - continue on-going research programs to refine safety limits

3. Berm Construction Techniques to Decrease Cost and to Minimize Maintenance
   - slope strengthening
   - slope stabilizing/protection
   - erosion protection methods

4. Seismicity Levels in the Beaufort and their Influence on Soil Stability

5. Thaw Settlement
   - analysis of multiple wells penetrating permafrost

6. Construction Equipment
   - deep water dredging
   - soil placement methods
   - large volume hauling capability
   - increasing length of construction season by ice breaking, ice strengthening

7. Wave Interactions at Berm/Caisson Structures and Optimization of Shape to Account for Spray, Superstructure Icing, and Ice Effects.

ACKNOWLEDGEMENTS

The authors would like to thank the officers of Gulf Canada Resources Inc. for permission to publish this paper, and the numerous workers in the fields of Arctic science and engineering upon whose published efforts much of this paper is based.

REFERENCES

Bercha, F.G. and J.V. Danys, 1976, "Investigation of Ice Forces on a Conical Offshore Structure," Ocean Engineering, Volume 3, No. 5, pp. 299-310.

Bercha, F.G. and J.V. Danys, 1975, "Prediction of Ice Forces on Conical Offshore Structures," Marine Science Communications, pp. 365-380.

Boon, D.J., 1980, "The Construction of an Artificial Drilling Island in Intermediate Water Depths in the Beaufort Sea," Offshore Technology Conference, Houston.

Bruce, J.C., "Steel Concepts for Arctic Exploration and Production," Proceedings CSSRA, Montreal, 1981.

Croasdale, K.R., "Ice Forces on Marine Structures," International Association of Hydraulic Research (IAHR) Third International Symposium on Ice Problems, Hanover, NH, August 18-21, 1975.

Croasdale, K.R. and R.W. Marcellus, 1978, "Ice and Wave Action on Artificial Islands in the Beaufort Sea," Canadian Journal of Civil Engineering, Volume 5, No. 1, pp. 98-113.

Croasdale, K.R. and R.W. Marcellus, 1981, "Ice Forces on Large Marine Structures," Sixth International Conference on Port and Ocean Engineering Under Arctic Conditions, Quebec, Canada, pp. 755-769.

Danys, J.V., 1977, "On Wind Induced Static Ice Forces on Offshore Structures," Fourth International POAC, Memorial University, St. John's Newfoundland, September 26-30, 1977, pp. 718-729.

DeJong, J.J.A. and J.C. Bruce, 1978, "Design and Construction of a Caisson Retained Island Drilling Platform for the Beaufort Sea," Offshore Technology Conference, Houston.

Edwards, R.Y. and K.R. Croasdale, 1976, "Model Experiments to Determine Ice Forces on Conical Structures," Symposium on Applied Glaciology, Cambridge, England.

EBA Engineering Consultants Ltd., 1981, Geotechnical Contribution to Design of a Caisson-Retained Island, Beaufort Sea, Engineering Report Submitted to Canadian Marine Drilling Ltd./Dome Petroleum Ltd., January, 1981, Proprietary to Gulf Canada Resources Inc. and Dome Petroleum.

Frederking, R. and B.D. Wright, 1980, "Characteristics of an Ice Rubble Field at Issungnak, February-March, 1980," Proceedings National Research Council Workshop on Ridges and Rubble Fields. Calgary, October.

Garratt, D.H and P.R. Kry, 1978, "Construction of Artificial Islands as Beaufort Sea Drilling Platforms," Journal of Canadian Petroleum Technology, April-June, pp. 73-79.

Gibson, R.E., 1958, "The Progress of Consolidation in a Clay Layer Increasing in Thickness with Time," Geotechnique, Volume 8, No. 2, pp. 171-182.

Hnatiuk, J., 1976, "Results of an Environmental Research Program in the Canadian Beaufort Sea," Eighth Annual Offshore Technology Conference, Houston, Texas.

Hnatiuk, J. and K.D. Brown, 1977, "Sea Bottom Scouring in the Canadian Beaufort Sea," Ninth Annual Offshore Technology Conference, Houston, TX.

Hnatiuk, J., 1982, "Exploration Methods in the Canadian Arctic," Conference on Arctic Energy Resources, Oslo, Norway.

Iyer, S.H., 1981, Mobile Arctic Caisson, A Synthesis of Design Considerations, Internal Report, Gulf Canada Resources, Inc.

Jazrawi, W. and J.R. Davies, 1975, "A Monopod Drilling System for the Canadian Beaufort Sea," S.N.A.M.E., Ice Tech. '75, Montreal.

Kent, D.D., B.W. Graham and R.H.B. Sangster, 1979, "Geotechnical Design of a Caisson Retained Island for Exploration Drilling in the Beaufort Sea," Proceedings, 1st Canadian Conference Marine Geotechnical Engineering, Calgary, Alberta, pp. 429-437.

Kovacs, A., W.R. Weeks and W.D. Hibler, 1971, "Pressure Ridge Characteristics in the Arctic Coastal Environment," First International Conference on Port and Ocean Engineering Under Arctic Conditions, University of Trondheim, Norway.

Kovacs, A. and D.S. Sodhi, 1979, "Shore Ice Pile-up and Ride-Up," Workshop on Problems of the Seasonal Sea Ice Zone, Naval Post Graduate School, Monterey, CA.

Kry, P.R., 1980, "Ice Forces on Wide Structures," Canadian Geotechnical Journal, Volume 17, No. 1, pp. 97-113.

Kry, P.R., 1979, "Implications of Structure Width for Design Ice Forces," International Union of Theoretical and Applied Mechanics, Symposium on the Physics and Mechanics of Ice, Copenhagen, Denmark, August 6-10.

Lewis, J.W. and K.R. Croasdale, 1978, "Modeling the Interaction Between Pressure Ridges and Conical Structures," IAHR Ice Symposium, Lulea, Sweden.

McGonigal, D and B.D. Wright, "Winter Ice Motion in the Southern Beaufort Sea from RAMS Buoy Data Analysis 1975-1978," Intermaritec 80, International Conference on Marine Sciences and Ocean Engineering, Hamburg.

Metge, M., A. Strilchuk and P. Trofimenkoff, 1975, "On Recording Stresses in Ice," Proceedings, 3rd International Symposium on Ice Problems, International Association for Hydraulic Research, Hanover, NH, pp. 459-468.

Meyerhoff, G.G., 1976, "Concepts of Safety in Foundation Engineering Ashore and Offshore," Conference on Behaviour of Off-Shore Structures, Volume 1, pp. 900-911.

Michel, B., 1978, Ice Mechanics, Les Presses de l'Universite Laval, Quebec.

Michel B. and N. Toussaint, 1977, "Mechanisms and Theory of Indentation of Ice Plates," Journal of Glaciology, 1981, pp. 285-300.

Neill, R., 1976, "Dynamic Ice Forces on Piers and Piles; An Assessment of Design Guidelines in the Light of Recent Research," Canadian Journal of Civil Engineering.

O'Conner, M.J. and Associates, 1982, 1981 Regional Borrow Investigation, Beaufort Sea, a report submitted to Gulf Canada Resources, Inc.

Parmerter, R.R. and M. Coon, 1973, "On the Mechanics of Pressure Ridge Formation in Sea Ice," Offshore Technology Conference, Houston, TX, Volume 1, pp. 733-742.

Ralston, T.D., 1977, "Ice Force Design Considerations for Conical Offshore Structures, Fourth POAC Conference, St. John's, Newfoundland.

Riley, J.G., 1975, "The Construction of Artificial Islands in the Beaufort Sea," Offshore Technology Conference, Houston, TX, Volume I, pp. 211-222.

Sackinger, W.M. and P.A. Sackinger, 1977, "Shear Strength of the Adfreeze Bond of Sea Ice to Structures," Fourth International POAC, Memorial University, St. John's, Newfoundland, pp. 607-614.

Sangster, R.H.B., 1979, "Some Aspects of Offshore Geotechnical Design and Construction in Canada," Proceedings, 1st Canadian Conference Marine Geotechnical Engineering, Calgary, Alberta, pp. 395-403.

Schwartz, J., 1970, "The Pressure of Floating Ice-Fields on Piles," International Association for Hydraulic Research, Ice and its Action on Hydraulic Structures, Iceland.

Spedding, L.G., 1979, "Landfast Ice Motion Observed in the Mackenzie Delta Region of the Southern Beaufort Sea in the 1972/73 Winter," Proceedings, 5th International Conference on Port and Ocean Engineering under Arctic Conditions, Trondheim, Norway, August 13-17, Volume 1, pp. 23-27.

Swan Wooster Engineering Co. Ltd., 1980, Progress Report - Caisson Retained Island, Beaufort Sea, Engineering Report submitted to Dome Petroleum Ltd., June, 1980.

Swan Wooster Engineering Co. Ltd., 1982, Environmental Impact Statement: Study of Production Structures, Beaufort Sea Oilfield Development, a report submitted to Dome Petroleum Ltd., Esso Resources Canada Ltd., and Gulf Canada Resources Inc.

Tryde, P., 1977, "Ice Forces," Journal of Glaciology, Volume 19, No. 81.

Wasilewski, B and J.C. Bruce, 1981, "Conceptual Design for a Mobile Arctic Gravity Platform," POAC, Quebec.

Weeks, W.F. and A. Assur, 1967, The Mechanical Properties of Sea Ice, U.S. Army, Corps of Engineers, Cold Regions Research and Engineering Laboratory, Hanover, NH, Monograph II-C3.

Wright, B.D., J. Hnatiuk and A. Kovacs, 1978, "Sea Ice Pressure Ridges in the Beaufort Sea," IAHR Symposium on Ice Problems, Lulea, Sweden.

ARCTIC TECHNOLOGY II

INTRODUCTION

Paul C. Xirouchakis

Department of Ocean Engineering
Massachusetts Institute of Technology
Cambridge, Massachusetts

In this session two important technological aspects of arctic engineering were addressed namely that of Arctic Concrete Structures and that of Ice Breaking. ·The subject of Concrete Structures in the Arctic was addressed by Professor Gerwick. Professor Gerwick in his paper "On Concrete Structures For The Arctic" gives a comprehensive review of the important aspects of technology involved in the design and construction of concrete structures for the Arctic. He shows how the material characteristics of prestressed and reinforced concrete structures can be selected so that they can withstand successfully the actions of the Arctic environment. He also demonstrates how the ice conditions pose distinct requirements on the selection of suitable structures that can resist (or absorb) the imposed loading. It is of interest to add to Professor Gerwick's discussion a brief summary of the concrete technology in the shipbuilding technology because it helps strengthen some of the important points that Professor Gerwick makes in his paper.

Early construction of reinforced concrete vessels during the two World War periods demonstrated the main advantages and limitations of concrete in shipbuilding. Low construction costs and excellent durability was achieved, while the deadweight to displacement ratio was significantly lower compared with steel ships and there was a size limitation. Reinforced concrete vessels also exhibited superior impact strength and resistance against brittle fracture qualities.

During the World War II prestressed concrete was introduced for the construction of simple box shaped barges. Prestressed concrete allowed larger size construction with thinner sections. This way the resistance to cracking was improved. Barge type construction took place after the war, which due to its simple form further reduced construction time.

During the seventies a number of feasibility studies were undertaken for large prestressed concrete tankers. The work by Moe (1974) indicated that the construction cost is expected to be lower for concrete than for steel. However, the estimate of cargo transportation cost was found somewhat higher for the concrete alternative due to the heavy structural weight. The study concluded that prestressed concrete appears to be more competitive for stationary or semi-stationary floating tanks. A technical feasibility and safety study by Gerwick, et al (1978), for a 300m long prestressed concrete vessel carrying liquefied petroleum gas (LPG) in free-standing tanks concluded that the proposed concrete design had adequate safety and concrete imposed no limitation in the ship's length.

When the main loads on the hull are due to wave-action a relatively heavy concrete structural weight results. This is so since an adequate level of prestressing is needed to maintain a dominant compressive stress field in the hull despite the cyclic hogging and sagging deformation of the girder. When a stationary or semi-stationary functional usage is intended, simpler barge type shapes can be employed since hydrodynamic resistance and maneuverability considerations become less important. This allows an optimized construction with simple geometrical shapes and as a consequence hull weight reductions. If in addition a volume limited design is considered, as in the case of carrying a low density cargo, then the heavier hull weight is not a limitation any more. Finally, the favorable concrete properties at cryogenic temperatures increase the concrete's competitive advantage for floating LNG and LPG processing and storage plants. In the recent past a growing interest has been demonstrated for this class of applications.

The two main areas for present and future research and development for concrete shipbuilding applications are in the direction of materials selection and structural design. Fiber reinforced concrete is promising improved tensile properties. Multiaxial prestressing and appropriate shaping of shell elements can further increase deadweight to displacement ratios. A very comprehensive literature review for concrete as a shipbuilding material can be found in the relevant committee report of the latest International Ship Structures Congress (1982).

When the main loads on the hull are due to ice action reinforced and prestressed concrete structures appear to gain definite competitive advantages for stationary or semi-stationary applic-actions. The conference paper by Professor Gerwick explains very clearly why this is so.

REFERENCES

1. Moe, J., A Feasibility Study of Prestressed Concrete Tanker Ships, ACI Journal, Vol.71, pp.617-626, 1974.

2. Gerwick, Jr., B.C., Mansour, A.E., Price,E. and Thayamballi, A., "Feasibility and Comparative Studies for the Use of Prestressed Concrete in Large Storage/Processing Vessels", SNAME Transactions, Vol.86, pp.163-196, 1978.

3. 8th International Ship Structures Congress, Report of Committee 111.2, Nonferrous and Composite Materials, Gdansk, 1982.

The second topic to be addressed in this session was given by Dr. Vance who gave a comprehensive historical review of the important aspects of ice-breaking technology. It provides a thorough assessment of the state of the art on methods to predict the resistance of ships and forces on offshore steel and concrete structures and gravel islands in the Arctic. The paper also defines the main directions for future work.

The understanding of the process of breaking ice is strongly dependent upon constantly improving our knowledge of the bearing capacity of floating ice plates. The critical survey article by Kerr, A. (1976) provides important information on the failure of floating ice plates under static or quasi-static loading. Information is also needed in the case of impact or impulsive loading.

This brief subject overview will supplement Dr. Vance's presentation by focusing on the latter type of loading on floating ice plates. This summary will discuss consecutively the applications, the classification of impacts and ice plates, the ice deformations and failure criteria, the ice plate perforation mechanisms, the classification of penetrators and the importance of fluid-interaction effects. On the basis of this discussion the design problem will be defined. A very comprehensive review article was reported by Backman, M.E. and Goldsmith, W. (1978) on the mechanics of penetration of projectiles into targets.

Information is needed in the case of impact or impulsive loading to describe the processes and interactions involved during submarine surfacing through ice, dynamic wheel loading during aircraft over-ice landing operations and explosive demolition of ice sheets. The ramming through ice of icebreakers is another important application. In the latter case significant horizontal force components are developed in addition to the lateral loads.

The various types of impacts can be described according to the angle of incidence of the impact, the configuration and material characteristics of the ice plate and/or penetrator and the range of initial velocities. One of the most important variable is the magnitude of the initial velocity.

The ice plates can be conveniently categorized by their thickness. Ice plates can be considered thin when it can be assumed that plane cross-sections remain plane after deformation. Thick ice plates are obtained when the rear surface significantly influences the response only after a substantial penetration depth is reached. Intermediate plates have a thickness in between the range bounded by thin and thick plates.

For very low impact velocities, elastic ice deformations can be assumed. Permanent deformations (most probably localized) must be considered for intermediate impact velocities. Finally, a hydrodynamic description is needed for hypervelocity impacts. More data are needed for the development of suitable ice constitutive relations under impact loading. In particular more work is necessary to describe the effects of viscosity in such conditions. Finally, more data are needed especially on multi-axial ice failure criteria and their dependence on strain and/or temperature.

Two basic ice plate perforation mechanisms have been observed. Flexual cracking, for relatively low impacts and thinner plates, which consists of a number of radial cracks followed by a circumferential crack. The second mechanism is the formation of a cylindrical or cylindrical-conical plug for higher impacts and thicker plates. The salinity and temperature of ice strongly influence the resulting perforation mechanism. The angle of incidence of the impact is another important variable.

There are kinetic energy penetrators, explosives and combinations of these two types. Depending on their shape, penetrators range from sharp to blunt categories. Finally, it is important to know, in each particular application, whether or not the penetrator can be assumed to behave as a rigid body.

Fluid-interaction effects are expected to be of importance when impulsive loads are exerted. This is especially so, when flexural cracking occurs. The compressibility of the water must be considered during the early stages of deformation. The incompressible fluid model can be used for the latter part of the response. A fundamental review article was reported by Krajcinovic, D. (1977) on transient response considerations of submerged elasto-plastic structures.

The design problem can be defined, in a simple way, as the determination of the boundary curves separating the various ice plate perforation (or partial penetration) mechanisms in a plot of the magnitude of initial impact velocity versus the ratio of ice plate thickness to penetrator diameter. Clearly, other important variables such as ice temperature and salinity are parameters of the problem.

This brief overview on the subject of ice-breaking technology was mainly focused on the failure of floating ice plates subjected to impact or impulsive loading. This is an important area for future research due to its practical significance. The conference paper by Dr. Vance further substantiates this argument. In addition it deals in a very comprehensive manner with other important aspects of ice breaking technology.

REFERENCES

1. Kerr, A.D., The Bearing Capacity of Floating Ice Plates Subjected to Static or Quasi-Static Loads, J. of Glaciology, Vol. 17, No. 76, pp.229-268, 1976.

2. Backman, M.E. and Goldsmith, W., The Mechanics of Penetration of Projectiles into Targets, Int. J. Engng. Sci., Vol. 16, pp.1-99, 1978.

3. Krajcinovic, D., Some Transient Problems of Submerged Elasto-Plastic Structures, Shock and Vibration Digest, pp.15-19, 1977.

CONCRETE STRUCTURES FOR THE ARCTIC

Ben C. Gerwick, Jr.
Professor of Civil Engineering,
University of California, Berkeley

Responding to the current intense interest in the exploration for and development of petroleum resources in offshore Arctic and sub-Arctic areas, considerable design and development efforts have recently been directed to prestressed concrete caissons. The successful use of concrete offshore platforms in the North Sea and concrete lighthouses in the Baltic Sea and Eastern Canada have encouraged consideration of concrete for structures in the extremely hostile environment of the Arctic where ice is a dominant design force.

Prestressed and reinforced concrete exhibits ductile behavior under impact at low temperatures. Properly designed and constructed, it is highly durable under the multiple attack phenomena of saline water, freeze-thaw cycling, ice abrasion, and thermal strains.

Thick shell or slab elements are typically employed as the peripheral ice wall. With substantial quantities of reinforcing and prestressing steel, including through wall stirrups, high resistance to local concentrated loads ("punching shear") and a ductile behavior under overload can be attained.

While early attention has focussed primarily on the configuration of the ice wall, the internal structural configuration must be designed to distribute and transfer the intense loads to the base and into the foundation soils. Concrete shear walls, diaphragms, and biaxially-prestressed slabs are structurally effective for this purpose.

Structural lightweight concrete can be employed to reduce the draft of structures in the shallow waters of the Beaufort Sea and has been successfully used on the caissons for the Tarsiut island in Canada.

Some areas for future development are indicated, including use of silica fumes to increase strength, steel-concrete composite sandwich construction, utilization of multiaxial stress conditions, and coatings to reduce friction and abrasion.

Concrete structures utilizing shell and slab elements are inherently economical solutions for structures in the Arctic environment. It is believed that their concomitant properties will enlarge their utilization for both sea-ice and iceberg resistant structures.

1. INTRODUCTION

The rapidly-growing interest in the Arctic offshore as a potential major petroleum province has intensified the development of concepts for exploratory and production platforms suitable to the unique environmental conditions. Sea ice dominates the design in the Beaufort and Chukchi Seas while icebergs are a principal design concern offshore Labrador and Newfoundland.

The successful use of prestressed concrete offshore platforms in the North Sea and for lighthouses in sub-Arctic conditions in Eastern Canada and the Baltic Sea has led to their consideration as offshore platforms in the Arctic. The basic considerations of ductile behavior under impact at low temperatures, ice abrasion resistance and durability, fatigue endurance, and overall economy appear well-suited to meet the requirements. On the other hand, their inherently heavy weight, hence draft, is a constraint on deployment in the shallow coastal waters typical of the Beaufort Sea and on their initial delivery around Point Barrow, where the Arctic ice pack limits the available ice-free waterway.

Lightweight concrete caissons were employed on the Tarsiut caisson-retained island in the Canadian Beaufort Sea. Short test sections of concrete seawall, crib, and slope protection units have been installed on the gravel islands off Prudhoe Bay. A small prestressed concrete tidal power plant was installed at Kislogybusk in the Barents Sea by Soviet engineers in 1968. Concrete pillars have been used to support quays in harbors in Greenland. These, plus the lighthouses referred to earlier, constitute our experience with concrete sea structures under Arctic conditions. Thus a major effort is being carried out currently to augment this relative paucity of data by laboratory and field tests, and by examination and evaluation of relevant experience from sub-Arctic and temperate environments.

For many of the offshore concepts now under study, the determining structural consideration is that of the so-called "punching shear", due to the impact from a multi-year ice floe or iceberg on a local zone of the periphery, where forces of 600 to 1000 tons/m^2

or more may be developed over a limited area.

Intense local loads such as these and global ice loads must be transmitted back through the structure and eventually into the foundation. Experience with icebreakers (steel hulls) has shown that although the hull plating is usually not punctured, it is the stiff supporting members, such as the frames, which buckle under the high local loads.

2. RELEVANT PROPERTIES OF PRESTRESSED AND REINFORCED CONCRETE.

a) Low temperature ductility under impact.

Reinforced concrete has long been used for river bridge piers subject to dynamic ice loading during breakup. The concrete itself grows stronger (and stiffer) as the water in the pores freezes. The reinforcing steel is subjected only to longitudinal stress. The massiveness and stiffness of typical concrete structures tends to limit dynamic amplification and lengthen the period of contact with an ice feature so that the load is primarily a quasi-static one.

Prestressing tendons made of cold-drawn wire have relatively constant properties over a wide range of temperatures down to that of LNG. This includes strength, ductility, and fatigue endurance. Thus prestressed and reinforced concrete structures can be considered to be free from problems of brittle fracture under impact at low temperatures.

b) Durability.

Properly designed and constructed concrete structures develop high resistance to corrosion of the reinforcement. The concrete mix must be relatively impermeable and must ensure an alkaline environment for the embedded steel. Excessive microcracking of the surface under thermal gradients or freeze-thaw attack must be prevented. In order to limit cracking and spalling while still keeping an adequate barrier of concrete, the cover should be carefully chosen, neither too little not too great. Typical values are 50mm,(+10m, -0mm), with a reduction up to 12.5mm for stirrups.

One area of potential corrosion needs to be addressed and that is the deck, where spray plus evaporation and the intentional application of chlorides to prevent glare ice may lead to high concentrations of chloride similar to those found in bridge decks in the Eastern U.S. For these conditions, the deck should have adequate drainage. Epoxy-coated reinforcement meeting ASTM specifications should be considered.

The freeze-thaw resistance of properly air-entrained concrete has been widely attested in the Northern U.S. and Canada. In the Arctic the number of cycles of freeze-thaw for the superstructure concrete are very limited. After a few cycles in September, the exposed concrete is below freezing for the winter. Below water level the temperature stays at about -1°C, generally insufficient to cause any deep freezing of the pore water in the concrete.

However, in a zone extending from waterline up 1 to 2 meters, the concrete is potentially subject to a significant numbers of cycles each fall, freezing when exposed to air, thawing when covered by the still warm seawater waves and storm surges. This concrete is potentially saturated to a high degree, since in most cases there will be water ballast on the inside: hence freeze-thaw resistance is of special concern.

This is also the zone of potential ice abrasion-erosion. Investigation of concrete ships from World War II moored in northern Norwegian waters and in Cook Inlet, Alaska, show two cases of degradation near the waterline, being attributed to a combination of freeze-thaw cycles of saturated concrete and ice abrasion. It should be noted that a number of concrete vessels and structures exposed in a similar manner show no such problems, and hence a preliminary conclusion is that the surficial damage is due to the permeable, non-air entrained concrete employed in these particular vessels. Nevertheless, these cases of damage indicate that the exposure is not just conjectural but is real. Hence special care is warranted for these areas. Inclusion of 5-6% air entrainment, design of the mix for low permeability, and using non-saturated aggregates at time of mixing, are advisable. Three relevant proprietary studies are underway at this time, involving both laboratory and field tests.

Table 1 Recommended mix design for typical Arctic Marine Structures.

Normal weight concrete

Coarse aggregate: 3/4" max. crushed rock or natural gravel.

Natural sand: about 40-45% of total aggregate.

Cement ASTM Type II, 650 lbs: Pozzolan, ASTM Type N or F, 100 lbs.

Water - to give W/C ratio about 0.40
Water-reducing admixture
Air entrainment 6%
(Air void spacing not greater than 0.2mm).

For structural lightweight concrete, use 5/8" or ½" max. coarse aggregate. Sand may be all natural, or part natural and part lightweight, or all lightweight.

Only sealed surface lightweight aggregates, coarse and fine should be used. Lightweight aggregates should not be saturated prior to mixing.

Durability must also consider the prestressing system. Ducts must be completely grouted, not only to prevent corrosion but to prevent freezing of voids which have become water-filled. Protection of post-tensioning anchorages must be assured.

The Treat Island tests of the U.S. Corps of Engineers have not only demonstrated the need for air entrainment but also the care required for post-tensioned anchors. Recessed pockets, concrete-filled but without the use of epoxy bonding, have proven to be the most durable.

A concern has been expressed over the effects of cracks (thermal, shrinkage or structural) in the severe freeze-thaw environment. This appears to be a matter of reinforcement percentage: if the steel percentage is too low, the freezing water in the crack will both widen and extend it, whereas if the reinforcement percentage across the initial crack is sufficient to exert a closing force equal to the tensile strength of concrete, then the freezing water extrudes in the crack without any cumulative effect. The above is based on field observations, including investigation of the Tarsiut caissons where structural cracking had occurred in the walls of some caissons due to unequal support: however, the high percentage of reinforcing had prevented any measurable further deterioration from freeze-thaw action.

c) The resistance of thick-walled slabs and shells to intense local loading, whether from collision by boats or impact of ice, has been the subject of numerous laboratory and analytical investigations. Because of the low span-to-depth ratio of the shell, and the presence of high percentages of reinforcing and/or prestressing steel, both in-plane and through-wall, the behavior of the shell or slab is significantly better than given by conventional building codes, which are relevant to high span-to-depth ratios and low steel percentages.

The behavior of such a shell (slab) under high concentrated loading is influenced to a high degree by the support conditions. The more rigid the supports, the more effective will be the internal arch or dome which forms within the shell thickness.

The term "punching shear" is really improper for such a case. In reality, the initial effect is primarily flexure, with severe cracking forming opposite the load. As the reinforcing steel reaches yield, the moments are transferred to the supports. Inclined diagonal struts form within the shell. The high intensity of load on them leads to diagonal laminar cracking within the strut, just as in a column. Ultimate failure will then take place by punching shear along rather flat diagonals or, if the struts have been sufficiently confined by stirrups, by crushing at the inner face of the haunch. In some slab tests, failure has also occurred in crushing under the load.

This indicates the need for high percentages of shear steel (stirrups) and for confinement of the compression zones. Closed stirrups often fail to develop yield strength prior to laminar cracking failure: there is insufficient room to properly bond them. Use of headed stirrups or specially-fabricated stirrups, which mechanically are fixed to the in-plane steel, are therefore much more efficient. If stirrups are used their lap should occur in the confined-section rather than at the face. As in column confinement, mechanical splicing or closure is most effective.

The nominal shear strengths so developed are significantly greater than those predicted by simple application of code formulae. Since the concrete, near ultimate, will have high through-wall tension, it seems advisable to rely entirely on the through-wall steel to resist the transverse tension.

d) Thermal gradients.

The problem of thermal gradients in service have already been successfully addressed in the North Sea where the delta T between the warm stored oil and cold water is usually taken as (35°C-6°C) or 29°C. In the case of exploratory caissons in the Arctic, temperature differentials between "warmed" water inside and cold water outside will be negligible. Above sea level, however, the temperature difference between "warmed" water inside and air outside may reach 40°C, sufficient to cause cracking on the cold face. This in turn could aggravate the attacks by freeze-thaw and ice abrasion. This problem can be successfully contained by providing a minimum of about 1% reinforcing steel, both directions, using closely-spaced bars, over the entire exposed face. The CEB/FIP Model Code defines this "tensile zone" over which the percentage of steel is to be computed, as $de = C + 8\,\emptyset$ where C is the cover and $\emptyset$ is the bar diameter. What this steel does is ensure a constant closure stress acting across the crack, thus preventing the crack from growing under freeze-thaw attack. Careful observation of the Tarsiut caissons has shown this to be effective.

The prevention of microcracks in the externally exposed surface is also of importance during construction. Proper control of the mix and curing will help to reduce microcracking due to shrinkage, but the effects of heat of hydration need special attention. The thick peripheral walls will see a significant increase in temperature to about 65°C due to heat of hydration, and this will not cool to ambient temperature for 2 to 3 weeks. Insulation must be applied to the forms and concrete surfaces after stripping the forms until the differential temperature has lowered to about 20°C.

Global temperature differentials will also arise, tending to create circumferential tension in the peripheral wall. These stresses should be considered in design, especially since the shear capacity is reduced by a membrane tensile stress. Prestressing has been successfully employed at Tarsiut to maintain a net compression in the external wall.

3. APPLICABLE STANDARDS AND GUIDELINES.

A number of recent publications plus several currently in preparation form an adequate basis for the design and construction of prestressed concrete structures for the Arctic. These include:

- API - Bulletin 2N, on Planning, Designing, and Constructing Fixed Offshore Structures in Ice Environments.
- ACI - 357, currently under revision to include Arctic applications.
- FIP - Recommendations for Concrete Sea Structures, 4th edition in preparation, to include Arctic applications.
- ABS - Rules for Fixed Offshore Structures currently under revision to include the Arctic.

Other standards, such as the DNV Rules, while primarily directed to temperate zones, are nevertheless highly relevant to the design and construction of such structures for the Arctic.

As in all new developments, there are also a number of very important aspects being promulgated in technical journals and presented at technical conferences. Some of the most important of these are listed in the "references" at the end of this paper.

4. LIGHTWEIGHT CONCRETE

The principal constraint to the use of concrete structures in the Beaufort Sea is their inherent heavy weight which increases the draft, thus making it difficult to pass Point Barrow and placing a limit on their ability to be used at shallow water drilling sites. This has led to an interest in the use of structural lightweight aggregate, which enables a reduction in weight and draft of about 20%. Other properties of high quality lightweight concrete are also attractive: lower modulus, greater impermeability, probably improved freeze-thaw behavior, better thermal compatibility with the embedded steel, and less microcracking.

Recent research at the University of California at Berkeley has shown that prestressed lightweight concrete possesses excellent structural properties under cyclic and repeated loads, even under extreme thermal strains such as those induced by an LNG spill.

Lightweight coarse aggregates were utilized in the Tarsiut caissons and an inspection after one year showed their condition to be highly satisfactory. There was no evidence of freeze-thaw attack or ice abrasion and structural cracks had not propagated or widened. Some spalling had occurred due to impact from barges, etc. Lightweight concrete tends to spall as thin, flat slabs, as contrasted to normal concrete, where the spalls are deeper but cover less extent. In areas subject to such impact, use of fenders or provision of wire mesh near the surface are indicated.

A number of proprietary research projects are currently verifying the properties of structural lightweight concrete for Arctic use.

The use of structural lightweight aggregate for these shallow water Arctic caissons facilitates construction as well as deployment and hence proves economical for these structures.

5. FAILURE MODE

A principal concern in any structure subject to possible overload, especially dynamic or impact loads, is to ensure a ductile mode of failure. While concrete per se is a brittle material, heavily reinforced and confined concrete behaves in a pseudo-ductile manner. In the case of the typical peripheral ice walls of concrete, having relatively high percentages of steel in both in-plane directions as well as heavy stirrups through the wall, the effect of high overload will be to produce cracking and yield in the tensile steel, followed eventually by spalling of the cover and eventually crushing. If heavy confinement has been provided, this will not be a brittle fracture, but will produce a major reduction in stiffness. Rotation will ensue and the wall between the supports will tend to act like a catenary, in membrane tension. Based on experience from the hull plates of icebreakers, we would expect the ice load to progressively transfer to the supporting diaphragms or bulkhead frames.

Thus by proper reinforcing of the exterior walls and its supporting structure, a ductile mode of failure can be assured.

The ice wall problem is very similar to the boat/barge collision problem on shafts of the North Sea concrete platforms, where similar approaches to design have proven effective.

5. CONFIGURATIONS

One of the very attractive features of concrete for Arctic offshore structures is its ability to be economically molded and configured into complex shell shapes. Offshore structures may be constructed as cast-in-place concrete or as an assemblage of precast elements, joined after erection by concreted joints and/or post-tensioning. The first method has been characteristic of the Condeep structures in the North Sea, whereas the Ninian Central Platform utilized precast concrete shell elements extensively.

For the current applications in the Alaskan Beaufort Sea, configurations being studied range from rectangular box-like structures to highly sophisticated doubly-curved conical shells, designed to cause the ice to fail in tension.

Floating structures in prestressed concrete have been studied and appear equally attractive for deeper water applications in the Arctic. One of these, the Exxon Floating Caisson Spar (ref. 1) was described at POAC 79 Conference in Trondheim. Other potential configurations might resemble Gulf Canada's steel floating drilling structure, currently being fabricated. Prestressed concrete hulls have even been considered for Arctic transport, both as submarine pressure hulls and as surface ice-breaking tanker hulls.(2)

While early attention of engineers has been directed primarily to the design of the peripheral ice wall, it is important to also address the internal configuration by which intense loads are transmitted through the structure and into the foundation. The primary means is typically through vertical shear walls but horizontal diaphragm slabs and circumferential walls help to distribute the load throughout the structure. When these high loads (often of the order of 100,000 tons or more) reach the base, they must be transmitted into the foundation. A detailed description of this critical interface problem is beyond the scope of this paper. However, whether this is accomplished by skirts, spuds, or ribs, etc., the base slab also becomes a highly-loaded complex structural element, for which multi-axial post-tensioning has proven very efficacious in similar applications in temperate climates and in preliminary designs for the Arctic.

7. FUTURE DEVELOPMENTS.

The current development of silica-fume additives (microsilica particles) offers the potential for significant increase in concrete strength, both normal weight and lightweight, as well as an increase in impermeability.

However, such higher compressive strengths must be properly confined if brittle behavior is to be prevented. Ductility can be attained by confinement with stirrups and also by the inclusion of randomly-oriented steel fibres.

The actual behavior of the thick-walled concrete elements is highly dependent on the multi-axial stress conditions existing. While the current nuclear code permits minor advantage to be taken of favorable multi-axial stresses, tests and three-dimensional analyses indicate that their effect may be very significant and enable more efficient designs to be developed for the concentrated loads inflicted by ice. The FIP is currently undertaking an in-depth study of these phenomena.

For many structural configurations, especially cones, the ice load on the structure is determined to a high degree by the friction between ice and the surface. Use of coatings to reduce this friction and to minimize abrasion is currently under investigation in several proprietary research projects.

Coatings should not only have low friction, high bond, elasticity, toughness, and abrasion resistance, but should not cause freeze-thaw damage due to trapping of water vapor beneath them. The water vapor in the concrete will migrate to the cold face, and tend to congeal under an impermeable coating, possibly leading to spalling when it freezes. The ideal coating, then would appear to be one which is permeable to water vapor.

While there will continue to be a healthy competition between concrete structures and steel structures, the optimal solution in some cases may well turn out to be a hybrid or sandwich structure, in which steel plates of moderate thickness enclose a concrete wall, with which they are constrained to act in composite behavior through the use of shear connectors such as welded studs. Tests in Japan and the U.K. have shown that such a hybrid design is extremely efficient in resisting high concentrated load.(3)

The use of concrete in composite action to stiffen a steel hull has recently been applied by Dome Petroleum on their SSDC-1, a converted VLCC mid-body being utilized as an exploratory drilling platform in the Canadian Beaufort Sea.

8. EVALUATION.

Steel-hulled ice breakers have given satisfactory performance over a number of years. What is the justification for the use of concrete for Arctic offshore structures?

First, it is generally believed that concrete structures will give satisfactory long-term service with minimum maintenance due to their inherent durability.

Secondly, is the ductility even at low temperatures and the redundancy which can be incorporated into concrete structures by proper design and construction.

Thirdly, is the ability to resist high local concentrated loads in shell action.

Finally, and perhaps the most important practical reason, is the lower cost. Although individual structures will vary considerably, when caisson-type structures are designed to equal performance criteria, prestressed concrete caissons will usually be found to cost only 50 to 70% as much as the comparable steel caissons.

This latter fact may be inherent in the environmental loading pattern which prevents the use of steel in its most favorable forms as framed members subject to axial tension and compression. In the Arctic, the caisson size and configuration is largely determined by the ice features and the foundation soils, hence is independent of the structural material employed. Prestressed and reinforced concrete is especially well suited to shell configurations.

As experience grows with the use of concrete structures in the Arctic, it is believed that in addition to economic justification, its many concomitant advantages will enlarge its utilization for sea ice and iceberg resistant structures.

Fig. 1 Prestressed lightweight concrete caissons for Dome Petroleum's Tarsiut Island.

REFERENCES

1. Gerwick, B. C. Jr., and Jahns, H., Conceptual Design of Floating Drilling, Production, and Storage Caisson for Arctic Waters, Engineering Structures, January 1981.

2. Gerwick, B. C. Jr., Prestressed Concrete Floating Terminal for Arctic Ocean Service, Proceedings, POAC 1975, Fairbanks.

3. Montague, P., and Sang Choo, Y., Double-Skin Construction of Large Pressure Vessels for Sub-Sea Systems, Offshore Technology Conference, OTC 4106, 1981.

4. Fédération Internationale de la Précontrainte, Guides to Good Practice, "Practical Construction" (Tendons, Prestressing, Grouting, Anchorages, Corrosion Protection.) FIP/2/1 Sept. 1975 "Grouting of Vertical Ducts", FIP/2/4 March 1978.

5. Gerwick, B. C. Jr., Causes and Prevention of Problems in Large-Scale Prestressed Concrete Construction, Journal, Prestressed Concrete Institute, May/June 1982.

6. Mehta, P. K., and Gerwick, B. C. Jr., Cracking-Corrosion Interaction in Concrete Exposed to the Marine Environment, Concrete International (ACI), October 1982.

7. Gerwick, B. C. Jr., Resistance of Concrete Walls to High Concentrated Ice Loads, Offshore Technology Conference Preprints, OTC 4111, 1981.

8. Proceedings, International Conference on Preformance of Concrete in the Marine Environment, ACI Special Publication SP 65, American Concrete Institute.

a. Stratfjord A Condeep.
Base raft constructed in basin.

b. Completion of main cells afloat.

Figure 2 Typical sequence in construction of North Sea Offshore Platform of Prestressed Concrete.

c. Construction of shafts.

d. Completion of shafts.

e. Transfer of pre-assembled integrated deck.

Figure 3 Ninian Central platform utilizes both precast and cast-in-place concrete construction.

Figure 4 Prestressed lightweight concrete caissons are towed to Canadian Beaufort Sea, to form Tarsiut caisson-retained island.

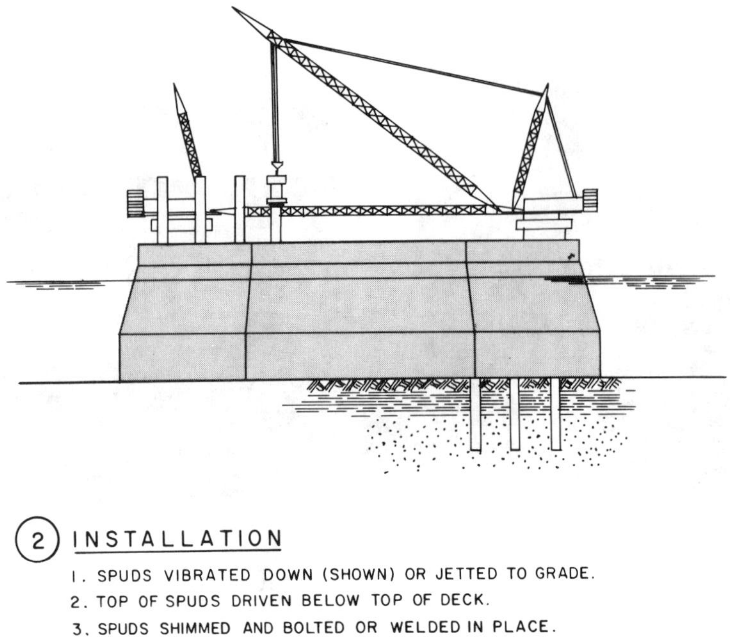

a.

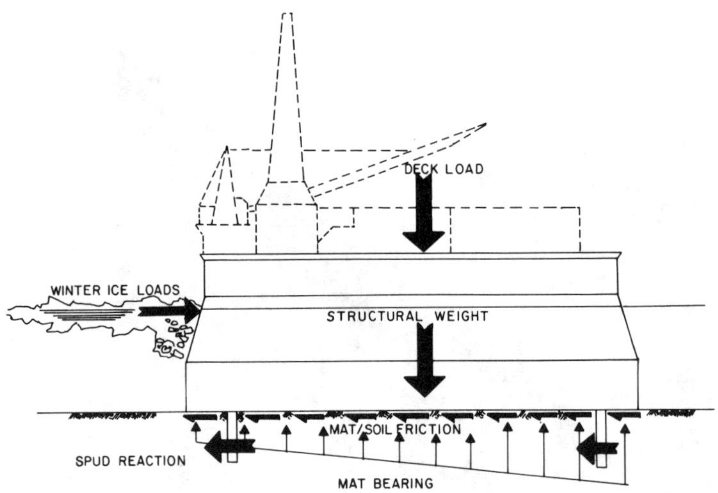

b.

Figure 5 SAMS concept utilizes a prestressed lightweight concrete caisson with added shear transfer by steel spuds.

A REVIEW OF ICEBREAKING TECHNOLOGY

George P. Vance

Mobil Research and Development Corporation
Offshore Engineering
Farmers Branch, Texas 75234

ABSTRACT

This paper traces the history of icebreaking technology from its early stages of educated guesses by ship designers to the sophisticated computer techniques utilized by current day Naval Architects and Structural Engineers. The paper covers the broad spectrum from the analytical techniques utilized to predict ship resistance to the methods used to predict the forces on offshore steel and concrete structures and gravel islands.

1. INTRODUCTION

The first documented attempt to analyze the process of breaking ice was presented by Runeburg (1) in 1888. In 1921 Kari (2) and in 1936 Simonson (3) published equations related to the downward force developed by the bow of an icebreaker. In 1938 Shimansky (4) attempted to incorporate the complete fore body shape of the vessel. The current evolution of icebreaking technology began with the work of Vinogradov (5) in 1958 when he attempted to employ the approach of conservation of energy to determine the resistance encountered by a vessel. White (6) in 1965 synthesized these past efforts to determine a more effective bow shape. In 1968 the Russian team of Kashteljan, Poznjak and Ryvlin (7) published an extensive work that described their efforts in combining theoretical and empirical attempts to accurately predict the resistance of a vessel in ice. In 1970 Lewis and Edwards (8) improved upon these techniques. In 1972 Enkvist (9) presented the results of the work carried out in Finland. That same year Milano (10) reinvistigated the energy approach to the problem and published a computer algorithm that incorporated all of the parameters of the bow shape.In 1974 Vance (11) published an improved empirical equation that incorporated the length of the ship. In 1978 a technique was derived,but not published, by Arctec and Mitsui, the program was entitled ICEREM. The latest effort in predicting icebreaking resistance was published by Naegle in 1980 (12) in which he examines the equations of motion involved in the icebreaking process.

The history of ice forces on structures is not as defined as that of vessel icebreaking. The problem is more diversified and varies from the determination of the vibration characteristics of slender structures in thin ice sheets to that of impact of large icebergs with large structures. In addition, the efforts in determining ice forces on structures is more recent and are still in the development stage.

Some of the earlier efforts were conduct- by Peyton(13) in the 1960's in relation to ice forces on oil production platforms in Cook Inlet, Alaska. In the 1970's Michell and Toussaint (14) published information on the forces involved in the penetration of an ice sheet by various shaped structures. In 1980 Crosdale (15) published the intial efforts conducted to determine the forces generated by pressure ridges interacting with a structure. In 1977 Ralston (16) published equations to determine the limiting forces involved with an ice sheet interacting with a cone shaped structure.

Although proprietary work is being conducted in such projects as Dome's Hans Island project and Gulf's Tarsiut Island project,very little has been published related to impact forces on structures. A simplified approach to the problem was published by Cammaret (17) in 1981. Additional work was also published in 1981 on pressure ridges by Prodanvic (18) and Gerwick (19). Very little has been published on the interaction of single and multiple leg structures with broken ice. This problem becomes important when investigating the problem of floating structures in a broken ice field.

In summary the ice forces on a structure can be broken down into the following broad categories:

Penetration of a solid ice sheet by a vertical sided structure

Penetrattion of a solid ice sheet by a cone shaped structure

Impact of a large ice mass with a structure

Interaction of a pressure ridge with a stucture

Interaction of a multi leg structure with broken ice.

The following sections will discuss some of the techniques mentioned above and provide some insight into their utility, however, the details of their application can be found in the authors original publication.

2. ICEBREAKING RESISTANCE

The methods discussed here are those that relate to icebreaking resistance in a homogenous ice field. The work conducted in a broken ice field and in pressure ridges, although just as important in the overall scheme of icebreaking resistance, has not been adequately verified with full scale data to have any particular technique accepted.

In level ice resistance prediction, the presently accepted analytical techniques are those proposed by Kashteljan (7), Lewis and Edwards (8), Milano (10), Vance (11) and Naegle (12) in addition to the proprietary technique available through Arctec and Mitsui. The governing equations and some typical results are presented here.

The Kashteljan (7) equation is given as:

$$R = 0.004 B h \sigma \mu_0 + 3.0 \gamma_i h^2 \mu_0 + 0.23 B^{1.65} h V / \eta_2 + R_w \qquad (1)$$

where:

- R- total resistance
- B- maximum waterline beam
- h- ice thickness
- σ- ice flexual strength
- μ_0- hull efficiency= $1 + 1/\eta_1$
- η_1- icebreaking coefficient (see (7))
- η_2- icecutting coefficient (see (7))
- γ_i- specific weight of ice
- V- ship speed
- R_w- open water resistance

In 1970 Lewis and Edwards (8) published the following equation:

$$R_i = 0.146 \sigma h^2 + 8.84 \rho_i g B h^2 + 5.905 \rho_i B h V^2 \qquad (2)$$

where:

- R_i-ice resistance
- ρ_i-mass density of ice

and the remaining parameters are as defined previously.

In 1972 Edwards and Lewis (20) published a purely empirical equation that included the effect of snow cover, however, that equation will not be presented here.

In 1974 Vance (11) presented an equation that included the effect of length:

$$R_i = C_s \rho_\Delta g B h^2 + C_b \sigma B h + C_v \rho_i V^2 L h^{0.65} B^{0.35} \qquad (3)$$

where:

- L -length between perpendiculars
- ρ_Δ-density difference between ice and water
- C_s-submergence coefficient
- C_b-breaking coefficient
- C_v-velocity coefficient

the remaining parameters are as defined previously. The coefficients for various ships are presented in the basic reference.

The algorithms of Milano (10) and Naegle (12) are to complicated to present in this paper and are available in the basic reference. The details of ICEREM are proprietary to Arctec and Mitsui. Comparison of the various techniques are presented in figures 1 through 4.

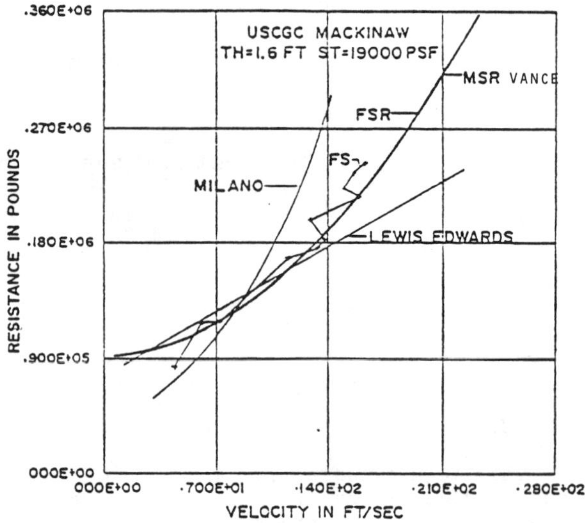

Fig. 1 Comparison of Mackinaw resistance predictions (Vance 1974, fig. 6.3)

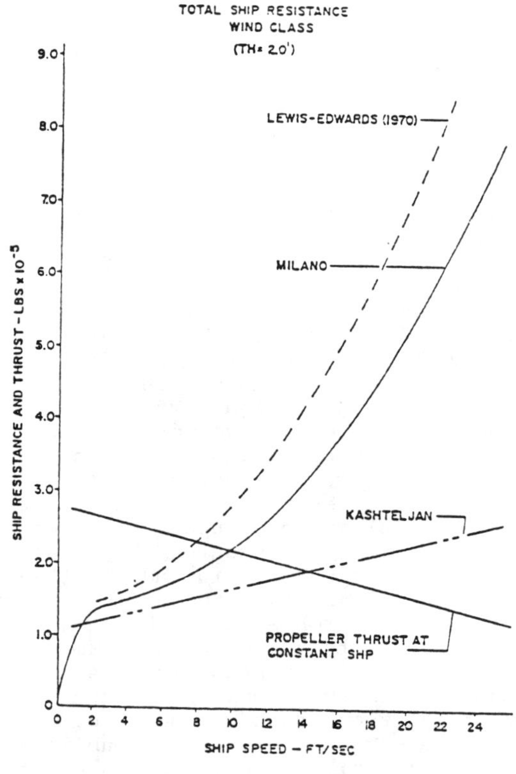

Fig. 2 Comparison of ice-resistance predictions for Wind class (Vance 1974, fig. 1.7)

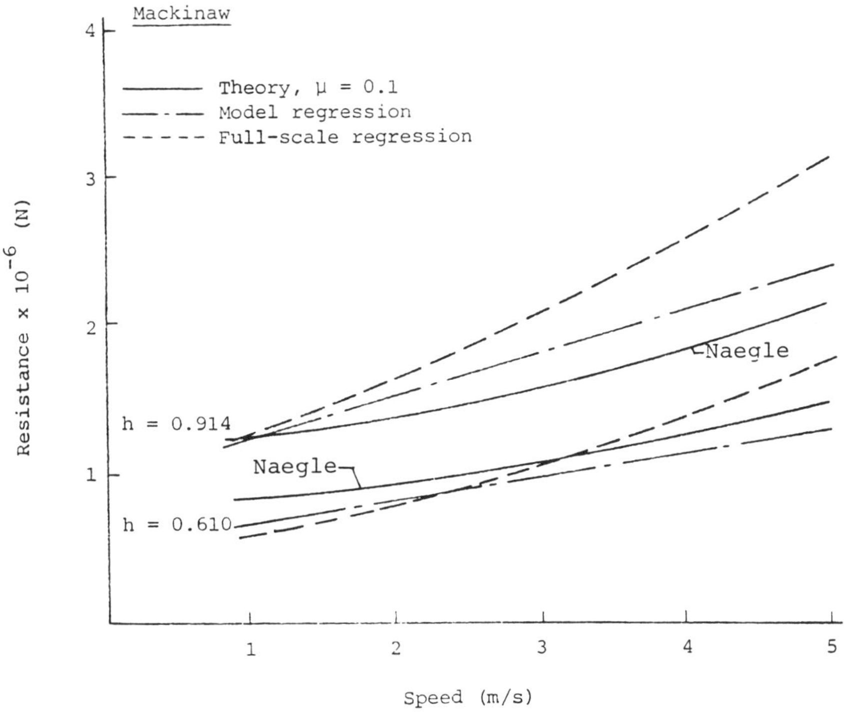

Fig. 3 Total resistance as a function of ship speed for Mackinaw (Naegle 1980)

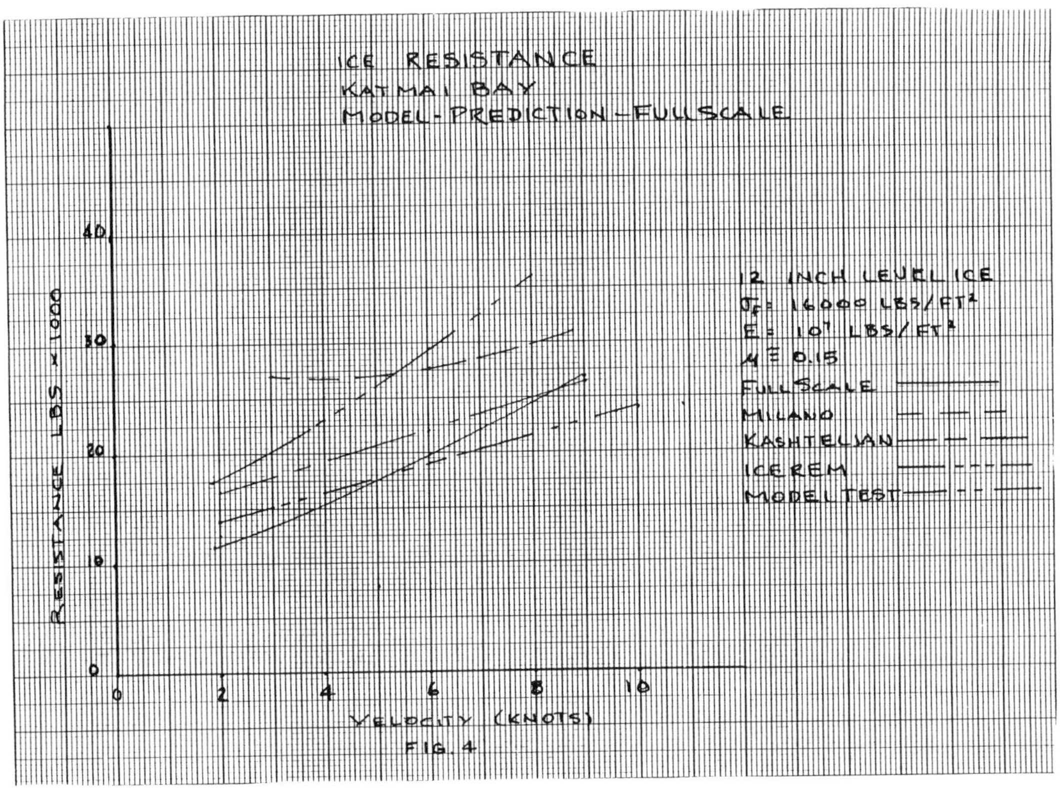

FIG. 4

3. FORCES ON A STRUCTURE - INDENTATION

The most comprehensive work on ice indentation was reported by Michel (21) and Toussaint (14). They presented the equation:

$$F = I\ M\ K\ \sigma_c\ t\ w \qquad (4)$$

where:

F- ice crushing force
I- indentation factor
M- shape factor
K- contact factor
σ_c- uniaxial compressive strength of ice
t- ice thickness
w- ice-structure contact width

M can be taken as 1 for a flat surface and 0.9 for a cylinder and 0.85 $(\sin\alpha/2)^{0.6}$ where α is the included angle in a wedge shaped structure. I, K and σ_c are functions of the strain rate. For narrow structures such as columns and pier heads the strain rate is taken as:

$$\dot{\varepsilon} = V/4w \qquad (5)$$

where V is the velocity of indentation, i.e. the ice velocity. For wider structures such as gravel islands Ralston (22) recommends that the strain rate be set equal to:

$$\dot{\varepsilon} = V/2w \qquad (6)$$

For $10^{-8} < \dot{\varepsilon} < 5x10^{-4}$ I can be taken as 2.97 and K as 0.6 and

$$\sigma_c = 7.5x10^5\ (\ \dot{\varepsilon}\ /5x10^{-4})^{0.32} \qquad (7)$$

For $5x10^{-4} < \dot{\varepsilon} < 10^{-2}$ I can be taken as 2.97 and K as 0.26 and

$$\sigma_c = 7.5x10^5\ (\dot{\varepsilon}\ /5x10^{-4})^{-0.126} \qquad (8)$$

For $\dot{\varepsilon} > 10^{-2}$ I can be taken as 1.57 and K as 0.5 with σ_c =5.176x10^5 pascals.

4. FORCES ON A STRUCTURE - CONES

In 1977 Ralston (16) published his plastic analysis of an ice sheet interacting with a conical surface. He presents the solution in the form of the following eqation:

$$R_h = (A_1\sigma_f\, t^2 + A_2\rho_w g\ t\ D^2 + A_3\ \rho_w gt(D^2 - D_T^{\ 2}))\ A_4 \qquad (9)$$

$$R_v = B_1\ R_h + B_2\ \rho_w gt(D^2 - D_T^{\ 2}) \qquad (10)$$

where:

R_h - horizontal force
R_v - vertical force
D - waterline diameter of cone
D_T - top diameter of cone
σ_f - flexual strength of ice
μ - coefficient of friction ice/cone
t - thickness of the ice sheet
g - gravitational acceleration
ρ_w - mass density of water

The values of A_1, A_2, A_3, A_4, B_1 and B_2 are given in figure 5. The predictions provided by these equations have been compared to model test results and show good agreement as shown in figure 6.

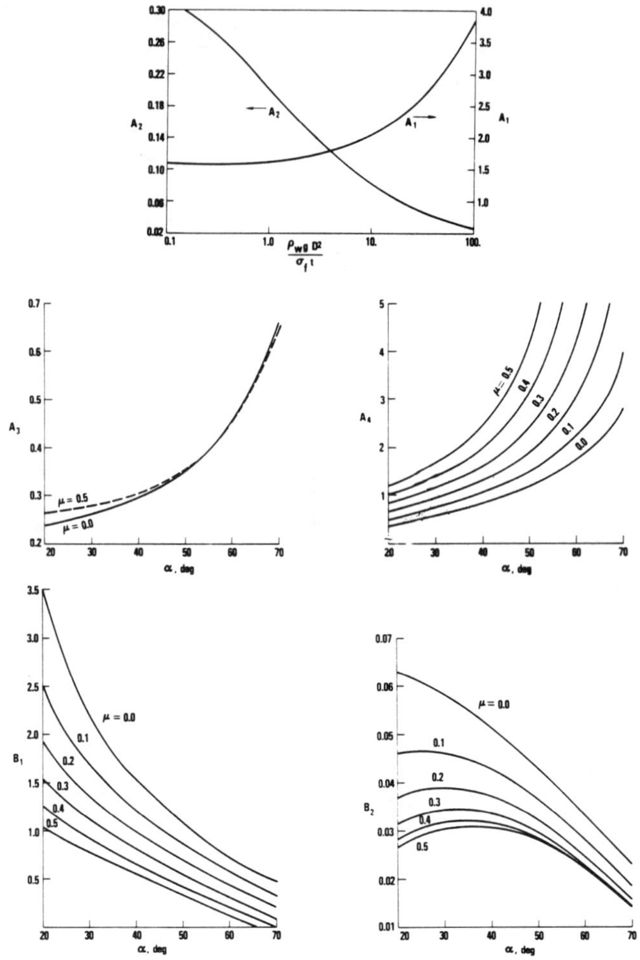

FIG. 5 ICE FORCE COEFFICIENTS FOR PLASTIC ANALYSIS.
(RALSTON 1977)

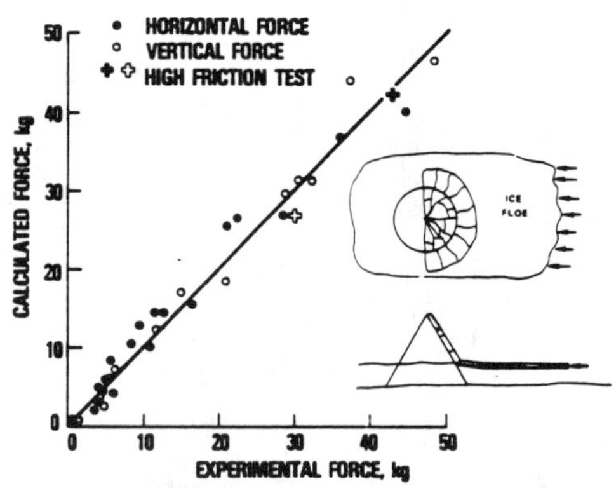

FIG. 6. COMPARISON OF THE PRESENT ANALYSIS WITH EDWARDS AND CROASDALE'S (1976) MODEL TEST DATA.
(RALSTON 1977)

5. FORCES ON A STRUCTURE - IMPACT

The impact of an ice mass with a structure involves many parameters that will effect the maximum force generated upon impact. Such factors as the shape of the ice mass, the shape of the structure, the compressive strength of the ice as well as the total mass and velocity of the ice will have an effect on the force generated. Cammaert (17) presented a simple approach that equated the kinetic energy of the ice to the energy dissipated during crushing of the ice. Under the assumption that only one half of the kinetic energy of the ice mass is available for crushing and that the ice mass is tabular and of greater width than the cylindrical structure and utilizing an added mass coefficient of 1.5, he arrives at a maximum force of:

$$F = 2D\,\sigma_c\,(2RX_m - X_m^{2})^{0.5} \qquad (11)$$

where:

- F - maximum impact force
- D - the total depth of the ice mass
- R - the radius of the cylindrical sturcture
- σ_c - the uniaxial crushing strength of ice
- w - the wieght of the ice mass
- V - the velocity of the ice mass
- g - gravitational acceleration

$$X_m - (0.198\ w\ V^2/D\ \ R^{0.5}g)^{0.5}$$

Very little model testing has been undertaken to verify this approach. Considerable work must be undertaken to examine the phenomenia in more detail.

6. FORCES ON A STRUCTURE - PRESSURE RIDGES

The problem of forces generated by pressure ridges interacting with a structure is significantly more complex that that of a homogeneous ice sheet. Various solutions have been presented in the literature but none have been truely verified. Vaudrey (23) presented a simple shear model where the maximum force on a column is generated by double shearing of the ice ridge by the column such that the force is given by:

$$F = 50\ S^2\ \sigma_s\ (1-P) \qquad (12)$$

where:

- F - maximum force on the column
- S - sail height of the ridge
- σ_s - shear strength of the ridge
- P - porosity of the ridge

A more sophisticated model,utilizing a finite difference technique was presented by Gerwick (19). This method takes into consideration various stages of failure of the ice sheet as well as the ridge failure itself. There are many other references relating to ridge forces on structures, however, there is little full scale or model scale verification of these various techniques. The properties of the pressure ridge itself must be better understood before any model can be developed and verified.

7. FORCES ON A STRUCTURE - BROKEN ICE

The area of ice forces on a structure from broken ice is even less developed than that of the pressure ridge. There have been presentations using the soil mechanics failure plane approach, an ice on ice friction approach and a shear failure approach, however, none have been developed to any great degree nor have they been verified by model or field test. The work is in the early stages of development and no comprehensive work has been published to date.

8. CONCLUSION

I have attempted to present a very brief historical review , along with some insight, into the development of icebreaking technology from its early begining with ship resistance to its current application to offshore structures. It is a fascinating and complex area of study. Although the techniques that I have presented are being used in the field today, they are by no means the final answer or for that matter, the correct answer. There are questions related to interaction effects, added mass effects, scale effects and hydrodynamic effects that have to be investigated in much more detail. The field is wide open in many areas, particularly in the development of analytical models and their verification by model and full scale testing. Much of this work is currently being carried out by the oil industry within their internal research and development programs in order to assure a safe and cost effective development of this countries arctic natural resources. However it will take the combined effort of industry, government and academia to answer all the questions that remain to be answered.

The author would like to express his appreciation to Mobil Oil Corporation for their cooperation in the publication of this paper and would like to point out that the opinions expressed in the paper are those of the author and do not necessarily reflect those of Mobil Oil Corporation.

REFERENCES

1. Runeburg, R., On Steamers for Navigation and Icebreaking, Proceedings of The Institute of Civil Engineers, vol. 97, Part 3, 1888.

2. Kari, A., The Design of Icebreakers, Shipbuilding and Shipping, December 1921.

3. Simonson, D.R., Bow Characteristics of Icebreaking, Journal of the American Society of Naval Engineers, 48 (2),1936.

4. Shimansky, J.A., Conditional Standards of Ice Qualities of a Ship, Transactions of the Arctic Institute of the Chief Administrator of the Northern Sea Route, Vol. 130, 1938.

5. Vinogradov, I.V., The Icebreaker, Schiff und Hafen, January 1958.

6. White, R.M., Dynamically Developed Force at the Bow of an Icebreaker. D.Sc. thesis-Massachusetts Institute of Technology, 1965.

7. Kashteljan, V.I., Poznjak, I.I., and Ryulin, A., Ice Resistance to the Motion of a Ship, Sudostroenie, Leningrad,1968.

8. Lewis, J.W., and Edwards, R.Y., Methods for Predicting Icebreaking and Ice Resistance Characteristics of Icebreakers, SNAME Transactions, Vol. 78, 1970.

9. Enkvist, E., On the Ice Resistance Encountered by Ships Operating in the Continuous Mode of Icebreaking, Swedish Academy of Engineering Science in Finland, Report No. 24, 1972.

10. Milano, V.R., Ship Resistance to continuous Motion in Ice, Ph.D. thesis, Stevens Institute of Technology, 1972.

11. Vance, G. P., A Modeling System for Vessels in Ice, Ph.D. thesis, The University of Rhode Island.

12. Naegle, J.N., Ice Resistance Prediction and Motion Simulation for Ships Operating in the Continuous Mode of Icebreaking, Ph.D. thesis, University of Michigan.

13. Peyton, H.R., Sea Ice Forces, National Research Council of Canada Technical Memo 92, 1968.

14. Michel, B., Toussaint, N., Mechanisms and Theory of Indentations of Ice Plates, Journal of Glaciology 19, No. 81, 1977.

15. Croasdale, K.R., Some Implications of Ice Ridges and Rubble Fields on the Design of Arctic Offshore Structures. NRC of Canada Workshops on Sea Ice, Ridging and Pileup., Tech Memo. No. 134,1980.

16. Ralston, T.D., Ice Force Designs for Conical Offshore Structures, Proceeding of POAC Conference, 1977.

17. Cammaert, A.B. and Tsinker, G.P., Impact of Large Ice Floes and Icebergs on Marine Structures, Proceedings of POAC Conference, 1981.

18. Prodanovic, A., Upper Bound of Ridge Crushing Pressure on Structures. Proceedings of POAC Conference, 1981.

19. Gerwick, B.C. and Rojansky, M., Failure Modes and Forces of Pressure Ridges Acting on Cylindrical Towers, Proceedings of POAC Conference 1981.

20. Edwards, R.Y. and Lewis, J.W., Full Scale and Model Tests of a Great Lakes Icebreaker, SNAME Transaction, Vol. 80, 1972.

21. Michel, B., Ice Mechanics, University of Laval Press, 1978.

22. Ralston T.D., Technical Seminar on Alaskan Beaufort Sea Gravel Island Design, Exxon Report, October 1979.

23. Vaudrey, K., Southern Bering Sea Production System Study, PMB Systems Engineering, Inc., 1982.

ARCTIC SCIENCE AND ENGINEERING I

INTRODUCTION

G. Leonard Johnson

Office of Naval Research
Arlington, Virginia

Oceanography/Meteorology

One of the key questions in understanding the cryosphere and its effect on both weather and long term climate variability in polar and sub-polar regions is what process(es) control(s) the geographic position of the edge of the ice. Certainly the ice edge is governed by response to winds, currents and vertical and horizontal heat fluxes to form a coupled system. At present, no acceptable theory exists to describe this coupled system, including the annual and interannual variations of the ice cover and their relation to atmospheric and oceanographic circulations. The MIZEX international effort as presented by Dr. Johannessen, et al is designed to address this complex question. The region of study has been fertile ground for Scandinavian oceanographers/meteorologists since the late eighteen hundreds and it is therefore appropriate that Dr. Ola Johannessen be the lead scientist as a continuation of this tradition.

Looking back in time the following leaders in physical oceanography and meteorology have worked in the Norwegian-Greenland Seas and formed the "Scandinavian School":

Norwegians:

H. Mohn (1835-1916): Derivation of current field in Norwegian Sea by mathematical calculations of temperature/salinity fields.

F. Nansen (1861-1930): explorer, scientist, humanist. First noticed that sea ice drifted 45° to right of wind direction and reasoned that ice might make the upper layer of water move even more toward right (Ekman Spiral). With Helland-Hansen, published the classic work on descriptive oceanography in the Norwegian/Greenland Sea.

V. Bjerknes (1862-1951): Developed circulation theorem (also used in oceanography) and made great contribution to modern weather forecasting (Mosby, 1983).

B. Helland-Hansen (1877-1957): equations to compute the relative field of currents from the observed field of mass - if one neglects friction - currents can be calculated at any depth if density distribution is known.

H. U. Sverdrup (1888-1957): Leader of Maud expedition in the Arctic Ocean, developed the basis for modern wind-current theory.

J. E. Fjeldstand (1985 -): Several theories, primarily for internal waves and heat transfer.

H. Mosby (1903-):Research in the Antarctic and Arctic especially bottom water formation.

Danish:

M. Knudsen (1871-1949): Determination of sea-water constants, "Knudsen's Tables".

Swedish:

V. W. Ekman (1874-1954): Ekman's current meter. Basic theory for ocean circulation taking into account the earth rotation (Ekman Spiral).

J. W. Sandstrom (1874-1947): Ocean dynamics theory developed with Helland-Hansen.

C. G. Rossby (1898-1957): Rossby waves.

It is hoped that MIZEX will provide the impetus for a renaissance in Arctic and sub-Arctic oceanography/meteorology to carry on the classic works of the late 1800's and early 1900's.

The second major area of interest in the Arctic is in paleoclimatology, paleoceanography and climate. The role of the Arctic Ocean on world climate is unknown and is likely of major significance. Key questions include: What was the role of this ocean on global climate in Mesozoic and early Cenozoic time? When was glaciation initiated in the Arctic Ocean and was it contemporaneous with the Antarctic? The Arctic Ocean, with its deep basins and sluggish bottom water and surface currents may have an undisturbed long geologic record of polar events which is not the case around Antarctica (Jackson, pers. comm, 1983).

Geology/Geophysics

Knowledge of the geologic structure and history of the Arctic is of more than regional interest, in that its plate tectonic evolution is closely tied to the North Atlantic and Pacific as well as to the continents that surround it. Arctic marine geology is at least an order of magnitude less known than any other area (Keen and Falconer 1978) and the econimic potential is high. The paper by Lawver et al reviews the current hypotheses of the origin of the Arctic. As the authors point out there are a number of first order geologic problems in the Arctic including the motion or lack thereof along the Nares Strait and the genesis of the Canada Basin, especially the northern section including Alpha Ridge. Hopefully the upcoming Canadian Field effort (CESAR) will provide some answers (Sweeney, 1983).

The Arctic requires innovative approaches to work within the rugged environmental constraints. Dr. Baggeroer's presentation will describe the cutting edge of geophysîcal techniques.

In my opinion major areas of interest in the next decade will be scientific drilling. Valuable information that drilling could provide to the tectonic framework is dating. Long continuous sediment cores and bedrock samples are desperately needed to provide timing constraints. No dates are available for the formation of the Lomonosov or Alpha Ridges. A date is badly needed to aid in plate reconstructions that involve the geology of Alaska and the Queen Elizabeth Islands as is apparent in the article by Lawver et al. Drilling could provide information of the composition and facies of the Cenozoic and Mesozoic deposits and crustal structure of the ridges. Drill core samples along the polar shelf would provide data on sediment thickness, and age distributions, history of margin vertical motions, pre-Aptian history, history of the Beaufort Shelf, location and facies of the Paleozoic continental margin (Sweeney, 1981).

Another area which I feel has a large growth opportunity will be the use of unmanned intelligent vehicles to perform a myriad of tasks from engineering services to scientific research beneath the ice.

Acknowledgements

Historical data were kindly supplied by E. A. S. Svendsen and A. Foldvik, University of Bergen. Special thanks are due Donna Parker for manuscript typing.

References

1. Grantz, A., Johnson, L., and Sweeney, J. F., The Arctic: In Perspectives in Regional Geological Synthesis, A. R. Palmer ed., D-NAG Spec. Publ 1, Geol. Soc. of Amer. Boulder, CO, 105-116, 1982.
2. Johannessen, O. M., Hibler, W. D., Wadhams, P., Campbell, W. G., Hasselmann, K., and Dyer, I., Marginal Ice Zones: A description of air-ice-ocean interactive processes, models and planned experiments, 1983, (this volume).
3. Keen, C. E., and Falconer, R. F. H., Continental Margins of the Arctic Ocean: in Continental Margins, Geological and Geophysical Research Needs and Problems, Nat'l Acad. of Science, Wash. D.C., pp 102-107, 1979.
4. Lawver, L. L., Grantz, A., and Meinke, L., the tectonics of the Arctic Ocean, 1983, (this volume).
5. Mosby, H., Historical Foreward. in: The Nordic Seas, B. Hurdle ed. Springer Press (in press) 1983.
6. Sweeney, J. F., Arctic Research for the 1980's, an Expose of Problems. Geoscience Canada, 8, NO. 4, pp. 1-5, 1983.

MARGINAL ICE ZONES: A DESCRIPTION OF AIR-ICE-OCEAN INTERACTIVE PROCESSES, MODELS AND PLANNED EXPERIMENTS

BY

Ola M.Johannessen : Geophysical Institute, University of Bergen, Norway.

William D.Hibler,III U.S.Army Cold Regions Research and Engineering Laboratory,Hanover, New Hampshire,U.S.A.

Peter Wadhams : Scott Polar Research Institute,Cambridge, England.

William J.Campbell : U.S.Geological Survey, Tacoma,Washington, and University of Puget Sound, U.S.A.

Klaus Hasselmann : Max-Planck Institute für Meteorologie,Hamburg, Federal Republic of Germany.

Ira Dyer : Massachusetts Institute of Technology,Cambridge, Massachusetts, U.S.A.

ABSTRACT

The marginal ice zones (MIZ) are regions where temperate and polar climate systems interact, resulting in strong horizontal and vertical gradients in the atmosphere and the ocean. These gradients lead to mesoscale processes which affect the heat, salt, and momentum fluxes at the ice margin. It is therefore important to increase our understanding of these processes in order to model the air-ice-ocean system in the MIZ, and to build up a predictive capability of the ice margin. Parameterization of these processes is also necessary in large scale modeling of the sea ice influence on the global climate system. This paper will review our knowledge of physical processes occuring in the marginal ice zones, point out problem areas and describe Marginal Ice Zone Program (MIZEX) to be initiated in 1983.

1. INTRODUCTION

Several important phenomena in air-ice-oceanography interaction were first discovered in the Arctic Ocean at the end of the last century. For example during Fritjof Nansens FRAM Expedition across the Polar Ocean in the years 1893-1896, Nansen discovered that the ice was drifting some 30° to the right of the surface wind. After Nansen's suggestion Ekman (1902) investigated this problem from a theoretical point of view and developed the Ekman wind driven theory which predicted that the current turned to the right with depth decreasing expotentially, resulting in the so-called Ekman spiral. The spiral was also first quantitatively verified from observation in the Arctic Ocean, Hunkins (1966) some 60 years after the theory had been formulated. Another interesting phenomenon, important for the deep circulation in the Atlantic Ocean was Nansen's theory of bottom water formation, in the Greenland Sea, published in the classical work by Nansen (1906) and Nansen and Helland Hansen (1909).

In more recent time during the last two decades a series of large projects culiminating in the Arctic Ice Dynamics Joint Experiment (AIDJEX), Pritchard(1980) yielded considerable understanding of the growth, motion and decay of sea in the interior of the Arctic Ocean. With these experiments concluded, and successful non-linear sea ice dynamic models in hand (Hibler 1979, Coon 1980), attention shifted to the problem of understanding the processes which occur near the open ocean boundaries of polar icefields (the marginal ice zone) and which determine the advance and retreat of the sea ice edge.

A workshop in Monterey in 1979, Andersen et al(1980), summarized the extent of the problems and the paucity of our knowledge in the Seasonal Sea Ice Zones. Other reviews which are dealing with physical oceanography in both the western and eastern Arctic and Subarctic Seas are reported by Coachman and Aagaard (1974) and recently by Rey (1982).

In this paper we have selected to deal with the significance of the marginal ice zones (MIZ), a description of air-ice-ocean interactive MIZ processes and models and a brief overview of a multinational Marginal Ice Zone Program (MIZEX) to be initiated in 1983, in the Fram Strait/Greenland Sea, Johannessen et.al. (1983).

2. SIGNIFICANCE OF THE MARGINAL ICE ZONE

The marginal ice zone is a significant region in two senses, firstly as a location for man's activities and secondly as an important geophysical boundary zone involving energy exchanges which require parameterization in larger-scale ocean-atmosphere models.

The MIZ is subject to fluctuations due to short-term forcing (e.g. cyclone passages, eddy generation) and to longer-term factors (seasonal and interannual). Successful modeling and prediction of variations in ice edge position and ice concentration would be of great value in furthering man's activities in the region. There are several areas of special interest.

2.1. Arctic Navigation

Present and future developments in offshore Arctic oil exploration, in seaborne transport of Arctic resources (e.g. liquefied natural gas, iron ore), and in the supplying by sea of rapidly growing Arctic communities all required a much better predictive capability for ice conditions. We need the ability to predict the blockage or opening of ports and channels, the opening or closure of shore leads or large leads within the pack, changes in the motion and concentration of the ice, and the development of any ice edge anomalies.

Typical ice-strengthened cargo vessels, for instance, can proceed even in multi-year ice so long as the ice field is open, but encounter difficulties in consolidated pack of any age. The richest fisheries in the North Atlantic lie close to the ice margin, and the development of Antarctic krill harvesting will produce an increase in ship activity close to the Antarctic ice edge; in both cases a predictive capability for the ice margin is highly desirable.

2.2. Biology

The biological regime in the MIZ is scarcely assessed at all, and a knowledge of the impact of ice margin processes of biological productivity would be of great value to the development of fisheries.

2.3. Naval Operations

The upper ocean in the MIZ is a region of extreme acoustic variability as well as having a high ambient noise level due to ice floe collisions. These effects interfere with the propagation of underwater sound.

2.4. Climate

An indication of the variability of the MIZ in the Greenland Sea, both from month to month and from year to year, is given by Figure 1,which shows mean and extreme limits for winter and summer over a 10-year period.

As a geophysical boundary zone the MIZ is unique in the complexity of the vertical and horizontal air-sea-ice energy interaction which take place there. In response to these the ice edge moves hundreds of kilometers north and south on a seasonal cycle. If the physical processes which occur on the mesoscale in the MIZ can be parameterized and included in large-scale models such that the ice edge motion can be understood,the results will be valuable not only to man's immediate activities but also to the study of the hypothetical response of the ice-covered oceans to major global disturbances. We would then be able to answer the question: Where would the ice edge lie if significant changes in certain energy fluxes occurred (e.g. effect of a dust veil due to volcanic eruption or meteoric impact, effect of a major increase in CO_2 or other atmospheric pollutants?)

There have already been some empirical studies which demonstrate a strong correlation between ice margin variations and interannual atmospheric variability (Walsh and Johnson 1980, Kukla and Gavin 1981, Vinnikov et al. 1981), and CO_2 sensitivity simultations by Bryan et al. (1982) and Manabe and Stouffer (1979) which indicate how strongly the insulating effect of sea ice affects the polar regions' climatic sensitivity To proceed further in these important areas of research, it is essential to have a better grasp of the physical processes which govern the ice edge position.

3. MIZ PROCESSES

3.1. Ice Dynamics And Thermodynamics

In the ice-covered oceans the growth, drift and decay of sea ice significantly modify the atmosphere-ocean interaction. The main effects are:

i) modification of the thermal fluxes at the air/sea interface
ii) modification of the buoyancy (salt) fluxes at the ocean surface
iii) modification of the surface albedo
iv) modification of the air-sea momentum exchange due to the ice interaction.

These modifications are particularly pronounced near the ice edge where the transition from ice to no ice occurs, and are further enhanced by the fact that the ice edge is in dynamic rather then static equilibrium. Specifically, in the presence of a free ice edge, advective effects can transfer ice to the MIZ to be rapidly melted.

The nature of the ice in the MIZ is different from the interior pack, because of its greater freedom of movement and also because it is broken up by incident waves and swell into discrete floes which are small (about 30 m diam) close to the ice edge and which are larger at deeper penetrations where the wave field has been attenuated. These floes contain fragments of the original pressure ridges which traversed them when they were in the interior of the pack, but it appears from submarine sonar profiles that considerable erosion of the ridge keels has occured. The combination of less ridging with a greater number of floe edges produces air-ice and ice-water drag coefficients which are different from the interior pack and which seem, from the slender evidence available, to be somewhat higher (Johannessen 1970, Smith et al. 1970). The characteristic floe size distribution also affects the ice dynamics by determining the rate of the floe collisions by which kinetic energy is redistributed within the icefield, and affects the thermodynamics by enhancing the melt rate in summer (through lateral melting around floe edges) and the growth rate in winter (through the incessant opening and closing of new open water areas). Figure 2 shows a typical scene at a compact ice edge composed of small floes broken up by wave action.

3.2. Oceanography

Oceanographic conditions in the MIZ are dominated by permanent and transient frontal systems, by eddies, and by upwelling events along the ice edge. Vertical fine structure (10 m) and mesoscale (100 m) structures formed by interleaving of Polar and Atlantic Water intrusions are also frequently observed in the Greenland Sea MIZ. These phenomena interact with the ice pack and the atmosphere. For example, salinity fronts off the ice edge develop strongly during summer due to meltwater input; eddies along the ice edge shed ice off into warmer water, thereby providing an ice export mechanism; surface boundaries of ocean fronts will limit ice extension particularly during winter because the ice will melt when forced across the boundary into warmer water by wind.Wind-driven upwelling along the ice edge is dependent on the ice roughness, the stability of the atmospheric surface boundary layer, and the ice interaction. Furthermore, there is a strong coupling between the ice pack and the oceanic mixed layer below it.

Fronts may be strong and permanent, such as the East Greenland Polar Front which separates the cold,low-salinity, southward-flowing East Greenland Current from the more saline water in the Greenland Sea, Fig. 3 more transient, such as ice edge meltwater fronts observed north of Svalbard, Fig. 4. The wave length of meanders observed on the meltwater front is in the order of 20-40 km (Johannessen et al. 1983), while the meander scale of the East Greenland Polar Front is longer (60-100 km). Several investigators, e.g. Perdue (1982) have established that the location of this front is correlated with the continental slope in the Greenland Sea, thereby implying bathymetric steering.

Along the fronts and the ice edge, eddies have been observed which develop from frontal meanders. In the MIZ region north of Svalbard the horizontal scale is approximately the Rossby internal radius of deformation of 10 km (Johannessen et al. 1983, NORSEX Group 1983). Figure 5 and airborne SAR image obtained in this area during the Norwegian Remote Sensing Experiments (NORSEX Group 1983), shows eddies of this type being shed from the ice edge. The image, obtained on a cloudy day, demonstrates the capability of aircraft radars in collecting sequential mesoscale synoptic information in the MIZ. Further down-stream in Fram Strait and the Greenland Sea, larger eddies of diameter 50 km or more are observed (Vinje 1977, Wadhams and Squire 1983) which appear to be generated through baroclinic instability of the polar front. Our present knowledge of the space and time scales of these high-latitude eddies, and their generation, energetics and role in lateral heat and mass exchange in the MIZ, is very sparse.

Transient wind-driven upwelling along the ice edge has been observed north of Svalbard: in winter by Buckley et al.(1979) and in the fall by Johannessen et al. (1983). In the winter, Fig. 6, water was upwelled from 150 m depth to the surface in a 10-km-wide zone along the ice edge, thereby generating two fronts, one coinciding with the edge and the other parallel to the edge and 10 km. off. During the fall upwelling event, where the vertical statification across the pycnocline (located at 20 m) was very strong, only a slight rise of the pycnocline, on the order of a few meters, took place during a 2 1/2-day 10 $m\text{-}s^{-1}$ wind events, Fig. 7. The upwelling is believed to be caused by changes in the wind stress across the ice edge due to the variation of the air drag coefficient between open water, broken ice floes and smoother ice, and to stability variations in the atmospheric surface boundary layer.

The planetary boundary layer and mixed layer under pack ice have been the subject of several studies, such as those of Hunkins (1966), McPhee and Smith (1976), Maykut(1977), Morison (1980) and Morison and Smith (1981). All of these have taken place in the interior pack, and it is expected that the boundary layer and mixed layer in the MIZ may behave quite differently. For instance, the few measurements of air and water drag coefficients for MIZ ice suggest that they are greater than over the interior pack or open water (Johannessen 1970, Smith et al. 1970), while recent theoretical work by McPhee (1981) suggest that in summer the water drag coefficient is very low because of the effect of meltwater on the boundary layer. In the mixed layer it is possible that one-dimensional models of behavior are no longer appropiate because of the large horizontal density gradients which may introduce vertical velocity shears, eddies or disruptions in the internal wave field.

3.3. Meteorology

The principal atmospheric processes that are important in the MIZ are those which control the exchanges of momentum, heat, moisture and radiative energy between the atmosphere and the ocean or ice surface.

Wind stress is the atmospheric momentum flux at the surface. It contributes to ice drift, wave and current generation, and mixing in the upper ocean. The magnitude and direction of the stress are determined mainly by the gradient of sea level pressure and the stratification and shear in the atmospheric surface layer and planetary boundary layer. Many features of the sea level pressure distribution are controlled by synoptic-scale and planetary-scale dynamics. Other features, however, are dependent on boundary layer exchange processes and mesoscale circulation features, thus creating a feedback between the pressure field and the stress field. Wind stress may also be coupled to changing surface roughness, notably in the case of wind-generated waves. Critical MIZ problems involve determining the wind stress under a variety of synoptic conditions, stratifications and roughness characteristics.

The sensible heat flux to and from the atmosphere strongly influence ice growth, the temperature of the ocean surface, and convection in the upper ocean. Its magnitude and direction are a function of the temperature difference between the atmosphere and the surface, and, as in the case of stress the details of the turbulent transfer are dependent upon atratification and shear in the surface layer and planetary boundary layer. In some cases, however, as when deep convection is generated during air mass modification, there may be strong vertical coupling between the surface and the mid-troposphere. Feedback may be involved if this convection further influences synoptic development. Relevant MIZ problems involve relating the sensible heat flux to the ambient synoptic conditions and boundary layer characteristics.

The physics of water vapor transfer (evaporation and condensation) are analogous to those of sensible heat, and all the preceding remarks apply. Precipitation, usually rain or snow, enters MIZ problems in several ways. Precipitation over the ocean decreases the salinity of the mixed layer and affects thermohaline convection, and over ice may cause melting or a change in surface wetness. Snow over ice surfaces usually causes an albedo change, and strongly influence the heat transfer by conduction; it may also slightly alter the surface roughness.

The solar and infrared radiative fluxes are large terms in the energy balance of both the ocean and the ice. Both quantities are critically dependent on cloudiness, and to a somewhat lesser degreee on atmospheric constituents, especially water vapor, carbon dioxide and aerosols. Albedo variations, especially over heterogeneous snow-ice-water surfacees, influence the short-wave fluxes, while variations of infrared emissivity affect the long-wave balance. The MIZ radiative problems is chiefly one of monitoring the fluxes at the surface together with the ambient cloud and moisture conditions.

There are numerous topographic considerations, mainly involving the Greenland land mass. These involve all scales, ranging from katabatic effects in coastal regions to large-scale, orographically induced cyclogenesis.

Most of the atmospheric processes listed above are coupled to one another through a variety of complicated mechanisms.

4. MODELS

A coupled mesoscale ice-ocean simulation of the MIZ has not yet been carried out. Some ice model simulations have been done for the Greenland Sea using Hibler's (1979) dynamic thermodynamic sea ice model to predict seasonal and interannual variations in the ice edge position (Hibler and Walsh 1982), and to predict week-to-week variations in ice drift and compactness (Tucker and Hibler 1982). The large-scale simulations (Hibler and Walsh 1982) yielded a seasonal cycle with excessive amounts of ice in the North Atlantic during winter and with somewhat excessive amounts of open water in the central Arctic during summer. The poor fit to the Atlantic ice margin in winter is likely partially due to the neglect of lateral oceanic heat transport, since the ocean portion of the model consisted of only a fixed depth, motionless mixed layer together with an upward oceanic heat flux. However, a similar model has successfully simulated the seasonal cycle of Wedell Sea pack ice (Hibler and Ackley 1981) indicating that there may be considerable asymmetry between the oceanographic characteristics in the different hemispheres. In general these results emphasize the need for carrying out more fully coupled ice-ocean simulations in the marginal ice zone regions.

To model ice drift, growth and decay it is important to understand the nature of the ice rheology. In the MIZ the ice cover is more fragmented than ice in the central pack, with substantial variations in compactness. These MIZ characteristics have an unknown effect on the ice dynamics. Of particular interest is the role of internal ice stress as compared to wind and water stresses on the ice drift. On the large scale the internal ice stress has a rectifying effect on motion in the marginal ice zone. In particular, under on-ice-winds, this stress tends to reduce further convergence after the ice has been sufficiently compacted. Off-ice winds, on the other hand, can cause motion with little ice resistance. Such features are characteristic of the plastic rheologies used in large-scale models (e.g. Hibler 1979, Coon 1980). However, superimposed on such a rectifying effect, random bumping or rotation of floes may produce an effective pressure term. Røed and O'Brien (1981) speculate that such an unconfined pressure may be a mechanism causing a jet-like motion at the ice edge. In addition, mesoscale simultations by Hibler et al.(1982) show the presence of wave effects during ice building. To methodically examine the role and effect of these rheology features on ice edge growth, drift and decay further numerical simulations are needed. Such studies can be carried out using the viscous plastic model developed by Hibler (1979). This numerical model provides for the simulation of a highly nonlinear ice interaction employing an arbitrary shear to compressive strength ratio, and an unconstrained pressure term of adjustable magnitude.

The specific problem of the ocean response to wind forcing in the MIZ has been approached through analytical work invoking a stationary and inactive ice cover, which readily allows ice edge upwelling (Gammelsrød et al. 1975, Clark 1978). However, direct observation of the MIZ reveals a highly mobile rather than inactive ice cover, and emphasizes the need for a coupled ice-ocean model. Mesoscale numerical models, coupling sea ice and ocean (Røed and O'Brien 1983) and including thermodynamic processes, are under development. They will be used in studying the influence of a moving ice cover on the oceanic circulation in the MIZ on short time scales of a few days to a few weeks.

There is also a need for more complete models of the atmospheric winds. One approach in this regard is to study the mesoscale wind and surface flux fields employing planetary boundary layer (PBL) models. Models for obtaining the surface flow, stress and heat fluxes with respect to large-scale parameters of pressure and temperature fields were developed during AIDJEX (Brown 1974, 1981). The model developed by Brown adapted to the ocean in connection with GOASEX and JASIN for remote sensing surface truth studies (Brown and Liu 1981). In these experiments, model fields were shown to agree with point measurements to $\pm 2 \text{ m s}^{-1}$ and $\pm 20°$. These surface truth comparisons are discussed by Brown et al (1981). The geostrophic flow (derived from the surface pressure field) is corrected for curvature effects and thermal wind. It is used as the boundary condition on a two-layer similarity solution for the PBL flow. Corrections are included for stratification effects in both layers, secondary flow in the outer layer, variable surface roughness and humidity effects.

The mesoscale eddies which occur along the ice edge have already been subjected to laboratory modeling (Griffiths and Linden 1981 a,b), but further numerical modeling is required in order to understand this phenomenon. We plan first to examine the dynamics of isolated mesoscale eddies found in the MIZ region through the use of an existing two-layer dynamical numerical model (Smith and O'Brien 1982). The roles of topography, vertical eddy structure, variable friction and lateral boundaries can all be addressed with the model, and there is the possibility of incorporating thermodynamics. More complete studies will likely involve the coupling of a nonlinear dynamic-thermodynamic sea ice model to an eddy resolving baroclinic ocean model.

On a smaller scale, ocean waves are important in breaking up the ice in the MIZ, and long swell may be effective up to 50-60 km inside the ice edge. Present models of the interaction of waves with an array of discrete ice floes (Wadhams 1982) are based on scattering mechanisms and are successful in predicting the wave decay rate so long as the pack is not consolidated. They cannot, as yet, predict wave refraction within the pack or the form of the energy spectrum reflected back out into the open water. The flexural response of floes to waves can also be modeled successfully (Goodman et al. 1980) and used to predict the maximum floe size that can occur at different penetrations into the ice pack under a specified incident wave spectrum. The actual nature of the flow size distribution within this maximum size limit is not predictable as yet, but has been measured empirically.

To summarize, several of the mesoscale MIZ processes are poorly described theoretically. Regional models exist which will couple a uniform depth mixed layer both to the ice and to the deep ocean but have not been numerically investigated. In addition there is a need for development of a model for the Greenland/Norwegian Seas, employing a more complete treatment of the mixed layer. Such studies, will help in the understanding of the physical processes which control the East Greenland and West Spitzbergen Currents and the ice edge position.

5. SCIENTIFIC QUESTIONS

The foregoing discussions of processes and models suggest that the following scientifio questions are of major importance and need to be addressed in the MIZEX experiment.

5.1. Ice Dynamics And Thermodynamics

* What are the roles of the internal ice stress, floe-floe interaction, wind and water stresses, inertial-tidal forces, and wave forces in MIZ ice-dynamics?
* Are lateral variations in vertical heat fluxes more important than lateral oceanic heat fluxes in determining the ice retreat?
* How does ice advection caused by general circulation such as the East Greenland Current,by eddies, ice bands and streamers, influence the retreat of the ice edge?

5.2. Ice Topography

* How does the ice thickness distribution vary with distance from the ice edge?
* What is the role of waves in the distributions of floe size and ice roughness?

5.3. Oceanography

* What is the three-dimensional structure of the fronts (East Greenland polar front and meltwater fronts) in the Fram Strait and Greenland Sea marginal ice zones? What is their temporal and spatial variability over a period of days? What is the relationship between the fronts, the ice edge and the bathymetry? How do instabilities, eddies and fine-structure occur in relation to fronts?
* What are the characteristics of the eddy field in the MIZ with respect to space and time scales, energies, generation mechanisms, propagation and role in lateral heat and mass exchange?
* How prevalent is upwelling along the ice edge? How does it relate to the wind-stress variation across the edge and is it important to the dynamics and thermodynamics of the ice edge region?
* How do the momentum , buoyancy, and heat fluxes in the oceanic mixed layer vary with varying ice conditions (melting rate, concentration, floe size, etc.)?
* How does meltwater input affects stratification and the upper layer circulation in the MIZ? Does the meltwater, for example, generate a jet-like current along the ice edge by analogy with coastal currents, with fresh water inputs from fjords and estuaries?
* How does the internal wave field differ under pack ice and in the open ocean?
* What is the role of vertical fine-structure in the transfer of properties across the front?

5.4. Meteorology

* How do the bulk aerodynamic coefficients change with the ice conditions and atmospheric surface layer stability in the MIZ?
* Is there a strong coupling between the ocean mixed layer and the atmospheric boundary layer in the MIZ?
* What is the horizontal distribution of the surface wind stress and heat flux and how are they influenced by the ice pack, broken ice, open water and the stability in the atmospheric surface layer?
* What is the relationship between synoptic scale meteorological patterns and wind flow adjacent to the MIZ?
* Are there jet-like structures in the surface winds near the ice margins and how are they driven?

6. MARGINAL ICE ZONE PROGRAM (MIZEX)

As a result of a workshop in Voss,Norway, in 1980, and subsequent meetings, a program emerged which has two complementary aspects. The overall problem of understanding the annual and interannual variability of the polar ice margins, and of relating these to the large-scale behavior of the atmospheric and ocean circulations, is to be addressed by a long-term monitoring and modeling program described in an associated document (Air-Sea-Ice Research Programs for the 1980's, Untersteiner 1982). Nested within this program will be a mesoscale experimental program to study specific physical processes occuring within the MIZ and to develop models of these processes. This is known as the Marginal Ice Zone Experiment (MIZEX), Wadhams et al. 1981, Johannessen et al. 1983. Essentially, MIZEX is related to the large-scale aspects of the Air-Sea-Ice (ASI) Programs as GATE was related to FGGE.

The vast extent of the Antarctic ice edge, its great zonal and seasonal variation, and the absence of land boundaries all make the Antarctic the ideal area for a MIZEX experiment. However, its remoteness increases greatly the cost of multiship experiments with aircraft remote sensing support, so in the first instance it was decided that MIZEX should take place in the Arctic area of greatest importance thermodynamically, i.e. the region north and west of Svalbard. Fram Strait handles most of the heat and water exchange between the Arctic Ocean and the rest of the world and therefore is a crucial area for studying energy interactions across the ice margin. The shallow Bering Sea, which is a MIZ of quite different character without large velocity shear, is to be studied in a parallel program beginning in early 1983 (Martin et al. 1982) and sharing many personnel, instruments and experimental concepts with the Fram Strait/Greenland Sea MIZEX.

Physical processes in the MIZ are different in winter than in summer, and experiments in both seasons are needed. The first major experiment is to take place during a six-week period from mid-June to the end of July 1984, and is to be preceded by a shorter study in 1983. The dates are chosen to cover the melt period and the transition to summer ice dynamics, and the 83 study is designed to test wheter the scales for the experimental arrays, and the cooperative measurement procedures, are appropriate for yielding the maximum amount of information. The winter experiment will follow in 1987.

The experiment is designed as a drifting one in which an area some 200 km square enclosing the ice edge is selected for intensive investigation. The center of the area is a ship moored to the ice some 30-50 km inside the ice edge and serving as the base for an array of transponders to measure ice deformation as well as for experiments on ice properties, the atmospheric boundary layer and the upper ocean. Other ships are dedicated to studies deeper inside the pack (requiring a heavy icebreaker), at the ice edge itself (where fronts, eddies and ice edge features will be mapped) and in the open water outside the ice ede. The work of these ships will be coordinated by a coordinator aboard one of the vessels, and the concept of following the downstream development of the MIZ ice will be combined with a fixed geographical grid for CTD measurements of ocean structure.

Regular remote sensing flights will map the "moving box" with synthetic aperture radar, passive microwave sensors and cameras, and will transmit imagery of the ice edge either directly to the ships by downlink or indirectly via the Tromsø Satellite Station in northern Norway, which will be the communication base for the experiment. As well as being a tool to assist in the experimental scheme, the remote sensing program is designed to increase our knowledge of the active and passive microwave signatures of sea ice in summer. The focus of the remote sensing experiments is on use of microwave sensors since they permit observation of ocean and ice surfaces through clouds. In spite of much research conducted in respect to microwave detection of sea ice during the last decade, i.e. BESEX, Gloersen et al. (1975), AIDJEX, Campbell et al. (1979), and NORSEX Group (1983), very little work has been done during the summer season. Many ambiguity problems are known to exist at this time of year due to snow melt and continual refreezing of ice surfaces. For example, passive microwave techniques yield good estimates of ice concentration when the ice is frozen (Svendsen et al 1983), but we are not sure how well this technique will work for wet ice. Another example is the SAR observations. This technique presently provides information about the ice edge and structure, as well as surface and internal waves in the ocean. However, we have not yet shown how useful the SAR is for estimating ice concentration and ice floe distribution during summer, and for locating fronts and eddies in the open ocean off the ice edge in cold water.

The scales of the arrays employed, and the set of measurements to be made, will be governed by the needs and results of MIZ modeling studies which are already being coordinated through a MIZEX Modeling Group. Figure 8 shows a possible area that will be traced out by the ships and arrays during the six-week period of the experiment, assuming initial deployment northwest of Svalbard. The initial region, area 1, is a zone of relatively low ice advection with normally a compact and well-defined ice edge, while once Fram Strait is encountered (area 2) the ice drift in the MIZ becomes much more rapid and the ice edge is likely to be more irregular and complex in form.

At every stage MIZEX is planned to be closely coordinated with other experiments in the region. As the ASI program develops, MIZEX will be able to define the major energy interactions which must be parameterized for use within largerscale grid of the ASI program. A Fram Strait Monitoring Program has been proposed, sponsored by the Comité Arctique and funded by industry, which, if successful, will begin in 1984 with a large number of fixed moorings and remote sensing flights across Fram Strait. MIZEX will cooperate closely with this program to achieve maximum scientific value and avoid duplication of facilities. Lastly, the ships of MIZEX can provide a unique platform for important biological and acoustical studies in the MIZ region, and plans for such programs are being submitted in associated documents by Dyer (1982) (acoustics) and Dunbar (1982) (biology).

7. ACOUSTICS AND MIZEX

Portions of the research program relate directly to goals of MIZEX, are described here. The overall acoustics research effort is described in a seperate document (Dyer 1982).

The advent of powerful array and other signal processing technology via the microprocessor has revolutionized acoustics research. Of particular value to MIZ research is the potential of tomographic sensing of the mesoscale eddy field and the potential of wide-area mapping of ice-cover roughness. Both are synoptic in character, and can be repeated either continuosly or on a schedule governed by the temporal scale of each process.

Acoustic tomography was first suggested by Munk and Wunsch (1978), and the first pilot experiment was carried forward by them and their collaborators in 1981. Preliminary results (Spindel 1980) have high promises of mesoscale eddy delineation in the open ocean. In the MIZ eddy scales are expected to be considerably less than those at lower latitudes and acoustic paths are expected to be of considerably different character than those encountered in the first tomography experiment. For these reasons acoustic research in the MIZ will focus upon path identification, signal stability and coherence, and signal degradation associated with ice-related noise and rough surface scattering. Such knowledge is essential to reasoned deployment of a MIZ tomographic system (Spiesberger et al. 198, Spindel 1980), and will begin to be acquired in the 1983 experiment. A preliminary tomographic experiment may well be sensible for the 1984 Greenland Sea MIZEX, but more likely further work will be required on path identification, and on environmental parameters affecting system design. We can reasonably expect deployment of a tomographic system for the follow-on MIZEX winter experiment, and useful but partial data on eddy structure for the 1984 summer experiment.

Wide-area mapping of ice roughness was first proposed by Dyer (1981), based on acoustic backscatter results of 1978 and 1980 Arctic experiments. Such data delineate areas of ice cover in excess of 10^5 km2, and a similar technique could readily cover the planned MIZ cell (4×10^4 km^2). The experiment in 1983 would be used to test a new drifting sensor technology required for array processing, to compare preliminary roughness synopses with direct measurements of ice roughness in a few localities, and to optimize system configuration based on environmental effects such as ice-related noise and rough bottom scattering. We do not know how well such a system can discriminate rough open water from rough ice, and ultimately doppler processing may have to be used to do so. We do believe, however, that smooth open water, such as in large leads and polynyas, can be discriminated. These questions could be addressed in the 1983 experiment, but more likely would be in the 1984 summer MIZEX, when a larger array with more useful resolving power would be deployed. We can reasonably expect the summer 1984 MIZEX to result in ice roughness synopses in coordination with aircraft-acquired imagery and direct ice measurements.

Ice roughness backscatter might yield estimates of rms roughness depth, correlation radius and/or number of rougness elements per unit area, depending upon the model applied to the data. Thus the 1984 MIZEX includes an appropriate ice backscatter modeling effort which, in collaboration with other MIZEX researches, would lead to a validated model.

Recent research also points to the potential of ice-cover forward scatter as a different way to extract roughness (Medwin et al. 1979). In certain regimes the forward scatter Biot boundary wave can be measured to extract the roughness-volume unit area. While not a synoptic technique, forward scatter experiments along a single line would serve as a valuable check or adjunct to the backscatter results. Such forward scatter experiments are planned for the 1984 MIZEX.

REFERENCES

Andersen, B., W.F. Weeks and J.L. Newton, 1980, The seasonal ice zone. Proceedings of an International Workshop. Cold Regions Sci.and Technol., 2.

Brown, R.A., 1974, Analytic Methods in Planetary Boundary Layer Modeling. London: Adam Hilger Ltd. and New York: Halstead Press,John Wiley & Sons, 150 pp.

Brown, R.A., 1981, Modeling the geophysical drag coefficient for AIDJEX: The model and the data. J.Geophys. Res., 86, C3, 1989-1994.

Brown, R.A. and T.Liu, 1981, An operational largescale marine planetary boundary layer model. J.Appl. Meteor.

Brown, R.A., V.J. Cardone, T.Guymer, J.Hawkins, J.E. Overland, E.J. Pierson, S.Peteherych, J.C. Wilkerson, P.M. Woiceshyn and M.Wurtele, 1981, Surface wind analysis for SEASAT. J. Geophys. Res.

Bryan, K., F.G. Komro, S.Manabe and M.J.Spelman, 1982, Transient climate response to increasing atmospheric carbon dioxide . Science, 215(1), 56-58.

Buckley, J.R., T.Gammelsrød, A.Johannessen, O.M.Johannessen and L.P.Røed, 1979, Upwelling: Oceanic structure at the edge of the arctic ice pack in winter. Science, 203, 165-167.

Campbell, W.J., J. Waynberg, J.B. Ramseyer, R.O. Ramseier, M.R. Vant, R.Weaver, A.Redmond, L. Arsenault, P.Gloersen, H.J.Zwally, T.T. Wilheit, C.Chang, D.Hall, L.Gray, D.C.Mecks, M.L.Bryan, F.T. Barath, C.Elachi, F.Leberl and T.Farr, 1981, Microwave remote sensing of sea ice in the AIDJEX main experiment. Boundary-Layer Meteor., 13, 309-337.

Clarke, A.J., 1978, On wind-driven quasi-geostrophic water movement at fast ice edges. Deep Sea Res., 25, 41-51.

Coachman, L.K. and K.Aagaard, 1975, Physical Oceanography of arctic and substratic seas. In Marine Geology and Oceanography of the Arctic Seas (ed. Y.Herman) Chap. 1, Springer Verlag. New York.

Coon, M., 1980, A review of AIDJEX modeling. In Sea Ice Processes and Models, R.S.Pritchard, Ed., University of Washington Press, 12-27.

Dunbar, M.J., 1982, MIZEX Biological Program. McGill University, Montreal.

Dyer, I., 1981, Review of arctic backscattering and noise experiments. Presented at workshop held at New London, 3 and 4 November 1981.

Dyer, I., 1982, MIZEX Acoustical Program. Massachusetts Institue of Technology, Boston.

Ekman, V.W., 1902, Om jordrotationens innverkan på vindstrømmar i havet.Nyt. Mag. f., Naturvid, Vol. 20, Kristiania.

Gammelsrød, T., M.Mork and L.P.Røed, 1975, Upwelling posibilities at an ice edge, homogeneous model. Mar. Sci. Comm., 1(2), 115-145.

Goodman, D.J., P.Wadhams and V.A.Squire, 1980, The flexural response of a tabular ice island to ocean swell. Ann. Glaciol., 1, 23-27.

Gloersen, P., R.O.Ramseier, W.J.Campbell, T.C. Chang and T.T. Wilheit, 1975, Variation of ice morphology of selected mesoscale test areas during the Bering Sea experiment. In U.S.S.R./U.S.A. Bering Sea Experiment: Proceedings of the Final Symposium on the Results of the Joint Soviet-American Expedition, Leningrad, 12-17 May 1974, K.Ya. Kondratiev, Yu. 1. Rabinovich and W.Nordberg, Eds., Gidrometeoizdat, Leningrad, 196-218.

Griffiths, R.W. and P.F.Linden, 1981 a, The stability of buoyance-driven coastal currents. Dynam. Atmos. Oceans, 5, 281-306.

Griffiths, R.W. and P.F.Linden, 1981 b, The stability of vortices in a rotating stratified fluid. J.Fluid Mech., 105, 283-316.

Hibler, W.D., III, 1979, A dynamic-thermodynamic sea ice model. J.Phys. Oceanog., 9, 815-846.

Hibler, W.D., III and S.F.Ackley, 1982, On modeling the Weddell Sea pack ice.Ann.Glaciol., in press.

Hibler, W.D., III and J.Walsh, 1982, On modeling seasonal and interannual fluctuations of arctic sea ice. Submitted for publication.

Hibler, W.D., III, I.Udin and A.Ullerstid,1982, On forecasting mesoscale ice dynamics and dynamics and buildup. Submitted for publication.

Hunkins, K., 1966, Ekman drift currents in the Arctic Ocean. Deep Sea Res., 13, 607-620.

Hunkins, K.Y. Kristoffersen, G.H. Johnson and A. Heiberg, 1979, The Fram 1 Expedition. EOS, 60, 1043-1044.
Helland Hansen, B. and F.Nansen, 1909, The Norwegian Sea. Its physical oceanography based upon the Norwegian research 1900-1904. Rept. Norw. Fish. Mar. Investigation. Mallingske, Christiania, 2. Pt. 1(2), 390 pp.
Johannessen, O.M., 1970, Note on some vertical profiles below ice floes in the Gulf of St. Lawrence and near the North Pole. J.Geophys. Res., 75, 2857-2862.
Johannessen, O.M., J.A.Johannessen, J. Morrison, B. Farrelly and E.Svendsen, 1983, NORSEX III: Mesoscale oceanographic conditions in the ice edge region north of Svalbard during Fall. J.Geophys. Res. in press.
Johannessen, O.M., W.Hibler III, P.Wadhams, W.J.Campbell, K.Hasselman and I.Dyer 1983, MIZEX. Program for mesoscale Air-Ice-Ocean Interaction Experiments in Arctic Marginal Ice Zones. II. A Plan for a summer Marginal Ice Zone Experiment in the Fram Strait/Greenland Sea, 1984. U.S.Army Cold Regions Research and Engineering Laboratory. Special Report. 1983.
Kukla, G.and J.Gavin, 1981, Summer ice and carbon dioxide. Science, 214 (4520),497-503.
McPhee, M., 1981, An analytic similarity theory for the planetary boundary layer stabilized by surface buoyancy. Boundary-Layer Meteor., 21, 325-339.
McPhee, M. and J.D.Smith, 1976, Measurement of the turbulent boundary layer under pack ice. J.Phys. Oceanogr., 6(5), 696-711.
Manabe, S. and R.Stouffer, 1979, Sensitivity of a global climate model to an increase of CO2, concentration in the atmosphere. J.Geophys. Res., 85, 5529-5554.
Martin, S. (Ed.), 1982, Field plan for the 1983 Bering Sea MIZ winter ice deformation experiment. Unpubl. ms., Department of Oceanography, University of Washington (Seattle).
Maykut, G., 1977, Estimates of the regional heat and mass balance of the ice cover. Proceedings of Symposium on Sea Ice Processes and Models. University of Washington, 6-9 September 1977, 65-74.
Medwin, H., J.Bailie, J.Bremhorst, B.J. Savage and I. Tolstoy, 1979, The scattered acoustic boundary wave. J. Acoust.Soc.Amer., 66, 1131-1134.
Morison, J.H., 1980, Forced internal waves in the Arctic Ocean. Ph.D.dissertation, Geophysics Program, University of Washington (Seattle).
Morison, J. and J.D.Smith, 1981, Seasonal variations in the upper Arctic Ocean as observed at T-3. Geophys. Res. Letters, 8(7), 753-756.
Munk, W. and C.Wünsch, 1978, Ocean acoustic tomography: A scheme for large-scale monitoring. Deep Sea Res., 26A, 123-161.
NORSEX Group, 1983, NORSEX 1: The Norwegian Remote Sensing Experiment in the marginal ice zone north of Svalbard during Fall.Science, in press.
Perdue, W.F., 1982, Oceanographic investigation of the east Greenland Polar Front in autumn. Master thesis, Naval Postgraduate School, Monterey, California, March.
Pritchard, R.S., 1980, Sea ice and models. Proceedings of the Arctic Ice Dynamics Joint Experiment, International Commission on Snow and Ice Symposium. Seattle and London: University of Washington Press.
Rey, L. 1982, The Arctic Ocean. The hydrographic Environment and the bate of pollutants. Macmillan Press LTD, London.
Røed, L.P. and J.J.O'Brien, 1981, Geostrophic adjustment in highly dispersive media. An application to the marginal ice zone. Geophys. Astrophys. Fluid Dynam., 18, 263-278.
Røed, L.P. and J.J.O'Brien, 1983, A coupled ice-ocean model of upwelling in the marginal ice zone. J.Geophys. Res. in press.
Smith, D.C., IV and J.J.O'Brien, 1982, The interaction of a two-layer isolated mesoscale eddy with bottom topography. Naval Postgraduate School, Monterey, California: submitted for publication.
Smith, S.D., E.H. Banke and O.M.Johannessen, 1970, Wind stress and turbulence over ice in the Gulf of St. Lawrence. J.Geophys. Res.,75.
Spiesberger, J.R. Spindel and K.Metzger,1980, Stability and identification of ocean acoustic multipaths. J. Acoust. Soc. Amer., 67, 2011-2017.
Spindel, R., 1980, Multipath processing for ocean acoustic tomography, IEEE EASCON '80, 165-170.
Spindel, R., 1981, Ocean acoustic tomography. Proc. Inst. Acoust., December, 3/1-3/11.
Svendsen, E., K.Kloster, B.Farrelly, O.M. Johannessen, J.A. Johannessen, W.J. Campbell, P.Gloersen, P. Davalieri and C.Matzler, 1983, NORSEX II: Evaluation of the Nimbus 7 scanning multifrequency microwave radiometer for sea ice research. J.Geophys. Res., in press.
Tucker, W.B. and W.D.Hibler, III, 1982, Preliminary results on modeling the East Greenland Sea ice. POAC Proceedings, Quebec.
Untersteiner, N., 1982, Air-sea-ice interaction research programs for the 1980's. Unpubl. rpt. Applied Physics Lab., Seattle, Washington.
Vinje, T.E., 1977, Sea ice conditions in the European sector of the marginal seas of the Arctic, 1966-75. Norsk Polarinstitutt Årbok 1975, 163-174.
Vinnikov, K.Ya. et al., 1981, Modern changes in climate of the Northern Hemisphere. Meteorologiya I Hidrologiya, no. 6, 5-17.
Wadhams, P., 1981, Sea ice topography of the Arctic Ocean in the region 70 W to 25 E. Phil. Trans. Roy. Soc., A302, 1464-1504.
Wadhams, P., 1982, The seasonal ice zone. Proc. NATO Advanced Study Inst. on Air-Sea-Ice Interaction, Maratea, 28 September - 10 October 1981. Plenum Press, Chapter 10, in press.
Wadhams, P. and V.A. Squire, 1983, An ice-water vortext at the edge of the East Greenland Current. J.Geophys. Res., in press.
Wadhams, P., S.Martin, O.M. Johannessen, W.D. Hibler III and W.J.Campbell, 1981, MIZEX, A program for mesoscale air-ice-ocean experiments in artctic marginal ice zones. 1. Research strategy. U.S.Army Cold Regions Research and Engineering Laboratory, Special Report 81-91.
Walsh, J.E. and C.M.Johnson, 1980, Interannual atmospheric variability and associated fluctuations in arctic sea ice extent.J.Geophys. Res., 84, 6915-6928.
Nansen, F. 1906, Northern waters: Captain Roald Amundsen's oceanographic observations in the Arctic Seas in 1901.Vid. Selskap Skrifter I. (Mat.Naturv. Kl. Dybwad,Christiania, 1(3). 145 pp.

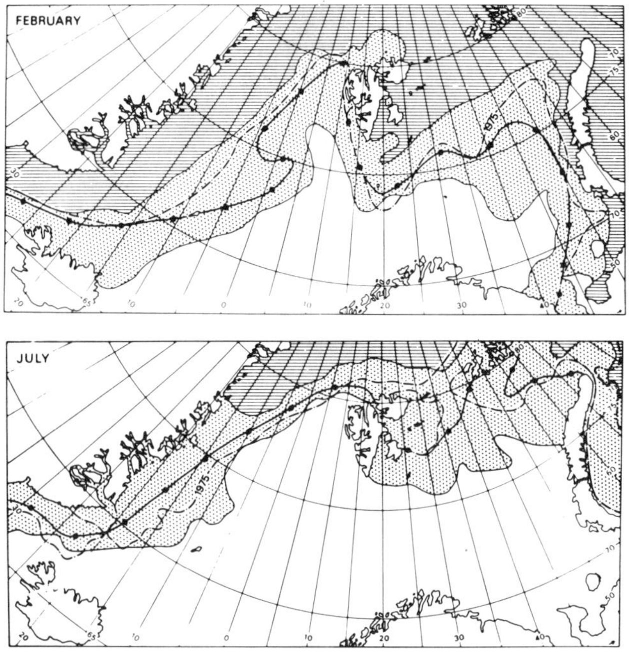

Fig. 1. Mean and extreme sea ice limits at the end of February and July for the years 1966/1974(Vinje 1977). The extreme range for 3/8 sea ice is bounded by the dotted areas, while the thick black line is the median limit for the decade and the dashed line is the 1975 limit.

Fig. 2. A typical compact ice edge, showing small floes about 20-40 m in diameter.

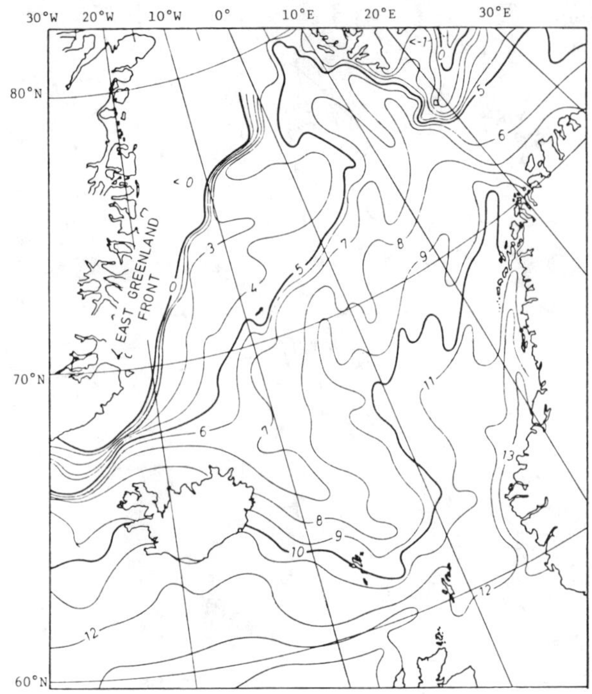

Fig. 3. Surface temperature in the Greenland and Norwegian Seas, summer 1958 (from Dietrich 1969).

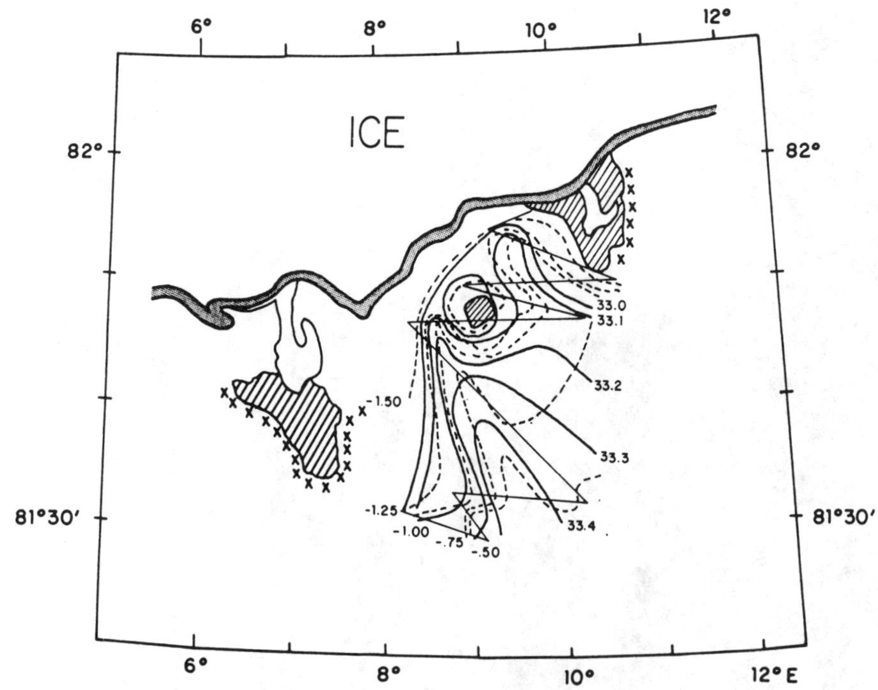

Fig. 4. Surface temperature and salinity mapped by R/V Polarsirkel on 1 October . Thin lines indicate ship track,and hatched areas indicate grease ice.Crosses indicate the frontal boundary interpreted from the SAR image the same day (from Johannessen et al. 1983).

Fig. 5. JPL Synthetic Aperture Radar (1.215 GH_z) image of the marginal ice zone on 1 October 1979 obtained during the NOSEX investigation. (After NORSEX Group, 1982).

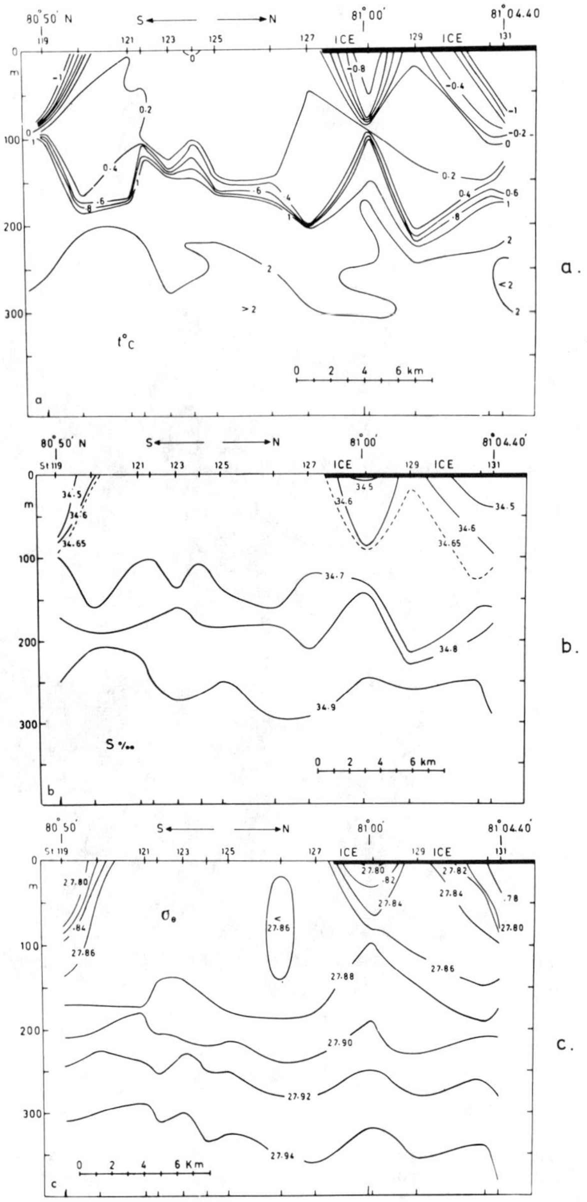

Fig. 6. Sections drawn from CTD casts made at 13°E on December 3,1977, showing upwelling near the ice edge (after Buckley et al.1979).(a) Temperature, degrees Celsius: (b)salinity,parts per thousand; (c) density in units of σ_Θ

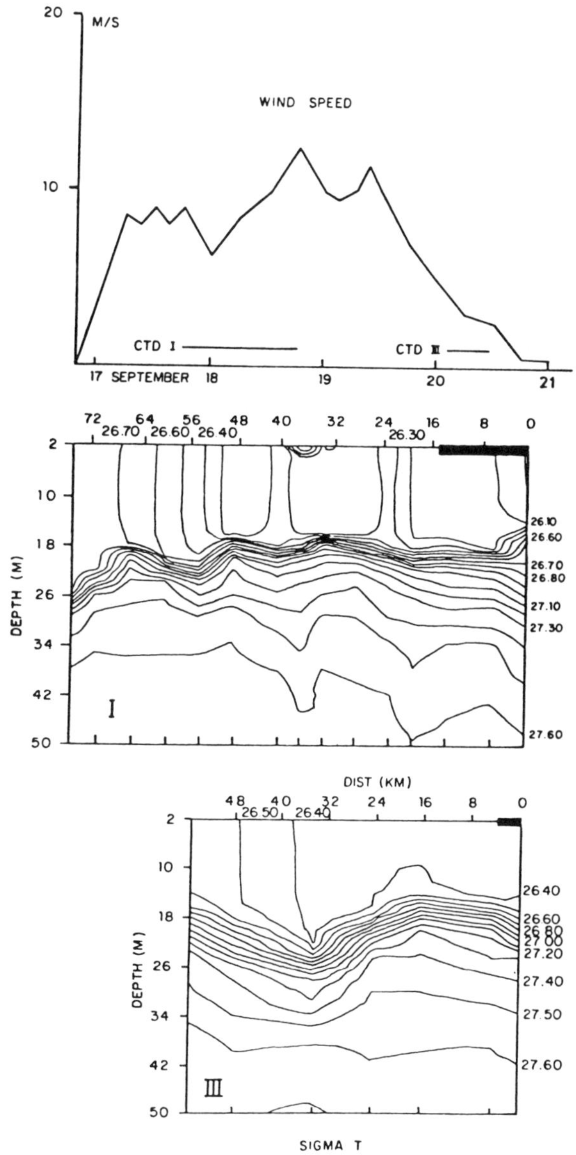

Fig. 7. Density structure from CTD sections I and III, and wind speed during a major easterly wind event. Stations indicated at the bottom of the sections. After Johannessen et al. 1983.

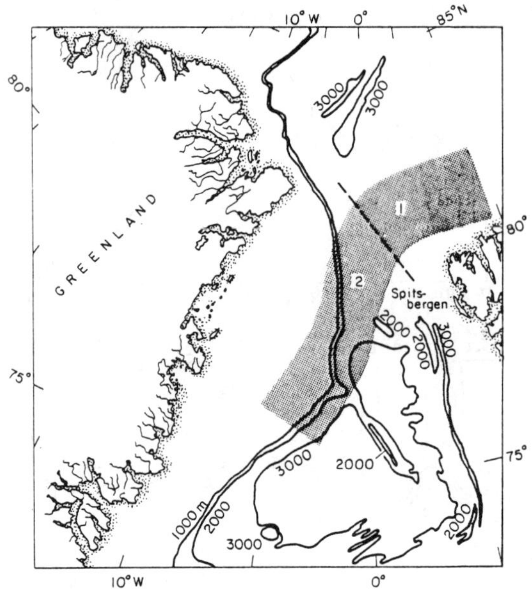

Fig. 8. A possible area swept out by the experimental box in six weeks, assuming a mean advection rate of 10 cm s^{-1}.

THE TECTONICS OF THE ARCTIC OCEAN

Lawrence A. Lawver

Department of Earth and Planetary Sciences,
Massachusetts Institute of Technology
Cambridge, MA 02139

Arthur Grantz

U.S. Geological Survey
Menlo Park, CA 94025

Linda Meinke

Department of Earth and Planetary Sciences,
Massachusetts Institute of Technology
Cambridge, MA 02139

ABSTRACT

Numerous theories concerning the tectonic evolution of the Arctic Ocean have been proposed. The nearly constant and continuous ice cover makes marine geological and geophysical reconnaisance of the Arctic extremely difficult. Aeromagnetic data collected by the Naval Research Laboratory and by Russian scientists have allowed us to understand the recent tectonic evolution of the Eurasian Basin. The Amerasian Basin has not produced unambiguous magnetic anomaly data and is still open to much speculation concerning its age and mode of evolution.

We review most of the current tectonic interpretations for the evolution of the Arctic. The three separate classes of models proposed for the Arctic include: oceanization of continental crust; entrapment of oceanic crust by the motions of the continents surrounding the Arctic; and production of oceanic crust in situ. While most tectonophysicists support the theory of creation of oceanic crust in situ in the Arctic they propose different directions and ages for the occurrence of seafloor spreading in the arctic particularly in the Amerasian Basin.

INTRODUCTION

Unlike the other oceanic regions of the earth, the Arctic ocean does not yet have a generally accepted tectonic history. In the north Atlantic region between the equator and 50°N, the plate motions of the North American, Eurasian and African plates have been determined from magnetic anomaly correlations (Pitman and Talwani, 1972). We know with a reasonable degree of certainty that North America has been moving away from Eurasia at 0.231 degrees/million years about a pole at 65.85^{o}N, 132.44^{o}E and from Africa at 0.258 deg/m.y. about a pole at 80.43^{o}N, 56.36^{o}E (Minster and Jordan, 1978). These rates and directions for the three plates can be determined or estimated for the last 150 Ma and the spreading center has always been along the Mid-Atlantic Ridge since the megacontinent Laurasia began to break up approximately 180 Ma.

Determining the past and present motions of continents and the age of oceanic crust has generally relied on the identification of marine magnetic anomalies, the trend of fracture zones that cross the spreading centers at a high angle, fault plane solutions for earthquakes along plate boundaries and information obtained by holes drilled into oceanic crust and sediments by the Deep Sea Drilling Project. Since it is continually covered with ice, no holes have been drilled in the deep basins of the Arctic Ocean. Consequently, we have no direct determination of the age of the initial sediments deposited on oceanic crust. Until the aeromagnetic data reported by Naval Research Laboratory personnel (Taylor et al., 1981; Vogt et al., 1982; Vogt et al., 1979; Vogt and Ostenso, 1970; Vogt et al., 1981) and by Karasik (1973, 1974), we have had no magnetic anomaly framework from which to build a tectonic model for the Arctic. Unfortunately, as will be discussed later, only one section of the Arctic can be tectonically resolved with the available aeromagnetic data.

Earthquake epicenters largely outline present day plate boundaries and since the establishment of the World-Wide Standardized Seismograph Network (WWSSN), the present-day active plate boundary between Eurasia and North America is now obvious. The bathymetry of the Arctic is still poorly known but the major features have probably been determined. Submarines and manned ice islands have contributed substantially to our knowledge of the bathymetry of the Arctic Ocean.

The most striking bathymetric feature of the Arctic Ocean is the Lomonosov Ridge which runs from close to the north end of Ellesmere Island nearly through the geographical north pole and ends near the Novosibirsk Islands. The main bathymetric features of the Arctic Ocean are shown in figure 1. The Lomonosov Ridge divides the Arctic Ocean into the Eurasian Basin and the Amerasian Basin. The Eurasian Basin contains the northern extension of the Mid-Atlantic Ridge spreading center. The ridge section in the Eurasian Basin is known as the Arctic Mid Ocean Ridge, the Nansen Ridge, the Gakkel Ridge or the Nansen-Gakkel Ridge. The Nansen-Gakkel Ridge splits the Eurasian Basin into the Nansen Basin along the Eurasian margin and the Fram or Amundsen Basin between the Nansen-Gakkel Ridge and the Lomonosov Ridge.

The Amerasian Basin is bounded by the Lomonosov Ridge on the north, the Queen Elizabeth Islands of Arctic Canada on the East, the MacKenzie Delta and the North Slope of Alaska on the south and the East Siberian Sea shelf on the west. The Alpha-Mendeleev Ridge divides the Amerasian Basin into the Canada Basin to the south and the Makarov Basin between the Alpha and Mendeleev Ridges and the Lomonosov Ridge to the north.

Tectonic Models

There have been three different classes of models proposed for the tectonic evolution of the Arctic Basin; oceanization of continental crust, entrapment of old oceanic crust and the formation of oceanic crust in situ. As recently as 1976, Pogrebitskiy (1976) and Pushcharovskiy (1976) have reiterated the Russian held

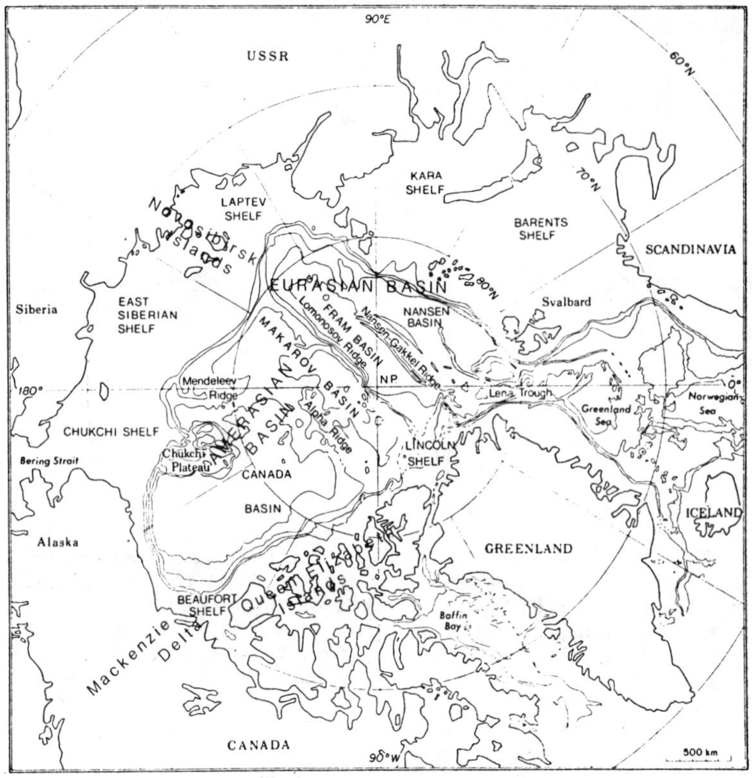

Fig. 1. Arctic seafloor. Bathymetric contour interval 1 km. Also shown is 500 m contour. Modified from Fig. 1 of Sweeney, 1981.

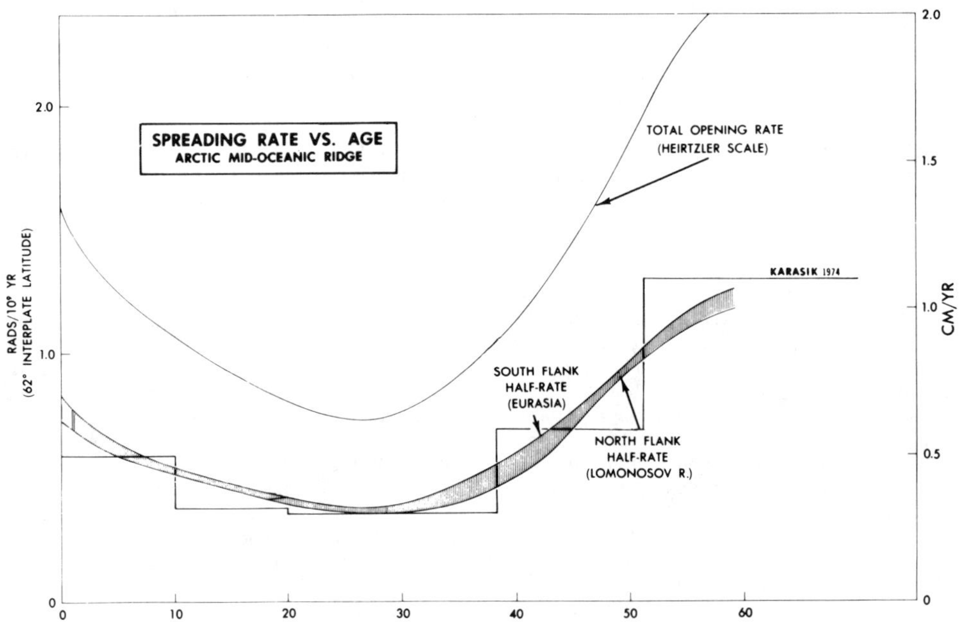

Fig. 2. Spreading half-rate (lower curves) and total opening rate (upper curves) as a function of time for segment of Eurasia Basin reported here. Curves were derived by graphically differentiating age versus distance curves fitted visually to identified magnetic lineations and their distance from axis. Approximate plate rotation rates (left scale, in radians per 10^6 years, assume constant interplate (rotation) latitude of 62°N for center of survey area. Note persistently slower spreading on north (Lomonosov Ridge) flank of Nansen Cordillera. Rate history (average for both flanks) deduced by Karasik (1974), shown by stepped curve, is in general agreement. Modified from Vogt et al., 1979.

view popularized in this country by Beloussov (1970) of oceanization of continental crust to explain the transformation of continental into oceanic crust.

The Lomonosov Ridge

Except for those people who wish to explain all of the Arctic Ocean as having resulted from subsidence and transformation of continental crust, most other researchers in Arctic tectonics agree about certain things. The first is that the Lomonosov Ridge is a continental fragment. Mair and Forsyth (1982) recently presented persuasive seismic refraction results that show the Lomonosov Ridge to be seismically similar to the Kara Sea and Barents Sea shelf region. The idea that the Lomonosov Ridge was rifted from the Barents shelf was originally proposed by Heezen and Ewing (1961). Even earlier, Ostenso (1960) had suggested that the Lomonosov Ridge was continental because it lacked the strong magnetic anomaly found over other oceanic ridges. Wilson (1963, 1965), Demenitskaya and Karasik (1969), Vogt and Ostenso (1970), Ostenso and Wold (1971), Pitman and Talwani (1972) and Ostenso (1972) have all added geophysical evidence to reinforce this idea. Ostenso (1972) presented a seismic profile obtained from the drifting ice island ARLIS-II, which showed the obvious differences between the Lomonosov Ridge and a normal seafloor spreading ridge. The elevated topography and the thick sediment cover on the ridge are much too shallow to be normal oceanic crust when judged with respect to the Parsons and Sclater (1977) age-versus-depth relationship for cooling oceanic crust.

Many authors have reiterated the idea that the Lomonosov Ridge is a continental fragment because reversing the spreading on the Nansen-Gakkel Ridge would close up the Eurasian Basin and fit the Lomonosov Ridge against the Barents-Kara shelf. Savostin and Karasik (1981) have recently suggested the presence of a third plate, the Spitsbergen plate, whose relative motions produce an even better closure of the Eurasian Basin. They calculated a present day pole of rotation for North America with respect to Eurasia to be 59.48°N, 140.83°E and a rate of opening to be .189 deg/m.y. This pole is 6° south of that found by Minster and Jordan (1978) and the rate of opening is 18% slower. Savostin and Karasik (1981) calculated the rate by fitting identified magnetic lineations along the Nansen-Gakkel Ridge.

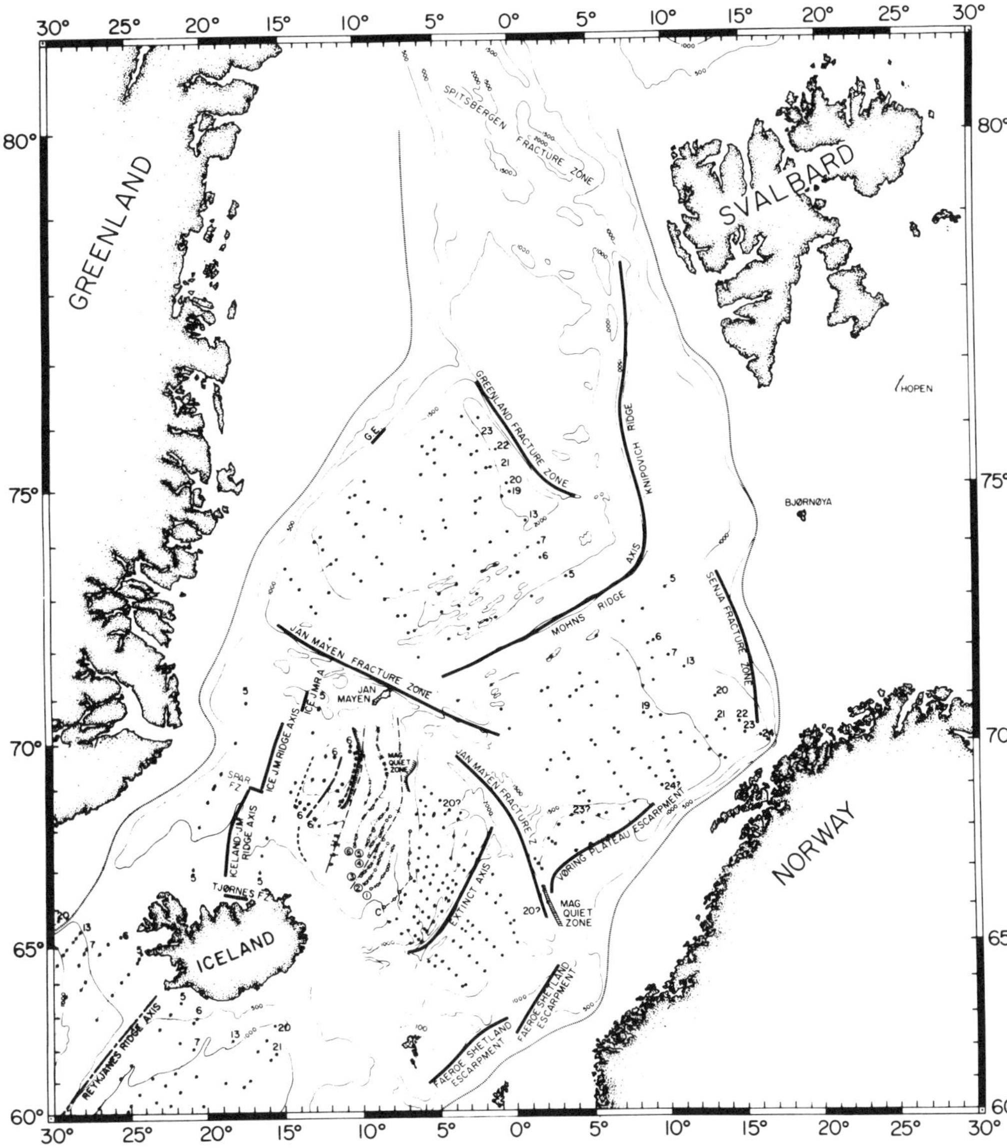

Fig. 3. Identified magnetic lineations in Norwegian-Greenland Sea. Taken from Talwani and Eldholm, 1977 with permission from the Geological Society of America Bulletin.

The Eurasian Basin

Vogt et al., (1979) positively identified anomaly 24 (55 Ma) in the Eurasian Basin but stated that it is difficult to make any correlations between Anomaly 24 and the shelf edge or the Lomonosov Ridge. They did find the Nansen-Gakkel Ridge to be remarkably uncomplicated with only a very few short transform faults and no evidence of jumped spreading centers. In addition, they found the spreading rate on the north flank of the Nansen-Gakkel Ridge to be consistently slower than that on the Eurasian side of the ridge. Figure 2 taken from Vogt et al., (1979) shows the variation in spreading rate on the Nansen-Gakkel Ridge with time and the difference in spreading rate on the two sides of the ridge. As can be seen from their figure, their results are very close to those found by Karasik (1974) although Karasik attempted to extrapolate the spreading history back to 70 Ma while Vogt et al. (1979) stopped it at Anomaly 24 time or 55 Ma.

Savostin and Karasik (1981) detail much of the geophysical evidence concerning the opening of the Eurasian Basin and the way that the opening is related to extension along the Mid-Atlantic Ridge in the Greenland and Norwegian Seas. Talwani and Eldholm (1977) identified the magnetic anomalies found in the Norwegian-Greenland Sea and suggested that the oldest identified anomaly is anomaly 24 (55 Ma). Unfortunately, north of the Greenland and Senja fracture zones, (see figure 3 taken from Talwani and Eldholm, 1977) there are no identified magnetic anomalies and the trend of the spreading ridge changes abruptly. Eldholm and Talwani (1977) suggest that motion along the Greenland-Senja Fracture Zone was transform until 38 Ma when seafloor spreading began along the Knipovich Ridge. Savostin and Karasik (1981) explain off axis earthquakes along the Knipovich Ridge as being indicative of a boundary of the Spitsbergen microplate. Figure 4 shows the geometry of three variants of their Spitsbergen microplate. It is conceivable, particularly since Srivastava (1978) claims that seafloor spreading in Baffin Bay did not commence until Anomaly 25 time (see Figure 5 taken from Srivastava, 1978) that rifting between Eurasia and North America at anomaly 24-25 time was extremely complicated. Rifting along the Mid-Atlantic Ridge north of the Greenland-Senja fracture zones might have consisted of extension on the Barents shelf much like Savostin and Karasik (1981) suggest is currently happening on the eastern boundary of their Spitsbergen microplate in two of their models. Motion at that time on the Barents shelf might have opened the North Cape Basin which Eldholm and Talwani (1977) discuss. At the same time, seafloor spreading in the Labrador Sea extended into Baffin Bay and must have been translated along Nares Strait into the Eurasian Basin from the west. The Labrador Sea and Baffin Bay extension died out at about anomaly 20 time (48 Ma) and seafloor spreading may have shifted from the suggested extension on the Barents shelf in the North Cape Basin to the Knipovich Ridge. The Nansen-Gakkel Ridge appears to have spread in a reasonably coherent manner during the period from anomaly 24 to anomaly 20 (Savostin and Karasik, 1981 and Vogt et al., 1979) as did the Mid-Atlantic Ridge in the Norwegian-Greenland Sea (Talwani and Eldholm, 1977). After anomaly 20, the Lomonosov Ridge moved away from Eurasia at the same rate as Greenland. Prior to anomaly 20 time, Greenland and Spitsbergen may have formed a small plate that moved independently of both North America and Eurasia. As will be discussed later, if Greenland is not moved back towards North America, in effect closing up Baffin Bay between anomalies 25 and 20, there is an unacceptable overlap of Greenland with Eurasia when the Eurasian Basin is closed.

The Amerasian Basin

While the Eurasian Basin appears to fit most plate tectonic models concerning the creation of oceanic crust at a seafloor spreading center, the

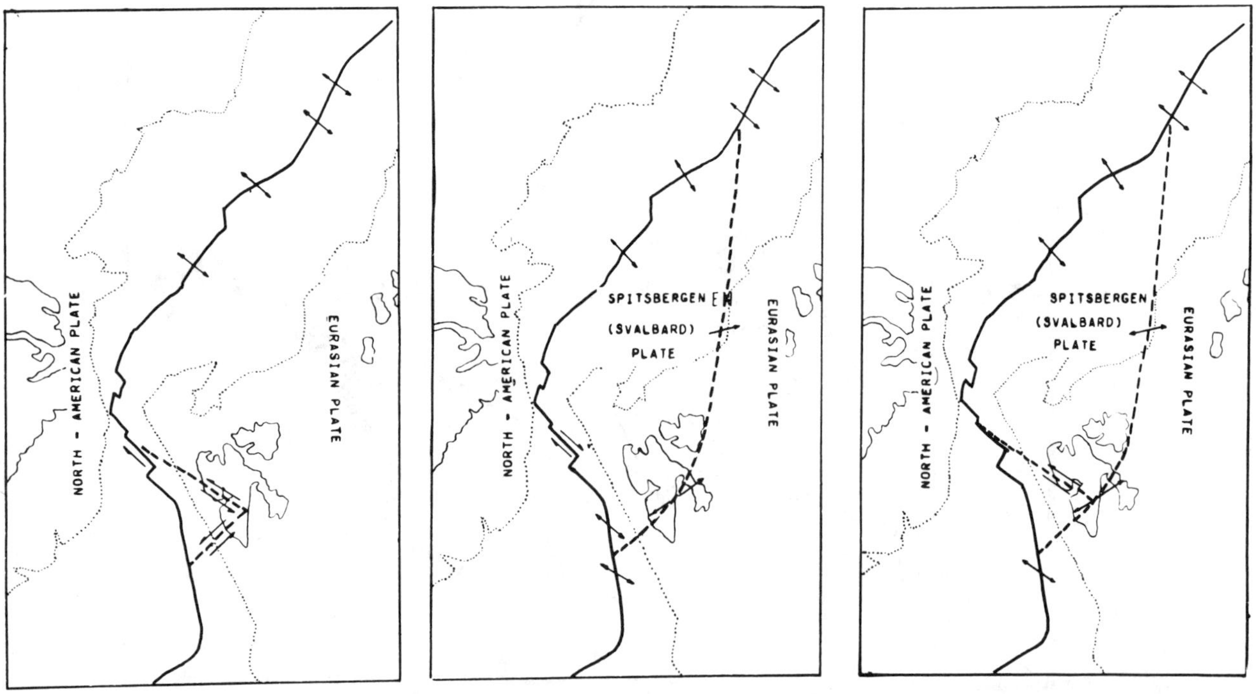

Fig. 4. Variants of the Spitsbergen microplate. Arrows show slip direction at plate boundaries. Taken from Savostin and Karasik, 1981 with permission from Elsevier Scientific Publishing Co.

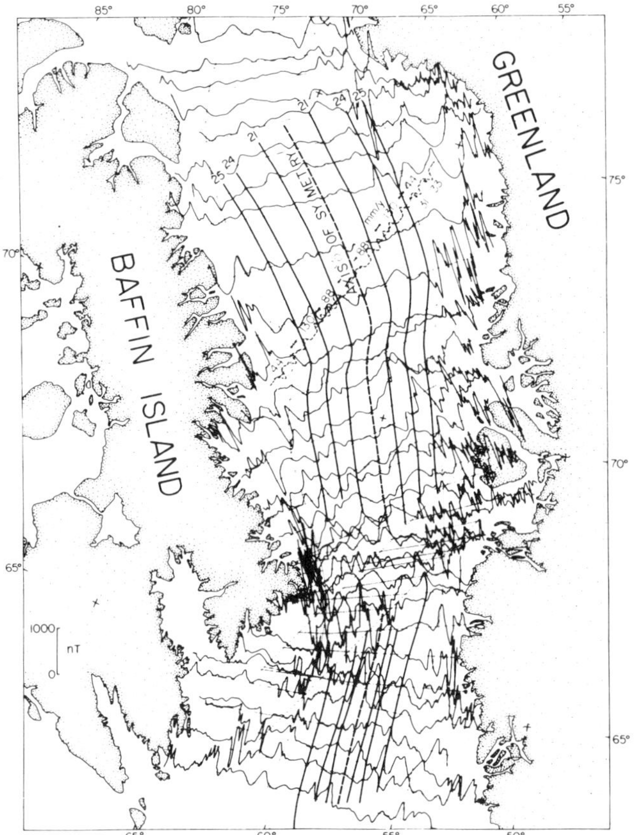

Fig. 5. Aeromagnetic profiles plotted along flight path across Baffin Bay and Davis Strait. Synthetic isochrons show the correlation of anomalies between tracks. Synthetic profile generated along a flow line through the northern Baffin Bay is shown by dotted line together with the rate of spreading in millimeters per year. The thick stippled area in the south marks the position of a linear feature whose direction coincides with the direction of the plate motions in this region. Taken from Srivastava, 1978 with permission from Blackwell Scientific Publications, Ltd.

origin of the Amerasian Basin remains uncertain. There are three classes of scenarios for the creation of the oceanic crust found in the Amerasian Basin. The first, already mentioned, is the view held by a few Russian oceanographers that the oceanic crust of the Amerasian Basin was oceanized in situ from continental crust. The second class of scenarios holds that the oceanic crust of the basin originated in the Pacific and was trapped in its present position by continental plate movements. The last idea is that the oceanic crust formed in situ sometime between the middle Paleozoic and the Tertiary.

The second model, which assumes that the oceanic crust in the Amerasian Basin was trapped there, has been most often expounded by Churkin (1969, 1973; Churkin and Trexler, 1980, 1981). Figure 6 shows one of Churkin and Trexler's (1980) paleogeographic reconstructions, according to which, what is now the Amerasian Basin consists of the Proto-Pacific (Kula) plate. The Kolyma block of Siberia is postulated to have been rafted into the region and to have docked with Siberia during the Early Cretaceous. Subsequently the Atlantic opened in Late Cretaceous time, and North America including the Brooks Range and its extension into northeastern Siberia, the Chukotsk Peninsula, moved northwest and sutured against the Kolyma block. This collision isolated an old piece of the Kula plate shown as (1) in figure 6 and thereby formed the Arctic Basin. Other pieces of the Kula plate, trapped later south of the Arctic would include the Yukon-Koyukuk region of west central Alaska (2) and the Bering Sea (3). While magnetic anomaly identifications in the Bering Sea are held to support the entrapment idea (Cooper et al., 1976), Langseth et al., (1980) argue that heat flow in the Bering Sea necessitates rifting in that basin after initiation of subduction along the Aleutian trench (Grow and Atwater 1970) and consequently that most of the Bering Sea is, in fact, a back-arc basin and not trapped Kula plate.

The third class of models concerning the formation of the Amerasian Basin considers that it is underlain by oceanic crust formed in situ. Two major features of the Amerasian Basin are pertinent to any of these models. The first feature is the Alpha-Mendeleev Ridge. The Alpha Ridge nearly intersects the Lomonosov Ridge near the northern coast of Ellesmere island while the Mendeleev Ridge is most prominent north of Wrangell Island. The two features are usually contoured as a continuous ridge (Johnson et al., 1979) but there are insufficient data between $80-85^{o}$N along the 180° meridian to ascertain whether they constitute a continuous prominent feature or if there is a saddle between them as shown on most charts. Near the Canadian and Siberian margins these ridges are as shallow as 1200 meters while the saddle may be as deep as 2200 meters. The other prominent bathymetric feature in the Amerasian Basin is the flat topped Chukchi Continental Borderland which is as shallow as 273 meters (Johnson et al., 1978) on Chukchi Cap.

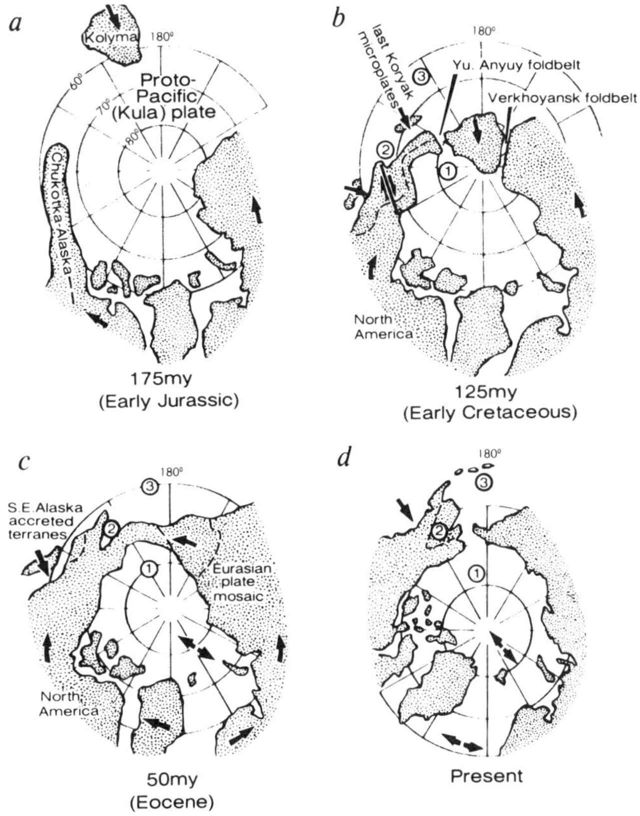

Fig. 6. Paleogeographic reconstructions of the Arctic showing northward drift, collision, and accretion. Projection is flat polar, and shelf areas are not shown so that the land areas can be recognized. These maps are not intended to show rigorous geographic locations but rather the process of terrane accretion and Kula plate capture. Taken from Churkin and Trexler, 1980 with permission from Elsevier Scientific Publishing Co.

Recent work by Grantz et al. (1979) on the Chukchi shelf suggest that the Chukchi Borderland may have rifted away from the mainland after the opening of the Canada Basin and as such should not be considered as a hinderance to most tectonic reconstructions.

The proposed seafloor spreading histories of the Amerasian basin depend heavily on the interpretation of the Alpha-Mendeleev Ridge. Of the four different modes of rifting suggested for the Amerasian Basin, the first utilized the ridge as a spreading center while another utilized it as a subduction zone island arc. Vogt and Ostenso (1970) summarize the early evidence suggesting the Alpha Cordillera as an inactive branch of the global mid-ocean ridge system. This hypothesis was originally suggested independently by Johnson and Heezen (1967), Beal (1968) and Vogt and Ostenso (no reference) according to Vogt and Ostenso (1970). These workers cite magnetic anomalies with wavelengths of 5 to 30 km, amplitudes exceeding 1000 gammas, and a reasonable degree of mutual correlation and biaxial symmetry as well as gravity and seismic refraction data to support their contention. Vogt and Ostenso (1970) postulated active spreading on the Alpha Ridge for the period 60-40 Ma with the spreading ceasing abruptly at 40 Ma. A recent paper by Taylor et al., (1981) reports that additional aeromagnetic lines flown perpendicular to the Alpha Ridge also show anomalies of high amplitude. These anomalies overlie much of the Alpha-Mendeleev Ridge and parts of the adjacent basins. The large amplitudes of the magnetic anomalies and the comparatively shallow depths of the Alpha-Mendeleev Ridge suggested to Herron et al., (1974) that the ridge cannot be an ancient spreading center, and support the hypothesis that it is a compressional feature.

Magnetic anomalies over the Makarov Basin, which lies between the Alpha-Mendeleev Ridge and the Lomonosov Ridge are tentatively identified by Taylor et al., (1981) to indicate that spreading there began in the Late Cretaceous at anomaly 34 (80 Ma) and continued until mid-Eocene (Anomaly 21; 53 Ma). Vogt et al., (1982) show the same anomalies and say they may be late Cretaceous spreading lineations (Anomaly 33? to Anomaly 28?) from 78 Ma to 65 Ma. Sweeney et al., (1982) report heat flow in the Makarov Basin as 60-70 mWm^{-2} which is somewhat less than the 90 mWm^{2} reported by Pushcharovskiy (1976) for the west slope of the Makarov basin. The Russian value, assuming the site of the measurement has not been disturbed, suggests that the basin is underlain by oceanic crust as young as 25 Ma, if one uses the Parsons and Sclater (1977) heat flow-versus-age relationship. Even the Canadian values give an uncorrected age of 42 to 60 Ma. These are minimum ages but they agree in a general way with the age of the Makarov Basin suggested by Taylor et al. (1979). Seafloor spreading in the Makarov Basin between 80 Ma and 53 Ma would have to have been connected to the spreading documented by Srivastava (1978) in the Labrador Sea. Srivastava's rotational poles for the opening of the Labrador Sea between anomaly 33 (78 Ma) time to anomaly 21 (53 Ma) do not preclude opening of the Makarov Basin, but there is no presently recognized structural link between the

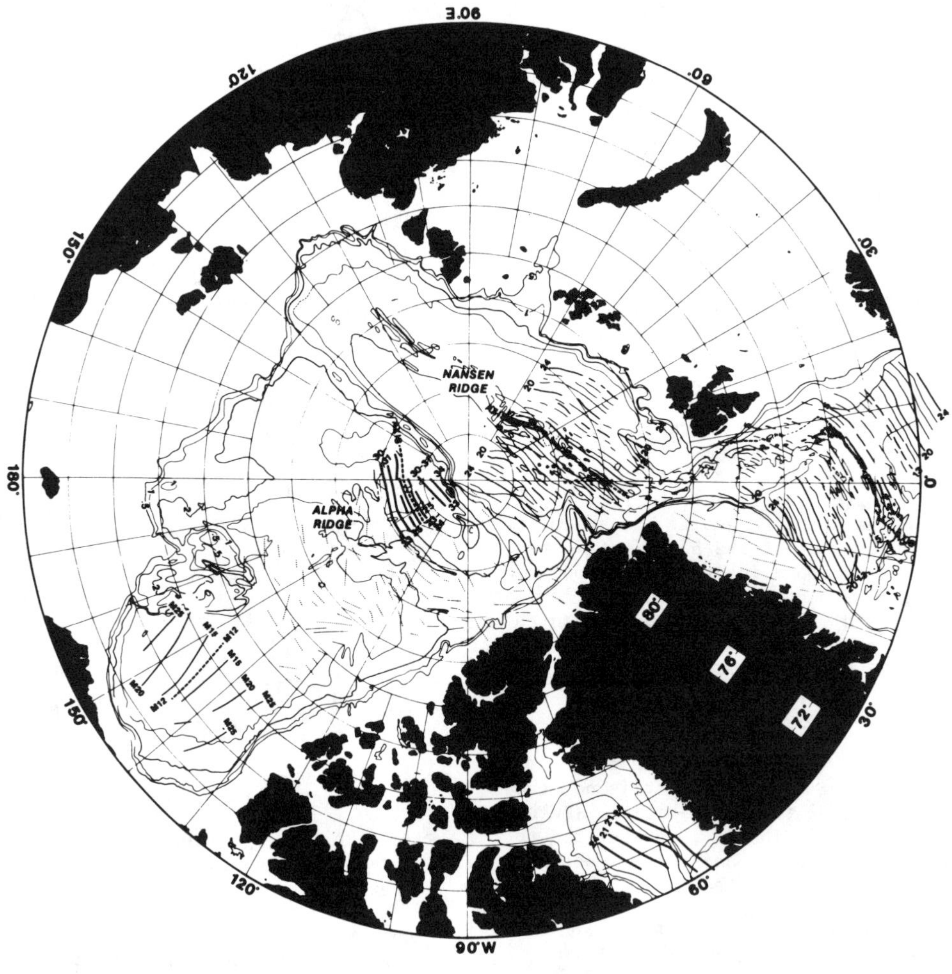

Fig.7. Polar projection of Arctic with major magnetic lineations from this study and Srivastava (1978), Phillips et al. (1981), and Vogt et al. (1979a). Bathymetric base from Johnson et al. (1979). Dashed lines represent inferred axes of sea-floor spreading. Taken from Taylor et al., 1981.

spreading centers in the Labrador Sea and in the Makarov Basin. Possibly subsequent plate motion has obliterated the necessary transform faults that might have connected the two basins.

If the Makarov Basin is underlain by late Cretaceous to early Tertiary seafloor and the Alpha Ridge is not an extinct spreading center, then the Canada Basin is the only other area of oceanic crust in the Amerasian basin. Taylor et al., (1981) report that a fan shaped set of low amplitude aeromagnetic anomalies over the Canada Basin (Figure 7) support the earlier idea of Carey (1958) that the Canada Basin opened about a pole near the Gulf of Alaska based on the bend in the geology of the North American Cordillera in Alaska, the so-called Alaska Orocline. Tailleur (1969), Tailleur and Brosgé (1970), and Rickwood (1970) elaborated on the idea, based on the need for a northern source land for the Carboniferous to Triassic clastic sedimentary strata of Alaska's north slope. These workers variously placed the pole in east central Alaska or the Mackenzie delta. Taylor et al., (1981) acknowledge that the lineations in the Canada Basin are very low amplitude and difficult to correlate but nevertheless identify the oldest anomaly as M25 (153 Ma) and the youngest which is the extinct spreading center as M12 (127 Ma). The early geological papers such as Tailleur and Brosgé (1970), Brosgé and Tailleur (1970) and Tailleur (1973) would agree with this or an older Jurassic age, while Rickwood (1970) and more recent marine geophysical papers by Grantz et al., (1979, 1981), Grantz and May (in press) and Lawver and Baggeroer (in press) would support a similar model but with most of the rotation occurring between 125 Ma and 80 Ma. Vogt et al., (1982) have recently written a lengthy paper that details a number of different scenarios for the Canada Basin. Their final model, which involves a ridge-ridge-ridge triple junction at 79°N, 140°W is much too complicated to discuss here but they do acknowledge that the paucity of good data does not rule out other models and leaves the question of the tectonic evolution of the Canada Basin unresolved.

There are at least three other models for the evolution of the Canada Basin besides the oroclinal bending suggested by Carey (1958) and Tailleur (1969) and the elaborate model suggested by Vogt et al., (1982). Herron et al., (1974), after Pitman and Talwani (1972) had determined a late Mesozoic pole of rotation for North America and Eurasia that lay near the north end of Greenland, realized that the tenets of plate tectonics required concurrent compression in the Arctic basin on the other side of the pole. Their scenario begins with paleozoic subduction of the Arctic Ocean along Arctic Canada, and the boundary between the Alaska-Chukotka continental region and the Arctic Basin having been a right lateral transform fault as shown in figure 8 taken from Herron et al. (1972). The Kolyma block eventually collided with Arctic Canada during the early Paleozoic. During the Jurassic magnetic quiet period (180-150 Ma), the Kolyma block was moved away from Arctic Canada and the Canada Basin was reopened. Concurrently the boundary between the north slope of Alaska and the Arctic Basin operated as a left-lateral transform fault. Opening during the Jurassic magnetic quiet period was held to explain the low-amplitude magnetic anomalies of the Canada Basin. Then when the Atlantic opened about a pole in northern Greenland in Late Cretaceous time, the Canada Basin was put under compression and the Alpha-Mendeleev Ridge was formed as an incipient island arc-subduction zone complex.

Compression on the Alpha-Mendeleev Ridge in late Cretaceous time is incompatible with the suggestion of Taylor et al., (1981) that there was spreading in the adjacent Makarov Basin during the Late Cretaceous and Paleocene. It is possible, however, that Late Cretaceous opening of the North Atlantic about a pole in northern Greenland may not have required compression in the Arctic Basin, since present day spreading in the Eurasian Basin does not seem to be

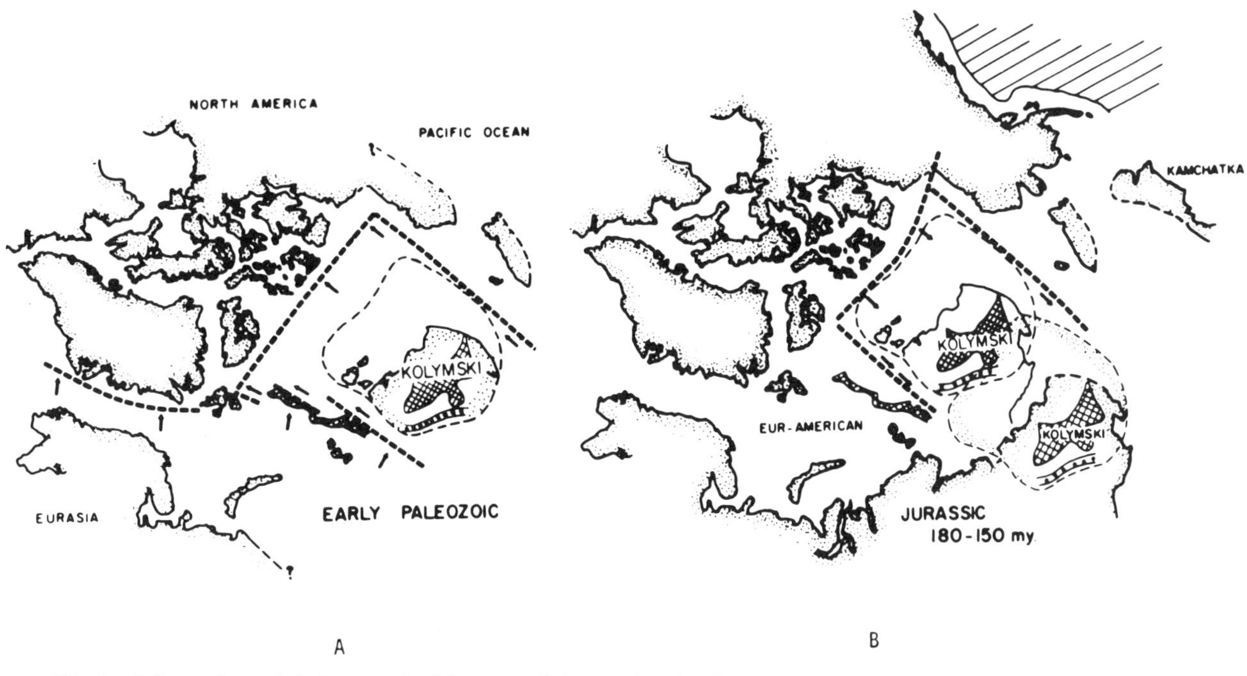

Fig 8. Schematic model for early history of Amerasian Basin. Heavy dashed lines with arrow = major plate boundaries and direction of movement relative to North America. A, closure of proto-Amerasian Basin in early Paleozoic time. This motion culminated with mid-Paleozoic folding of Franklinian (Parry Island and Ellesmere) geoclinal sediments and shedding of clastic rocks onto Canadian Arctic Archipelago, Brooks Range area of Alaska, and Wrangel Island north of Chukotskiy (easternmost tip of Siberia). B, our model for opening of Amerasian Basin during Jurassic magnetic quiet period, as Kolymski broke away from North America. Taken from Herron et al., 1974 with permission from Geology.

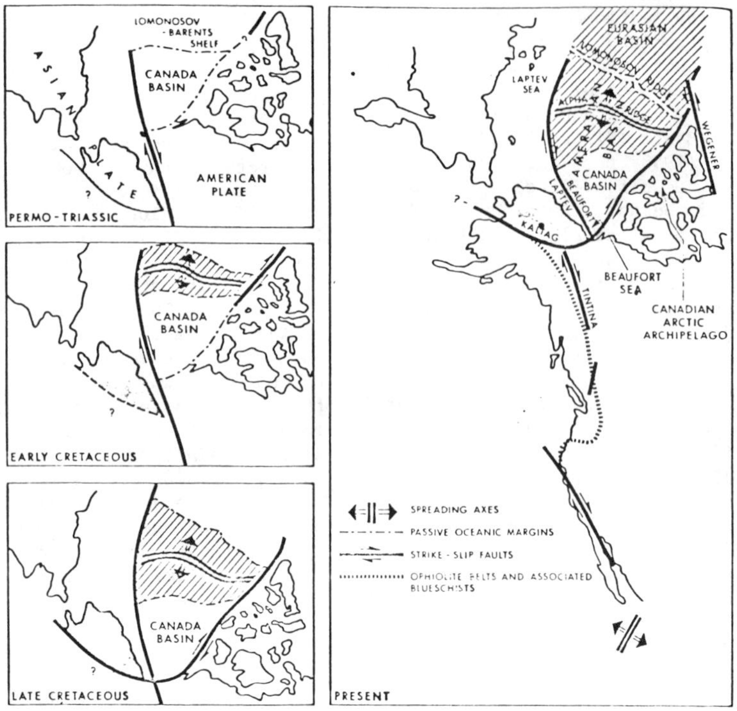

Fig. 9. Model for the evolution of the Amerasian Basin and associated displacements along Tintina and Kaltag faults. Taken from Jones, 1980 with permission from Nature.

producing compression on the opposite side of its pole, in northeast Siberia and the northwest Pacific Basin. In addition, Srivastava's (1978) work in the Labrador Sea indicates that the opening of the North Atlantic was geometrically far more complicated than Pitman and Talwani (1972) modeled it to be.

Jones (1980) presented a model for the opening of the Canada Basin that was similar to Herron et al. (1974) but the direction of opening is at right angles to their model. Jones' model shown in figure 9, requires the Alpha-Mendeleev Ridge to have been a Jurassic-Cretaceous spreading center with the southern Canada Basin being unspecified Paleozoic oceanic crust. His model requires a great deal of motion along the Kaltag Fault of northern Alaska. There is no evidence however that motion along the Kaltag Fault can be connected through the Yukon (Brosgé, personal communication, 1983) and then along the Arctic Canada continental margin as a transform fault.

The last model for the evolution of the Canada Basin is that proposed by Mair and Forsyth (1982) and is a variation on the model first suggested by Carey (1958). The Canada Basin first opened about pole 1 shown in figure 10 and then the Chukchi Continental Borderland rotated away from northern Alaska about pole 2.

Computer Models

By using pole 1 of Mair and Forsyth (1982)(a modification by Grantz et al., (1979) of a suggestion first made by Rickwood (1970))we tested the bathymetric fit of the opposite sides of the Canada Basin. Our computer fit for plate rotations using the 1000 meter bathymetric contours taken from the Heezen and Tharp (1975) chart of the Arctic region gave a good match-up between Arctic Canada and the Siberian shelf as far east as the Alpha-Mendeleev Ridge. We also used Savostin and Karasik's (1981) poles to "close-up" the Eurasian basin. The most striking problem with closing up the Arctic as we did, was the unacceptable overlap of Greenland with Eurasia mentioned previously. If Srivastiva's (1978) pole of opening for the Labrador Sea-Baffin Bay is used, we feel that we will get a much better match between Greenland and Eurasia, indicating that Greenland must have moved away from North America along Nares Strait.

While our simple plate rotations based on bathymetry do not resolve the age of the Canada Basin,they certainly support Carey (1958) and his model for the rotational rifting of northern Alaska away from the Canadian Arctic Islands. If the Alpha Ridge is not a spreading center, which seems reasonably certain because its depth is inconsistent with the age-versus-depth curves of Parsons and Sclater (1977), then we must suggest another explanation for it. If the Alpha-Mendeleev Ridge formed when the Makarov Basin first began to spread and the Lomonosov Ridge was still attached to Eurasia, the Alpha-Mendeleev Ridge may be a continental fragment that partially foundered and otherwise deformed prior to rifting of the Makarov Basin. When actual seafloor spreading began in the

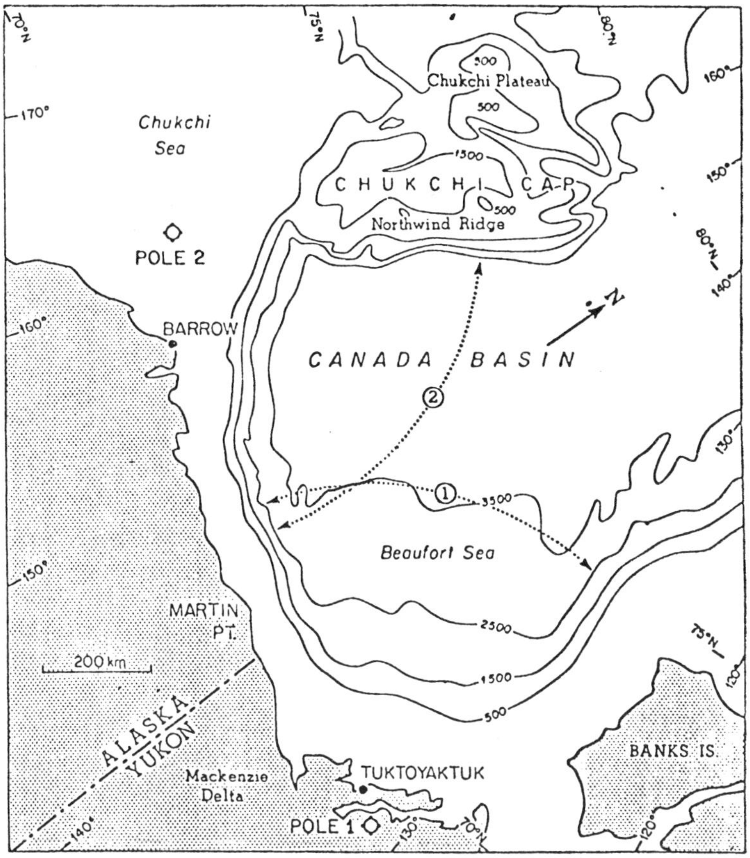

Fig 10. Bathymetry is in meters. Pole 1 in the Mackenzie Delta is initial pole of opening of the Canada Basin while pole 2 is a subsequent pole of opening for the rotation of the Chukchi Cap away from the North Slope of Alaska. Taken from Mair and Forsyth, 1982 with permission from Elsevier Scientific Publishing Co.

Makarov Basin it may have isolated the Alpha-Mendeleev Ridge as a semi-continental fragment.

An alternative explanation might exist if the Canada Basin tectonic province includes the Alpha Rise and the Amerasian side of the Makarov Basin. As the Canada Basin opened there may have been a hot spot that started near the ridge crest on the Siberian side and which produced an anomalous volcanic mass of material similar to the Rio Grande and Walvis Rises in the South Atlantic (Rabinowitz and La Breque, 1979). The hot spot slowly migrated across the ridge and ended near Arctic Canada. These last two suggestions for the origin of the Alpha-Mendeleev Ridge are highly speculative.

Conclusion

Most tectonophysicists would agree that the Nansen-Gakkel Ridge in the Eurasian Basin is the presently active plate boundary between Eurasia and North America. They would also agree that the Lomonosov Ridge is a continental fragment that has been rifted away from the Barents and Kara Sea shelves by spreading on the Nansen-Gakkel Ridge. The early evolution of the Nansen-Gakkel Ridge and how it connected to the Mid-Atlantic Ridge is still controversial particularly concerning transform motion along the Nares Strait (Dawes and Kerr, 1982). The tectonic origin of the Alpha-Mendeleev Ridge, and the age and mode of origin of the Canada Basin are unknown.

Acknowledgments

This work was supported by the Division of Polar Programs of the NSF(8117402 DPP) for L. Lawver and L. Meinke and by the U.S. Geological Survey from A. Grantz. The background for this review paper grew out of numerous discussions that the authors have had with their colleagues, particularly those in the Alaskan and Pacific-Arctic marine geology branches of the U.S.G.S. We wish to thank J. Peoples for typing the paper and E. Harding for her tolerance and help.

REFERENCES

Baggeroer, A.B., and Falconer, R., Array refraction profiles and crustal models of the Canada Basin, J. Geophys. Res., vol. 87, pp. 5461-5476, 1982.

Balkwill, H.R., Evolution of Sverdrup Basin, Arctic Canada, Am. Assoc. Pet. Geol. Bull., vol. 62, pp. 1004-1028, 1978.

Beal, M.A., Bathymetry and structure of the Arctic Ocean, Ph.D. Dissertation, Oreg. State Univ., Corvallis, 1968.

Beloussov, V.V., Against the hypothesis of ocean-floor spreading, Tectonophysics, vol. 9, pp. 489-511, 1970.

Brosge, W.P., and Tailleur, I.L., Depositional history of northern Alaska, in Proceedings of the geological seminar on the North Slope of Alaska, eds., W.L. Adkison and M.M. Brosge, Pacific Sec., Am. Assoc. Petroleum Geologists, Los Angeles, pp. D1-D17, 1970.

Carey, S.W., The orocline concept in geotectonics, R. Soc. Tasmania Pap. Proc., vol. 89, pp. 255-288, 1955.

Churkin, M., Jr., Paleozoic tectonic history of the Arctic basin north of Alaska, Science, vol. 165, no. 3893, pp. 549-555, 1969.

Churkin, M., Jr., Western Boundary of the North American Continental plate in Asia, Geol. Soc. Am. Bull., vol. 83, pp. 1027-1036, 1972.

Churkin, M., Jr., Geologic concepts of Arctic Ocean Basin: in Arctic Geology, ed. Pitcher, M.G., American Assoc. Petrol. Geol. Memoir, vol. 19, pp. 485-499, 1973.

Churkin, M. and Trexler, J.H., Jr., Circum-Arctic plate accretion-isolating part of a Pacific plate to form the nucleus of the Arctic Basin, Earth Planet. Sci. Lett., vol. 49, pp. 356-362, 1980.

Churkin, M., Jr., and Trexler, J.H., Jr., Continental plates and accreted oceanic terranes in the Arctic, in The Ocean Basins and Margins, eds. A.E.M. Nairn, M. Churkin, and F.G. Stehli, vol. 5, pp. 1-20, Plenum, New York, N.Y. 1981.

Coles, R.L., Hannaford, W. and Haines, G.V. Magnetic anomalies and the evolution of the Arctic, in Arctic Geophysical Review, ed. J.F. Sweeney Publ. Earth Phys. Branch, Dep. Energy, Mines Resour., Ottawa, vol. 45, pp. 51-66, 1978.

Cooper, A.K., Marlow, M.S., and Scholl, D.W., Mesozoic magnetic lineations in the Bering Sea marginal basin, J. Geophys. Res., vol. 81, pp. 1916-1943, 1976.

Dawes, R.R. and Kerr, J.W., Nares Strait and the drift of Greenland: A conflict in plate tectonics, Meddelelser om Gronland, Geo. Science, vol. 8, 1982.

DeLaurier, J., Magnetic anomalies and the evolution of the Arctic, in Arctic Geophysical Review, ed. J.F. Sweeney, Publ. Earth Phys. Branch, Dep. Energy Mines Resour., Ottawa, vol. 45-4, pp. 87-90, 1978.

Demenitskaya, R.M. and Karasik, A.M., Mirovaya Sistema Spedinno-okeanskikh Knrebtov i ee Severnyi Fragment, Noveishie Drizheniya, Vulkanize i Zemletriaseniya Materikov i Dna Okeanov, Nauka, pp. 249-257, 1969.

Dutro, J.T., Jr., Geology of Alaska bordering the Arctic Ocean, in The Ocean Basins and Margins, eds. A.E.M. Nairn, M. Churkin, and F.G. Stehli, vol. 5, pp. 21-36, Plenum New York, N.Y., 1981.

Eittreim, S. and Grantz, A., CDP Seismic sections of western Beaufort continental margin, Tectonophysics, vol. 59, pp. 251-262, 1979.

Eldholm, O. and Talwani, M., Sediment distribution and structural framework of the Barents Sea, Geol. Soc. Am. Bull., vol. 88, pp. 1015-1029, 1977.

Forsyth, D.A., Mair, J.A. and Fraser, I., Crustal structure of the Central Sverdrup Basin, Can. J. Earth Sci., vol. 16, pp. 1581-1598, 1979.

Geol. Survey of Canada, Low level aeromagnetic charts 8071G-8074G, 8089G-8092G, 8105G-8109G 8118G-8122G, 7255G, 8129G-8132G, 8141G-8144G, 4 miles to an inch, 1967-1968.

Grantz, A., and May, S.D., Rifting history and structural development of the continental margin north of Alaska, Proc. Hedburg Conf. on Continental Margin Processes, Am. Assoc. Pet. Geol., Mem., (in press) 1982.

Grantz, A., Eittreim, S. and Dinter, D.A., Geology and tectonic development of the continental margin north of Alaska, Tectonophysics, vol. 59, pp. 263-291, 1979.

Grantz, A., Eittreim, S., and Whitney, O.T., Geology and physiography of the continental margin north of Alaska and implications for the origin of the Canada Basin, in The Ocean Basins and Margins, eds. A.E.M.Nairn and M. Churkin, vol. 5, pp. 21-36, Plenum, New York, N.Y., 1981.

Grantz, A. and May, S.D., Origin of the Canada Basin as inferred from the Seismic Geology of off-shore Northern Alaska, Proceedings of the Alaska Geological Society, Symposium on the origin of the Canada Basin, Anchorage, Alaska, 1983. (in press)

Grow, J.A. and Atwater, T., Mid-tertiary tectonic transition in the Aleutian Arc, Geol. Soc. Amer. Bull., vol. 81, pp. 3715-3722, 1970.

Hamilton, W., The Uralides and the motion of the Russian and Siberian platforms, Geol. Soc. Am. Bull., vol. 81, pp. 2253-2576, 1970.

Heezin, B.C. and Ewing, M., The mid-oceanic ridge and its extension through the arctic basin, in Geology of the Arctic, vol. 1, ed. G.O. Raasch, pp. 622-642,University of Toronto Press, Toronto, Ontario, 1961.

Heezen, B.C. and Tharp, M., Map of the Arctic Region, World 1:5,000,000, Sheet 14, Am. Geograph. Soc., New York, N.Y., 1975.

Herron, E.M., Dewey, J.F. and Pitman, W.C., III, Plate tectonics model for the evolution of the Arctic, Geology, vol. 2, pp. 377-380, 1974.

Hillhouse, J.W. and Gromme, S., Cretaceous remagnetization of Paleozoic sedimentary rocks in the Brooks Range, Alaska, in Geology of the National Petroleum Reserve in Alaska, U.S. Geol. Surv. Prof. Pap., (in press) 1982.

Hunkins, K. and Tiemann, W., Geophysical data summary for Fletchers Ice Island (T-3) May 1962-Oct. 1974, Lamont-Doherty Geol. Obs., Tech. Rep. CU-1-77, pp. 219, 1977.

Johnson, G.L., Taylor, P.T., Vogt, P.R., and Sweeney, J.F., Arctic Basin morphology, Polarforschung, vol. 48, pp. 20-30, 1978.

Johnson, G.L., Monahan, D., Gronlie, G. and Sobczak, L., Bathymetric Chart of the Arctic, Gebco Sheet 5·17, Canadian Hydrographic Service, Ottawa, 1979.

Jones, P.R., Evidence from Canada and Alaska on plate tectonic evolution of the Arctic Ocean Basin, Nature, vol. 285, pp. 215-217, 1980.

Karasik, A.M., Anomalous magnetic field of Eurasian basin of the Arctic Sea, Dokl. Akad. Nauk. SSSR, vol. 211, No. 1 (in Russian), 1973.

Karasik, A.M., Plate tectonics and Eurasian basin of the Arctic Sea, in Problems of geology of the Earth's polar regions, Izd. NIIGA, Leningrad (in Russian), 1974.

Karasik, A.M., Yevrazisikii bassein cevernogo ledovitogo okeana's pozitisi tektoniki plit., in Problemi Geologii Polynarnikh Oblastei Zemli, Nidra, Leningrad, pp. 24-31, 1974.

Karasik, A.M., Basic specifics of the history of development and the structure of the Arctic basin bottom according to aeromagnetic data, in Marine Geology, Sedimentology, Sedimentary Petrography, and Geology of the Ocean, Academy of Sciences of the U.S.S.R., Ministry of Geology of the U.S.S.R., National Committee of Geologists of the Soviet Union, Leningrad, pp. 178-193, (Transl. (1981): NASA GSFC-07.), 1980.

Kerr, J.W., A plate tectonic contrast in northern Canada, Geophys. Assoc. Can., Spec. Pap., vol. 20, pp. 457-484, 1980.

Khramov, A.N., Paleomagnetic directions and paleomagnetic poles - data for U.S.S.R., Acad. Sci. U.S.S.R., Sov. Geophys. Commun. WD3, pp. 1-43, 1975.

Kovacs, L.C., Motion along Nares Strait recorded in the Lincoln Sea: aeromagnetic evidence, in Nares Strait. Medd. Gronland, eds. P.R. Dawes and J.W. Kerr, Geosci., vol. 8, pp. 105-120, 1982.

Kovacs, L.C., and Vogt. P.R., Depth-to-magnetic source analysis of the Arctic Ocean region, in Structure of the Arctic, eds. G.L. Johnson and J.F. Sweeney, Tectonophysics, vol. 89, pp. 255-294, 1982.

Kovacs, L.C., Bernero, L.C., Johnson, G.L., Pilger, R., Taylor, P.T., and Vogt, P.R., Residual magnetic anomaly chart of the Arctic Ocean region, Geol. Soc. Am. Map Ser., 1982 (in press).

Kristoffersen, Y. and Talwani, M., Extinct triple junction south of Greenland and the Tertiary motion of Greenland relative to North America, Geol. Soc. Am. Bull., vol. 88, pp. 1037-1049, 1977.

LaBrecque, J.L., Kent, D.V. and Cande, S.C., Revised magnetic polarity time scale for late Cretaceous and Cenozoic time, Geology, vol. 5, pp. 330-335, 1977.

Langseth, M.G., Hobart, M.A. and Horai, K., Heat flow in the Bering Sea, J. Geophys. Res., vol. 85, pp. 3740-3750, 1980.

Larsen, R.L. and Hilde, T.W.C., A revised time scale of magnetic reversals for the Early Cretaceous and Late Jurassic, J. Geophys. Res., vol. 80, pp. 2586-2594, 1975.

Lawver, L. and Baggeroer, A., A note on the age of the Canada Basin, Proceedings of the Alaska Geological Society, Symposium on the origin of the Canada Basin, Anchorage, Alaska, 1983, (in press)

Mair, J.A. and Lyons, J.A., Crustal structure and velocity anisotrophy beneath the Beaufort Sea, Can. J. Earth Sci., vol. 18, pp. 724-741, 1981.

Mair, J.A. and Forsyth, D.A., Crustal structures of the Canada Basin near Alaska, the Lomonosov Ridge and adjoining basins near the North Pole, Tectonophysics, vol. 89, pp. 239-253, 1982.

McElhinney, M.W., Paleomagnetism and plate tectonics of eastern Asia, in The Western Pacific: Island Arcs, Marginal Seas, ed. P.J.Coleman, Geochemistry, Univ. of Western Australia Press, Nedlands, W.A., pp. 407-414, 1973.

Milen, A.R., A seismic refraction measurement in the Beaufort Sea, Bull. Seismol. Soc. Am., vol. 56, pp. 775-779, 1966.

Minster, J.B. and Jordan, T.H., Present-day plate motions, J. Geophys. Res., vol. 83, pp. 5331-5354, 1978.

Morgridge, D.L. and Smith, W.B., Geology and discovery of Prudhoe Bay Field, eastern Arctic slope, Alaska, Am. Assoc. Pet. Geol. Mem., vol. 16, pp. 489-501, 1972.

Newman, G.W., Mull, C.G., and Watkins, N.D., Northern Alaska paleomagnetism, plate rotation, and tectonics, Alaska Geol. Soc. Symp., Anchorage, Alaska, pp. 16-19, (abstr.), 1977.

Ostenso, N.A., Aeromagnetic survey of the Arctic Ocean Basin, Proc. 13th Alaskan Sci. Cong., pp. 115-148, 1960.

Ostenso, N.A., Seafloor spreading and the origin of the Arctic Ocean Basin, in Implications of continental drift to the Earth Sciences, Academic Press, New York, N.Y., pp. 165-173, 1972.

Ostenso, N.A. and Wold, R.J., Aeromagnetic Survey of the Arctic Ocean: techniques and interpretations, Mar. Geophys. Res., vol. 1, pp. 178-219, 1971.

Parsons, B. and Sclater, J.G., An analysis of the variation of ocean floor bathymetry and heat flow with age, J. Geophys. Res., vol. 82, pp. 803-827, 1977.

Phillips, J.D., Feden, R., Fleming, H.S. and Tapscott, C., Aeromagnetic studies of the Greenland/

Norwegian Sea and Arctic Ocean, Bull. Geol. Soc., Am., (in press) 1982.

Pitman, W.C., III and Talwani, M., Sea-floor spreading in the North Atlantic, Geol. Soc. Am. Bull., vol. 83, pp. 619-646, 1972.

Pogrebitskiy, Yu. Ye., The geodynamic system of the Arctic Ocean and its structural evolution, Sov. Geol., vol. 12, pp. 3-22, Translated in Int'l. Geol. Rev., vol. 20, pp. 1251-1266, 1978.

Pushcharovskiy, Yu. M., Tectonics of the Arctic Ocean Basin, Geotectonics, vol. 10, pp. 85-91, 1976.

Rabinowitz, P.D. and J. La Breque, The Mesozoic South Atlantic Ocean and Evolution of Its Continental Margins, J. Geophys. Res., v.84, 5973-6002, 1979.

Rickwood, F.K., The Prudhoe Bay Field, in Proceedings of the geological seminar on the North Slope of Alaska, eds. W.L. Adkison and M.M. Brosge, Am Assoc, Petro. Geol., Pacific Sec., Los Angeles, L1-L11, 1970.

Ruppel, B.D. and McHendrie, A.G., Free-air gravity anomaly map of the eastern Chukchi and southern Beaufort Seas, USGS Map MF-785, 1976.

Savostin, L.A., and Karasik, A.M., Recent plate tectonics of the Arctic Basin and of north-eastern Asia, Tectonophysics, vol. 74, pp. 111-145, 1981.

Scrutton, R.A., On Sheared passive continental margins, Tectonophysics, vol. 59, pp. 293-305, 1979.

Sobzak, L.W., Bathymetry of the Arctic Ocean north of 85°N latitude, Tectonophysics, vol. 42, pp. T27-T33, 1977.

Srivastava, S.P., Evolution of the Labrador Sea and its bearing on the early evolution of the North Atlantic, Geophys. J.R. astr. Soc., vol. 52, 313-357, 1978.

Sweeney, J.F., Arctic seafloor structure and tectonic evolution, Paleo-reconstructions of the continents, Am. Geophys. Union Geodyn. Ser., vol. 2, pp. 55-64, 1981.

Sweeney, J.F., Irving, E. and Geuer, J.W., Evolution of the Arctic Basin, in Arctic Geophysical Review, ed. J.F. Sweeney, Earth Phys. Branch, Dep. Energy Mines Resour., Ottawa, vol. 45, no. 4, pp. 91-100, 1978.

Sweeney, J.F., Weber, J.R., and Blasco, S.M., Continental ridges in the Arctic Ocean, LOREX constraints, Tectonophysics, vol. 89, pp. 217-237, 1982.

Tailleur, I.L., Rifting speculation of the geology of Alaska's North Slope: Oil and Gas Jour., vol. 67, no. 39, pp. 128-130, 1969.

Tailler, I.L. and Brosge, W.P., Tectonic history of northern Alaska, in Proceedings of the geological seminar on the North Slope of Alaska, eds. W.L. Adkison and M.M. Brosge, Am. Assoc. Petroleum Geologists, Pacific Sec., Los Angeles, E1-E19, 1970.

Tailleur, I.L., Probable rift origin of Canada Basin, in Arctic Geology, ed. M.G. Fitcher, Am. Assoc. Petr. Geol., Mem. 19, pp. 526-535, 1973.

Talwani, M. and Eldholm, O., Evolution of the Norwegian-Greenland Sea, Geol. Soc. Amer. Bull., vol. 88, pp. 969-999, 1977.

Taylor. P.T., Kovacs, L.C., Vogt, P.R., and Johnson, G.L., Detailed aeromagnetic investigation of the Arctic Basin, 2., J. Geophys. Res., vol. 86, pp. 6323-6333, 1981.

Vogt. P.R. and Einwich, A.M., Magnetic anomalies and sea-floor spreading in the western north Atlantic, and a revised calibration of the Keathley (M) geomagnetic reversal chronology, in Initial Reports of the Deep Sea Drilling Project, Government Printing Office, Washington, D.C., pp. 857-876, 1979.

Vogt. P.R. and Ostenso, N.A., Magnetic and gravity profiles across the Alpha Cordillera and their relation to Arctic sea-floor spreading, J. Geophys. Res., vol. 75, 4925-4937, 1970.

Vogt. P.R., Taylor. P.T., Kovacs, L.C., and Johnson, G.L., Detailed aeromagnetic investigations on the Arctic Basin, J. Geophys. Res., vol. 84, 1071-1089, 1979.

Vogt, P.R., Bernero, C., Kovacs, L.C., and Taylor, P., Structure and plate tectonic evolution of the marine Arctic as revealed by aeromagnetics. Proc. 28th Int. Congr. Geol. of Oceans Symp., Paris, July 7-17, 1980, Oceanol. Acta, SP:25-40, 1981.

Vogt, P.R., Kovacs, L.C., Bernero, C. and Srivastava, S.P., Asymmetric geophysical signatures in the Greenland-Norwegian and southern Labrador seas and the Eurasian basin, in: Structure of the Arctic, eds. G.L. Johnson and J.F. Sweeney, Tectonophysics, vol. 89, pp. 95-160, 1982.

Vogt, P.R., Taylor, P.T., Kovacs, L.C. and Johnson, G.L., The Canada Basin: Aeromagnetic constraints on structure and evolution, Tectonophysics, vol. 89, pp. 295-336, 1982.

Wilson, J.T., Continental drift, Sci. Amer., vol. 208, pp. 86-100, 1963.

Wilson, J.T., Evidence from Ocean Islands suggesting movement in the Earth, Phil. Trans. Roy. Soc., London, No. 1088, pp. 145-167, 1965.

SEISMIC EXPLORATION IN THE ARCTIC OCEAN

Arthur B. Baggeroer and Gregory L. Duckworth

Massachusetts Institute of Technology
Cambridge, Massachusetts 02139 USA

ABSTRACT

Because of its harsh environment, the Arctic still remains one of the least explored areas of the world. Research beginning in the early 1950's, including the passage of the submarine *Nautilus* beneath the ice cap, began to define the bathymetry and major features of the Arctic Ocean. The occupation of several long-lasting ice islands in the 1950's, 60's, and 70's started the accumulation of seismic data through reflection and limited refraction surveys, along with the acquisition of spot gravity, bathymetry, shallow core, and extensive physical oceanographic data. Aeromagnetic surveys in the mid and late 1970's set the tectonic framework for many major features of the region. From the late 70's to the present the establishment of several short term camps on the ice floes has greatly expanded the wealth of seismic reflection and refraction data available from the central Arctic. For these experiments seismic sources such as airguns and helicopter deployment of explosive charges at large offsets were received on multi-channel hydrophone arrays and ocean bottom seismometers. Recent advances in digital hardware and processing technology have allowed much more detailed analyses of these data on a routine basis than were possible in the past. Finally, the recent discovery of oil in continental margin regions has led to the beginning of what promises to be extensive exploration of the shelf areas by the oil industry. The large experimental and research resources of these companies will certainly contribute greatly to our knowledge of Arctic crustal structure. In this survey we hope to summarize some of the considerations and methods involved in Arctic geophysics, focusing mainly on seismic technology.

1. INTRODUCTION

The Arctic is one of the least known regions of the world's oceans. Major questions still remain concerning its origins and the processes that led to its tectonic evolution. The principal reason for this state is the paucity of seismic data available from the Arctic. While this can in part be attributed to the harshness of the cold environment upon men and their machines, it is the ice cover that is principal limiting factor. It is not stable or contiguous enough to reliably support the use of land techniques, yet only the most sturdy of ice breakers can maneuver within it, thereby preventing the use of routine marine procedures. As a result, seismic exploration in the Arctic has been a peculiar mix of techniques used in both land and marine environments.

Early explorers of the Arctic took soundings which indicated that the Arctic was not a shallow sea, but a full fledged ocean. It was, however, the accurate bathymetric records over long track lengths taken by nuclear submarine traverses that started to delineate its major physiographic features [Beal, 1968]. While data from ice stations and spot soundings have become available in greater quantity through several scientific research programs, the submarine data still provide the principal synoptic bathymetric information about the deep Arctic [Johnson, et al., 1979]. These research programs have, however, determined a great deal about the tectonic structure of the Arctic through seismic, gravity, magnetic, heat-flow, and coring data. The quest for oil and gas has led to extensive exploration of the continental margins [*Business Week*, 1983]. In the winter crews push offshore a few tens of kilometers on the shorefast ice, while in the summer marine surveys operate when the ice recedes. Nevertheless, long exploration lines across the margin and down the shelf-break are sparse [Eittreim and Grantz, 1979; Sundvor and Eldholm, 1979].

Before discussing the seismic exploration techniques used in the Arctic it is useful to examine its major features [Vogt and Avery, 1974]. Figure 1 indicates the Arctic and its adjacent continental margins. Tectonically, it can be divided into two basins - the Amerasian Basin bounded by western Siberia, Alaska and the Canadian archipelago, and the Eurasian Basin bounded by Greenland, Svalbard, European Russia and eastern Siberia. These basins are separated by the Lomonosov Ridge. The Eurasian Basin is relatively well understood and is hypothesized to be the result of seafloor spreading initiated 60 m.y.b.p. along the Gakkel (Nansen, Arctic Mid-Ocean) Ridge separating the Pole Abyssal Plain (Fram Basin) and the Barents Abyssal Plain (Nansen Basin) [Jackson, et al., 1982; Kristoffersen, et al., 1982; Kristoffersen, 1982]. This is the northern extension of the Mid-Atlantic Ridge. The continental margin north of Greenland and Svalbard is relatively narrow. It is, however, extended in each of these locations by the Morris Jesup Rise and the Yermak Plateau respectively. These are conjectured to have originated as a "hot spot" of thickened crust similar to Iceland that formed at the ridge and rifted apart about 35 m.y.b.p. [Feden, et al. 1979; Kovacs and Vogt,

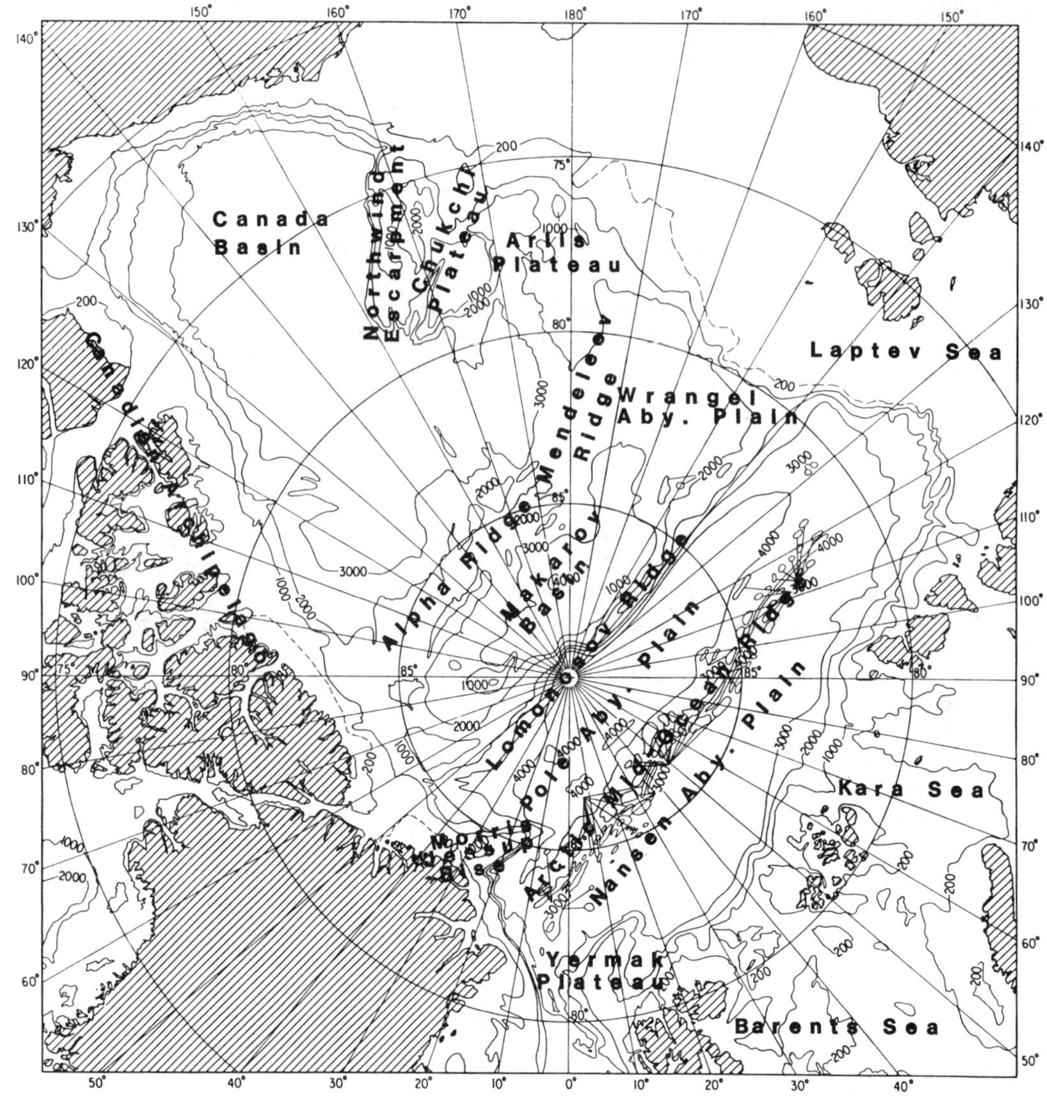

Fig. 1) Bathymetric and geographical feature map of the Arctic Ocean and adjacent continental margins.

1982]. The margin of the Barents, Kara and Laptev Seas is one of the widest in the world [Vogt and Ostenso, 1973] and is the site of significant Soviet oil and gas exploration.

The Amerasian Basin is much more complex tectonically [Hall, 1973; Vogt and Avery, 1974; Clark, 1981]. North of Alaska and Canada is the Canada Basin, an oceanic basin with heavy (over 4 km) sedimentation. This basin is hypothesized to have been formed between 80 and 135 m.y.b.p. [Johnson and Vogt, 1981; Baggeroer and Falconer, 1982; Lawver and Baggeroer, 1982; Vogt, et al., 1982; Sweeney, 1982; Sweeney, to appear]. Further to the north are the Alpha Ridge which is still of uncertain origin, and the Makarov Basin. Protruding from the continental margin of Siberia are the Northwind Ridge, the Chukchi Plateau and the Mendeleev Ridge. The Alaskan and Canadian continental margins, now the site of intense seismic exploration for oil, are average in their extent, while the Siberian margin is extremely wide. The entire tectonic structure for the Amerasian basin is still conjectural with much of the debate resting upon the acquisition of seismic data [Grantz, et al., 1981; Herron, et al., 1974]. The Lomonosov Ridge which crosses the central Arctic divides the Eurasian and Amerasian Basins. It is generally felt to be a continental fragment associated with the early rifting of the Eurasian Basin; however, its juxtaposition to the Makarov Basin is a part of the tectonic uncertainty of the Amerasian Basin [Sweeney, et al., 1982; Mair and Forsyth, 1982].

Figure 2 illustrates the tracks of a series of drifting buoys in the Arctic over a 7-8 month period [ICEX, 1979]. The prevailing current in the eastern Arctic sweeps from Siberia to the Fram Straits between Svalbard and Greenland. The western Arctic is dominated by the Circum-polar Current and the Beaufort Gyre [Coachman and Aagaard, 1974]. While there are exceptions, the ice conditions in the eastern Arctic are much more dynamic than those in the western Arctic. This has impacted the seismic exploration at several ice camps when the ice floes upon which the camps were sited broke up during the course of the experiments [Hunkins, et al., 1979; Baggeroer and Dyer, 1982].

Figure 3 indicates the average maximum and minimum ice cover [Central Intelligence Agency, 1975;

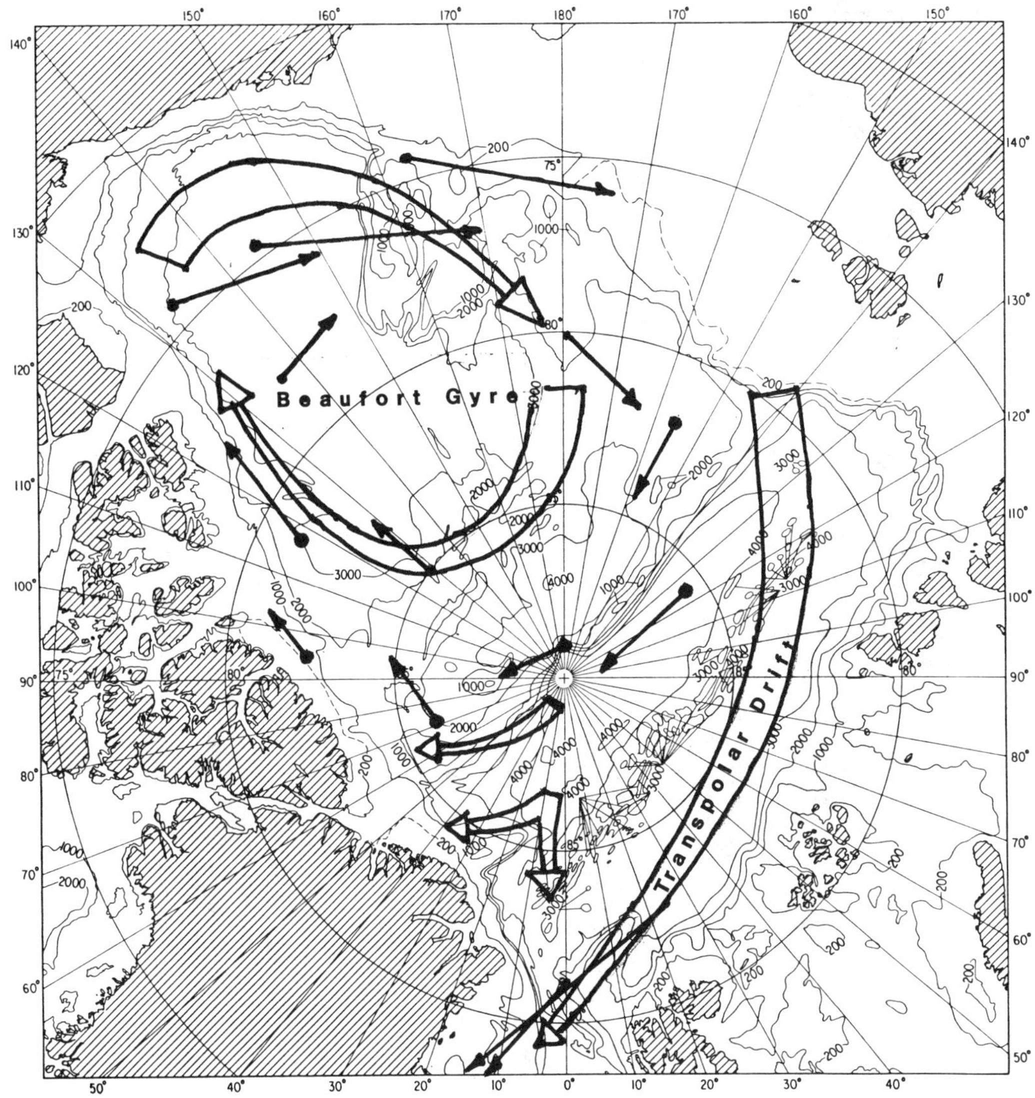

Fig. 2) Map of ice drift patterns in the central Arctic from the ICEX buoys showing initial and final locations from 7 month drift [from Thorndike, ICEX report, 1979]. General drift systems shown by large arrows from CIA Polar Regions Atlas, 1976.

Wadhams, 1981]. Pack ice in various states of unrest covers the central Arctic year round; however, its seasonal retreat does permit open water exploration methods near some continental margins. There is wide variation in the extent of open water both from year to year and from one locale to another. Moreover, it is now not predictable enough so that advance preparations can be made to exploit an unusually good season [Pritchard, 1980; Coon, et al., 1977]. The maximum retreat of the ice edge essentially defines the areas where conventional marine seismic exploration can be done.

Seismic exploration within the ice pack of the Arctic Ocean has been both difficult and sparse when compared to that done in the rest of the world's oceans. Most of the research has been done from manned ice stations. Figure 4 indicates the location of most of the U.S. and Canadian ice stations where some form of seismic exploration has been performed [Central Intelligence Agency, 1975]. The methods used have included seismic reflection and refraction with varying degrees of sophistication. In addition, spot bathymetric soundings have been made. [e.g. Kristoffersen, 1982] By far the most extensive set of geophysical data has been a series of airborne magnetometer surveys [Vogt, et al., 1979]. These have been used to identify magnetic anomalies in the Eurasian Basin of the eastern Arctic and suggest tectonic constraints for the Amerasian Basin of the western Arctic. In addition, heat flow and gravity measurements as well as shallow coring have been made at several ice stations [Ostenso and Wold, 1977; Lachenruch and Marshall, 1969; Kutschale, 1966].

Seismic exploration on the continental margins in the Arctic is dependent upon the characteristics of the ice cover. In addition to the seasonal coverage, landfast ice and shallow water often prevent the use of the towed arrays usually employed in marine seismic surveys. Permafrost and unusually large surface wave noise introduce unique problems in processing seismic data. Finally, much of the seismic exploration in the Arctic is done near Eskimo settlements and their hunting areas. This has led to the imposition of constraints upon the seismic sources used.

In this paper the technology of seismic exploration in the Arctic is surveyed. The discussion has been divided into seismic exploration for i) the deep ocean of the cen-

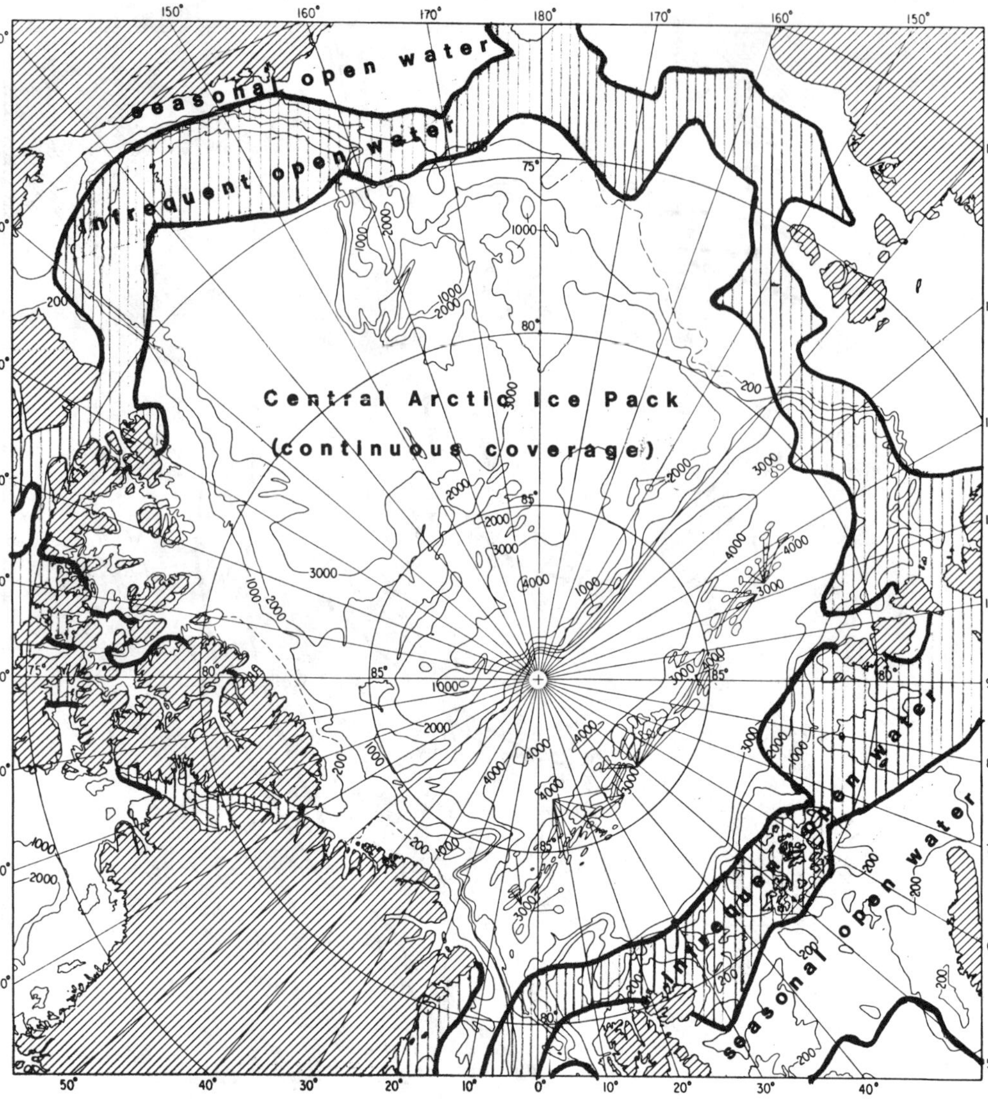

Fig. 3) Seasonal absolute minimum, average minimum, and average maximum ice coverage in the Arctic Ocean. The absolute minimum curve delineates the regions of exploration possible by conventional marine techniques.

tral Arctic which has primarily been the concern of the oceanographic research community and ii) the adjacent continental margins which have been the focus of the oil exploration industry. It examines just the Arctic Ocean where the ice cover is a major factor in exploration and not more southerly regions that are seasonally completely ice free and easily accessible. Land seismic exploration in cold regions also is not covered [see Roethlisberger, 1972]. In the nature of a survey paper and due to space constraints, the details of equipment and the supporting mathematics used in the various exploration methods are not included. The authors have attempted to review the open literature relevant to seismic exploration in the Arctic. Much of it is scattered in a large number of journals and even larger number of reports. We have attempted to cite a representative sampling of this literature, but it is by no means complete. The authors are aware that a large proprietary literature as well as some relevant classified information exists, particularly in the oil exploration companies; consequently, we are still left with the perception that much more remains unpublished and unnoticed. Finally, the authors admit to biases instilled by their own experiences in the Arctic.

2. DEEP OCEAN SEISMIC EXPLORATION

Almost all the seismic exploration in the central Arctic has been done from either ice islands or ice floe camps. The ice islands are actually icebergs that are thought to have calved from glaciers near Ellesmere Island. These include the Fletcher (T-3) and Arlis programs of the U.S. as well as several Soviet programs. Typically, they have been occupied for several years. These islands have often been several kilometers in extent and 10's of meters thick. This has permitted the landing of heavy, long range aircraft such as the C-130 Hercules, the most practical means of supplying camps in the central Arctic. Due to the thin ice and extensive ridging encountered on ice floe camps, smaller aircraft, such as the DC-3 or Twin Otter, that can land on refrozen leads of 1-2 year old ice must be used, and the Hercules employed for air droppable supplies only. Surprisingly, ice islands are relatively rare in the Arctic, with a recent inventory identify-

Fig. 4a) Drift track map of Russian ice stations and U.S. ice islands. Single channel reflection experiments have been carried out on most of the U.S. efforts.

ing only 17 suitable for long term occupancy [Brewer Research, New London, CT, personal communication]. They tend to circulate in the Beaufort Gyre before being caught in the Transpolar Drift where they then exit into the North Atlantic. (T-3, occupied intermittently since 1952, is predicted to exit sometime in early 1983 [op. cit.])

Camps on ice floes have been established for short term experiments, usually during the season from early March when the daylight necessary for aircraft operations first appears to late May when the ice conditions deteriorate and runway problems are encountered. In mid May ice fog becomes common and can limit operations up to 50% of the time as well. Ice floe camps have included the AIDJEX, CANBARX, LOREX and Fram I-IV Series [Trowbridge, 1976; Dyer, et al., 1982; Sweeney, 1982; Hunkins, et al., 1979; Baggeroer and Dyer, 1982; Manley, et al., 1982; Kristoffersen, 1982]. These camps are seasonal and drift with the ice pack 100-200 km during their 2 month lifetimes. The drift has an average trend governed by the currents; however, locally it is very erratic and not very adaptable to the constant course lines usually shot in seismic operations.

Both ice islands and ice floe camps impose constraints upon traditional seismic exploration methods. The drifts are both random in direction and slow, so the coverage is both uneven and sparse. Ice floe camps usually have limited local power (12-25 kW) which restricts the available sources for normal incidence reflection experiments. The deployment of charges at large offsets from the receivers for refraction experiments requires a helicopter or small fixed wing aircraft and is much more tedious than the steam-away technique used in temperate waters. Finally, there are problems with 60 Hz hum from seawater ground loops which are more pervasive than that encountered in shipboard operations due to the distributed nature of the camp , 30 Hz pickup due to mechanical vibration of the diesel/electric generator, and low frequency (< 20 Hz) strumming created by vortex shedding of the water flowing past suspended hydrophone cables.

2.1. Seismic Reflection

Seismic reflection experiments have been done on a sporadic basis in the central Arctic [e.g. Kutschale, 1966; Ostenso and Wold, 1977]. Figure 4 indicates the tracks of some of the ice islands and camps which have gathered seismic reflection data. Until recently these have been single channel operations. Sources for these have included sparkers, boomers, airguns and small explosives ranging from blasting caps to 1 kg SUS charges. All the non-explosive sources work reasonably well in an enclosed hut, although the extreme cold can cause airguns to be very tempermental if they are not properly maintained. In addition, camp power limitations often set the maximum repetition rate [Kristoffersen, 1982]. The receivers have been been either suspended hydrophones or ice mounted geophones. Each have problems unique to Arctic operations. The hydrophones are prone to strum oscillations which can dominate the observations at frequencies below 20 Hz. This has led to extraordinary applications of cable fairing and compliant suspension systems borrowed from sonobuoy technology. In addition, turbulence due to currents interacting with ice keels generates a substantial amount of noise at very low frequencies (< 1 Hz) [Hunkins, 1981]. The geophones are very prone to vibrational

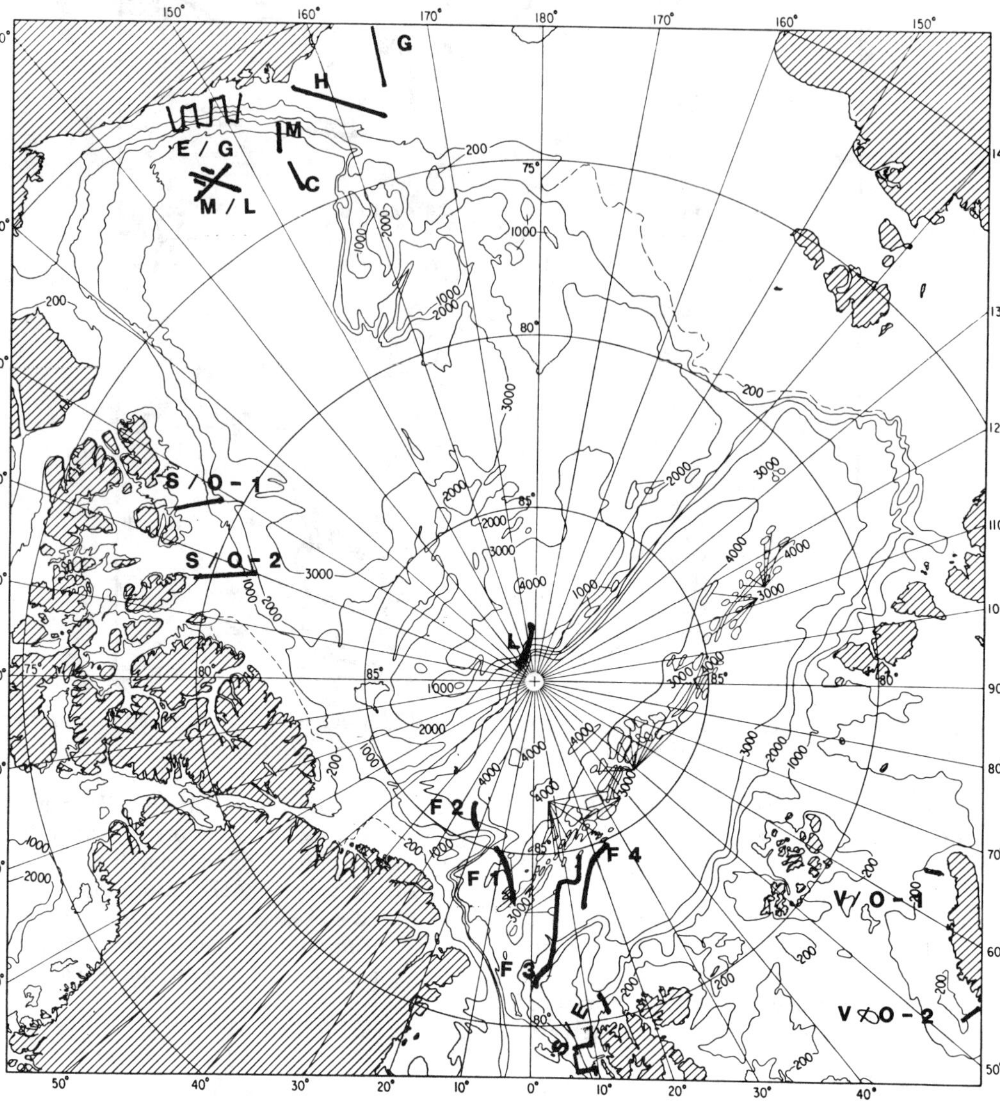

Fig. 4b) Drift tracks of temporary ice camps deployed by the United States and Canada.

C: CANBARX	[Baggeroer and Falconer, 1982]	refraction.
E/G:	[Eittreim and Grantz, 1979]	reflection and refraction
F1: Fram I	[Hunkins, et al., 1979]	reflection and refraction
F2: Fram II	[Baggeroer and Dyer, 1982]	reflection and refraction
F3: Fram III	[Manley, et al., 1982]	reflection and refraction
F4: Fram IV	[Kristoffersen, 1982]	reflection and refraction
G:	[Grantz and Forsyth, 1978]	refraction
H:	[Hunkins, 1962]	refraction
L:	[Sweeney, et al., 1982]	refraction and reflection
M:	[Milne, 1966]	refraction
M/L: AIDJEX	[Mair and Lyons, 1982]	refraction
S/E:	[Sundvor and Eldholm, 1979]	reflection and refraction
S/O:	[Sanders and Overton, 1965]	refraction
V/O:	[Vogt and Ostenso, 1973]	refraction

pickup from the camp power generator and man made camp noise, e.g. winches used in other experiments, snowmobiles, helicopters, and portable heating units. This has required efforts to decouple these two for noise free observations, such as isolation mounting of the generator units. Until recently, data have been recorded using analog methods. With the high levels of strum and/or hum prevalent in drifting ice experiments the limited dynamic range of analog recording is a serious limitation. This is particularly unfortunate since both noise sources have signatures that are very amenable to suppression using signal processing techniques if the dynamic range of the recording system is adequate.

Single channel systems have produced seismic records of limited quality. They are often not powerful enough to penetrate the structure significantly [e.g. Ostenso and Wold, 1977]. Except when the sediment cover is very thin, such as near the mid-ocean ridge, they are seldom adequate to define the basement structure, especially in the Canada Basin where sediment thicknesses can exceed 4 km. Figure 5 shows a successful example from Fram I using a 40 cubic inch airgun which penetrated to basement and is typical of a single channel profile from an ice station [Jackson, et al., 1982]. These systems have been used extensively on the ice islands T-3 and Arlis which were occupied for long drift tracks within the central Arctic.

Multi-channel systems with digital recording have been used only recently for seismic exploration in the Central Arctic [Prada, et al., 1981; Kristoffersen, 1982]. The direction of drift is seldom either linear or constant, so the commonly used (CDP) shooting procedure is awkward at best. Fortunately, most of the operations have been in deep water where the constraints of the CDP shooting are least restrictive. The slow drift rates, however, lead to very limited coverage compared to open water profiling. During the Fram II experiment a multichannel array of hydrophones recorded seismic reflection data from 1 kg SUS charges as it drifted in the Pole Abyssal Plain. While these charges were certainly powerful enough, the shooting interval of approximately 4 hours corresponding to a distance of a kilometer at a nominal drift rate of 3 nm per day was not sufficient to produce a profile that could be easily interpreted [Marshall, 1982]. In addition, since coverage is at the whim of the ice movement, any equipment down-time for maintenance can leave extensive gaps in the profiles. During the Fram IV experiment two arrays of hydrophones, one with 22 hard-wired phones and the other with 18 sonobuoys using an RF link, recorded data from a 120 cubic inch airgun firing every 4 to 60 minutes, corresponding to a 50 meter shot spacing at the variable drift rates encountered. This produced a profile that defines the basement structure very well in this region of relatively thin sediment cover even with very limited signal processing [Kristoffersen, 1982].

The signal processing associated with deep water Arctic is not significantly different from that used for open water. The major difference is that the hydrophones are typically deployed at 60 to 93 m depths instead of just under the surface. For acoustic propagation experiments shallower depths increase the transmission losses significantly in the 5-80 Hz band of interest because of the Lloyd's mirror reflection from the (effectively) free surface. In addition, the spectral peak of an explosive charge is dependent upon its weight and depth of detonation.

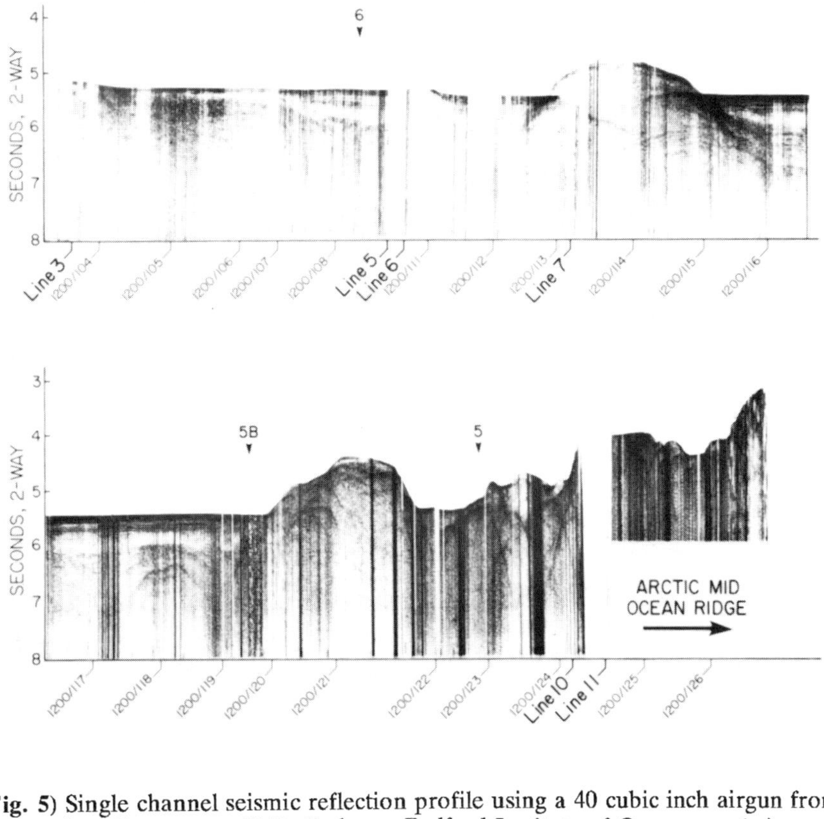

Fig. 5) Single channel seismic reflection profile using a 40 cubic inch airgun from Fram I. (profile courtesy H.R. Jackson, Bedford Institute of Oceanography)

Since the charge size is set by the power needed for adequate crustal penetration and logistics considerations, the depth is fixed by the spectral characteristics desired. For the large charges (25+ kg) used in central Arctic crustal studies, a depth of approximately 250 m is required for a 15 Hz spectral peak. This increased source and receiver depth requires the use of source deconvolution algorithms with long operators to suppress the transient surface ghosts. Beyond deconvolution operations, stacking has been also been done on the multi-channel data. Both common source and common depth point stacks have been done; however, timing the shooting interval appropriate for the group length is awkward because of the random ice motion. Generally, nominal stacking velocity models have been used since the water depths have been large compared to the array lengths. Velocity spectral estimation from reflection data has not been done due to the poor resolution inherent in the use of short array apertures. Work is now being carried out to increase practical array sizes.

An alternative to detailed seismic reflection for reconnaissance is to measure the backscattering from large explosions as a function of two way travel time and angle with an array. This has proven to be an expedient method for identifying regions of high bathymetric relief both in the Arctic and in more temperate waters [Dyer, et al., 1982]. Unfortunately, it does not yield detailed bathymetry.

The important points to emphasize about seismic exploration in the central Arctic are i) it has been very limited in quality since it is only recently that array and digital technologies commonplace in open water for many years have been used and ii) the coverage has been very sparse. The implications of these points are that traditional exploration methods are simply not going to acquire the data about the central Arctic that are needed to resolve the scientific issues. Consequently, some major technology developments, particularly those using remote sensing systems and/or submarine mounted systems, must be considered for future seismic exploration [Francois, 1977].

2.2. Seismic Refraction

Seismic refraction experiments have been performed during the scientific programs on several of the ice islands and ice floe camps in the Central Arctic [Forsyth, 1978; Hunkins, et al., 1962; Milne, 1966; Sander and Overton, 1965; Vogt and Ostenso, 1973; Mair and Lyons, 1981; Baggeroer and Falconer, 1982; Jackson, et al., 1982; Duckworth, et al., 1982]. Figure 4 indicates the sites of most of the recent refraction shooting programs. This type of seismic exploration has received some degree of emphasis by marine geophysicists because of its important role in defining the deep crustal structure required to reconstruct the tectonic evolution of the Arctic.

The presence of the ice pack introduces major constraints on the acquisition of seismic refraction data. The availability of open water for source charge deployment limits the source/receiver offsets; consequently, dense, uniform shooting coverage is seldom possible. Moreover, the shots or sensors must be deployed by light aircraft or helicopter which have limited payloads. A significant amount of time is also consumed during the shooting of a line in searching for open water, landing the aircraft, rigging the charges, and deployment of the shot break recorder for shot instant and bathymetry data. The net

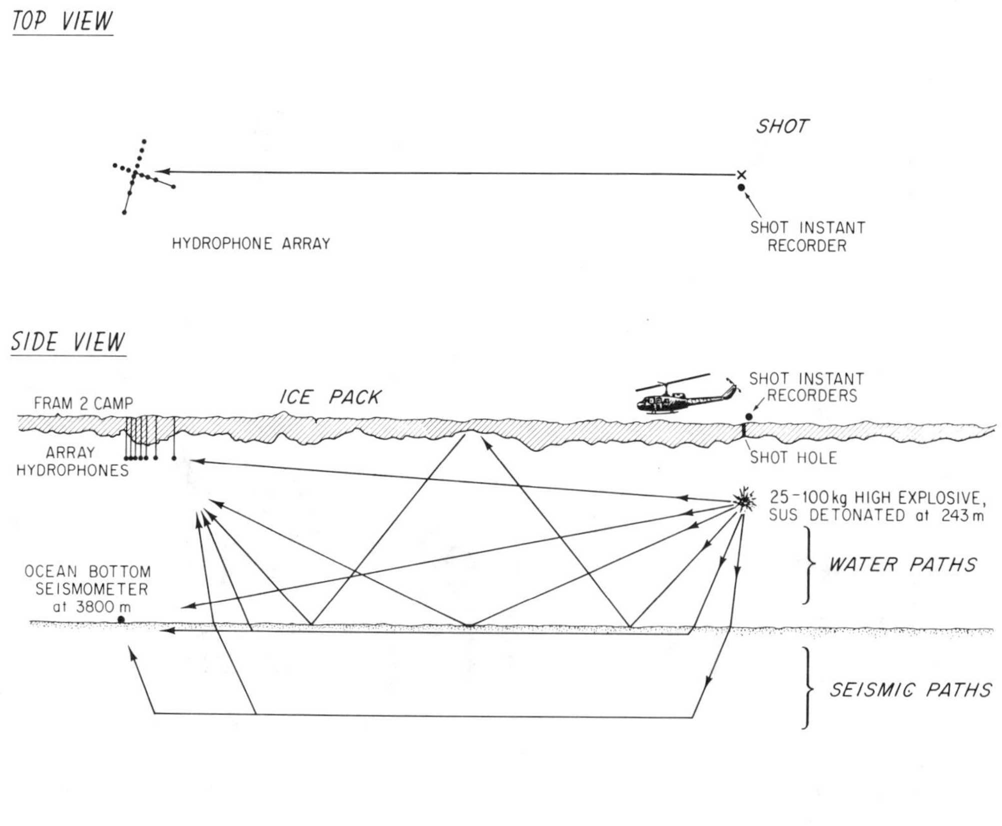

Fig. 6) Refraction shooting geometry for Arctic ice covered operations.

effect is that a very good seismic refraction line in the Arctic has approximately fifteen source/receiver offsets over 60+ km, whereas a good line in open water shot between two boats or a boat and a sonobuoy has one hundred or more shots with essentially continuous coverage of 100 to 300 m spacing [Stoffa, et al., 1981; Detrick and Purdy, 1981].

Refraction shooting on the ice pack has usually been done by deploying shots on a line from a fixed receiver location as indicated in Figure 6. The shots are typically 25-100 kg of TNT detonated at 93 or 243 m by depth activated SUS charges. Shot instants are obtained by a portable a shot break recorder simultaneously recording a time code synchronized to the data acquisition time base. The receivers have included hydrophones, ice mounted geophones, geophone arrays, ocean bottom seismometers [Heffler and Barrett, 1979] and hydrophone arrays. Virtually all the data have been recorded by analog means with the exception of the hydrophone arrays which have used gain ranging, digital recording [Prada, et al., 1981; Kristoffersen, 1982]. An alternative shooting procedure is to move the receiver instead of the source [Mair and Lyons, 1982]. For ice mounted geophones this is convenient if there are a few elements in the receiver; however, it is awkward for more than one hydrophone and impossible for ocean bottom seismometers.

Most refraction work in the Arctic has used single source/single receiver and first arrival analysis as the primary method for structural determination [Jackson, et al., 1982; Kristoffersen, et al., 1982]. Recent experiments, however, have used multi-channel arrays with high dynamic range digital recording. This permits the use of much more sophisticated signal processing which can exploit both the first arrivals and the subsequent multiple paths previously relegated to reverberation in deducing a velocity model for the seismic structure [Baggeroer and Falconer, 1982; Baggeroer and Duckworth, 1982; Duckworth, et al., 1982].

The sparseness of source/receiver offsets in Arctic refraction profiles makes the theoretical inversion of the first arrival function using the time-distance Herglotz-Wiechert formulae practically impossible [Aki and Richards, 1980; Diebold, et al., 1981]. This is further complicated by uncertain bathymetric information along the line. The impact of this is that the signal processing must extract as many constraints on the seismic model through secondary arrivals as possible. Interpretation of the secondary arrivals is feasible only when array data permitting velocity analysis are obtained. Even then it is subject to a significant amount of uncertainty. Fortunately, the acquisition of high quality, digital data and the application of recently developed signal processing methods has advanced the interpretation of Arctic seismic refraction data significantly.

Figures 7a-7c are indicative of the state of the art in deep ocean seismic data acquisition and interpretation. Figures 7a show the waveforms recorded on two array channels separated by 600 m from a Fram II refraction shot at 27.2 km. The array was a 24 channel 800 m x 800 m cross with hydrophones suspended at 93 m beneath the ice. The data were recorded in an SEG-Y format on a multi-channel, gain ranging, digital acquisition system with a 120 dB dynamic range. The acquisition system used a general purpose minicomputer which also allowed limited data analysis and review during the experiment when data were not being taken. In Figures 7a some of the prominent events have been labeled for reference and are described in the captions. It should be noted that the data are exceptionally coherent and noise free for refraction data. This can be attributed to the quiet platform provided by the ice, the high dynamic range of the acquisition system, and the relatively large charge sizes employed (25 kg of TNT). The differences between the two traces in Figs. 7a-1 through 7a-5 are real variations due to differential propagation paths for the sensors.

Along with the waveforms in Figures 7a is the high resolution velocity spectrum determined from all the array channels [Baggeroer and Falconer, 1982; Duckworth, et al., 1982]. One can observe that the important events stand out much more clearly than on the waveforms themselves. The high phase velocity deep crustal events from the Moho and layer 3 appear as the first arrivals. These are followed by events associated with layer 2 at approximately 18 seconds after the shot. There are also high phase velocity multiples of the earlier deep events present. The direct water wave and sediment multiple events follow with progressively increasing phase velocities. One can also identify arrivals with shear wave propagation paths interspersed among these events. The important point is that the multi-channel data and the velocity spectra identify many more events in the data from a single source/receiver offset than just the first arrival detection customarily done in refraction analysis. In addition, the phase velocity, or its inverse, horizontal slowness, are directly estimated from the data when arrays are used. These identifications are especially useful in filling in the gaps in the offset coverage imposed by the ice cover.

The velocity spectra can be used to estimate a velocity/depth section by first transforming to tau-slowness spectra and then migration of this parameterization to a velocity model. Figure 7b illustrates the tau-slowness spectrum obtained by rotating the coordinates of the data in Fig. 7a. The same events noted in Fig. 7a are labeled in Fig. 7b. In the tau-slowness domain several "stacking" operations can be done to fill in the gaps in the section. First, the velocity spectra are produced in 4 Hz analysis bands. If dispersion phenomena are neglected, the spectra from the different frequency bands can be composited. The events from the shots at different offsets can can also be composited on the tau-slowness spectrum since the offset dependence is removed by the tau-slowness transformation [McMechan and Ottolini, 1980]. Finally, multiple events for which travel paths can be accurately identified can be converted to "pseudo primary events" by appropriate reduction in their travel time and offset and added to the primary tau-slowness spectrum.

The stacked tau-slowness spectrum can be inverted for crustal velocity structure in several ways [e.g. Kennett, 1981; Baggeroer and Duckworth, 1982]. In this example, the spectrum is the input to an iterative migration algorithm which generates a velocity-depth section [Clayton and McMechan, 1981]. The section derived from the data in Figure 7a as well as other data on the same line is indicated in Figure 7c. Error bounds on the section are also noted to suggest the degree of confidence that can be associated with the model. This was done using the method of Bessonova, et al. [1974]. As one can observe, the deep crustal section is well identified using just this 6 shot refraction line. While there are many ways to the interpret seismic refraction data, and Arctic data are no exception, the previous sequence of figures is indicative of the

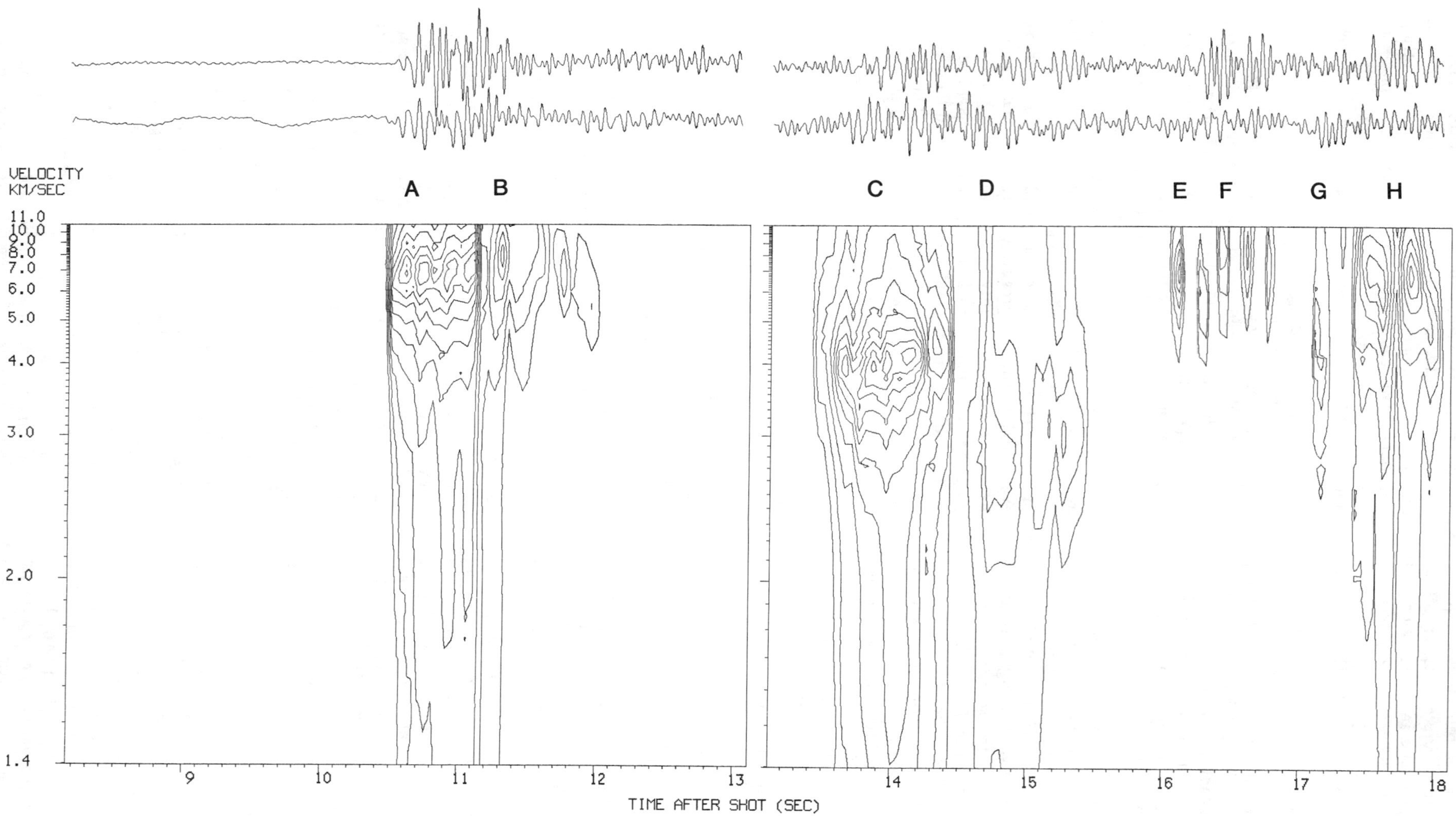

Fig. 7a-1) Waveform and velocity spectrum for Fram II seismic refraction experiment with source/receiver offset = 27.2 km. The waveforms are lowpassed at 30 Hz and are taken at sensors 0.6 km apart to indicate coherency. The traces are uncorrected for moveout, and the top trace is the reference for the velocity spectrum. For the spectrum, the contour intervals are 3 dB. The waveforms are plotted with a gain of 20 in amplitude. Event A is the layer 3 primary. The four peaks are due to surface reflections at the source and receiver. Event B is the Moho compressional arrival.

Fig. 7a-2) Waveform and velocity spectrum of Fig 7a-1 continued. The waveforms are plotted with a gain of 20 in amplitude. Event C is a layer 3 shear primary arrival. Event D is a deep sediment compressional primary. Events E and F are layer 3 and Moho pegleg multiples. Arrival G is a deep layer 2 second multiple, and event H is the layer 3 free surface multiple. Note that the events that are clearly delineated on the spectrum are not easily picked on the waveforms themselves.

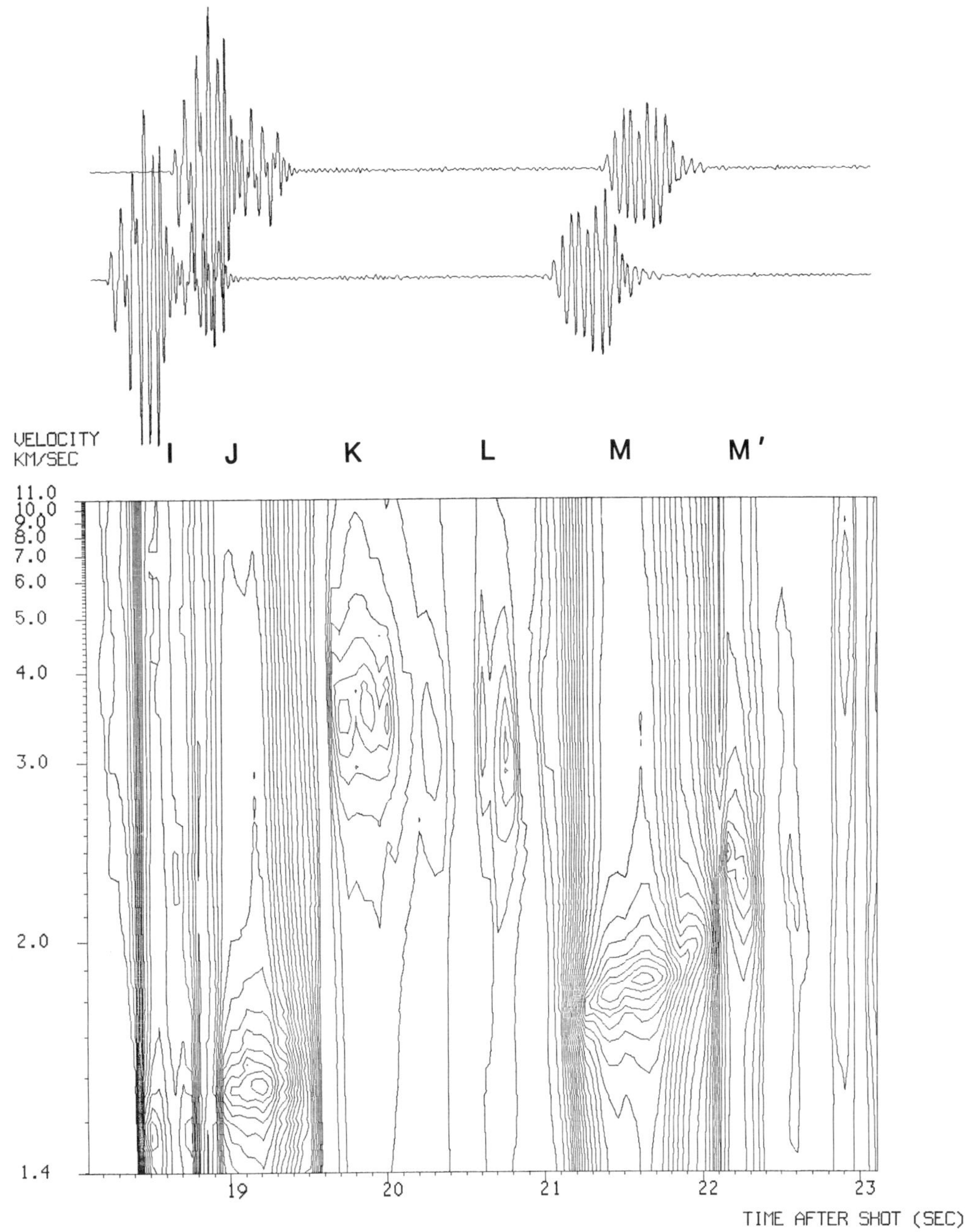

Fig. 7a-3) Waveform and velocity spectrum of Fig 7a-1 continued. The waveforms are plotted with a gain of 0.5 in amplitude. Event I is the direct water arrival. Note that at 27.2 km the offset is exactly half the distance to the first convergence zone and the arrival is relatively weak. Arrival J is the primary refraction from the upper sediments. Event K is the compressional arrival from the deeper sediments, and L is a shear multiple from layer 3. Complex M is the second free surface multiple from a sediment refraction path. A high sediment velocity gradient causes the rise in phase velocity as the near source/receiver free surface ghosts arrive. Event M' is an upper layer 2 shear path, converted to shear above the sediment-layer 2 transition.

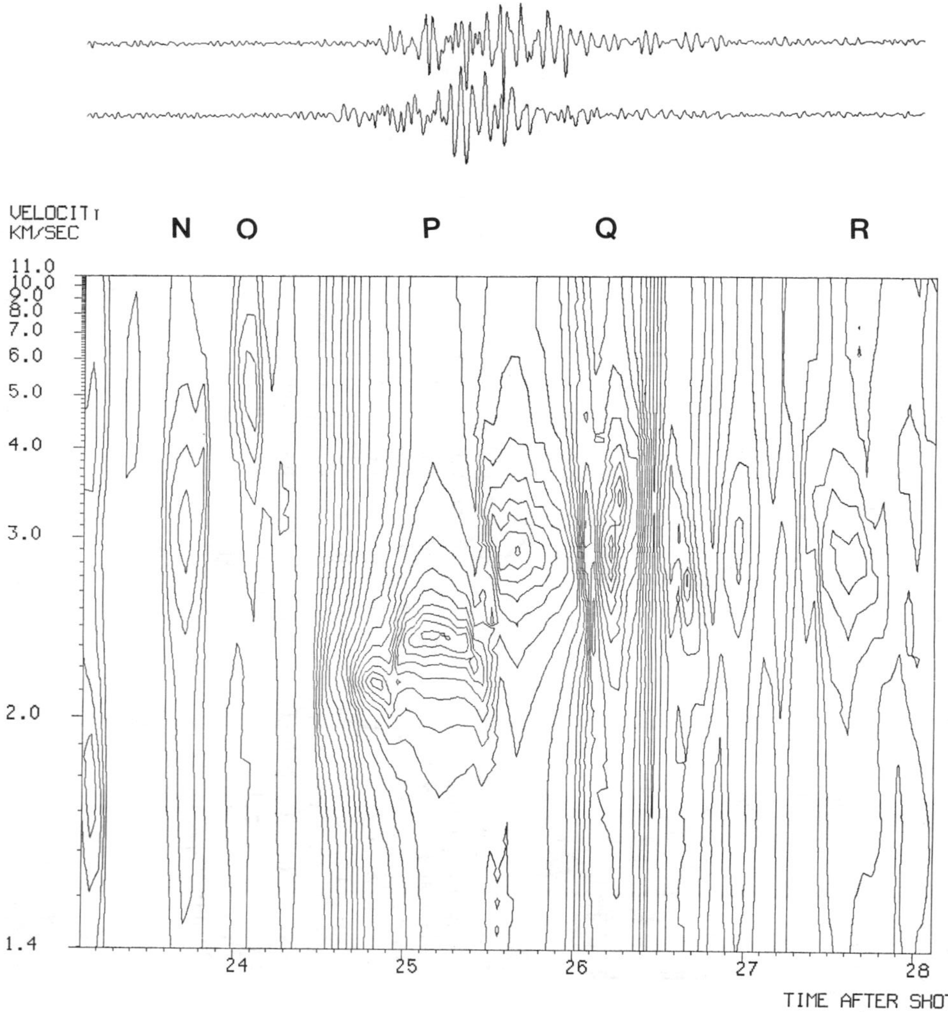

Fig. 7a-4) Waveform and velocity spectrum of Fig 7a-1 continued. The waveforms are plotted with a gain of 2.0 in amplitude. Event N is an internal multiple of undetermined path. Arrival O is the layer 2 compressional third free surface multiple. The complex P is the third free surface multiple of the sediment refraction. The event labeled Q is a deep layer 2 shear path third free surface multiple. Weak arrival R is a shallower layer 2 shear third free surface multiple.

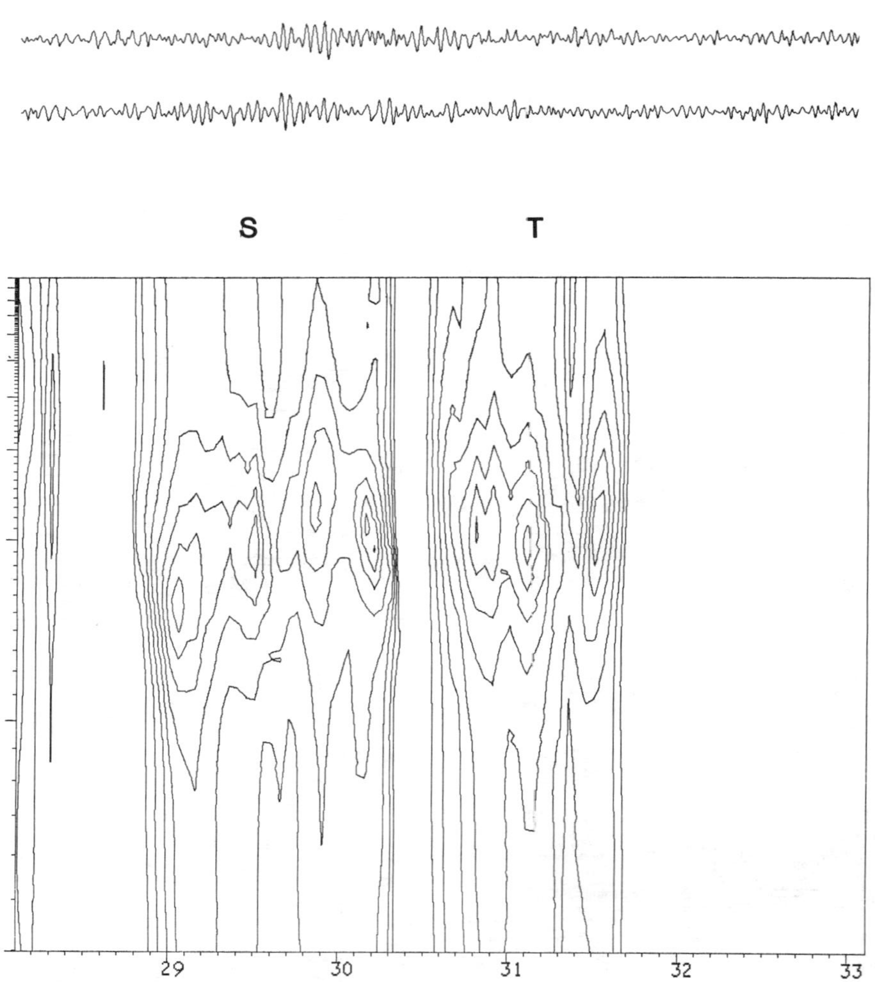

Fig. 7a-5) Waveform and velocity spectrum of Fig 7a-1 continued. The waveforms are plotted with a gain of 5.0 in amplitude. Event S is probably a layer 2 shear path. Complex T is the sediment compressional refraction fourth free surface multiple.

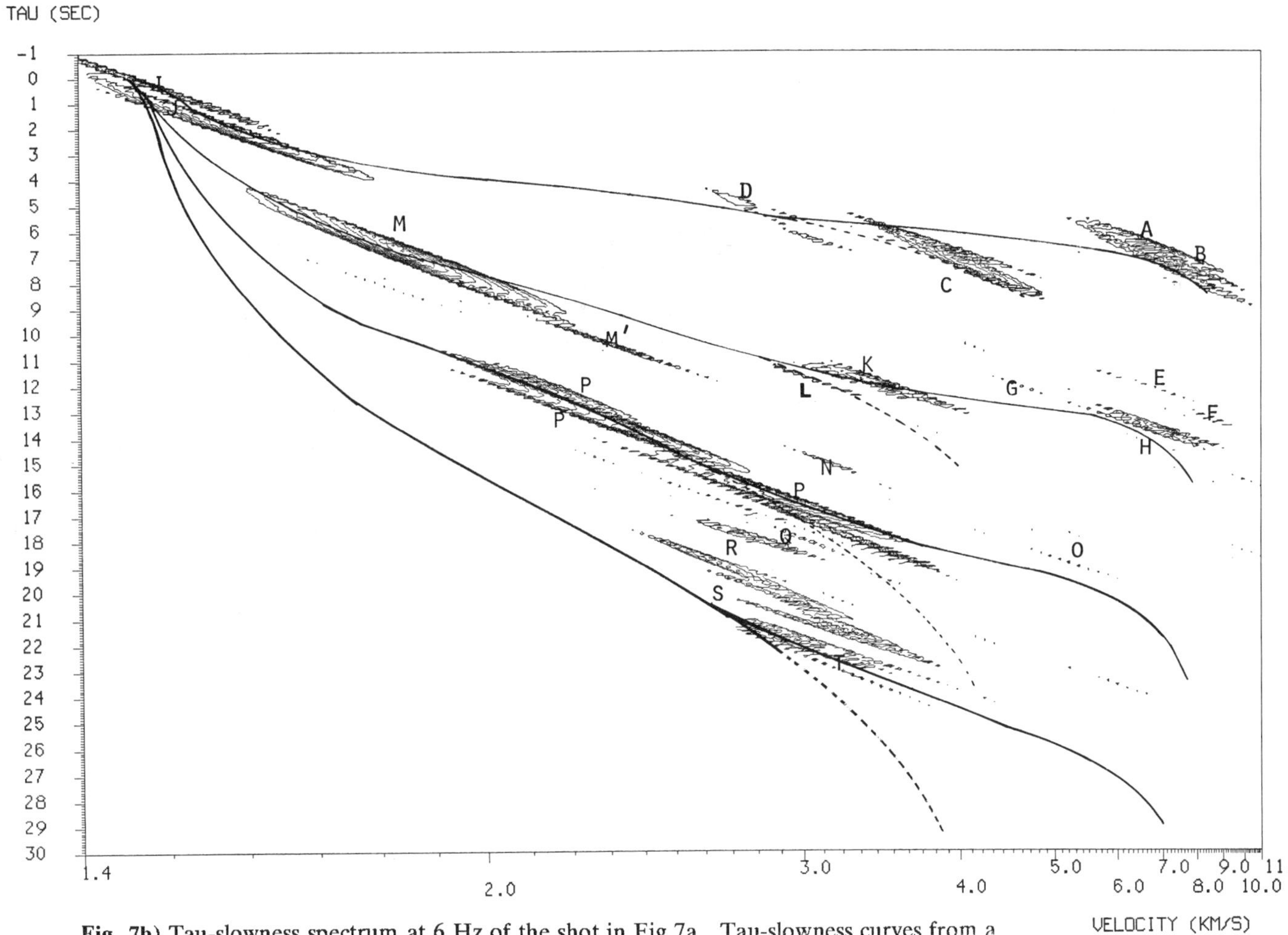

Fig. 7b) Tau-slowness spectrum at 6 Hz of the shot in Fig 7a. Tau-slowness curves from a possible model obtained from this and other shots is shown superposed on the spectrum. Solid lines are curves for the compressional model, dotted lines are curves for the shear model. The events indicated in Figs. 7a are labeled.

current level of sophistication possible with high quality multi-channel array data obtained from an ice floe camp.

In addition to refraction methods, seismic exploration has encompassed several other techniques. Conventional seismic refraction data often do not lead to good estimates of the upper sedimentary section because they are sparsely shot over the offsets where the sedimentary arrivals appear and are simultaneous with the onset of the very high intensity direct water arrivals. Ocean bottom seismometers can mitigate this to a degree, but they have not been used extensively enough to obtain good sediment velocity models. However, the near bottom sediments also have a dominant influence on long range, low frequency acoustic propagation in the water column. These effects can be exploited to produce useful models of the upper part of the seismic section [Kutschale, 1982; Baggeroer and Duckworth, 1982]. The major limitation of this approach is that most of the supporting propagation models are underpinned by the assumption of horizontal homogeneity. This is reasonable for sources and receivers simultaneously located within an abyssal plain, but is not appropriate when significant bathymetric relief intervenes.

For seismic sections extending beyond the Moho the charge weights of up to 200 kg used in refraction experiments are seldom adequate. Teleseismic events from earthquakes can be used for this purpose, with both body wave and surface wave events being used to infer deep crustal and mantle seismic structure [Stewart, 1980].

3. CONTINENTAL MARGIN EXPLORATION

Seismic exploration on the Arctic continental margin has been very extensive. The preferred method of exploration is seismic reflection using conventional marine techniques when the ice retreats during the summer. Some seismic refraction data using sonobuoys and a single ship have also been acquired. Extensive work has been done on the Beaufort Sea off Alaska and Canada, in the Canadian Archipelago in the Amerasian Arctic, and off Svalbard and in the Barents Sea of the Eurasian Arctic.

3.1. Data Acquisition

Marine seismic data are generally of much higher quality, and the cost per mile of survey much lower, than that of lines shot on the ice pack. Nevertheless, conventional marine seismic exploration has several problems unique to the Arctic environment. For instance, the seasonal two to three months of ice retreat is not predictable

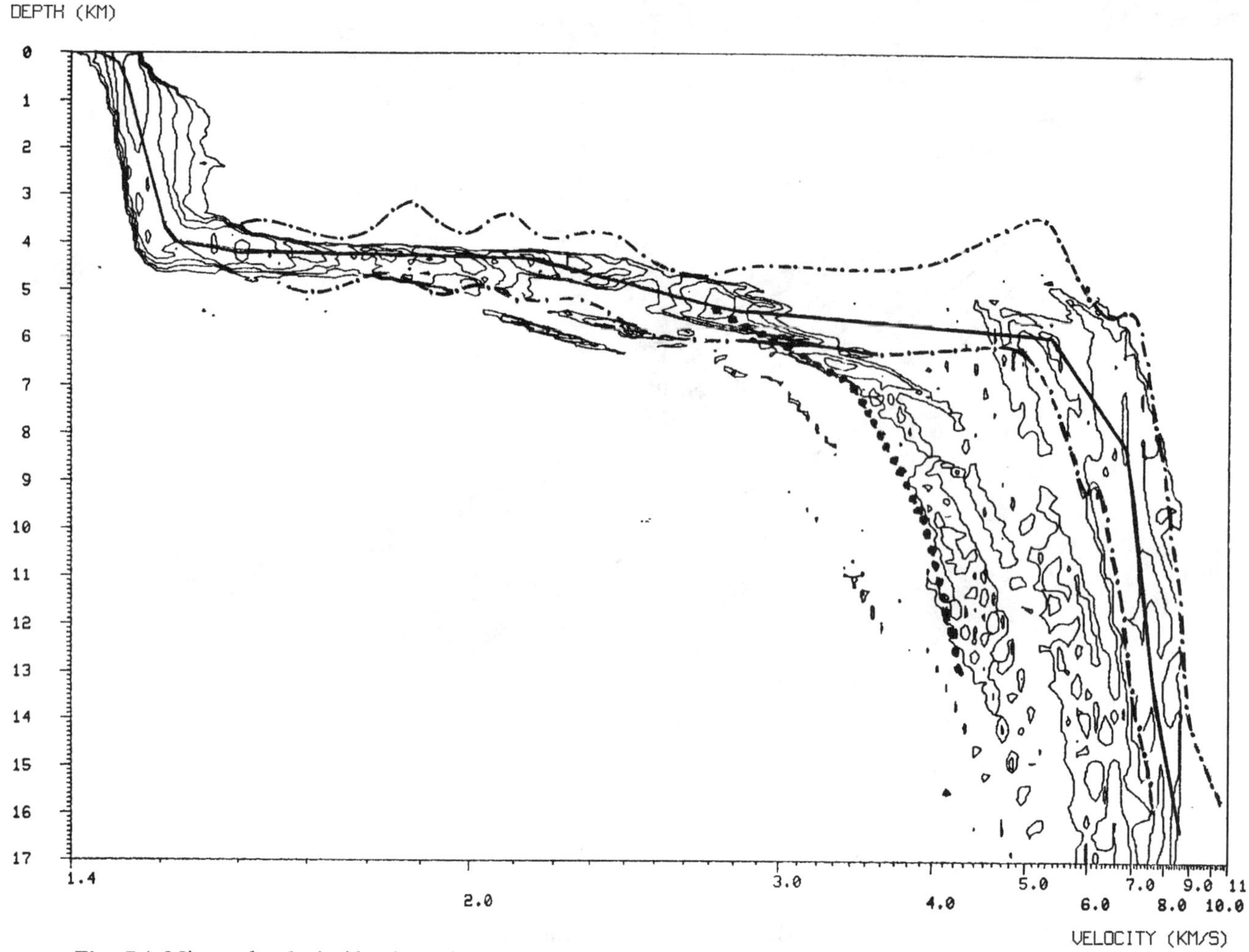

Fig. 7c) Migrated velocity/depth section for Fram II line which includes data in Figs. 7a and 7b. Only the compressional data is shown migrated correctly in this plot. The solid curve is the compressional profile. The dotted curve is the shear profile. The dot-dashed curves are the extremal bounds for the compressional profile.

either in time or extent so planning is uncertain at best. In spite of these problems a significant number of lines have been shot on the Arctic continental margins. Most are proprietary data; however, several have been published in the open literature.

The marine surveys are shot in essentially the same manner as in temperate zones but with an alert watch for drifting ice. Surveys as far as 200 km offshore have been obtained. While the receiver spreads are the same, explosive sources have been prohibited because of their potential environmental impact on marine mammals hunted by the Eskimos.

There is a band separating land seismic exploration, where dynamite and/or vibrators can be used, and offshore regions where marine exploration is feasible on a seasonal basis. This is caused both by very shallow water, which is a problem not unique to the Arctic, and multi-season shorefast ice, which is. This narrow gap in coverage can pose several problems in tying the offshore profiles to onshore lines and well logs. The approach to this problem has been to use vibrators and geophones on landfast and seasonal ice out to water depths overlapping the offshore marine lines. Line geometries with 2400 percent split spreads with 220 ft group intervals and 4800 percent with 110 ft group intervals have been shot with four vibrator sources at a 6 mile per day rate [Mertz, et al., 1981]. Ice roughness, particularly the pressure ridges, are a major impediment to high production. This has stimulated inventive application of diverse transport systems including hovercraft for mobility about the rough ice. Unfortunately, shorefast ice is also typically very rough. Another factor is the presence of leads of open water, or thin ice covering recently opened leads. By their nature, vibrators concentrate a large mass in a small area. It is rumored that more than one has fallen through thin ice. Airguns have been used and compared to vibrator records using the same receiving array. The records are similar and relative comparisons are subjective [Mertz, et al., 1981]. An alternative, but far less desirable approach is to use "mud guns", airguns specially modified to operate in shallow water or mud. The time required to deploy this source and the degradation in record quality has made it far less popular than than the vibrator or airgun. In addition, it is useful to maintain a common source signature throughout a line so artifacts are not introduced.

3.2. Signal Processing

The Arctic environment introduces several unique problems in processing seismic records. Permafrost un-

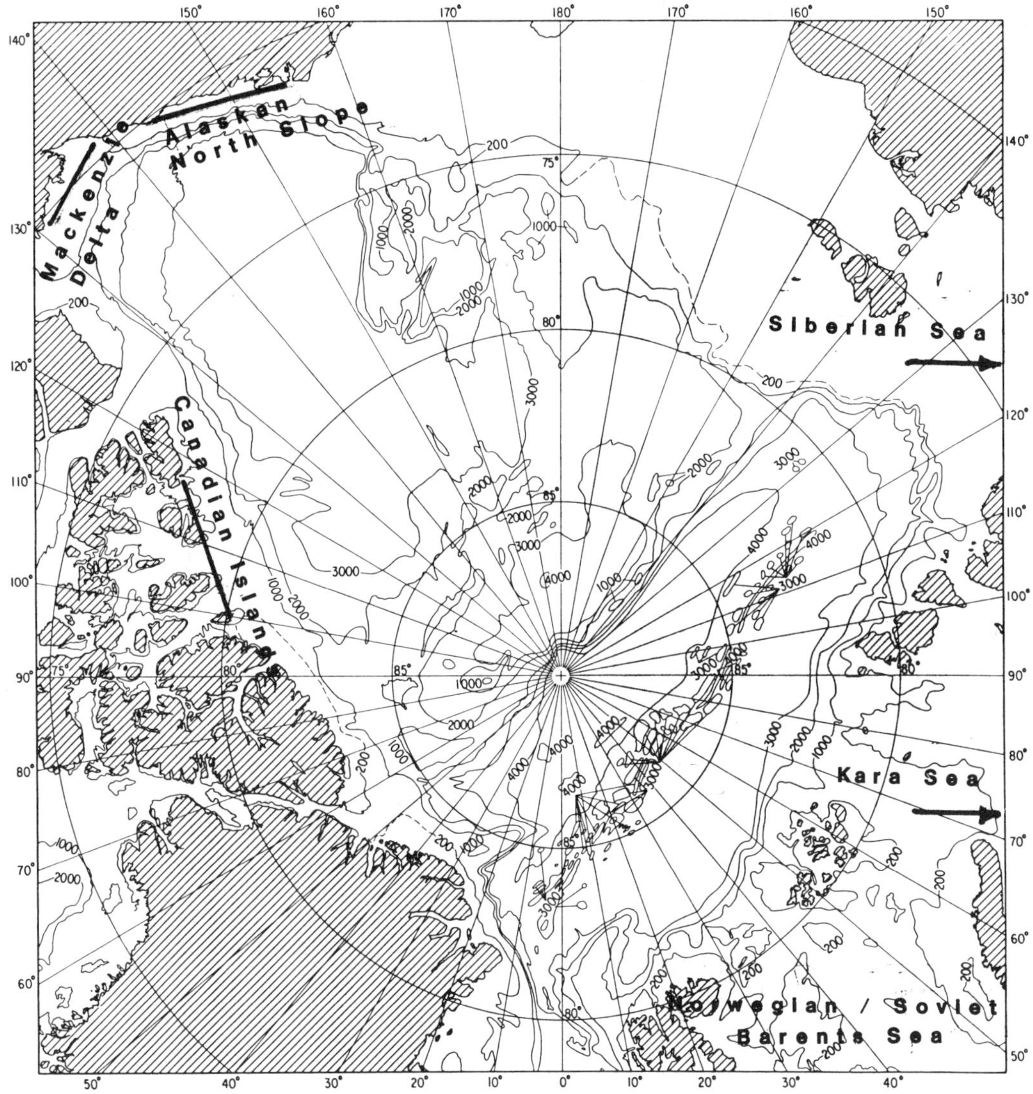

Fig. 8) Labels show the regions in the Arctic currently undergoing exploration for oil and natural gas.

derlies extensive offshore regions of the Arctic. Some of the permafrost is a relic of sealevel subsidence occurring 20,000 to 40,000 years ago to a depth approximately 100m below the current level. Nearshore permafrost can extend to depths of 600 m. Offshore, the permafrost is warmer and thinner due to the warmer boundary condition of the -1.55 degree Celsius seawater [Lewellen, 1973; 1974; 1975; Rogers, et al., 1975]. In addition, the upper section has seasonal variations as well as long term trends of regression or advance. Permafrost has a compressional wave speed between 2400 and 4000 m/s. It usually is not an easily identifiable and distinct layer, but it has its own internal velocity structure. Its occurrence can be very regional, so strong horizontal variations in the velocity structure are common. These changes create a number of problems in the signal processing for making static corrections and/or determining stacking velocities.

Ice floe operations using vibrators introduce flexural wave noise that can be exceptionally strong because of the direct coupling of the source to the ice, instead of the water. The propagated ice wave interference can be 40 dB greater than the reflected signals on a single trace, which imposes some strong demands on the noise cancellation or velocity filtering algorithms [Crary, 1954; Ewing, et al., 1957]. This seems to be suppressed to some extent if there is an open water lead or thin ice between the source and receiver to decouple this motion. The flexural wave noise can be suppressed to an extent by velocity filtering; however, significant residuals remain in many cases because of their intensity and the local variability of the ice thickness which affects surface wave propagation speeds.

An additional complication arises when the ice keels are deep enough to contact the bottom [Wadhams, 1981]. This provides a direct, high velocity coupling to the seabed when compared to that when an intervening water layer is present. This introduces additional horizontal inhomogeneities into the data.

Subsea permafrost with its high velocities can lead to the presence of an exceptionally efficient waveguide producing strong reflected and/or refracted reverberation. The permafrost depth and its sound speed gradient influence the refraction paths significantly, so care must be taken in deconvolution and muting algorithms.

Finally, there appear to be numerous locales of shallow gas-bearing layers in several exploration areas. These introduce a highly reflective horizon followed by a low velocity layer [Boucher, et al., 1981; Grantz, et al., 1979]. This can lead to strong reverberation as well as shallower penetration of the underlying structure due to the lowered transmission coefficient.

4. SUMMARY

Seismic exploration in the Arctic which had just begun 30 years ago is now proceeding with great vigor. An intense amount of activity is now underway in the quest for oil and gas, and the development of array processing and computer technology has made this, and more academically oriented tectonic studies in the central Arctic very fruitful. During this period many inventive approaches have been developed to overcome the impediments of the Arctic, especially the unique effects introduced by the ice and limitations imposed by the ice cover. We hope to have highlighted some of these approaches in this short survey. In spite of the great amount of research already done, the Arctic still remains the least known of the world's oceans. While the technology is evolving which permits men to work in its harsh climate, further advances must be made to acquire the comprehensive seismic data needed to develop the resources of the Arctic, and to understand its origins and evolution.

5. ACKNOWLEDGEMENTS

This work was supported by the Office of Naval Research, Arctic Program Office. Their support for the authors' work on the field programs CANBARX, Fram II and Fram IV is gratefully acknowledged. Parts of the data and the text were prepared on the MIT Digital Signal Processing Computer supported by ARPA.

The authors benefited greatly from discussion with many people about seismic exploration in the Arctic. It is a pleasure to acknowledge the assistance of I. Dyer of MIT, G.L. Johnson of ONR, J. Sweeney of the Pacific Geoscience Center, Canada, R. Jackson of the Atlantic Geoscience Center, Canada, Y. Kristoffersen of the Norsk Polarinstitutt, Norway, C. Frazier, J. Mendiguren and W. Ostrander of Chevron Oil and D. Bennyhoff and J. Yancey of ARCO.

6. REFERENCES

Aki, K., and Richards, P.G., *Quantitative Seismology: Theory and Methods,* W.H. Freeman, San Francisco, 1980.

Baggeroer, A.B., and Dyer, I., The FRAM II Experiment in the Eastern Arctic, *EOS,* **63**, pp. 220-222, 1982.

Baggeroer, A.B., and Falconer, R.K.H., Array refraction profiles and crustal models of the Canada Basin, *J. of Geophysical Research,* **87**, pp. 5461-5476, 1982.

Baggeroer, A.B., and Duckworth, G.L., Seismic structure modeling in the Arctic Ocean, *Proceedings of Ocean's 82 Conference, Washington, D.C.,* IEEE Press, New York, 1982.

Beal, M.A., Bathymetry and structure of the Arctic Basin, Ph.D. Thesis, Oregon State University, Corvallis, Oregon, 1968.

Bessonova, E.N., Fishman, V.M., Ryaboyi, V.Z., Setnikova, G.A., The Tau Method of Inversion of Travel Times- I. Deep Sounding Seismic Data, *The Geophysical Journal of the Royal Astronomical Society* **36**, pp. 377-398, 1974.

Boucher, G., Reimnitz, E., and Kempema, E., Seismic evidence for an extensive gas-bearing layer at shallow depth, offshore from Prudhoe Bay, Alaska, *Cold Regions Science and Technology,* **4**, pp. 63- 71, 1981.

Business Week, The Arctic Energy Rush, pp. 52-57, Jan 24, 1983.

Coachman, L.K., and Aagaard, K., Physical oceanography of Arctic and Subarctic seas, in *Marine Geology and Oceanography of the Arctic Seas,* ed. Y. Herman, pp. 1-72, Springer-Verlag, New York, 1974.

Central Intelligence Agency, *Polar Regions Atlas,* Supt. of Documents, U.S. Government Printing Office, Washington, D.C., May 1978.

Clark, D.L., Geology and geophysics of the Amerasian Basin, in *The Ocean Basins and Margins, Volume 5, The Arctic Ocean,* ed. A.E. Nairn, M. Churkin, Jr., and F.G. Stehli, Plenum Press, New York, 1981.

Clayton, R., and McMechan, G., Inversion of refraction data by wavefield continuation, *Geophysics,* **46**, pp. 860-872, 1981.

Coon, M.D., Hall, R.T., and Pritchard, R.S., Prediction of Arctic Ice Conditions for Operations, *Proceedings of the Offshore Technology Conference,* Houston, TX, 1977.

Crary, A.P., Seismic studies on Fletcher's Ice Island, T-3, *Trans. of the American Geophysical Union,* **35**, pp. 293-300, 1954.

Detrick, R.S., and Purdy, G.M., The structure of the Kane Fracture Zone from seismic refraction surveys, *J. of Geophysical Research,* **85**, pp. 3759-3777, 1980.

Diebold, J.B., and Stoffa, P.L., The travel time equation, tau-p mapping, and inversion of common midpoint data, *Geophysics,* **46**, pp. 238-254, 1981.

Duckworth, G.L., Baggeroer, A.B., and Jackson, H.R., Crustal structure measurements near Fram II in the Pole Abyssal Plain, in *Structure of the Arctic,* ed G.L. Johnson and J.F. Sweeney, *Tectonophysics,* **89**, pp. 173-215, 1982.

Dyer, I., Baggeroer, A.B., Zittel, J., and Williams, R.J., Acoustic backscattering from the basins and margins of the Arctic Oceans, *J. of Geophysical Research,* **87**, pp. 9477-9488, 1982.

Eittreim, S., and Grantz, A., CDP seismic sections of the western Beaufort continental margin, *Tectonophysics,* **59**, pp. 251-262, 1979.

Ewing, M., Jardetsky, W.S. and Press, F., *Elastic Waves in Layered Media,* McGraw-Hill, 1957.

Feden, R.H., Vogt, P.R., and Fleming, H.S., Magnetic and Bathymetric Evidence to the "Yermak Hot Spot" Northwest of Svalbard in the Arctic Basin, *Earth and Planetary Science Letters* **44,** pp. 18-38, 1979.

Forsyth, D.A., Review of Arctic crustal studies, in *Arctic Geophysical Review, Earth Physics Branch,* ed. J.F. Sweeney, **45,** pp. 75-85, 1978.

Francois, R.E., Arctic underwater operational systems, in *Arctic Systems,* ed. P.J. Amaria, A.A. Bruneau and P.A. Lapp, Plenum Press, New York, 1977.

Grantz, A., Eittreim, S., and Dinter, D.A., Geology and tectonic development of the continental margin north of Alaska, in Crustal Properties across Passive Margins, ed. C.E. Keen, *Tectonophysics,* **59,** pp. 263-291, 1979.

Grantz, A., Eittreim, S., and Whitney, O.T., Geology and physiography of the continental margin north of Alaska and implications for the origin of the Canada Basin, in *The Ocean Basins and Margins, Volume 5, The Arctic Ocean,* ed. A.E. Nairn, M. Churkin, Jr. and F.G. Stehli, Plenum Press, New York, 1981.

Hall, J.K., Geophysical evidence for ancient sea-floor spreading from Alpha Cordillera and Mendeleyev Ridge, in *Arctic Geology,* ed. M.G. Pitcher, American Assoc. of Petroleum Geologists, Tulsa, Okla, 1973.

Heffler, D.E., and Barrett, D.L., OBS Development at the Bedford Institute of Oceanography, *Marine Geophysical Research* **4,** pp. 227-245, 1979.

Herron, E.M., Dewey, J.F., and Pitman, W.C., Plate Tectonics Model for the evolution of the Arctic, *Geology* **2,** pp. 377-380, 1974.

Hofer, J., and Varga, W., Seismological experience in the Beaufort Sea, *Geophysics,* **37,** pp. 605-619, 1972.

Hunkins, K., Herron, T., Kutschale, H., and Peter, P., Geophysical studies of the Chukchi Cap, Arctic Ocean, *J. of Geophysical Research,* **67,** pp. 234-247, 1962.

Hunkins, K., Arctic Ocean eddies and baroclinic instability, *Rept. CU-2-81,* Lamont-Doherty Geological Obsv., 1981.

Hunkins, K., Kristoffersen, Y., Johnson, G.L., and Heiberg, A., The FRAM I Expedition, *EOS,* **60(52),** 1979.

ICEX, *Ice and Climate Experiment, Report of the Science and Applications Working Group,* A. Thorndike, Chairman, Goddard Space Flight Center, Greenville, MD, Dec. 1979.

Jackson, H.R., and Reid, I., Crustal refraction and seismic reflection from FRAM I, *EOS,* **61,** pp. 276, 1981.

Jackson, H.R., Reid, I., and Falconer, R.K.H., Crustal structure near the Arctic Mid-Ocean Ridge, *J. of Geophysical Research,* **87- B3,** pp. 1773-1783, 1982.

Johnson, G.L., Monahan, D., Gronlie, G., and Sobczak, L., General Bathymetric Chart of the Oceans (Arctic Ocean), Canadian Hydrographic Service, Ottawa, Canada, 1979.

Kennett, B.L.N., Slowness techniques in seismic interpretation, *J. of Geophysical Research,* **86-B12,** pp. 11575-11584, 1981.

Kovacs, L.C. and Vogt, P.R., Depth-to-magnetic source analysis of the Arctic Ocean Region, in *Structure of the Arctic,* ed. G.L. Johnson and J.F. Sweeney, *Tectonophysics,* **89,** pp. 255-294, 1982.

Kristoffersen, Y., Husebye, E.S., Bungum, H., and Gregersen, S., Seismic investigations of the Nansen Ridge during the FRAM I experiment, *Tectonophysics,* **82,** pp. 57-68, 1982.

Kristoffersen, Y., The Nansen Ridge, Arctic Ocean; some geophysical observations of the rift valley at slow spreading rate, in *Structure of the Arctic,* ed. G.L. Johnson and J.F. Sweeney, *Tectonophysics,* **89,** pp. 161-172, 1982.

Kristoffersen, Y., U.S. ice drift station FRAM IV: Report on the Norwegian field program, *Rapportserie Nr. 11,* Norsk Polarinstitutt, Oslo, 1982.

Kutschale, H., Arctic Ocean geophysical studies: the southern half of the Siberia Basin, *Geophysics,* **21,** pp. 683-710, 1966.

Kutschale, H., Pulse propagation in the Arctic Ocean, Lamont- Doherty Geological Observatory, 1982 *in preparation.*

Lachenruch, A.H., and Marshall, B.V., Heat flow in the Arctic, *Arctic,* **22,** Arctic Inst. of North America, pp. 300-311, 1969.

Lawver, L.A., and Baggeroer, A.B., Heat flow and depth to basement in the Canada Basin, *Proceedings of the Alaska Geological Society,* 1982.

Lewellen, R.I., The occurrence and characteristics of nearshore permafrost, northern Alaska, in *Permafrost: The North American Contribution to the Second International Conference, National Academy of Science, Washington, D.C.,* 1973.

Lewellen, R.I., Offshore permafrost of Beaufort Sea, Alaska, in *The Coast and and Shelf of the Beaufort Sea,* ed. J.C. Reed and J.E. Sater, The Arctic Institute of North America, Arlington, Va., 1974.

Lewellen, R.I., Subsea permafrost research techniques, *Geoscience and Man,* **23,** pp. 29-34, 1977.

Mair, J.A. and Lyons, J.A., Crustal structure and velocity anisotropy beneath the Beaufort Sea, *Canadian Journal of Earth Science,* **18,** pp. 724-741, 1981.

Mair, J.A. and Forsyth, D.A., Crustal structures of the Canada Basin near Alaska, the Lomonosov Ridge and adjoining basins near the North Pole, in *Structure of the Arctic,* ed. G.L. Johnson and J.F. Sweeney, *Tectonophysics,* **89,** pp. 239-253, 1982.

Manley, T.O., Codispoti, L.A., Hunkins, K.L., Jackson, H.R., Jones, E.P., Lee, V., Moore, S., Morison, J., Packard, T.T., and Wadhams, P., The FRAM III Expedition, *EOS,* **63,** pp. 627-636, 1982.

Marshall, W.F., Seismic reflection at the FRAM II ice station, S.M. Thesis, Dept. of Ocean Engineering, MIT, Cambridge, Mass, 1982.

McMechan, G.A. and Ottolini, R., Direct observation of a tau-p curve in a slant stacked wavefield, *Bulletin of the Seismological Society of America,* **70,** pp. 775-789, 1980.

Mertz, R.W., Brooks, L.D., and Lansley, M., Deepwater vibrator operations, Beaufort Sea, Alaska, 1979 winter season, *Geophysics,* **46,** pp. 172-181, 1981.

Milne, A.R., A seismic refraction in the Beaufort Sea, *Bulletin of the Seismological Society of America,* **56,** pp. 775-779, 1966.

Ostenso, N.A. and Wold, R.J., A seismic and gravity profile across the Arctic Ocean basin, *Tectonophysics,* **38,** pp. 1-24, 1977.

Prada, K., von der Heydt, K., and O'Brien, T.F., A versatile multichannel data acquisition system for seismic and acoustic applications, *Proceedings of the Ocean's 81 Conference, Boston, MA,* pp. 43-47, IEEE Press, New York, 1981.

Pritchard, R.S. (Editor), *Sea Ice Processes and Models,* Univ. of Washington, Seattle, WA., 1980.

Roethlisberger, H., *Seismic exploration in cold regions,* Cold Regions Research and Engineering Laboratory, 1972.

Rogers, J.C., Harrison, W.D., Shapiro, L.H., Ostercamp, T.E., Gedney, L.D., and VanWormer, J.D., Nearshore permafrost studies in the vicinity of Point Barrow, Alaska, *Report No. 75-6,* Geophysical Institute, University of Alaska, Fairbanks, Alaska, 1975.

Sander, G.W. and Overton, A., Deep seismic refraction investigation in the Canadian Arctic Archipelago, *Geophysics,* **30,** pp. 87-96, 1965.

Stewart, I.C.F., Arctic lithospheric structure from delays of teleseismic P wave reflections, *Tectonophysics,* **69,** pp. 37-62, 1980.

Stoffa, P.L., Buhl, P., Diebold, J.B., and Wentzel, F., Direct mapping of seismic data to the domain of intercept time and ray parameter: a plane wave decomposition, *Geophysics,* **46,** pp. 255-267, 1981.

Sundvor, E., and Eldholm, O., The western and northern margin off Svalbard, in *Crustal Properties across Passive Margins,* ed C.E. Keen, *Tectonophysics,* **59,** pp. 239-250, 1979.

Sweeney, J.F., Weber, J.R. and Blasco, S.M., Continental ridges in the Arctic Ocean: LOREX constraints, in *Structure of the Arctic,* ed. G.L. Johnson and J.F. Sweeney, *Tectonophysics,* **89,** pp. 217-237, 1982.

Sweeney, J.F., Structure and development of the polar margins of North America, *Dynamics of Passive Margins Geodynamic Series, Vol. 6,* American Geophysical Union, pp. 17-29, 1982.

Sweeney, J.F., Evidence for the origin of the Canada Basin margin by rifting in early Cretacous time, to appear in *Origin of the Canada Basin,* ed. Mulle and Reed, *Alaska Geological Society Journal*

Trowbridge, R., The Arctic Ice Dynamics Joint Experiment (AIDJEX), *Naval Research Review,* pp. 8-17, May 1976.

Vogt, P.R., and Ostenso, N.A., Reconnaissance geophysical studies in Barents and Kara Seas - Summary, in *Arctic Geology,* ed. M.G. Pitcher, American Association of Petroleum Geologists, Tulsa, OK, 1973.

Vogt, P.R., and Avery, O.E., Tectonic history of the Arctic basins: partial solutions and unsolved mysteries, in *Marine Geology and Oceanography of the Arctic Seas,* ed. Y. Herman, pp. 83- 118, Springer-Verlag, New York, 1974.

Vogt, P.R., Taylor, P.T., Kovacs, L.C., and Johnson, G.L., Detailed aeromagnetic investigation of the Arctic Basin, *J. of Geophysical Research,* **84,** pp. 1071-1089, 1979.

Vogt, P.R., Taylor, P.T., Kovacs, L.C., and Johnson, G.L., The Canada Basin: aeromagnetic constraints on structure and evolution, in *Structure of the Arctic,* ed. G.L. Johnson and J.F. Sweeney, *Tectonophysics,* **89,** pp. 295-336, 1982.

Wadhams, P., The ice cover in the Greenland and Norwegian Seas, *Reviews of Geophysics and Space Physics,* **19,** pp. 345-393, 1981.

INTRODUCTION

N. Untersteiner

Polar Science Center
Applied Physics Laboratory
University of Washington
Seattle, Washington

Sea ice is a unique geophysical material, consisting of an ensemble of pieces ranging from gravel-sized fragments to unbroken plates many kilometers in diameter. In seemingly random patterns, the individual pieces, or "floes", are separated by lanes of open water, or joined in ridges where thin ice was crushed and compressed to form welts of thick ice extending both upward, and downward into the water.

To describe the mechanical and thermodynamic behavior of such a chaotic material in terms of a manageable number of parameters requires a statistical treatment. The desired statistical parameters can only be derived from a very large number of observations. The technical means of acquiring such data became available about 2 decades ago: the upward looking sonar mounted on a submarine, and the satellite-borne passive microwave imaging device. The following two papers, "Arctic Sea Ice Morphology and Its Measurement" by P. Wadhams and "Aspects of Arctic Sea Ice Observable by Sequential Passive Microwave Observations from the Nimbus-5 Satellite" by W. J. Campbell, P. Gloersen, and H. J. Zwally, are informative by themselves as well as in their juxtaposition.

Except for a small amount of ice thickness data laboriously obtained by drilling holes in the ice, virtually all available information about the statistics of ice thickness and ridge keels and their regional differences has come from the 28,000 km of published sonar profiles. There are two difficulties with the interpretation of sonar data: one is the definition of undeformed or level ice based on a somewhat arbitrarily chosen slope angle, and the other is the distinction between open water and thin ice. Even so, the sonar method per se is physically straightforward and whatever ambiguities exist in the data are known. The most severe limitation of these data is their scarcity and the fact that nuclear-powered submarines are rarely available for unclassified missions.

In contrast, passive microwave observations from satellites have been taken, nearly without interruption, for over a decade and cover both arctic and antarctic sea ice fields. At the chosen wavelength of 1.55 cm, the ratio of emissivities for water, first-year (FY) and multiyear (MY) ice is about 10:19:21. Owing to the large difference in brightness temperature between open water and ice of any kind, passive microwave data have provided an impressive time series of data on the global extent of sea ice. Ice of different thickness and age radiates with emissivities within a range of only 10%, making the more subtle task of distinguishing between ice of different types far more difficult. Perhaps the greatest problem--and one that is also being encountered with certain other types of satellite data--lies in the average view that a satellite takes over a sizable element of the earth's surface. It is difficult to estimate the real, small-scale variations on the ground from which the average is composed, and thus to relate "ground truth" to its appearance from the height of a satellite.

Visual observations from aircraft gathered during many years, as well as considerations of the overall ice budget of the Arctic Basin, indicate that the basin should contain a higher MY fraction than that derived from microwave data, but such information is hardly reliable. Observers on the ice surface can clearly make the MY-FY distinction, but their surveys are necessarily small scale and cannot cover the 25 km pixels of a satellite image. Data that could help calibrate the MY/FY fractions from microwave data are the sonar profiles of ice draft.

Both the GURNARD cruise (April 1976) and the SOVEREIGN cruise (October 1976) occurred at a time when ESMR on Nimbus-5 was in operation. While the sonar-derived ice thickness alone is not necessarily a measure of the age of the ice, a comparison and analysis of apparent brightness temperatures and ice draft along identical profiles might be of great help toward resolving existing uncertainties.

ARCTIC SEA ICE MORPHOLOGY AND ITS MEASUREMENT

Peter Wadhams

Scott Polar Research Institute

University of Cambridge, Cambridge CB2 1ER, England

1. INTRODUCTION

At present there is no reliable air-borne technique for the measurement of sea ice thickness, although impulse radar has yielded promising results over fresh-water ice and over undeformed sea-ice sheets [1], and we shall also show in this review that airborne laser profilometer data can be interpreted to yield a reasonable estimate of the pressure ridge depth distribution in sea ice. Traditional techniques of hole drilling can yield a reasonable estimate of the mean ice draft over a region with a few hundred randomly sited holes [2] but an impossibly large number of holes is required to yield useful information about the distribution of ice thickness. The only reliable method for the synoptic measurement of ice thickness distribution and bottom topography is still the submerged upward-looking sonar.

An upward-looking sonar can be mounted on the sea bed [3], allowing ice to pass overhead and thus generating a time series of ice draft at a point. This is especially useful for site-specific surveys in shallow areas subject to scouring where a time series measured over, say, a year can be analysed using extreme value techniques to predict the deepest ice keel that will pass that location in a longer interval. To obtain the geo-graphical variation in ice morphology, however, one must use a mobile sonar. Successful measurements have been made from sonars mounted in unmanned vehicles [4,5,6] and small submersibles, but such vehicles are only really useful for local surveys, and the only means of obtaining data over a large area is the full-sized submarine. For the central Arctic this implies a nuclear submarine, although conventional submarines have been used to obtain local data near the ice edge, e.g. during wave surveys [7].

Since the first voyage of U.S.S. Nautilus to the North Pole in 1958 there have been more than 30 exploratory cruises beneath the Arctic ice by U.S. nuclear submarines and three by Royal Navy submarines. These have covered most parts of the Arctic Basin and peripheral seas such as the Greenland Sea, Baffin Bay and the Northwest Passage, in most cases with winter and summer profiles. Table I is a list of cruises where the data have been released and analysed; the total analysed track length now exceeds 28,000 km.

These datasets provide a wealth of statistical information. Among the parameters which we shall discuss in this review are:- mean ice draft, ice draft distribution, the distribution of level ice, pressure ridge spacings and frequencies, and the occurrence and width distribution of leads. These are the parameters of major geophysical interest. In addition, the spectral characteristics of the ice bottom are of importance in calculating the back-scattering of sound by an ice cover.

2. INSTRUMENTS

Initially, most U.S. surveys were carried out with a simple fathometer mounted on the upper casing of the submarine, and operating a standard electrically-sensitive paper chart recorder. The beamwidth of the fathometer was large (10° - 30°) and therefore the ice bottom profile was smoothed such that fine detail of the topography was lost. The chart record has to be digitised; normally in this process the envelope of the sonar record is traced by the curve follower, i.e. the point on the ice bottom which is first encountered by the beam footprint. This causes troughs between closely spaced ridge crests to be lost. The overall effects of a finite beamwidth are thus:-

a) over-estimate of mean ice draft;

b) under-estimate of pressure ridge numbers;

c) under-estimate of the slope of a pressure ridge, and distortion of its shape, especially rounding of the crest [17];

d) correct estimate of the absolute draft of a pressure ridge, so long as it is not 'lost' by merging with a deeper one;

e) loss of information on fine scale spatial roughness.

Having been digitised, the record has to be corrected for the varying speed of the submarine, and also in many cases for the fact that a circular rotating stylus arm is used in the sonar recorder. If no independent measurement of sea level is available, this has to be removed from the

Table I. Data from under-ice submarine cruises

Ship	Month/Year	Region	Track length	Reference
Nautilus	Aug 1958	Transarctic	2700 km	[8]
Sargo	Feb 1960	M'Clure Strait	730	[9]
Sargo	Feb 1960	Canada Basin	5000	[10]
Seadragon	Aug 1960	" "	3400	[10]
Seadragon	Aug 1960	M'Clure Strait	540	[9]
Seadragon	July-Aug 1962	Canada Basin	5000	[10]
Skate	July 1962	Eurasian Basin	1800	[10]
Queenfish	Feb 1967	Davis Strait	900	[11]
Dreadnought	March 1971	W Eurasian Basin	600	[12]
Trepang	March 1971	Denmark Strait	300	[13]
Gurnard	April 1976	S Beaufort Sea	1400	[14]
Sovereign	Oct 1976	W Eurasian Basin	4000	[15]
Sovereign	April 1979	" "	2000	[16]

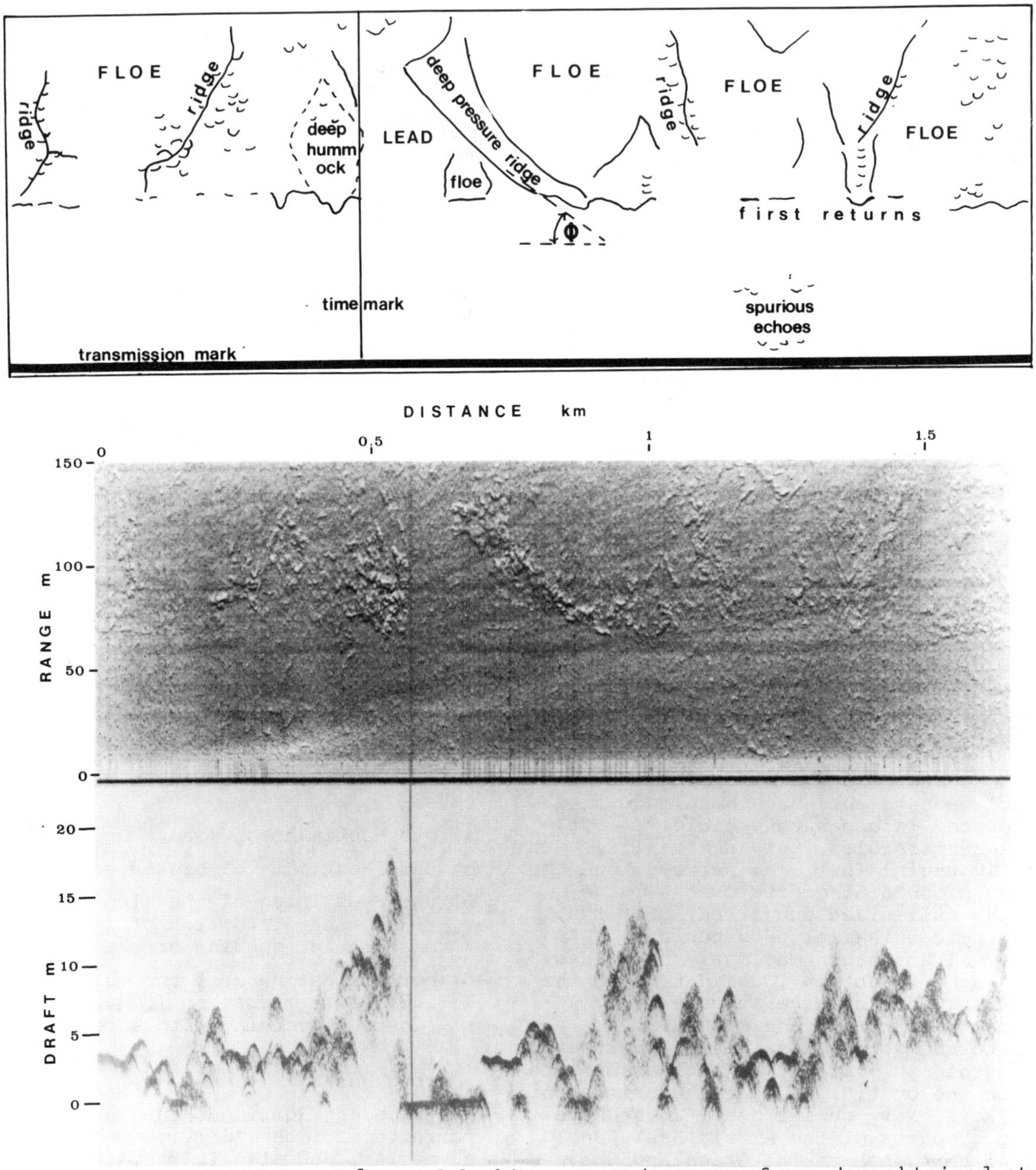

FIGURE 1. Concurrent sidescan and upward-looking sonar imagery of sea ice obtained at 83°N 13°W on October 19th 1976. The upper sketch is an interpretation of the sidescan sonar image. (After [19]).

record by reference to open leads; these have to be frequent enough to permit the zero correction to be valid. Even when a zero correction has been made, the porpoising of the submarine adds a random error to every depth value.

More recent U.S. cruises have solved most of these problems by using a narrow-beam (about 3°) sonar, with digital recording of depth and a zero reference provided automatically by a coupled pressure transducer. Records from this type of sonar can be regarded for all practical needs as perfect representations of the ice underside, failing only in the resolution of very fine scale topographic variations (since the beam footprint is about 4 m in diameter).

It would be very desirable if on every cruise a sidescan sonar profile were recorded concurrently with the upward-looking profile. This would give information on the orientation and structure of pressure ridges as the submarine crosses them, which otherwise are unknown. An experimental deployment of sidescan sonar was made from HMS Sovereign in 1976 [18, 19]. The instrument was a normal 100 kHz sidescan towfish mounted on the upper casing, and figure 1 shows some typical imagery from one channel only (the second channel failed) with concurrent records from an upward looking sonar. The individual ice blocks in pressure ridges can be seen, and it is clear that pressure ridges, which may appear on the top surface to be narrow linear features, are far more irregular and wide when seen from below, covering much or most of the ice bottom with roughness elements. While some quantitative estimates can be made from ordinary sidescan imagery (e.g. the relief of ice blocks from the length of their shadows, typically 2-3 m on fig. 1, or the percentage of thin ice in leads) the most desirable form of sonar would be a sector scanning instrument which combines the quantitative ability of the upward-looking sonar with the plan view afforded by sidescan. Such instruments exist commercially (e.g. 'Bosun' [20]) but have never been fitted to a submarine. They would yield valuable information on the two-dimensional spectral characteristics of the ice bottom.

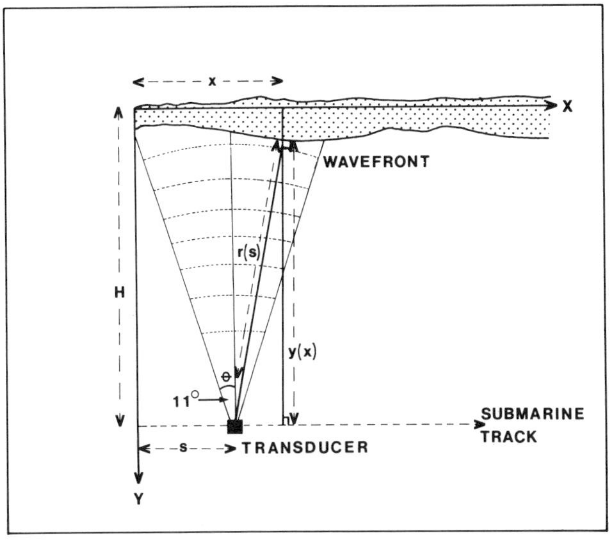

FIGURE 2. Geometry of a wide-beam upward-looking sonar.

3. BEAMWIDTH CORRECTION

Where a sonar beam is wide in the fore-and-aft plane only, a correction for the effect of beamwidth can be made using reconstruction equations, such as those of Harrison [21]. In the notation of figure 2, the digitised ice profile envelope is given by (s,H-r(s)), if the digitisation of sea level is assumed correct, while the corresponding corrected depth point is (x,H-y(x)). The parameters x and y are related to s and r by

$$x = s - r\,dr/ds, \quad (1)$$

$$y = r\left[1-(dr/ds)^2\right]^{\frac{1}{2}} \quad (2)$$

These equations are valid so long as dr/ds is single-valued, i.e. is not a cusp between two overlapping hyperbolae. When the equations are invalid a corrected point cannot be generated; this occurs most frequently at the trough between two close peaks. The equations are also inapplicable if the reconstruction involves an angle greater than the beamwidth if the transducer, i.e. if $y/r < \cos\Theta$. In these cases the reconstructed point is placed on the edge of the beam, i.e.

$$y = r\cos\Theta, \quad (3)$$

$$x = s + r\sin\Theta \text{ if } dr/ds < 0, \qquad = s - r\sin\Theta \text{ if } dr/ds > 0, \quad (4)$$

A major drawback of profile reconstruction, or deconvolution, is that it cannot regenerate the full depth of a trough between two peaks when that trough is obscured by two overlapping hyperbolae. A possible, although exceedingly time-consuming, technique would be to digitise each of the two hyperbolae rather than the envelope of the profile; even so the bottom

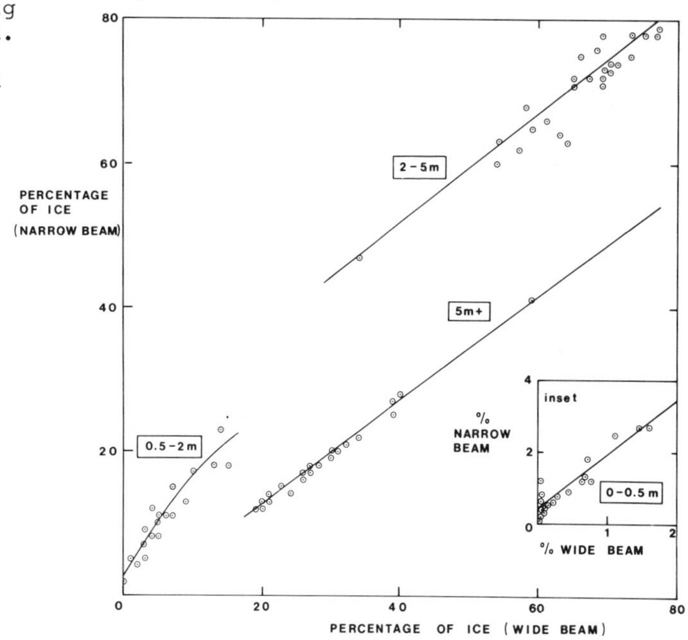

FIGURE 3. Result of applying a simulated wide beam to 50 km sections of a narrow-beam ice profile: effect on the percentages of ice occurring in four different draft ranges (after [15]).

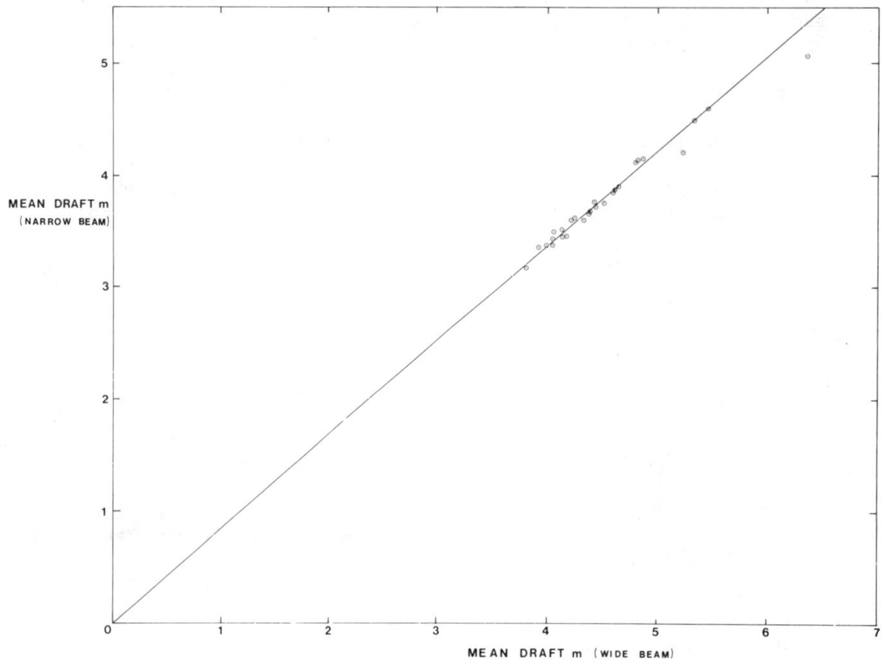

FIGURE 4. Result of applying a simulated wide beam to 50 km sections of a narrow-beam ice profile: effect on mean draft (after [15]).

of the trough is lost, although the reconstruction can proceed further down the slopes.

A second method of correction for beamwidth effects, which avoids the problem of 'lost troughs' is to proceed in the reverse direction, that is, to take a narrow-beam ice profile and to simulate the effect of passing a wide beam over it. Calibration curves for the various statistical parameters can be drawn by applying the simulation to profile sections with widely differing ice characteristics; the corrected values for the parameters are then read off the curves. This was done for the 1976 *Sovereign* cruise to the Eurasian Basin [15] by using narrow-beam data collected by U.S.S. *Gurnard* in the Beaufort Sea [14]. Figures 3 and 4 show the effect of applying a simulated *Sovereign* sonar beam (22° overall beamwidth, run at a depth of 74.7 m) to the *Gurnard* profile analysed in 50 km sections. The results were consistent enough to enable the *Sovereign* data to be corrected - in particular, the effect on mean ice draft appeared to be quite simple, the narrow-beam mean draft $\bar{h}_n$ being related to the wide-beam mean draft $\bar{h}_w$ by

$$\bar{h}_n = 0.840\,\bar{h}_w \qquad (5)$$

It must be emphasised that such a simulation should be run afresh for every cruise analysed since the combination of beamwidth and submarine depth will be different each time, so the results of figs 3 and 4 should not be regarded as universal.

In all the interpretations of sea ice profiles which follow, the limitations imposed by sonar beamwidth and the inadequate possibilities for its correction should always be borne in mind.

4. MEAN ICE DRAFT

The simplest statistic to be derived from ice profiles is the mean draft ($\bar{h}$). Even so, its calculation is not straightforward, and when comparing mean drafts from different cruises several factors must be considered. Firstly, there are the beamwidth effects discussed above. Next there is the problem of statistical reliability. The mean draft must be measured over a sufficient length (L) of profile to generate a statistically valid value, but a length that is not so great that it runs from one ice regime into a quite different one. Typically lengths of 50 km or 100 km are used to compute $\bar{h}$, and if a 'spot value' is required, e.g. for purposes of contouring, the position of the 'spot' is taken to be the centre of gravity of the submarine's path within that section. In coastal zones of the Arctic the mean draft varies rapidly with distance from shore, especially where shear is occurring (e.g. Beaufort Sea, north Greenland). Here even a 50 km section fails to resolve the rapid variation in mean draft, but if shorter sections are used the problem of sample size becomes acute, since as L decreases the accuracy of an $\bar{h}$ estimate becomes lower. We can find the accuracy of $\bar{h}$ by examining repeated samples taken over a region of the Arctic where ice conditions vary extremely slowly. Two zones have been studied where these conditions apply: the central Beaufort Sea (Gurnard sections 3-25 [14]), which gave $\bar{h} = (3.67 \pm 0.06)$m for L = 50 km; and the Eurasian Basin (Sovereign sections 14-31 [15]) which gave $\bar{h} = (4.51 \pm 0.05)$m for L = 100 km. We assume that these standard deviations are typical of the respective track lengths, and therefore 0.05-0.06 m is the magnitude of error involved when we quote a mean draft for a 50-100 km section.

We now briefly review mean drafts as observed in various regions of the Arctic.

4.1 Beaufort Sea in spring.

The April 1976 cruise of Gurnard [14] yielded the high value of 5.09 m for a 50 km section running northward from the 100 m isobath off Barter Island. The 17 km that lay nearest to shore reached 5.58 m. The second 50 km gave $\bar{h}$ = 4.22 m, and northward of this the statistically homogeneous region described above was reached. At the end of its cruise the submarine again approached the shear zone north of Point Barrow (300 m isobath) where an $\bar{h}$ of 4.61 m was recorded.

4.2 Eurasian Basin and north Greenland

The October 1976 cruise of Sovereign sampled much of the western Eurasian Basin and the heavily ridged offshore zone north of Greenland and Ellesmere Island [15, 22]. The greatest observed $\bar{h}$ was 7.49 m at 85°N, 70°W, north of Ellesmere Island, although the entire track across the north of Greenland had high values in the range 5.1 - 6.7 m. According to numerical models [23] this is the zone of heaviest ridging in the Arctic Ocean, since the wind and current stresses both tend to pile up ice from the Trans Polar Drift Stream against the downstream land boundaries. Moving north on the 70°W meridian towards the North Pole the mean ice draft remained high out to 400 km from the coast of Ellesmere Island, after which a very rapid transition occurred to a statistically homogeneous regime which extended across the rest of the western Eurasian Basin. The mean draft dropped to 4.51 m, with variations from section to section which were small enough for a run test to demonstrate that the sections came from the same population. The difference of 0.84 m between the Gurnard mean drafts in the central Beaufort Sea and the Sovereign mean drafts in the Eurasian Basin is highly

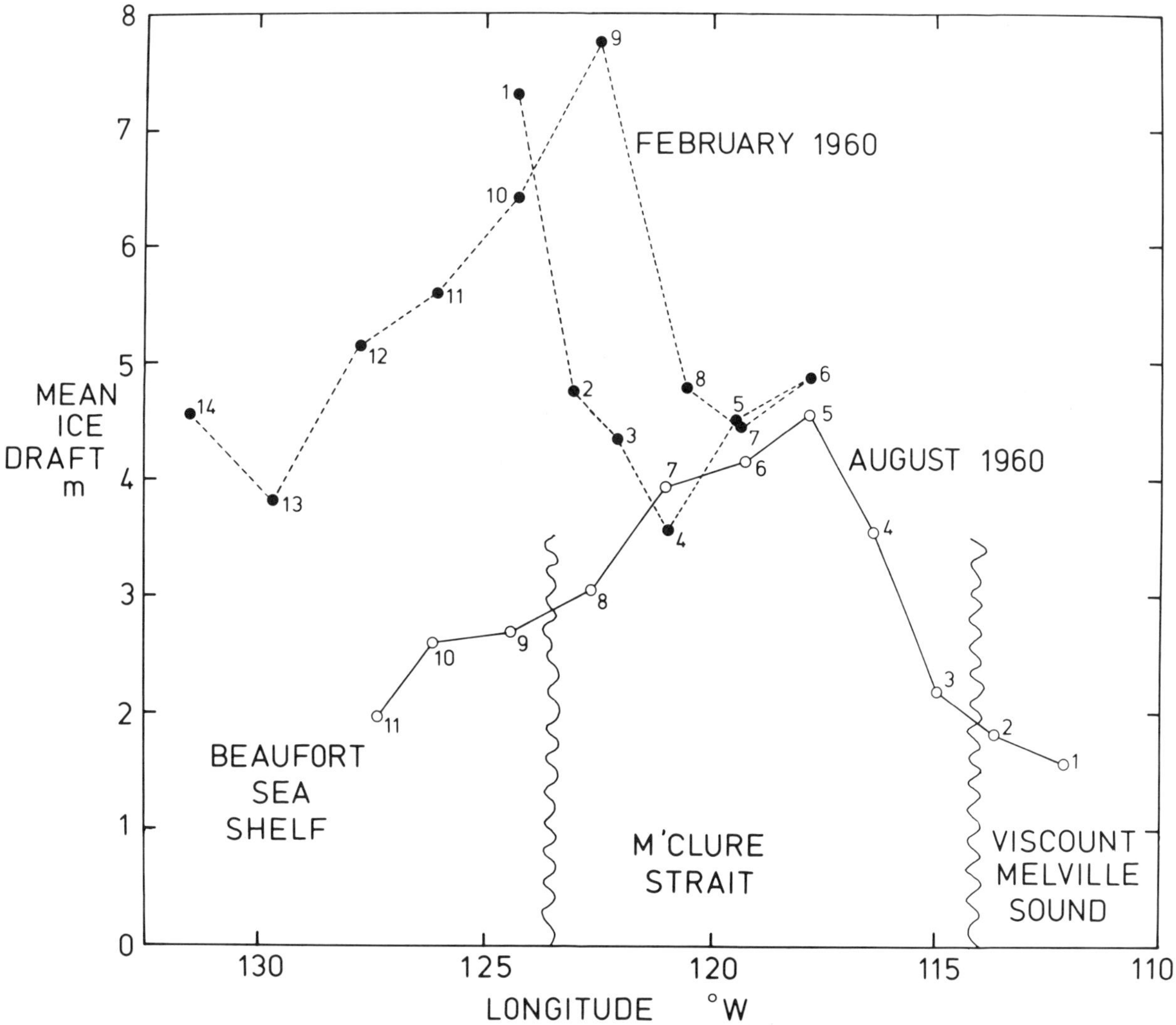

FIGURE 5. Mean ice draft in M'Clure Strait as a function of longitude (after [9]). Sections 1, 9 and 10 were obtained on the northern side of the Strait, the other sections were approximately along the mid-line.

significant, and indicates a clear tendency towards higher mean ice thicknesses in the Eurasian than in the Canada Basin.

4.3 M'Clure Strait.

Two cruises through M'Clure Strait were carried out in 1960, by U.S.S. *Sargo* in February and U.S.S. *Seadragon* in August [9]. Figure 5 shows the mean ice draft as a function of longitude, using a 50 km section length. The very highest value of 7.77 m (the highest mean draft yet observed in the Arctic) was obtained off the southwest tip of Prince Patrick Island just within the mouth of the Strait in winter. This zone is already known from airborne surveys to be one of heavy ridging, and is the source area for 'multi-year hummock fields' (R. Hudson, personal comm.,), large coherent ice masses composed of continuous ridging, which eventually break out and drift into the southern Beaufort Sea. The accumulation of ridging is local, not widespread like that across the north of Greenland, but it is exceptionally intense and the reason for its generation is not known. The rest of the winter profile shows a change in $\bar{h}$ across the Beaufort Sea shear zone (sections 10-14) which resembles the *Gurnard* observations north of Alaska. The mean ice draft within M'Clure Strait in winter lies in the range 4-5 m. In the following summer $\bar{h}$ within the Strait was very similar, only 22 cm less on average, implying ice which is melting *in situ*. There is a rapid decline towards the open water ice edge in Viscount Melville Sound, and also out into the Beaufort Sea, where the Spring break-up of the coastal zone has generated a large amount of open water.

4.4 Fram Strait.

Although the 1976 *Sovereign* cruise sampled Fram Strait, a much more complete survey of the ice conditions there was carried out by the same submarine in April-May 1979 [16]. Figure 6 shows spot values of mean ice draft for 50 km sections, which are sufficiently well distributed to permit contouring. The most obvious feature is the very rapid decline of $\bar{h}$ in the marginal ice zone (MIZ) as the ice edge is approached. Further, $\bar{h}$ is low on the eastern side of Fram Strait relative to the western side, implying a different ice composition - in fact ice approaching Fram Strait from the north-east comes from north of Franz Josef Land and Siberia and tends to be younger than ice approaching from the north and north-west

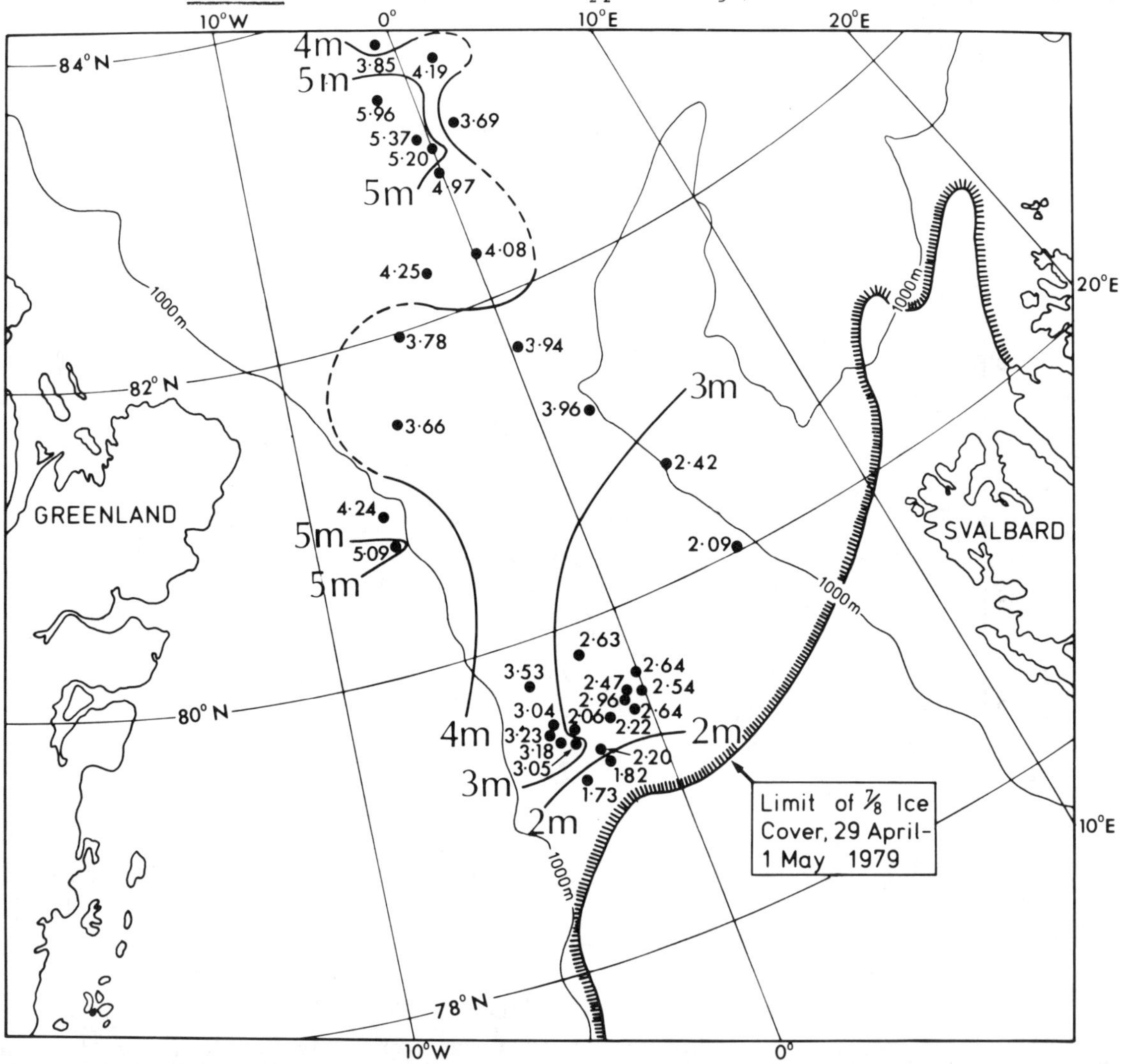

FIGURE 6. Contours of mean ice draft in Fram Strait during April-May 1979 derived from 50 km sections profiled by HMS "Sovereign" (after [16]).

which has crossed the entire Arctic Basin. Note the high $\bar{h}$ values close to north Greenland, reinforcing the earlier Sovereign data.

4.5 Denmark Strait.

A profile obtained in March 1971 between 67° and 70°N was reported by Kozo and Tucker [13]. They found that the modal ice draft (which we expect to be slightly lower than $\bar{h}$) varied from 0.8 m at the ice edge to 2.9 m some 200 km inside. These values are significantly lower than those observed at 79°N in fig. 6, indicating considerable melting and/or divergence of the ice cover during its southward passage in the East Greenland Current.

LeSchack and Chang [24] produced a contour map of r.m.s ice draft over much of the Arctic Basin, which was reproduced in Hibler [23]. Apart from the difficulty of comparing r.m.s. with mean values, this chart possesses some curious features which probably derive from the fact that it was based on a mixture of winter and summer data. For instance, it shows no shear zone in the Beaufort Sea. In other respects, however, it agrees reasonably well with the results described above.

Mean ice draft can be converted approximately to mean ice thickness by multiplying by the ratio 1.12, assuming a water density of 1020 and an ice density of 910 kg m^{-3}. Ackley et al [25] describe more complex procedures which allow for ice density variation with depth and for a contribution from snow cover; they must be applied to the whole ice profile. Mean ice draft, however, is a useful concept to retain, since it is a direct measure of the mass of ice per unit area of sea surface.

5. ICE DRAFT DISTRIBUTION

The probability density function P(h) of ice draft is defined such that P(h) dh is the probability that a random point on the ice underside has a draft between h and (h+dh). P(h) again has to be defined over a length scale L; a longer track length gives greater reliability to the shape of the distribution but may mask a real transition between ice regimes. P(h) also varies with time, the most rapid variation being in the percentage of thin ice present, since the ice cover can change in days or hours from a state of net divergence to one of net convergence. Since thin ice also tends to be contained in a relatively small number of polynyas, its contribution to a measured P(h) involves a sampling problem which is greater than that for other types of ice. The importance of P(h), or rather of g(h), the ice thickness distribution function [26], is that it is both an input and an output to models of Arctic ice dynamics and thermodynamics [27, 23]; in particular, the concentration of ice less than 1 m thick is the chief determinant of ocean-atmosphere heat flow [28].

Figure 7 shows a typical set of P(h) functions obtained from 100 km track sections running northward from 85°N 70°W (section 11) to the North Pole (section 16) in October 1976 [15]. Each distribution shows a peak at less than 1 m draft due to young ice in recently opened leads and polynyas. There is then a gap with relatively little ice, followed by a major peak at about 3 m due to undeformed first- and multi-year ice, followed by a tail which contains ice from the sides and bottoms of pressure ridges. The reason for the gap after the young ice peak is that any ice less than 1 m thick found in an otherwise mature icefield exists only in leads and polynyas, and is easily crushed during periods of convergence to form pressure ridges. Most young ice is therefore transported straight to the 'tail' of the distribution rather than being allowed to evolve peacefully to add to the undeformed ice peak. The major peak itself may appear split into two or more closely spaced peaks due to the slightly different preferred depths of first-year and multi-year ice. In many cases, however, the peak is broad and the two categories are not resolved.

When averages of many such distributions

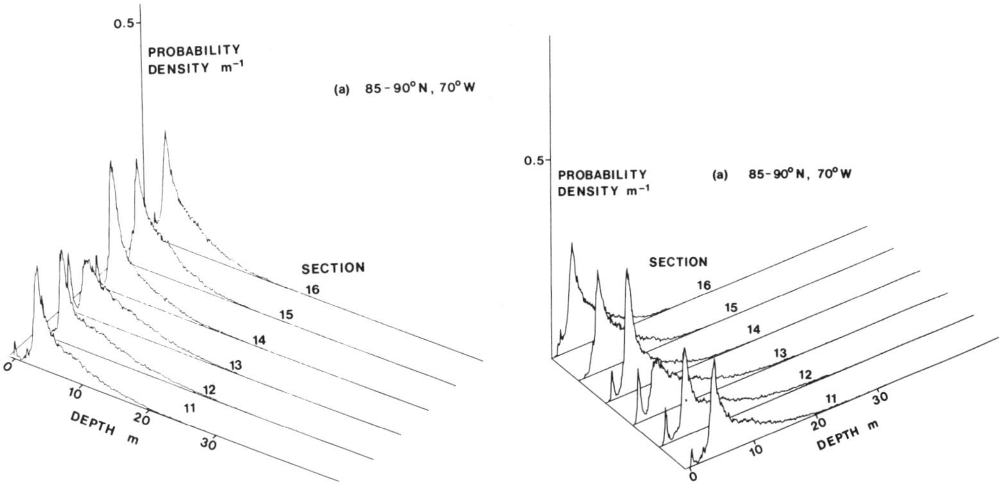

FIGURE 7. Probability density functions of ice draft derived from 100 km sections running north from 85°N70°W (section 11) to the North Pole (section 16).

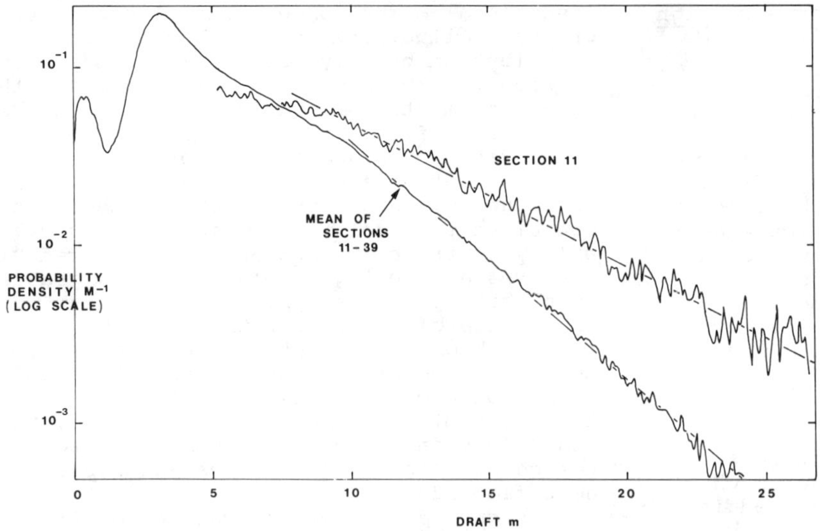

FIGURE 8. Semilogarithmic plot of P(h) for a 100 km section and for 2900 km of ice profile from the Eurasian Basin (after [15]).

from the Arctic Basin are made, it is found that the tail gives a close fit to a negative exponential distribution. Figure 8 shows this for Sovereign data [15] and it was also found for Gurnard [14] and M'Clure Strait [9] data. The reason for the draft distribution of deformed ice being a negative exponential is unknown. We shall see, however, that the pressure ridge draft distribution also fits a negative exponential and that the two results can be tied together on the assumption that ridges tend to take the shape of an isosceles triangle in cross-section. The two results therefore involve only a single mystery.

The cumulative probability G(h), defined by

$$G(h) = \int_0^h P(h)\, dh \tag{6}$$

is also a useful concept; it was employed in the AIDJEX model [27]. Figure 9 shows a typical form for G(h), obtained from two intersecting 200 km sections in the southern Beaufort Sea [14]. From this type of curve we can see, for instance, that the median

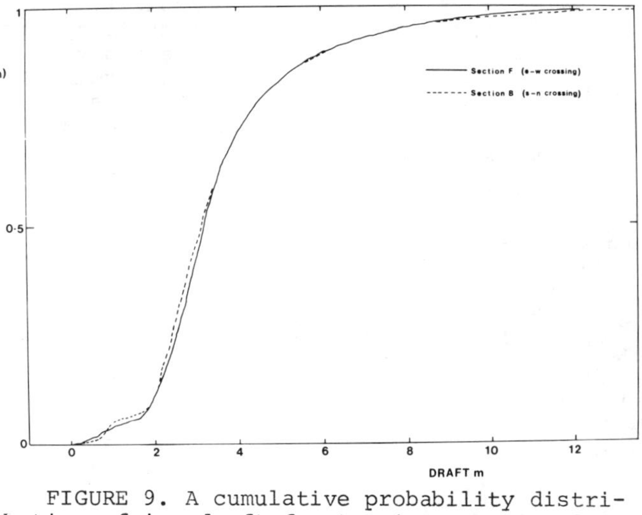

FIGURE 9. A cumulative probability distribution of ice draft for two intersecting 200 km profiles from the Beaufort Sea (after [14]).

depth (G(h) = 0.5) is reached at 3.2 m and that 99% of the ice is less than 12.2 m thick.

6. LEVEL ICE

We saw from fig. 7 that the major peak in P(h) is usually smoothed out to the point where the relative contributions of undeformed first-year and multi-year (second-year or older) ice to the overall cover cannot be resolved. It is important to attempt a separation of these ice types, since an ice cover with extensive multi-year ice is more of a hazard to navigation and to offshore structures than an ice cover of similar mean draft but composed entirely of first-year ice. The obvious property of undeformed ice is its comparative flatness, and by trial and error Williams et al [12] found that if 'level ice' were defined as ice with a local gradient of less than 1 in 40, then statistically generated level ice percentages agreed well with the results of visual estimates. The idea is that by isolating 'level ice' we can find the preferred drafts of undeformed first- and multi-year ice and the relative contributions of each to the ice cover.

The 1 in 40 criterion can be applied in different ways. On one definition (D1, say), a point (x, h(x)) is defined as a level ice point if its draft differs from that of a point 10 m away to either side by less than 25 cm. This has the disadvantage that it includes ice on opposite flanks of a pressure ridge at a depth where the ridge width is 10 m, or ice on the same flank of successive ridges separated by 10 m. A more restrictive definition D2, suggested by A.S. Thorndike (personal comm.) is that no point within 10 m of a level ice point may differ in draft by more than 25 cm from that of the level ice point. The two definitions are thus

$$\text{D1} \quad |h(x+10)-h(x)| \leqslant 0.25 \text{ or } |h(x-10)-h(x)| \leqslant 0.25, \tag{7}$$

$$\text{D2} \quad |h(x \pm d)-h(x)| \leqslant 0.25, \quad d=0 \text{ to } 10 \text{ m} \tag{8}$$

Wadhams and Horne [14] found that D2 was so restrictive that it found very little ice, since normal undulations in undeformed ice exceed 25 cm within a 20 m gauge length. It is effective, however, in picking out the preferred drafts of level ice, since although it does not see all the level ice, all that it sees is level ice. D1 is more effective in assessing the relative contributions of different undeformed ice types to the overall ice cover.

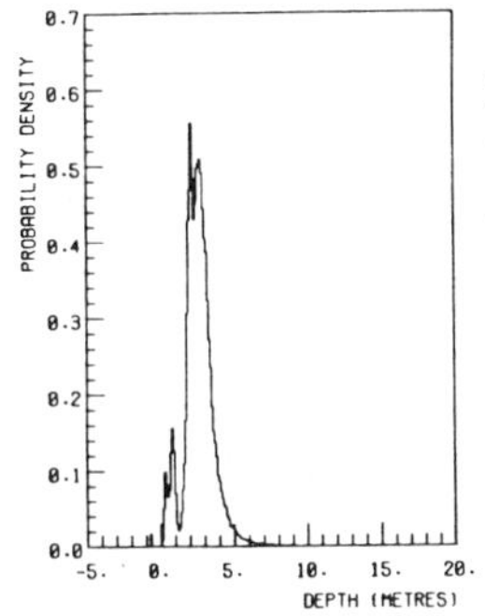

FIGURE 10. Probability density function of level ice draft from the "Gurnard" profile in the Beaufort Sea (after [14]).

Figure 10 shows a probability density function of level ice draft using definition D2 and taken from the whole 1400 km Gurnard profile [14]. The technique has separated both the young ice peak and the undeformed ice peak into two components. The two thin ice peaks are at 0.3-0.4 m and 0.8-0.9 m drafts (representing two separate episodes of divergence in the recent past), while the main peak is split into 2.1-2.2 m and 2.7-2.8 m categories. It is reasonable to assign the 2.1 m and 2.7 m peaks to first-year and multi-year ice respectively. Estimates of thermodynamic growth rate by Thorndike et al [26] suggest that first-year ice which forms on 1 October will reach a thickness of only 1.8 m by 10 April (date of the Gurnard profile), but 2.7 m corresponds quite well to their estimates and those of Maykut and Untersteiner [29] for the equilibrium thickness of multi-year ice in April. It is interesting to note that Wadhams [15] found only a single peak in the level ice draft distribution from Sovereign data in late October, since the ice cover at that time of year would not contain first-year ice.

We expect the percentage of the sea surface occupied by level ice to vary inversely with the mean ice draft, since a high mean draft implies a large amount of deformation. This can be used as a means of separating ice regimes, analogous to the use of (T,S) diagrams to separate water masses. Figure 11 shows mean ice draft plotted against level ice percentage (using definition D1 from Sovereign data [15]). The overall data show a negative linear correlation, and there is a clear separation between ice regimes. The 'outer offshore zone' comprises the region further from shore (200-400 km) than the 'Greenland offshore zone' itself, which is the part of the track traversing the northern coasts of Greenland and Ellesmere Island.

7. PRESSURE RIDGES

In an ice profile obtained by upward-looking sonar it is impossible to identify every ice feature without additional information. For instance, a single ridge keel may have a complex ice block structure which gives it an appearance indistinguishable from that of two independent keels which happen to cross one another above the submarine track. Nevertheless, as it is undeniable that pressure ridges exist and are an important component of the ice cover, it is necessary to adopt an arbitrary criterion for counting independent keels. Ideally, the same criterion should be used for all analyses since this permits comparison of results. The most common definition is that an independent keel is one in which the troughs (points of minimum draft) on either side of the keel crest (point of maximum draft) each rise at least half way towards the local level ice bottom before beginning to descent again. The 'local level ice bottom' is difficult to find in heavily ridged areas, so it is defined arbitrarily as being at a draft of 2.5 m. This is similar to the Rayleigh criterion for resolving spectral lines. It has been used in analyses of sonar and laser profiles by Leppäranta [30]; McLaren et al [9]; Lowry and Wadhams [31], Tucker et al [32], Wadhams [15, 17, 22, 33], Wadhams and Lowry [34], Weeks et al [35] and Williams et al [12]. Earlier definitions, such as that of Hibler et al [36], involved troughs which descend a fixed distance from the peak; Hibler [37] and Rothrock and Thorndike [38] have discussed the problems involved in this type of definition.

Hibler et al [39] showed that if ridges occur at random along a track, the distribution of spacings between them is given by

$$P_r(x)\ dx = \mu \exp(-\mu x)\ dx \qquad (9)$$

where μ is the mean mumber of ridges per unit length of track and $P_r(x)\ dx$ is the probability that a given spacing lies between x and (x + dx). Mock et al [40] found good agreement with this relation for surface sails, except for an excess at small spacings. More recently, Lowry and Wadhams [31] derived a modified relation which allows for the possibility of ridges overlapping so that the trough between two crests disappears. The shallower of the two closely spaced ridges is thus lost from the statistics. This is the so-called 'ridge-shadowing effect'. If two ridges have reliefs h,h' $(h>h')$ and are taken to be triangular in cross section with slope α in the along-track direction, then the closest approach x_{crit} of the two ridges that still permits h' to be independent is

$$x_{crit} = h \cot \alpha \qquad (10)$$

A relation based on this idea was found to give better agreement with the observed spacing distribution.

Figure 12 shows keel spacing distributions from the southern Beaufort Sea [14]. At moderate spacings (40-250 m) the distributions agree well with (9). At small spacings the keel shadowing effect reduces

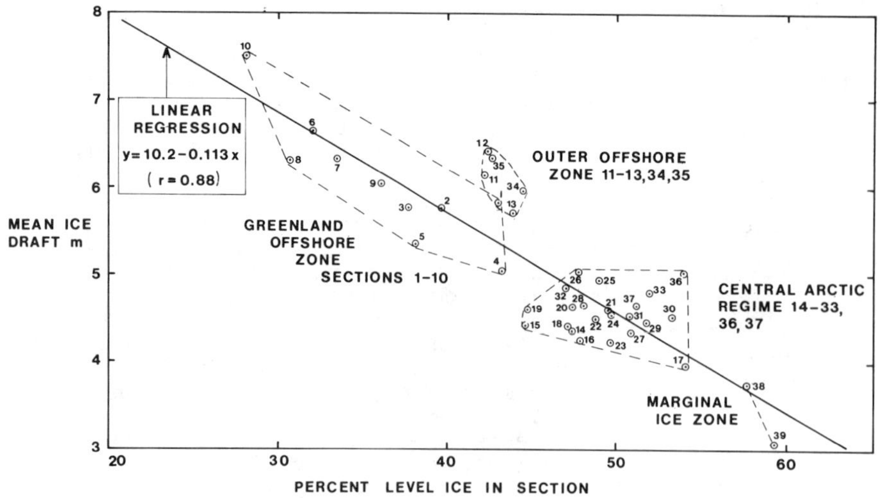

FIGURE 11. Relation between mean ice draft in a 100 km profile and the percentage of level ice in the profile (after [15]).

the number of occurrences, while at very large spacings the number of occurrences is greater than predicted by (9). This is because of the presence of leads and polynyas, which interpose additional smooth stretches of ice into the formerly random icefield and thus generate an anomalous number of large keel spacings.

In many independent sets of observations it has been found that the distribution of keel drafts fits a negative exponential distribution, i.e.

$$n(h)\,dh = B \exp(-b h)\,dh \quad (11)$$

where $n(h)$ is the number of keels per km of track per m of draft increment, and B,b are parameters that can be derived in terms of the experimentally observed mean keel draft $\bar{h}$, mean number of keels per km μ, and low-value cut-off draft h_o:

$$b = (\bar{h} - h_o)^{-1} \quad (12)$$

$$B = \mu b \exp(b h_o) \quad (13)$$

This relationship fits keel drafts observed by narrow-beam sonar in the Beaufort Sea [14] and M'Clure Strait [9] as well as sail heights observed by laser profiling [15, 22, 30, 32, 33, 35].

Given the validity of the empirical result (11), two important consequences follow. Firstly, we can relate keel draft distribution to ice draft distribution $P(h)$. Beyond the maximum draft which can be attained by undeformed ice h_{max} say, the whole of $P(h)$ is due to contributions from ice keels. If keels tend to the shape of isosceles triangles with a mean along-track slope angle α, then it is easy to show that

$$P(h)\,dh \Big|_{h>h_{max}} = \frac{2\,B \cot \alpha}{b} \exp(-b h)\,dh \quad (14)$$

The observations of Wadhams and Horne [14] gave 11.3° as the best value for α to enable (11) and (14) to agree. This appears low in relation to observed slope angles, but the random angle of encounter between the submarine and the ridge is sufficient to make α approximately half of the real ridge slope [17].

The second important consequence is that (11) makes it easy to predict extreme keel depths. If S km is the distance drifted per year by the ice cover which passes over a given location, then the total number of keels per year of depth D or greater which will drift past that point is

$$N_D = S \int_D^{\infty} n(h)\,dh = S\mu \exp\left[-\frac{D - h_o}{\bar{h} - h_o}\right] \quad (15)$$

$T_D = 1/N_D$ is the return period in years for a keel of draft D or greater at the point concerned. The chief problem, then, in predicting extreme depths is the measurement or estimation of S. Extreme depth prediction is discussed further in section 9.

Some observations of keel drafts, especially those made with wide-beam sonars, have fitted a distribution proposed by Hibler *et al* [39], of form

$$n(h) \propto \exp(-\lambda h^2) \quad (16)$$

This distribution was proposed on theoretical grounds by assuming that keels are randomly oriented linear features which are geometrically congruent in cross-section (i.e. cross-sectional area is proportional to the square of their relief), and that their depth distribution can be obtained by a variational calculation holding the volume

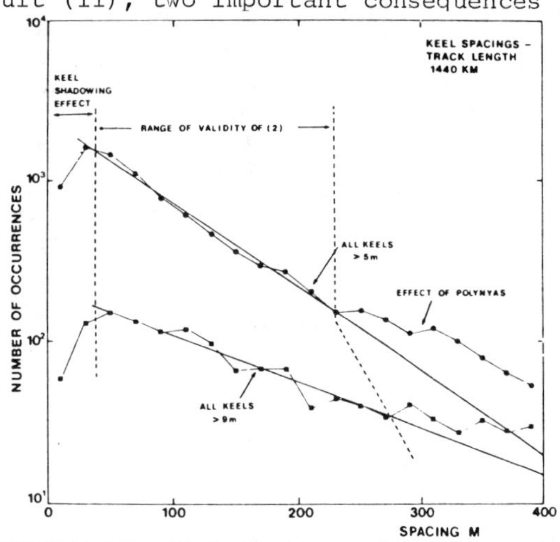

FIGURE 12. Distribution of keel spacings from 1400 km profile in Beaufort Sea, plotted for keels deeper than 5m and 9 m (after [14]).

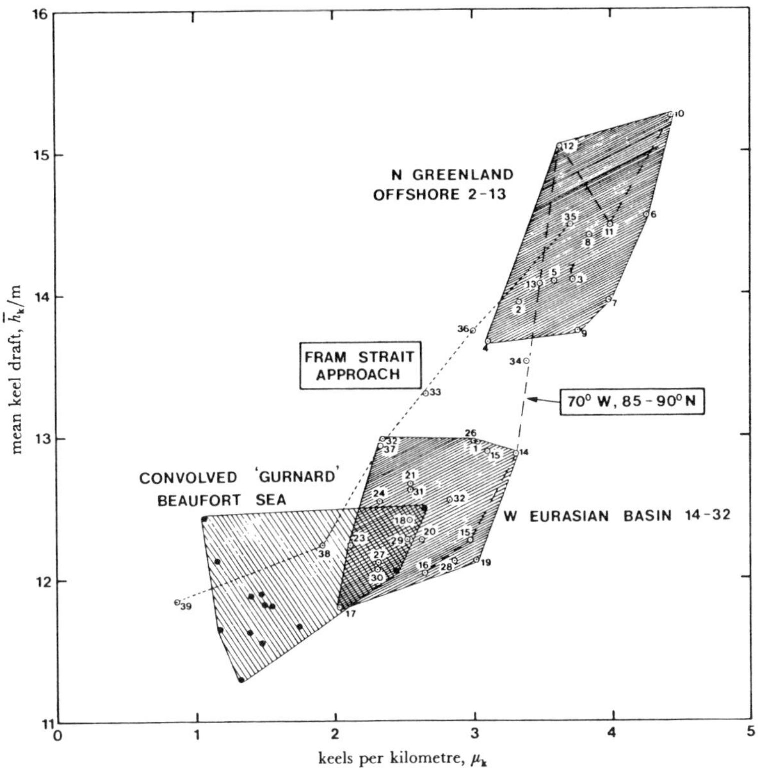

FIGURE 13. Relation between mean keel draft and number of keels per kilometre, for keels deeper than 9m. Numbered points refer to 100 km sections from the *Sovereign* profile, black spots to 50 km sections from the *Gurnard* profile convolved with a wide beam (after [15]).

of deformed ice as constant. The *Sovereign* data [15] fitted (16), but the relationship appears to hold mainly for wide-beam results, suggesting that the wide beam 'smooths out' the complex topography of keels so as to yield deformation elements which resemble the simple congruent features assumed in the theory.

Some datasets obtained from the Arctic show a positive correlation between the mean keel draft and the keel frequency. Figure 13 shows this relation for the *Sovereign* data, and also shows that the *Gurnard data*, when convolved with a wide beam footprint so as to mimic the *Sovereign*'s sonar, also gives points which lie on the same curve. Many other data sets show only a weak correlation, however.

The deepest keel ever observed in the Arctic had a draft of 47 m (Lyon quoted in [41]), while a 43 m keel was observed by *Sovereign* [15]. Wadhams [17] analysed the distribution of keels deeper than 30 m in the region profiled by *Sovereign* and found, as expected from (15), that they were concentrated in areas of high ridge frequency. A significant clustering in groups was also found, however, suggesting that very deep keels are formed by singular events of major deformation, and that they subsequently split into a number of keel linkages which drift downstream in fairly close company.

8. LEADS AND POLYNYAS

Although the function P(h) gives information about thin ice occurrence which is vital to heat flow calculations, there are other applications for which it is desirable to know how the thin ice is distributed in leads and polynyas. One set of applications concerns sea ice mechanics, since each lead is a potential pressure ridge, and the distribution of leads is a measure of the ice cover's capacity to sustain deformation through convergence or shear. Other applications are practical - leads and polynyas offer a submarine the ability to surface and also are possible landing sites for aircraft which support ice operations.

The definition of a lead adopted for the computer analysis of ice profiles [9, 14, 15] is that it is a continuous stretch of ice at least 5 m in length, within which no point has a draft of more than 1 m. A polynya containing an ice floe more than 1 m deep is thus counted as two polynyas. We must use the terms 'lead' and 'polynya' interchangeably, since an upward-looking sonar alone gives no information about the shape of the thin ice region which is passing overhead.

The distribution of lead widths has not been found to fit any simple function. *Sovereign* data [15] fitted a relationship where the frequency of leads was proportional to (diameter)$^{-2}$, but this has not been found to be of general validity. The normal way to represent leads has therefore been by means of a trafficability diagram such as figure 14, where the ordinate represents the average distance between encounters with a lead of minimum width given by the abscissa. From this it is possible to see at a glance the average spacing of leads which exceed the critical width for some operational requirement, e.g. the surfacing of a submarine.

Wadhams [15] found that in October leads occupied 1-5% of the ice cover in the western Eurasian Basin, with the percentage rising to 10-20 in the marginal ice zone and also in the heavily ridged zone north of Greenland. This association of heavy ridging with large amounts of thin ice was ascribed [42] to the acceleration of already deformed ice as it moved around the northeast coast of Greenland to approach Fram Strait. Typical

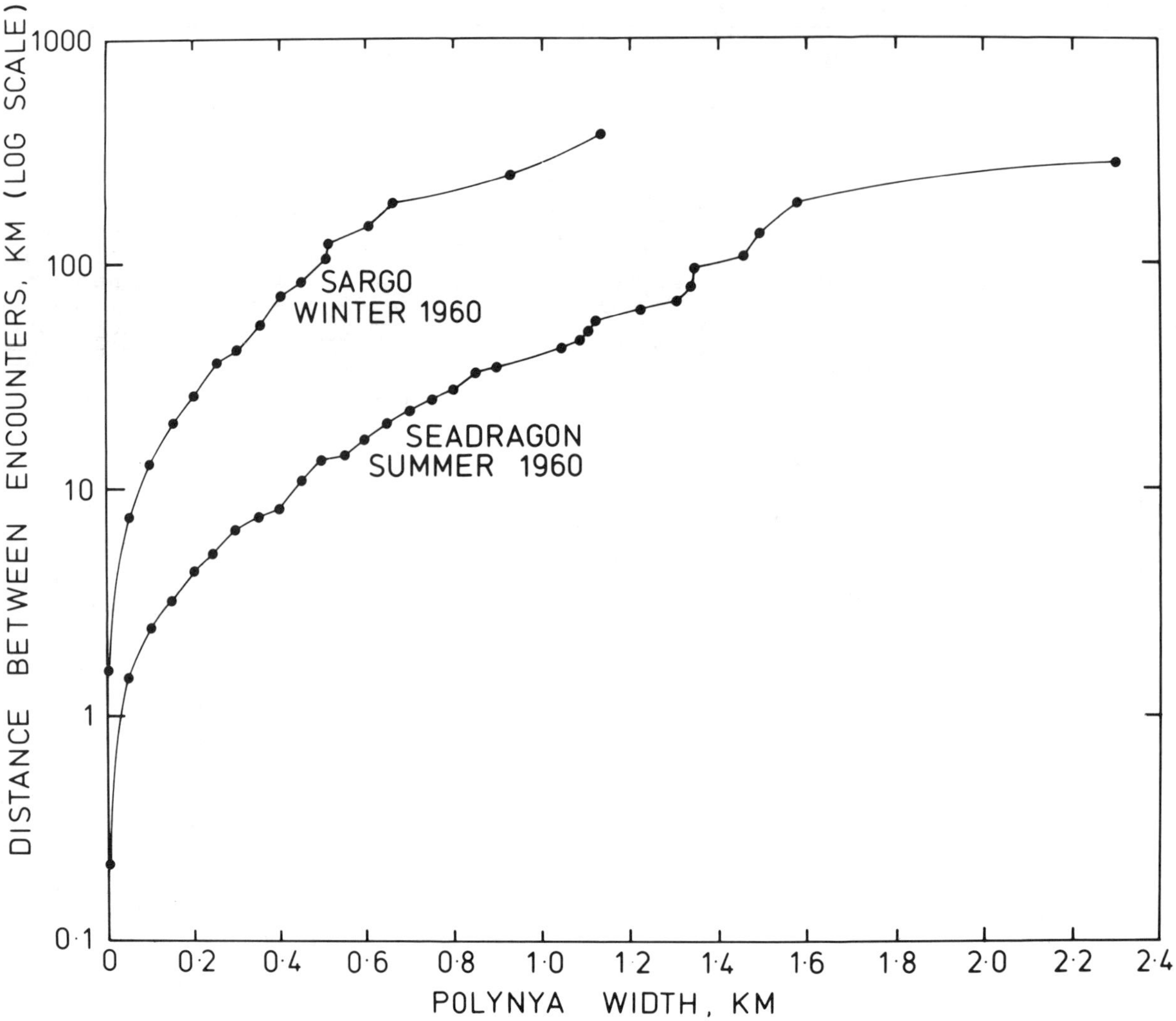

FIGURE 14. Trafficability diagrams for encounters with polynyas in M'Clure Strait and the nearby Beaufort Sea shelf during winter and summer (after [9]).

lead frequencies were 0.2 - 0.9 per km, with 90% or more of leads being less than 50 m wide. Figure 14 shows the difference between the lead distributions in winter and summer in M'Clure Strait and the neighbouring Beaufort Sea shelf. Leads are almost a factor of 10 more frequent at all widths in summer.

9. EXTREME DEPTH PREDICTION

An important application of upward-looking sonar profiles is their use in the prediction of the deepest keel that will be encountered in a time interval or a length of track greater than that observed. This is similar to the use of wave measurements at a site to predict a '100-year wave'. In shelf areas, the need for such predictions is concerned with the recurrence period for scouring of the sea bed by deep keels. In the central Arctic, the relevance is to the safety of submarine operations and to scientific problems such as the water depth to which stirring and internal wave generation by deep keels takes place.

The simplest extreme value problem is one-dimensional, i.e. what is the deepest keel that will cross a given spot in a given

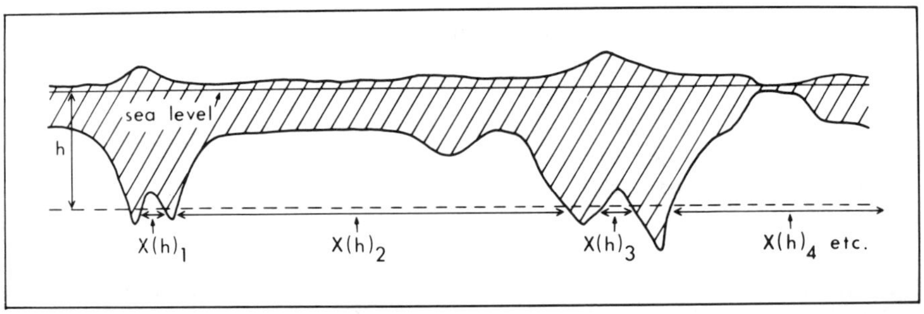

FIGURE 15. Definition of the depth crossing statistic X(h).

interval of time? Wadhams [43] proposed three techniques for tackling this problem. They are:-

(i) Use of the negative exponential distribution, where it has been shown to be the frequency distribution followed by the keel drafts. This method has already been discussed in section 7 (equation 15).

(ii) Use of a depth crossing technique. A statistic X(h) is defined (figure 15), which is the set of distances between the upward crossing of depth horizon h by the ice profile, and the subsequent downward crossing. $\bar{X}(h)$ is computed, and for the *Gurnard* data was found to fit a positive exponential distribution as a function of h. Extrapolation to some greater depth D tells us that the return period T_D for keels of depth D or greater is just

$$T_D = \bar{X}(D) / S \qquad (17)$$

where S is defined as in (15). Although in the example studied, $\bar{X}(h)$ fitted a simple distribution, the advantage of this technique in general is that there is no need to define or count independent pressure ridges.

(iii) Use of a probability plotting technique. The record is divided into uniform sampling intervals (say 50 km) and the deepest ice point in each interval found. The resulting points are ranked in order of depth and plotted on exponential extreme-value probability paper [44] using the Weibull plotting formula

$$F = \frac{1}{P(X>x)} = \frac{n+1}{m} \qquad (18)$$

where F = return distance in units of 50 km
n = number of depth values ranked
m = rank of a given depth point, m = 1 being deepest of all.

The results from *Gurnard* fit a straight line (figure 16), from which we can infer the return distance for extremely low probabilities. This differs from a technique proposed by Tucker *et al* [32] using normal probability paper, which was found not to give a straight line.

Predictions from these three techniques agreed well, and in the Beaufort Sea coastal zone gave a depth of 40 m for a return period of 10 years, and of 55 m for 1000 years.

A more difficult two-dimensional prediction problem concerns the return period for the exceedence of a given depth anywhere along a line of unit length drawn at right angles to the mean ice drift. The application here is to the frequency of scouring of a seabed pipeline laid to shore at right angles to the net ice drift in the shear zone. Wadhams [43] showed that a parameter which must be measured to solve this problem is ℓ, the mean length of a keel in km, measured along its crest, which continuously exceeds a depth D. No such parameter can be derived from upward-looking sonar profiles, but the distribution of this statistic is identical to the distribution of the widths of ice scours themselves, since it is only the sections of keel deeper than D

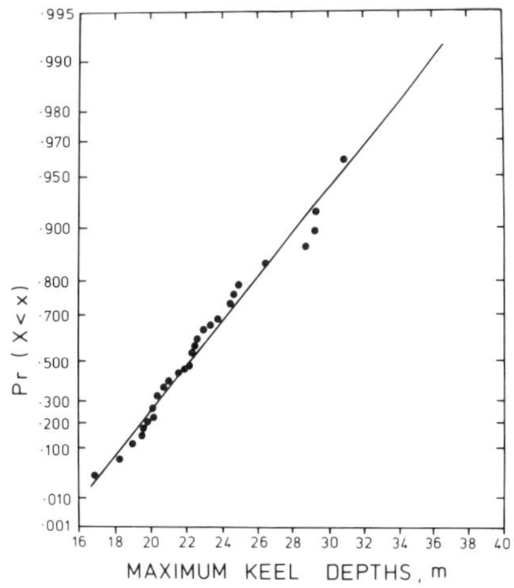

FIGURE 16. Fit of *Gurnard* data to Gumbel extreme-value plotting (after [43]).

(where D is water depth) that scour the bottom. Using the existing observation [45] that the depth d of scouring below the undisturbed seabed follows a negative exponential distribution exp (-k d), Wadhams derived the prediction formula

$$d = \frac{1}{k} \ln \left[S \mu T f \exp \left(- \frac{D - h_o}{\bar{h} - h_o} \right) \right] \qquad (19)$$

where d is the maximum depth of scour achieved per km in a recurrence interval T yrs, and f = Π/2ℓ, where ℓ is measured over the same region as the submarine profile. Values for d of 4-8 m were predicted for 100 year recurrence at depths of 15-65 m in the Canadian Beaufort Sea coastal zone, necessitating deep burial of any pipeline.

10. SPECTRAL CHARACTERISTICS OF THE ICE SURFACE

The spectral characteristics of ice surfaces were investigated in studies by Hibler [46] and Hibler and LeSchack [47]. Hibler discussed the two-dimensional spectral analysis of airborne laser profiles obtained by a star-shaped pattern of flights, i.e. intersecting tracks over the same area differing in orientation by equal increments. He showed that it is possible to determine the anisotropy of ridge structure and of higher-frequency roughness elements (interpreted as drifting snow). He defined a 'roughness tensor' which is a measure of the degree and orientation of the anisotropy and which also indicates the sense in which convergence has occurred in the icefield. Sonar data of comparable quality are not available, but Hibler and LeSchack [47] found that two tracks crossing at right angles each displayed a single significant spectral peak at different wavelengths, enabling the presumed orientation of a local array of aligned pressure ridges to be determined.

More recently Rothrock and Thorndike [38] proposed a fractal approach to the geometric properties of the ice underside. They found that one-dimensional spectral

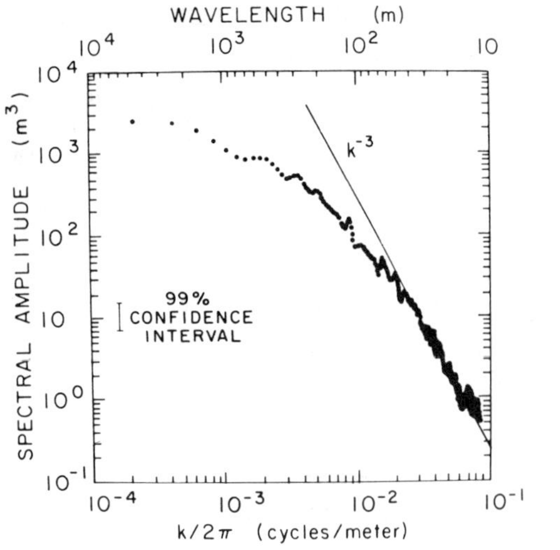

FIGURE 17. Power spectrum of bottom roughness of a 60 km ice profile from the Beaufort Sea. k is wave number (after [38]).

amplitudes of ice roughness vary as k^{-3}, where k is wave number, in the limit of high wave numbers (figure 17). An exponent of -3 separates surfaces which are 'rough' (continuous but not differentiable) from those which are 'smooth' (continuous and differentiable), and in addition a k^{-3} relation implies that there is no natural length scale to the ice roughness, i.e. that an ice surface looks the same when viewed under any degree of magnification.

The fractal approach is a promising route towards ice surface simulation. It would be very useful to be able to numerically simulate ice surfaces such that the shapes and roughness scales of ridges, hummocks, level ice and leads are correctly represented. These simulations could then be used to examine the scattering of sound by the ice bottom surface, or the scattering of microwaves by the upper surface on a statistical basis. An application of the latter is to the interpretation of return pulses from airborne or spaceborne radar altimeters over sea ice.

11. UNDERSIDE-TOPSIDE COMPARISONS

In this review we have concentrated on ice morphology as measured from below, since this offers the only reliable means of determining the ice thickness distribution. The chief drawback of under-ice profiling is that the instrument platform, a nuclear submarine, is not available where and when wanted. The airborne laser profilometer, however, is an instrument which can be cheaply flown in a light aircraft and which can profile the top surface of the ice with high accuracy. It would be very valuable, then, if we could develop algorithms which give valid information about ice thickness distribution from the results of laser profiles alone.

Attempts at direct point-for-point comparison between freeboard and draft using submarines and aircraft have not been very successful, because of the difficulty of navigational control. Even a few metres of error between the submarine track and the aircraft track yield top and bottom profiles which cannot be adequately correlated, and even the use of transponders and beacons cannot guarantee a sufficient degree of superimposition. The only reliable point-for-point technique is hole drilling, and by the aid of this laborious method Ackley et al [25] were able to test models of top-bottom transformations whereby the surface topographic variations were multiplied by a factor to take account of the buoyancy effect and then subjected to a smoothing filter to allow for the observed fact that roughness elements are longer on the underside than on the topside.

A statistical comparison is a more promising method, since ice conditions vary slowly enough that an error of a km or so in space and of a few hours in time is not crucial to the interpretation of 100 km section lengths. This experiment was carried out by HMS Sovereign and a Canadian Argus aircraft in 1976 [15, 17, 18, 22, 31, 34]. Significant correlations were found both for draft/elevation and for keels/sails.

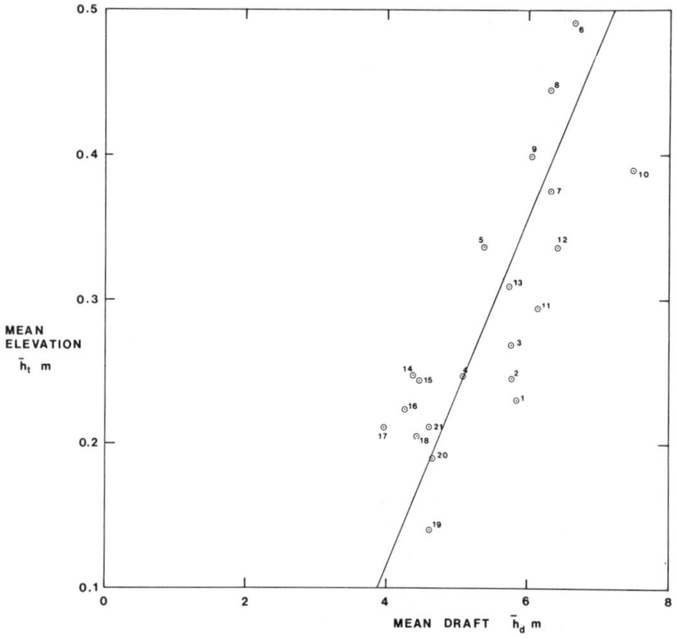

FIGURE 18. Mean draft versus mean elevation for corresponding 100 km sonar and laser sections (after [15]).

Figure 18 shows that the mean draft ($\bar{h}_d$) of a 100 km sonar section was positively correlated with the mean elevation ($\bar{h}_t$) of the corresponding laser section by

$$\bar{h}_d = 8.35\bar{h}_t + 3.04 \quad (r = 0.79) \tag{20}$$

In the analysis of laser profiles it is necessary to remove aircraft porpoising, and except in regions of extensive open water this can only be done to give a 'zero level' which is the mean freeboard of undeformed ice, not sea level. Therefore (20) implies that the intercept at $\bar{h}_t = 0$ represents the mean draft of undeformed floes (3.04 m) while the gradient is the ratio of added draft to added elevation when pressure ridges are formed, i.e. $(\rho_i)_{sails} / (\rho_w - (\rho_i)_{keels})$. Setting $(\rho_i)_{sails} = (\rho_i)_{keels} = \rho_i$, (20) yields $\rho_i = 0.915$. Both this and 3.04 are reasonable numbers for a polar icefield, suggesting that (20) is a valid relation for inferring mean ice draft from mean 'elevation' as derived from a corrected laser profile.

For sails and keels, two linear correlations were found, of form

$$\bar{h}_k = 9.509\ \bar{h}_s - 1.834 \quad (r = 0.85) \qquad (21)$$

and

$$\mu_k = 0.2421\ \mu_s + 1.688 \quad (r = 0.76) \qquad (22)$$

Here $\bar{h}_k$ is the mean keel draft relative to a 9 m cut-off, μ_k is the number of keels per km, again with a 9 m cut-off, $\bar{h}_s$ is the mean sail height above the undeformed ice surface relative to a 1 m cut-off, and μ_s is the corresponding mean number of sails per km. It should be noted that the mean ice draft $\bar{h}_d$ has been corrected for the effect of beamwidth, while the keel numbers and mean keel drafts have not, since there is no valid technique for doing so; the keel parameters in (21) and (22) are therefore relative to a wide-beam sounder record.

Using (21) and (22) it is possible to convert a distribution of sails into a distribution of keels. Wadhams [48] went further in suggesting that having generated a keel distribution from a sail distribution it is even possible to generate the form of the thick end of the ice draft distribution by using (14). Eventually this process of successive inference causes unacceptable errors, but in principle it is now possible to make reasonable statements about ice keels and ice drafts from laser measurements alone. It is hoped that future experiments of the same kind will improve the correlations further.

12. FLOE SIZE DISTRIBUTIONS

We end this review by mentioning the problem of measuring sea ice morphology in regions where ice characteristics are quite different from the central Arctic. In the marginal ice zone (MIZ) near an open ocean boundary the standard techniques described so far will yield large amounts of thin ice and large numbers of 'leads'. This is simply because wave penetration from the open ocean has broken up the icefield into discrete floes. There is an extensive literature on the processes of wave attenuation and wave-induced breakup (summarised in [7, 42]). The overall conclusions are that wave-induced fracture is important in the outermost 100 km of an icefield; and that in any sea state there is a maximum floe diameter which is typically 40-80 m in the outermost parts of the icefield but greater at deeper penetrations. At any given penetration the number density of floes as a function of diameter has been found to fit a negative exponential distribution but it has not yet been possible to relate the parameters of the distribution to other physical factors.

A second type of icefield is one far from any ocean boundary but subject to the throttling effect of a passage through a strait (Bering Strait, Nares Strait) or equivalent barrier, e.g. the narrowing of the ice drift in the Trans Polar Drift Stream as it approaches Fram Strait with the barrier of northeast Greenland on its right hand. In these situations an arcuate fracture pattern develops in the ice cover generating a distribution of very large floes (tens of km in diameter) which also follow a negative exponential distribution of diameter [49, 50].

The measurement of floe size distributions in the MIZ is best carried out by vertical aerial photography from a helicopter or low-flying aircraft, although infra-red line scan has also been used successfully and SAR is valuable for surveys where fog or cloud obstruct visibility. The distribution of very large floes is best studied from satellite imagery, of which NOAA imagery is especially useful because of the high northern latitude to which it reaches. The relationship of floe size distribution to physical forcing, and the effect of floe size distribution upon the large-scale rheology and dynamics of an ice cover, are important subjects of current research.

ACKNOWLEDGEMENTS

It is a pleasure to acknowledge continuing support from the Arctic Program, Office of Naval Research.

REFERENCES

1. Rossiter, J.R., Review of impulse radar sounding of sea ice, Proc. Intl. Workshop on Remote Estimation of Sea Ice Thickness, St. John's, Nfld, Sept 25-26 1979 (ed. J.R. Rossiter, D.P. Bazeley), Memorial Univ., St. John's, pp 77-108, 1980.

2. Rothrock, D.A., The ice thickness distribution, In Proc. NATO Advanced Study Inst. on Air-Sea-Interaction in the Presence of Ice, Maratea, Sept 1981, Plenum Press, in press.

3. Hoare, R.D., Danielewicz, B.W., Pilkington G.R. and O'Rourke, J.C., An upward-looking sonar system to profile ice keels for one year, Oceans '80, Conf. on Ocean Engng in the '80s, Seattle, 8-10 Sept 1980, Inst. Electrical & Electronic Engrs, New York, pp 123-126, 1980

4. Francois, R.E. and Nodland, W.K., Unmanned Arctic Research Submersible (UARS) system development and test report, Univ. Washington, Applied Physics Lab., Tech. Rept. APL-UW 7219, 88pp, 1972.

5. Francois, R.E. and Nodland, W.K., Arctic acoustic measurements at 50 kHz, Univ. Washington, Applied Physics Lab., Tech. Rept. APL-UW 7313, 35pp, 1973.

6. Francois, R.E., High resolution observations of under-ice morphology, Univ. Washington, Applied Physics Lab., Tech. Rept. APL-UW 7712, 30pp, 1977.

7. Wadhams, P., Wave decay in the marginal ice zone measured from a submarine, Deep-Sea Res., vol 25, pp23-40, 1978.

8. Lyon, W.K., Ocean and sea-ice research in the Arctic Ocean via submarine, Trans. N.Y. Acad. Sci., srs 2, vol. 23, pp 662-674, 1961.

9. McLaren, A.S., Wadhams, P. and Weintraub, R., Winter and summer ice draft distributions in M'Clure Strait measured by submarine profiling, Arctic, submitted.

10. LeSchack, L.A., Arctic Ocean sea ice statistics derived from the upward-looking sonar data recorded during five nuclear submarine cruises, LeSchack Associates Ltd., 116-1111 University Blvd W., Silver Sprind Md., Tech. Rept. for Office of Naval Research Contract N00014-76-C-0757/ NR 307-374, Sept. 1980.

11. Wadhams, P., McLaren, A.S., and Weintraub, R., Ice characteristics in Davis Strait measured by submarine profiling, paper in prepn. for Arctic.

12. Williams, E., Swithinbank, C.W.M., and Robin, G. de Q., A submarine sonar study of Arctic pack ice, J. Glaciol., vol. 15, pp 349-362, 1975.

13. Kozo, T.L. and Tucker, W.B. III, Sea ice bottomside features in the Denmark Strait, J. Geophys. Res., Vol. 79, pp 4505-4511, 1974.

14. Wadhams, P. and Horne, R.J., An analysis of ice profiles obtained by submarine sonar in the Beaufort Sea, J. Glaciol., vol. 25, pp 401-424, 1980.

15. Wadhams, P., Sea-ice topography of the Arctic Ocean in the region 70°W to 25°E., Phil. Trans. Roy. Soc. London, srs A, vol. 302, pp 45-85, 1981.

16. Wadhams, P., Ice thickness characteristics in Fram Strait, paper in prepn. for Nature.

17. Wadhams, P., Characteristics of deep pressure ridges in the Arctic Ocean, Proc. 4th Intl. Conf. on Port & Ocean Engng Under Arctic Condns, St. John's, 26-30 Sept 1977 (ed. D.B. Muggeridge, G.R. Peters), Memorial Univ., St. John's, vol. 1, pp 544-555, 1978.

18. Wadhams, P., A British submarine expedition to the North Pole, 1976, Polar Record, vol. 18, pp 487-491, 1977.

19. Wadhams, P., Sidescan sonar imagery of sea ice in the Arctic Ocean, Can. J. Remote Sensing, vol. 4, pp 161-173, 1978.

20. Burke, R., and Robson, J., An evaluation of the Bo'sun multi-beam sonar system, Intl. Hydrog. Rev., vol, 52, pp 53-69, 1975.

21. Harrison, C.H., Reconstruction of subglacial relief from radio-echo sounding records, Geophysics, vol. 35, pp 1099-1115, 1970.

22. Wadhams, P., A comparison of sonar and laser profiles along corresponding tracks in the Arctic Ocean, In Sea Ice Processes and Models, ed. R.S. Pritchard, pp 283-299, Univ. Washington Press, Seattle, 1980.

23. Hibler, W.D. III, A dynamic thermodynamic sea ice model, J. Phys. Oceanogr., vol. 9. pp 815-846, 1979.

24. LeSchack, L.A. and Chang, D.C., Arctic under-ice roughness, Tech Rept., Development & Resources Transportation Co., 116-1111 University Blvd W, Silver Spring, Md, 1977.

25. Ackley, S.F., Hibler, W.D. III, Kugzruk, F., Kovacs, A. and Weeks, W.F., Thickness and roughness variations of Arctic multiyear sea ice, AIDJEX Bull., vol. 25, pp. 75-96, 1974.

26. Thorndike, A.S., Rothrock, D.A., Maykut, G.A., and Colony, R., The thickness distribution of sea ice, J. Geophys. Res., vol. 80, pp 4501-4513, 1975.

27. Coon, M.D., A review of AIDJEX modelling, In Sea Ice Processes and Models, ed. R.S. Pritchard, pp 12-27, Univ. Washington Press, Seattle, 1980.

28. Maykut, G.A., Energy exchange over young sea ice in the central Arctic, J. Geophys. Res., vol. 83, pp 3646-3658, 1978.

29. Maykut, G.A. and Untersteiner, N., Some results from a time dependent thermodynamic model of sea ice, J. Geophys. Res., vol. 76, pp 1550-1575, 1971.

30. Leppäranta, M., Statistical features of sea ice ridging in the Gulf of Bothnia, Tech. Rept. no. 32, Winter Navigation Res. Bd., Helsinki, 1981.

31. Lowry, R.T. and Wadhams, P., On the statistical distribution of pressure ridges in sea ice, J. Geophys. Res., vol. 84, pp 2487-2494, 1979.

32. Tucker, W.B. III, Weeks, W.F. and Frank, M., Sea ice ridging over the Alaskan continental shelf, J. Geophys, Res., vol. 84, pp 4885-4897, 1979.

33. Wadhams, P., Sea ice topography in the Beaufort Sea and its effect on oil containment, AIDJEX Bull., vol. 33, pp 1-52, 1976.

34. Wadhams, P. and Lowry, R.T., A joint topside-bottomside remote sensing experiment on Arctic sea ice, Proc. 4th Can. Symp. Remote Sensing, Quebec, 16-18 May 1977, pp 407-423, Canadian Remote Sensing Soc., Ottawa, 1977.

35. Weeks, W.F., Tucker, W.B. III, Frank, M. and Fungcharoen, S., Characterization of the surface roughness and floe geometry of the sea ice over the continental shelves of the Beaufort and Chukchi Seas, In Sea Ice Processes and Models, ed. R.S. Pritchard, pp 300-312, Univ. Washington Press, Seattle, 1980.

36. Hibler, W.D. III, Mock, S.J. and Tucker, W.B., Classification and variation of sea ice ridging in the western Arctic Basin, J. Geophys. Res., vol. 79, pp 2735-2743, 1974.

37. Hibler, W.D. III, Statistical variations in Arctic sea ice ridging and deformation rates, In Ice Tech 75, Symp. on Icebreaking and Related Technologies, Montreal, April 1975, pp J1-J9, Soc. Naval Architects Marine Engrs, New York, 1975.

38. Rothrock, D.A., and Thorndike, A.S., Geometric properties of the under side of sea ice, J. Geophys. Res., vol. 85, pp 3955-3963, 1980.

39. Hibler, W.D. III, Weeks, W.F. and Mock, S.J., Statistical aspects of sea ice ridge distributions, J. Geophys. Res., vol. 77, pp. 5954-5970, 1972.

40. Mock, S.J., Hartwell, A.D. and Hibler, W.D. III, Spatial aspects of pressure ridge statistics, J. Geophys. Res., vol. 77, pp 5945-5953.

41. Weeks, W.F., Kovacs, A. and Hibler, W.D. III, Pressure ridge characteristics in the Arctic coastal environment, Proc. 1st Intl. Conf. on Port & Ocean Engng. Under Arctic Condns., ed. S.S. Wetteland and P. Bruun, pp 152-183, Tech. Univ. of Norway, Trondheim, 1971.

42. Wadhams, P., The ice cover in the Greenland and Norwegian Seas, Revs Geophys. Space Phys., vol. 19, pp 345-393, 1981.

43. Wadhams, P., The prediction of extreme keel depths from sea ice profiles, Cold Regions Sci. & Technol., in press.

44. Gumbel, E.J., Statistics of Extremes, 375 pp, Columbia Univ. Press, New York, 1958.

45. Lewis, C.F.M., The frequency and magnitude of drift ice groundings from ice scour tracks in the Canadian Beaufort Sea, Proc. 4th Intl. Conf. on Port & Ocean Engng. Under Arctic Condns., St. John's, 26-30 Sept 1977, ed. D.B. Muggeridge and G.R. Peters, pp568-579, Memorial Univ., St. John's, 1978.

46. Hibler, W.D. III, Two dimensional statistical analysis of Arctic sea ice ridges, In Sea Ice, ed. T. Karlsson, pp 261-275, Natl. Res. Counc. of Iceland, Reykjavik, 1972.

47. Hibler, W.D. III and LeSchack, L.A., Power spectrum analysis of undersea and surface sea-ice profiles, J. Glaciol., vol. 11, pp 345-356, 1972.

48. Wadhams, P., The estimation of sea ice thickness from the distribution of pressure ridge heights and depths, Proc. Intl. Workshop on Remote Estimation of Sea Ice Thickness, ed. J.R. Rossiter, D.P. Bazeley, pp 53-76, Memorial Univ., St. John's, 1980.

49. Sodhi, D.S., Ice arching and the drift of pack ice through restricted channels, U.S. Army Cold Regions Res. & Engng. Lab., Hanover, N.H., Rept. 77-18, 11 pp, 1977.

50. Vinje, T.E., Sea ice studies in the Spitzbergen-Greenland area, Landsat Rept. E77-10206, US Dept of Commerce, Natl. Tech. Info. Service, Springfiled, Va., 1977.

ASPECTS OF ARCTIC SEA ICE OBSERVABLE BY SEQUENTIAL PASSIVE MICROWAVE OBSERVATIONS FROM THE NIMBUS-5 SATELLITE

William J. Campbell
Ice Dynamics Project - U.S. Geological Survey
University of Puget Sound
Tacoma, WA 98416

Per Gloersen and H. Jay Zwally

Goddard Laboratory for Atmospheric Sciences
Goddard Space Flight Center
Greenbelt, MD 20771

Observations made from 1972 to 1976 with the Electrically Scanning Microwave Radiometer on board the Nimbus-5 satellite provide sequential synoptic information of the Arctic sea ice cover. This four-year data set was used to construct a fairly continuous series of three-day average 19-GHz passive microwave images which has become a valuable source of polar information, yielding many anticipated and unanticipated discoveries of the sea ice canopy observed in its entirety through the clouds and during the polar night. Short-term, seasonal, and annual variations of key sea ice parameters, such as ice edge position, ice types, mixtures of ice types, ice concentration, and snow melt on the ice, are presented for various parts of the Arctic.

1. INTRODUCTION

The first synoptic view of the Arctic and Antarctic sea ice covers and ice sheets was obtained by the Electrically Scanning Microwave Radiometer (ESMR-5) on the Nimbus-5 satellite. From the time of its launch in December 1972, 19-GHz passive microwave data was acquired and used to construct images of the polar regions on a fairly continuous basis for four years. The series of images has yielded many discoveries of the polar ice as observed in its entirety through clouds and during the polar nights. The combination of complete spatial coverage, provided by the scanning microwave sensors, and the temporal detail, provided by sequential coverage at short-time scales, has yielded unique information on the cryosphere that can not be obtained by surface or aircraft measurements.

During the 1970s an extensive series of sea ice remote sensing experiments were conducted to gain an understanding of the microwave properties of sea ice and provide an accurate interpretation of the complex satellite microwave images. The experiments involved remote sensing aircraft, icebreakers, manned and unmanned sea ice drifting stations--all operating in various combined modes coupled when possible with satellite observations.

The principal instrument used for most of the studies is the ESMR-5. Before the ESMR-5 instrument was used in space, an airborne version was flown on the NASA CV-900 Airborne Laboratory extensively over the Arctic Ocean in 1970, 71, and 72. The 1970 data (Wilheit et al., 1972) showed that it was possible to distinguish sea ice from the open ocean both through the clouds and in the dark. The observed brightness temperatures of sea ice ranged between 210 and 250 K, whereas the brightness temperature of open sea water range between 130 and 160 K. This great difference in brightness temperature between water and ice readily allowed their discrimination in the imagery, and pointed the way to the ability to observe the fractional area covered by leads and polynyas, even when the individual leads or polynyas are not resolved in the imagery. The early data also showed that strong microwave emissivity differences occur on the ice surface itself. The ground-truth measurements and mesoscale microwave mosaic maps (10,000 km^2) acquired during the 1971 experiments allowed Gloersen et al. (1973) to show that the observed emissivity differences of sea ice at a wavelength of 1.55 cm are associated with the age of the ice. Multiyear ice had lower emissivities (cold brightness temperature of approximately 210 K) and first-year ice had higher emissivities (warm brightness temperature of approximately 235 K). This important result suggested that passive microwave imagery could provide an all-time capability of distinguishing between old (thick) and new (thin) ice as well as determining the area of open leads and polynyas within the ice pack.

Additional research was conducted during the joint US/USSR Bering Sea Experiment (BESEX), performed in February-March 1973, and during the most extensive sea ice experiment ever mounted, the Arctic Ice Dynamics Joint Experiment (AIDJEX), leading to a series of papers on various aspects of the variations of the microwave emissivities. Information on extent, concentration, type, and motion of Arctic sea ice has since been deduced from ESMR-5 imagery, aircraft observations, and surface-truth experiments (e.g. Gloersen et al., 1974a, 1974b, 1975a, 1975b, 1975c, 1978; Ramseier et al., 1974; Campbell et al., 1974, 1975a, 1975b, 1975c, 1977, 1978, 1980, 1981; Ramseier et al., 1974, 1975; Vant 1976; Zwally and Gloersen, 1977; Carsey, 1982; and Crane et al., 1982).

The purpose of this paper is to discuss our present understanding of the physical basis for the measurement of ice parameters by the ESMR 5 sensor and to discuss the significance of some of the sea ice observations. Sequences of ESMR-5 images are used to describe several of the observed sea ice phenomena. Such observations include repeatability of the radiances from one annual cycle to the next, variations in type and extent, interpretation of low radiance areas within the pack, comparison with in situ and airborne observations, large-scale ice dynamics inferred from variation of ice concentrations during as annual cycle, differences in such dynamics from one annual cycle to the next, and inference of ice melt patterns from the ESMR-5 radiances during the summer.

2. SEA ICE MICROWAVE PROPERTIES

2.1. Equations and Assumptions.

Both the surface of the earth and its atmosphere radiate in the microwave region (1 millimeter to 1/2 meter). In the microwave wavelength region and for the physical temperatures encountered, the Rayleigh-Jeans approximation to the Planck radiation law pertains, and the radiated power (usually expressed as brightness temperature, T_B) is therefore proportional to the physical temperature, T. However, most real objects emit only a fraction of the radiation that a perfect emitter would emit at the same physical temperature. This fraction defines the emissivity, ε, of the object, so that the brightness temperature is,

$$T_B = \varepsilon\ T. \qquad (1)$$

Variations in the brightness temperature observed over the surface of the earth are primarily due to variations in the emissivity of the surface material and secondarily to variations in physical temperature. For example, the emissivity of sea water at the 1.55 cm wavelength of ESMR-5 is about 0.44 compared to 0.92 for first-year sea ice and 0.84 for multiyear sea ice.

If the field-of-view of the sensor on the satellite includes a mixture of two materials, for example, sea water and sea ice with emissivities ε_W and ε_I and physical temperatures of T_W and T_I, respectively, then the observed brightness temperature is approximately a linear combination of the respective brightness temperatures according to the fractional area, C, of the ocean covered by sea ice,

$$T_B = (1-C)\ \varepsilon_W\ T_W + C\ \varepsilon_I\ T_I \quad . \qquad (2)$$

Sea ice concentration is defined as C, and the fractional area covered by sea water is, therefore, (1-C). If the emissivities and physical temperatures are known or estimated, sea ice concentration is determined by Equation (2) as a function of the observed brightness temperature,

$$C = \frac{T_B - \varepsilon_W\ T_W}{\varepsilon_I\ T_I - \varepsilon_W\ T_W} \quad . \qquad (3)$$

Equation (3) is plotted in Figure 1 for the value of $\varepsilon_W\ T_W$ = 130K, two values of ε_I, and several values of T_I (labeled T_S in Figure 1). Uncertainties in T_I and ε_I both contribute to uncertainties in the ice concentration calculated from Equation (3).

The full radiative-transfer equation applicable to passive microwave observations at a given wavelength within the sea ice canopy includes terms that account for the attenuation of the radiation in the atmosphere, atmospheric emission, and reflection of background radiation from space (Gloersen, et al., 1978). The principal effect of the additional terms, which tend to be small in polar regions, is included in Equation (3) by interpreting $\varepsilon_W\ T_W$ to be the brightness temperature of sea water as observed through a typical polar atmosphere. Variations in the atmosphere, as well as variations in the ocean emissivity due to surface roughness, produce variations of the observed $\varepsilon_W\ T_W$, with a standard deviation of about 5 K. Substantially larger deviations, however, are caused by the presence of rainfall in the sensor field-of-view. In the absence of atmospheric effects, variations of $\varepsilon_W\ T_W$ are negligible because T_W tends to be nearly constant at 271 K in the presence of sea ice, and ε_W at 1.55 cm is inversely proportional to T_W, with a proportionality constant of approximately 130 K.

Over consolidated sea ice, the atmosphere is usually very dry, with cloudiness generally consisting underlying stratus containing mostly ice crystals. Such clouds have low opacity in the microwave region and could be neglected except for their influence on the sea ice canopy surface temperature, T_I. Local variations in T_I on this account have been observed to be as much as 7 K (Gloersen et al., 1973). Annual variations in T_I are typically 30 K.

If the sea ice pack consists only of first-year ice (FY) and open water, the sea ice concentration (C) can be determined by equation (3) to about $\pm$ 15%, using estimated ice physical temperatures (T_I) (Comiso and Zwally, 1982; Zwally et al., 1983). In estimating T_I, the appropriate value is the temperature of the radiating layer, which for first-year sea ice is the saline ice just below the snow-ice interface. Consequently, T_I is usually somewhat warmer than the air temperature T_{air}, except near the melting point, and is approximately given by $T_I = T_{air} + 0.75\ (T_W - T_{air})$, (ibid).

If the sea ice pack consists only of a mixture of first-year (FY) and multi-year ice (MY) with emissivity ε_{IM} and temperature T_{IM}, then the observed brightness temperature is

$$T_B = F\ \varepsilon_{IM}\ T_{IM} + (1-F)\ \varepsilon_I\ T_I, \qquad (5)$$

and solving for F gives

$$F = \frac{T_B - \varepsilon_I\ T_I}{\varepsilon_{IM}\ T_{IM} - \varepsilon_I\ T_I} \quad , \qquad (6)$$

where f is defined as the fraction of ice cover that is multiyear ice. If open water is also present, the observed brightness temperature is

$$T_B = (1-C)\ \varepsilon_W\ T_W + C\ \{F\ \varepsilon_{IM}\ T_{IM} + (1-F)\ \varepsilon_I\ T_I\} \ . \qquad (7)$$

As can be seen in Equation 7, the values of both sea ice concentration (C) and multiyear fraction (F) cannot be obtained from only the single wavelength T_B obtained by the ESMR-5. In Figure 4 of Section 2.4, nomograms are presented for the three variables, C, F, and T_B, and for various seasonal values of the ice temperatures and emissivities to assist in the interpretation of the microwave images.

2.2. Sea Ice Types and Emissivities.

In the previous section, we mentioned differences in the emissivity of sea ice. The emissivity of sea ice depends on its type, varying from about 0.84 for multiyear (MY) sea ice to about 0.92 for first-year (FY) ice in the months when no melting occurs. These sea ice emissivities are at 1.55 cm wavelength and nadir viewing angle consistent with field measurements (Gloersen et al., 1973; Meeks et al., 1974; Campbell et al., 1978) and model calculations (Gloersen and Larabee, 1981), within the uncertainty of the actual radiating temperature. Thus, the average emissivity in a given field-of-view containing a mixture of ice types assumes an intermediate value depending on the composition of the mixture.

When the freeboard portion of the sea ice nears the melting point, the distinguishing features of first-year (FY) and multiyear (MY) ice disappear (see next section), and the emissivities become uniformly high. At the melting point, melt ponds appear on the surface of the ice, especially under cloudy conditions, again lowering the average emissivity in a given field-of-view to values depending on the area of the melt ponds and whether the melt ponds are ice-free or ice-covered (Gloersen et al., 1978).

A model accounting for the differences in the emissivities of the two principal ice types has been described in detail elsewhere (Gloersen and Larabee, 1981), but the results are summarized here since they are useful in the interpretation of the T_B images. Four principal factors enter into the observed T_B's: 1) the vertical temperature profile in the ice, 2) the surface reflectivity, 3) the optical depth, and 4) the volume scattering cross-section. These considerations have also been discussed in detail for the deep snow cover on Greenland and Antarctica (Chang et al., 1979); the basic elements of that discussion also pertain here.

The optical depth and the surface reflectivity are determined by the complex index of refraction in the freeboard portion of the ice, the scattering cross-section, and the density of the snow cover. Physically, the principal difference between first-year (FY) and multiyear (MY) ice is that in MY the freeboard portion is devoid of any liquid component (at below freezing temperatures), since the brine has drained out of the brine cells during the preceding melt period. Thus, the imaginary part of the index of refraction of the freeboard portion of the MY is similar to that of fresh water ice, which is the order of 0.0003 at 250 K (Cummings, 1952). As a consequence, the optical depth, although modified by the contribution to the extinction coefficient by volume scattering from the empty brine cells, is much greater than the elevation of the ice above sea level. Below sea level, the brine cells do contain brine, the imaginary part of the index is at least 0.1, and the optical depth becomes the order of 1 cm. Therefore the resultant optical depth in MY is approximately the thickness of the freeboard layer. In FY the brine component in the freeboard portion of the ice (present at temperatures down to at least 230 K; cf. A. Assur, 1960) restricts the optical depth to about 1 wavelength below the ice surface.

2.3. Repeatability of Nimbus-5 ESMR Observations

Recognizing that the microwave properties of sea ice depend on the ice temperature, as indicated above, we have examined the time history of two specific ice types in the Arctic basin utilizing the three-day averaged images of ESMR data over the three-year period. In order to circumvent effects of varying concentration in our first-year (FY) ice sample, we have selected the maximum radiance for each three-day ESMR-5 image in the area of the Chukchi Sea (Figure 2), known to be largely first-year ice. For the multiyear (MY) ice sample, the area chosen is north of Greenland and Ellesmere Island (Figure 2); here the minimum radiance in the area is chosen in order to avoid including effects of any first-year ice that might have been produced in leads and on the reasonable premise that the ice concentration in that area is generally close to unity.

The results are summarized in Figure 3. It should be kept in mind that the brightness temperatures in these images are represented by a color scale with 5 K steps; therefore, the points in Figure 3 are artificially bimodal, and their true values lie between the plotted points. Nevertheless, the points in Figure 3 show an remarkable repeatability in each of the seasons for the three years studied. In the January-February period, FY sea ice appears to have a brightness temperature of 237 K and MY of 203 K. For March and April, the FY value became 243 K, and the MY became 203 K. In May and June, both FY and MY radiances increased, a result due to both the increase in the sea ice temperature and the temperature-dependent emissivity during the onset of surface melting, as previously discussed. From mid-June through mid-August, FY and MY ice appear to radiate with approximately the same brightness temperatures, and are thus indistinguishable in this time period when each ice type has moisture in the freeboard structure.

From mid-August through September there is a cooling trend and a bifurcation of the points again occurs. It is in this period that much of the FY ice for the ensuing winter season is newly formed; the lower set of points in Figure 3 represents the changing properties of the pack remaining after the summer, i.e., the MY (including second year) sea ice. It will be noted that these points approach an average minimum value of 193 K, probably indicative of the of melt ponds in the ice at that time.

In early October, then, as snow begins to cover the new ice and it thickens, becoming FY, the Chuckchi Sea area assumes the steady-state value of 243 K as in March-April. At the same time, the melt ponds in the old ice are largely refrozen, and MY ice assumes the steady-state value of 208 K.

2.4. Seasonal Nomograms of Sea Ice Concentration

These observations have enabled us to construct nomograms for four distinct ice seasons, as shown in Figure 5, to aid in the interpretation of the sequential ESMR-5 images. We have lumped the January-April and October-December periods into a single ice season we call "winter", using average values of 240 K for fully consolidated FY ice and 205 K for consolidated MY. Thus, this nomograph can be used with assumptions as to ice type mixture in a given locality to infer ice concentration. This is accomplished by selecting a value for the MY fraction, F, and utilizing the appropriate level on the nomograph, which is a linear interpolation between the radiances of the two pure ice types, FY and MY, for various ice concentrations. The "spring" ice season is defined as the May-early June period (see Figures 3 and 4). In the "summer" period (mid-June mid-August) the nomogram indicates that the FY and MY are indistinguishable. Finally, the "fall" period (mid-August through September) shows again a difference between FY and MY sea ice types, with average values of T_B for each which differ from the "winter" and "spring" nomograms. Bear in mind that these seasons we have chosen and which are labelled winter, spring, summer, and fall in Figure 4 only approximate the normal calendar seasons. (Figure 4 and other color figures may be found on pages 219-222).

3. OBSERVATIONS OF SEA ICE EXTENT AND TYPE

The most distinct sea ice feature that appears in the ESMR-5 images is the ice edge because of the large emissivity difference between open water and sea ice, as discussed earlier. A number of studies using Arctic and Antarctic ESMR-5 data have shown that the ice edges are far more complex than depicted in various ice climatological atlases. Similar observations of this complexity have occasionally been made, when lack of cloudiness permitted, with visible and infrared imagers on board spacecraft (see, for instance, Campbell et al., 1975b). However, ESMR-5 provided the first synoptic view of the entire edge of the polar ice packs.

The first Antarctic ESMR-5 ice edge maps were also presented in Gloersen et al. (1974a) for select days in the winter of 1972-3. A thorough discussion of all the Antarctic ice edge data for the entire useful lifetime (1973-6) of ESMR-5 is given in Zwally et al. (1983).

The first Arctic ESMR-5 ice edge maps were also presented in Gloersen et al. (1974a) and show the entire ice edge with detailed discussion of the positions in the Greenland and Barents Seas for selected days in the winter of 1972-3. Campbell et al. (1977) give ESMR-5 ice edge positions for 1973-6 for the southern Beaufort Sea, and Campbell et al. (1975c and 1981) give the same for the Bering and Okhotsk Seas.

Between the maximum and minimum ice extent extremes, the ice concentration and edge location vary in a rapid and complex way. A time-lapse motion picture of ESMR-5 images (Campbell et al. 1980) has shown this most dramatically, particularly in the East Greenland, Barents, and Bering Seas. Features have been observed with changes in position up to 100 km/week in these areas.

Approximate Arctic minimum and maximum sea ice extents derived from ESMR-5 data are shown in Figure 5 for the years 1973, 1974, 1975, and 1976. These were obtained by selection of approximate extrema from the series of three-day ESMR-5 images. Experience has shown that the ice edge so determined is usually geographically located to an accuracy of approximately 50 km. The largest qualitative difference is the large tongue of sea ice extending from east of Greenland toward Svalbard in 1973 which is absent in 1974, appears less prominently in 1975, and is absent again in 1976. 1975 appears to be an anomalous year in which the minimum boundary touches the Alaskan north shore and is detached from Cape Cholyuskian whereas in the other three years shown the opposite condition obtains. (More will be said about this anomaly later.)

The discovery of how to distinguish first-year (FY) from multiyear (MY) sea ice by means of microwave remote sensing (Wilheit et al., 1972) was confirmed during the 1971 AIDJEX Pilot Experiment where the first simultaneous _in situ_ and airborne microwave measurements were made (Gloersen et al., 1973). The first application of this discovery on a global basis using satellite microwave data was made using ESMR-5 images (Gloersen et al., 1974a; Campbell et al., 1974).

Two ice types and their mixtures are observed in the ESMR-5 images to make up the Arctic ice pack: multiyear ice, with an average thickness of approximately 4 m, covering most of the area between the North Pole and Canada-Greenland and having radiance values ranging from 205 to 215 K; first-year ice, with an average thickness of approximately 1-1/2 m, covering most of the Beaufort, Chukchi, East Siberian, Laptev, Kara, and Barents Seas and having radiance values between 235 and 240 K; and various mixtures of these two ice types in other areas. There are apparently significant variations between the spatial distributions of these ice types from one year to the next as can be seen in Figures 10, 13, and 19. Since there are probably also significant temporal changes in the physical temperature distributions in the ice canopy, we avoid the temptation of over-interpreting these radiance distribution variations. Clearly, the ice within the minimum ice boundary for one year becomes the multiyear, or mixed first-year/multiyear ice, for the following year. The degree of mixing depends on how much ice divergence and refreezing takes place in a given location, and how much compression of the newly-formed ice into hummocks results from subsequent ice convergence. With the uncertainties in the interpretation of multiyear fraction (F) from the single-channel ESMR-5 radiances, as a result of not having independent information on surface temperatures or ice concentration, a cross-check on the interpretations of these ESMR-5 data is highly desirable. Such a comparison between spacecraft data obtained during part of the Main AIDJEX in April 1975 and data acquired at the same time from airborne radiometers has been discussed in detail in P. Gloersen et al. (1978). Briefly, the spacecraft data used for this comparison are illustrated in Figure 13, with the path of the aircraft underflights superimposed in white. The corresponding airborne ESMR data (with the cross-track image data averaged into a line profile) along these tracks are shown in Figure 14. The aircraft flights were carried out on nearly cloudfree days, so that an onboard thermal infrared radiometer could be used to determine surface temperature along the tracks. The ice was observed to be fully consolidated along the flight tracks (Figure 14), and the observed surface temperatures were extrapolated to other areas of the pack and used to interpret the ESMR-5 radiances in terms of F in Figure 13 (note scale on right hand side of figure). Comparison of the data in Figures 13 and 14 reveals that the relative changes in radiance (and so the first-year/multiyear ice mixture) is the same along the tracks for both ESMR-5 and the airborne ESMR. This experiment provided additional confidence in the ability to distinguish ice type via ESMR-5.

The transects discussed above traversed the Main AIDJEX manned drift station array. The nature of the mixing of the MY and FY ice can be seen in the meso-scale airborne microwave image acquired five days after the transect data (Figure 15). The image shows large multiyear floes clearly separated by various size areas of first-year ice, with F approximately equal to 0.6. Although interpretation of the microwave radiances for ice types had previously been confirmed by in-situ sea ice property measurements (Gloersen et al. 1973), it was reaffirmed by similar observations during the Main AIDJEX (Campbell et al., 1978). The same area enclosed within the AIDJEX triangle contained considerable open water during the following August (See Figures 16 and 17, which will be discussed in the next section).

4. OBSERVATIONS OF LOW RADIANCE AREAS WITHIN THE ICE PACK

Because of its importance in energy-balance studies, for example, there has been considerable speculation during the last few decades concerning the amount of open water that occurs within the polar sea ice covers. One of the most important results of the ESMR ice observations is that Gloersen et al. (1978) have shown that large areas of reduced ice concentration occasionally occur within the Arctic ice pack during all seasons.

In each of the four years of the ESMR-5 data set, the most pronounced cases of large areas of low microwave radiances (< 195 K) occurring within the pack are in August and September. The interpretation of these low radiance areas as reduced ice concentrations and a discussion of alternate interpretations are given in Campbell et al. (1980) and Crane et al. (1982). In Figures 6 and 7, a series of 16 ESMR images of the central Artic show the appearance and disappearance of several large areas of low microwave radiances which occurred during the period 14 August through 28 September 1974. All dates on the images indicate the first day of the 3-day period for which overlapping ESMR data are averaged in the image map. In Figure 9, a series of three ESMR images of the same area in the late summer of 1975 show the variations of similar large areas of low microwave radiances that have a very different distribution than those observed a year earlier.

During the summer, three phenomena can account for low microwave radiances from sea ice - low ice concentration, substantial liquid water in the surface layer, and ice-free melt ponds. Indeed, if in a given location ice-free melt ponds cover 30% of the area, then the microwave radiance from it would be the same as if it had 30% open water in leads and polynyas, or 70% ice concentration. Thus, in the absence of appropriate surface truth information, the open water versus ice-free melt pond ambiguity in the interpretation of ESMR-5 images is unavoidable. Similar surface truth problems exist in other areas of remote sensing, but in sea ice studies obtaining suitable surface truth is especially difficult during the melt season, because large spatial variations in ice concentration and the freeze/thaw of melt ponds occur at small time scales, and the only available surface truth information comes from sources which are seldom operating--manned drifting stations and flights of remote sensing aircraft equipped with microwave radiometers. Drifting stations give relevant data for extended periods when they exist, but the information is only for a point or a small area in a vast region - hardly the synoptic data on surface conditions needed to interpret synoptic satellite images. Remote sensing aircraft can acquire microwave images, which when coupled with visual and infrared observations from the aircraft, yield mesoscale synoptic surface information (Gloersen et al., 1973; Campbell et al., 1974, 1978). The advantage of aircraft passive microwave imagery lies in its ability to resolve individual leads and polynyas and unambiguously determine the presence or absence of open water. Aircraft can also acquire microwave data along extensive transects of the ice cover, which can also be used to infer surface characteristics over large areas (Gloersen et al., 1978). But even during the best of years, such aircraft flights over the Arctic Ocean occur during only a few days. During the operating lifetime of ESMR, therefore, very little summer surface truth data were acquired. As we discuss the satellite data for the four summers observed by ESMR (1973-76), we use the available surface and aircraft data to aid in their interpretation.

4.1. Case of August - September 1975

The most extensive summer surface truth measurements of sea ice during the ESMR lifetime were made as part of the Arctic Ice Dynamics Joint Experiment (AIDJEX) during the summer of 1975. During August 1975, the NASA CV-990 flew a series of microwave mapping missions over the AIDJEX manned drifting station and a series of transects over extensive areas of the Arctic Basin (see, for example, Figure 13). The geographical position of the AIDJEX triangle of drifting stations (Figure 12) at the time of these observations was in the Southern Beaufort Sea to the north of the MacKenzie Delta, and is shown in the ESMR-5 image for 30 August 1975 in Figure 9. The physical conditions of the ice surface which existed at the AIDJEX stations Big Bear (BB), Snow Bird (SB), Blue Fox (BF), and Caribon (C) for the period 10 August to 4 September 1975 were observed by Hanson (personal communication) and are given in Figure 8 along with the 2 m air temperature at all camps. The observation periods of the CV-990 aircraft and the ESMR-5 (Gloersen et al., 1978) are also noted in the figure. Two of the aircraft ESMR microwave maps of the AIDJEX area (Campbell et al., 1978) are given in Figure 16 (22 August 1975) and Figure 17 (27 August 1975).

According to the AIDJEX surface observations, the surface-layer moisture increased steadily during the period of the three CV-990 flights. Microwave measurements of an alpine snow pack by Edgerton et al. (1971) and snowpack model computations by Chang and Gloersen (1975) show that the emissivity of snow increases rapidly with the onset of melting and then decreases slowly as the free water content increases, but the slope of this decrease is not well known. The increase of surface-layer moisture in the AIDJEX area during the CV-990 overflights agrees with the aircraft microwave results, which show a decreasing trend of the T_B of the ice floes; note that the average T_B of the floes shown in Figure 16 (22 August 1975) is approximately 240 K while five days later (Figure 17) it was approximately 230 K. There are no aircraft measurements to indicate how cool the T_B's became due to the progressive ice-surface melting after the CV-990 flights, that is after 27 August 1975. However, there is a limit to how much free water can remain in the surface layer, because percolation through the layer and into the melt ponds occurs rapidly. Therefore, because the last CV-990 flight occurred immediately after the largest seasonal melt increment (Figure 8), we assume that the 10 K decrease in T_B to 230 K was the maximum due to increasing surface-layer moisture and estimate that the minimum T_B of the wet sea ice is about 230 K. Furthermore, because the melt ponds during the 19 to 27 August melt period retained a slush surface, we have assumed that the T_B of the slush-covered melt ponds remains as high as the surrounding wet sea ice; if not, then the estimated minimum T_B of wet sea ice would be greater than 230 K. The value of T_B = 230 K is also the signature of dry MY just below the melt point. Therefore, any T_B appreciably lower than 230 K at this time of the year can be due neither to wet sea ice nor dry MY just below the melt point.

The change in T_B due to surface wetness or formation of dry MY after freezing is significantly less than the change observed by ESMR-5 for parts of the Arctic Basin during this period of late August 1975. Appreciably lower T_B's can be caused by either large areas of open water within the pack, due to ice divergence, or by extensive areas of ice-free melt ponds. Of these two possibilities, we believe according to the discussion below that the former mechanism is responsible for the large areas of low T_B observed in the summer ice pack shown in the ESMR-5 images (e.g., Figures 6, 7, and 9).

Considering the 1975 observations (Figure 9) in some detail, the large areas of low microwave radiance appear as green and light yellow (T_B < 195 K) zones within the ice pack. The color-coded scale chosen to make the images resulted in some of the 5 K changes having stronger visual effect than others, for example, the 195 → 190 K changes in the images are more obvious than the 190 → 185 K changes. The scales are chosen to represent quantitatively a range of T_B's, and no particular color change is intrinsically more important than another.

Not only are the absolute values of the T_B's in the low radiance areas important, but these regions almost invariably have more pronounced T_B gradients around them than the gradients elsewhere in the pack. To show how rapidly the T_B gradients can change, in Figure 11 are plotted the variations in the ESMR-5 T_B's along the transect shown in Figure 9. Note that these profiles pass through the AIDJEX area. Since the surface observations (Figure 8) indicate that the melt ponds were not ice-free during this period and since the ESMR-5 T_B's, except for 24-26 August, are below the temperature where surface-layer free water could have a significant effect, we attribute these T_B variations to ice concentration variations. This thesis based on surface and satellite observations is also supported by the following aircraft observations.

The CV-990 aircraft ESMR images for 22 August 1975 (Figure 16) and for 27 August 1975 (Figure 17) both show low radiance areas in the vicinity of Blue Fox (BF). A histogram analysis (Campbell et al., 1978) of these images indicates that the percent of open water within the AIDJEX triangle was 8.5% on 22 August and 9.6% on 27 August, and that on 18 August it was zero. On both 22 and 27 August, visual sightings from the NASA CV-990 aircraft were made of the leads and polynyas in the Blue Fox area, which not only verified their existence but also proved that they were free of thin ice because sun glint from capillary and small gravity waves within them was observed. Indeed, on 27 August the aircraft made a northward turn from the Blue Fox area and a large area of open water with an ice concentration of approximately 60-70% was observed to extend to at least 25 km north of the station.

We now compare these 27 August visual and aircraft observations of open water to the ESMR-5 satellite image for 27-29 August (Figure 9). Because the footprint of the ESMR-5 is approximately 30 km, the individual leads and polynyas generally are not resolved as fully-open water but appear as reduced ice concentrations. The average ice concentration of the Blue Fox corner of the AIDJEX triangle and the area just north for 25 km was approximately 75% on 27 August. In comparison, the average ESMR-5 value of T_B, averaging over four 30 km footprints in the same areas, was 210 K, which gives an ice concentration of 80% using the nomogram shown in Figure 4. The ESMR-5 image for 30 August - 1 September (Figure 9) indicates that the ice concentration in the area of Blue Fox continued to decrease after 27 August. Note the dark green area with T_B = 190 K immediately north of the northeast corner of the AIDJEX triangle, for which the nomogram gives an ice concentration of 60% consistent with the nearby aircraft observations.

Having verified that the low microwave radiances observed by the ESMR-5 in the AIDJEX area in August 1975 were due to open water in leads and polynyas, we now turn to the very large low T_B areas which existed north of the AIDJEX area and extended in a ring centered around the North Pole at approximately 86°N latitude (Figure 9). The radiances ranging from 170 to 195 K observed in these areas, which move quite rapidly, orders of magnitude faster than typical ice drift velocities, indicate ice concentrations from about 40 to 65%. Although no surface data or aircraft flights were obtained over these transient and interesting events, we believe these low radiance areas were also due to low ice concentration just as we have shown existed at the same time in the AIDJEX vicinity.

Returning to the question of the possible contribution of melt ponds to the low radiances, it was observed that in the AIDJEX area the percentage of the ice surface covered by melt ponds was approximately 30% in August 1975, although these melt ponds were not ice-free. Melt ponds in the Arctic Basin grow in extent rapidly in the early summer, but by August they normally have reached their maximum extent and thereafter only grow in depth. As noted earlier, only limited drifting station and aircraft data exists on the spatial and temporal behavior of many sea ice phenomena, including melt ponds. A figure of 30% for total melt pond area of the time of maximum melt is about average. Drifting Station Alpha was in the area of 85°N and 130°W in August 1958, in the region where large low radiances area appear in Figures 6, 7, and 9, and the average maximum pond areas was then observed to be approximately 30%.

Assuming that the melt pond area in August 1975 over the central Arctic Basin was 30%, if the ponds were ice-free everywhere, then the decrease in T_B compared to the frozen state would be about 30 K. When the melt ponds are completely ice-free, the surrounding surface of the ice floes is also wet. Previously we established that the T_B values of wet sea ice floes (Figure 16 and 17) had minimum values in the range of 230 to 240 K. Therefore, the combined affect of ice-free melt ponds with 30% coverage and a wet surrounding surface would lower the average sea ice T_B to approximately 200 to 210 K. However, the ESMR-5 images for August show low radiance areas are considerably colder than 200 K, ranging normally in the region 170 - 195 K, with some values as low as 150 K. Therefore, although preferential thawing of melt pond surface within the low T_B zones could account for part of the decrease in T_B, this mechanism could not account for the major T_B decrease. In order that melt pond thawing could account for all of the T_B decrease in these low radiance areas, which move and change shape rapidly, most of the central Arctic Basin would have to have ice-free melt ponds covering 40-60% of the area, and they would have to thaw and refreeze rapidly and preferentially over large and small areas. Thawing and refreezing of melt ponds generally occurs over large areas, because the controlling mechanisms are associated with weather systems and the existence of extensive cloud cover. These phenomena generally have much larger dimensions than those of the low radiance areas observed by ESMR-5, and the summer weather systems generally do not move in the way the low radiance areas move in the sequential ESMR-5 images, along the specific routes described later.

We have observed from a NOAA-4 image that on 30 August 1975 the entire ice pack north of the AIDJEX to the Pole was covered by a thick stratus cloud layer. Therefore, if the melt ponds were ice-free due to backradiation from the clouds, then they should be ice-free over very large areas and not just in the low radiance areas shown in Figure 9. Also, as was shown earlier, the melt ponds in the AIDJEX area at this time were not ice-free, and, although this area had slightly less cloud cover than the area north to the Pole, the general meteorological regime in both areas was the same. Thus, we find no reason to assume ice-free ponds in the area of the large low radiance events which occurred to the north of the AIDJEX area.

The history of what becomes of the ice after summer when all melt ponds are thoroughly frozen, say in December, gives additional evidence that the low radiance areas shown in Figure 5 are primarily due to the presence of open water. Melt ponds which do not connect to the ocean typically have salinity of several parts per thousand or greater when they are fully developed, and when they do connect to the ocean they have salinities ranging up to that of the ocean. Unfortunately, we do not know the ratio of ocean-connected to unconnected melt ponds over the Arctic at the height of the summer season. Meeks et al. (1974) found that at 13 GHz the brightness temperature of FY ice remained constant for all ice thicknesses greater than 1 cm that had salinities equal to or greater than 2 ppt. Therefore, if we assume that the typical melt pond has a salinity sufficiently high to form ice with salinity equal to or greater than 2 ppt then when it refreeezes, it will have the signature typical of FY ice no matter what the age of the ice it rests upon. Observations of the radiometric signatures of refrozen melt ponds were made at various times of the year during the Main AIDJEX. Observations for Summer 1975 (Campbell et al., 1978, Gloersen et al., 1978) confirm that frozen melt ponds with a thin layer of slightly moist ice do indeed have the microwave signature of FY sea ice. When this new ice on the melt pond becomes FY ice as the winter progresses, the signature continues to be that of FY ice. It is worth noting that even if the melt pond water were fresh, there would be no empty brine pockets upon refreezing, and so the scattering in the freeboard portion typical of multiyear ice would be absent. The implication, then, is that areas which show a nearly pure MY sea ice signature in the winter had few if any melt ponds the previous summer, however, additional surface measurements on the emissivity of refrozen melt ponds are needed to confirm this conclusion. Therefore, assuming that the refrozen melt ponds do have emissivity of FY ice, then if the low radiance areas were due to preferential melting of pond ice that covers 50% or more of a given area, then when those areas refreeze their microwave signature would range from total FY to a maximum of 50% of MY (F=0.5). In other words, the signatures could only become that of a mixture of MY and FY ice with no values of $F > 0.5$. On the other hand, if, as is shown below, in the areas where the low radiances occurred in August-September, the winter ESMR-5 signatures revealed the presence of greater than 50% MY to as much as 100% MY, then the summer low radiances could not be due to preferential melting of large areas of 50% melt pond distribution.

In Figure 5 are shown envelopes of the low radiance areas ($\leq$ 197.5 K) which occurred within the Arctic Basin during August and September for each of the four summers of ESMR-5 , 1973-76. Note that for 1975, the distribution of the envelope is quite dissimilar to that of the other three years, appearing as a ring at 85°N latitude, whereas for all the other years the envelopes are arranged along the Transpolar Drift Stream, which is a major component of the average Arctic Basin ice circulation that transports sea ice northward from the East Siberian and Chukchi Seas towards the Pole and continues southward into the area between Greenland and Svalbard. The other major component of the average Arctic Basin circulation is the Beaufort Sea Gyre, which can clearly be seen in the drift data from the Arctic Data Buoy Network shown in Figure 18.

Consider again the envelope for the low ice radiances observed in August-September 1975 (Figure 5). The area covered by this envelope is large, and if we assume that the ice velocities within and around this envelope was similar to those shown in Figure 18 (a safe assumption since a plot of the trajectories of all drifting stations since Mansen's Fram voyage yields a pattern of average ice circulation showing the Transpolar Drift Stream and the Beaufort Sea Gyre with velocities similar to the buoy velocities) then the major part of the ice that was in the envelope during August - September 1975 was still in it when the ESMR-5 image shown in Figure 10 was acquired on 19 December 1975. We have verified by analysis of sequential images preceding the December 1975 image that no significant ice divergence, that is change in C, occurred in the central Arctic Basin. Therefore, these images can be used directly to indicate ice type distribution using the nomogram given in Figure 4. A comparison of the 1975 low ice radiance areas in Figure 5 with the same areas in ensuing winter (Figure 10) reveals that a wide range of ice type mixture, from full FY to full MY, occurred in these low radiance areas in December. In particular, the high F values from 0.5 to 1.0 could not have come from MY ice covered with 50% refrozen melt ponds.

Considering all of the arguments and observations presented above, we conclude that the low radiance areas observed within the Arctic icepack on the ESMR-5 images of August-September 1975 are due to large amounts of open water.

4.2. Cases of August-September 1973, 1974, and 1976

A comparison of the August-September low radiance areas with December ice type distribution by ESMR-5 in the same areas for 1973 and 1974 (Figures 5 and 10) reveals as in 1975, that a full range of ice type radiances occurred where they could not if large amounts (i.e., 50%) of refrozen melt ponds were present. A 1976 comparison is not made since the EMSR-5 images for December 1976 were not available.

In an analysis of long-term Soviet ice observations in the Arctic Basin, Yakovlev (1977) gives a map showing a "zone of occasional polynya" that covers a triangularly shaped area with the corners at 85°N, 150°E; 72°N, 180°; 72°N, 160°W. This polynya zone very closely coincides with the low radiance envelopes for August-September 1973, 74, and 76 (Figure 5) that lie between the North Pole and the Siberian coast. Thus, there exists long-term observational evidence that supports our premise that the low radiance zones are areas of low ice concentration.

Note again that for 1973, 74, and 76 the low radian envelopes in August-September all lie along the Transpolar Drift Steam. Only in 1975, the AIDJEX year, did a dissimilar distribution occur. Consider that in the August-September period when the large low radiance areas appear in the ice pack interior, the pack is at its minimum extent, much of the FY and younger ice types have disappeared, and the pack is far less rigid and compact than during other times of the year. Unfortunately, the Arctic Data Buoy Network was not in operation during this period to provide ice drift information. Thus, we are unable to deduce the actual ice motions, which occurred in the region of the low radiance envelopes, and compare them with sequential images to discover how the observed distributions of ice concentration and type relate to the ice dynamics. However, we do have ice dynamic observations in the southern Beaufort Sea in summer 1975 from AIDJEX.

The drift of the AIDJEX manned drifting station array (Figure 12) was very different from the drift that was expected when the stations were established in the spring of 1975. Indeed, the array was positioned with the expectation that it would drift westwards with the long-time average drift of the southern Beaufort Sea Gyre and then, when it arrived in the area north of Pt. Barrow, it would drift northwestward along the western part of the Gyre. The opposite drift actually happened. Thus, it can be said that the 1975 circulation of the Beaufort Sea was anomalous compared to the average circulation.

During August-September 1975, then, the southeastward drift of the southern Beaufort Sea ice was anomalous compared to the average summer westward drift. Furthermore, for this same period the distribution of the envelope of low radiance areas (Figure 5) was anomalous compared to that of the other three appearing in the same image along the Transpolar Drift Stream. The year 1975 was also the only one of the four years 1973-76 that a large shore lead that normally opens in August-September along the northern coast of Alaska and the McKenzie Delta region did not open (Figure 5). The low radiance envelopes for this 1975 period are arrayed in two modes: the largest envelope starts in the central part of the Beaufort Sea Gyre and extends westward into the Transpolar Drift Stream, thus being positioned perpendicular to the envelopes for 1973, 74, and 76 which lie along the Transpolar Drift Stream; a series of three smaller envelopes north of Greenland, Svalbard, and Novaya Zemlya that also lie along the Transpolar Drift Stream.

Based on the evidence above, we surmize that: (1) the low radiance (Figure 5) areas are primarily due to lowered ice concentration in all years; (2) during August-September 1973, 74, and 76, the long-time characteristic average ice circulation existed and a zone of lowered ice concentration, on the order of 60%, was formed along the Transpolar Drift Stream; and (3) during August-September 1975, an anomalous circulation occurred in the Beaufort Sea with a large part of the ice pack being pushed southward toward the Alaskan-Canadian coast, which resulted in the formation of an extensive zone of lowered ice concentration, again on the order of 60%, running from the central Beaufort Sea westwards into the Siberian Sea.

There is additional evidence in the ESMR-5 images for August-September 1974 (Figures 6 and 7) that the low radiance areas are primarily due to lowered ice concentration. Note in the image for 20 August, that two small low radiance areas appeared at 81°N and 90°E and at 79°N and 73°E. The brightness temperatures in these two areas are at least as low as those in the larger radiances appearing in the same image along the Transpolar Drift Stream. Following the evolution of the two small areas, near the central top of each image, in Figures 6 and 7, they gradually expand, while maintaining their low radiances, until by 13 September they have opened into the ocean at their southern extremities still maintaining their low radiances, indicative of low ice concentration. Note also that these two areas grew while maintaining their centroid and did not migrate rapidly like the similar features in the Transpolar Drift Stream. Since these two polynyas were located at the edge of the continental shelf, we believe that they are formed by a local oceanographic effect, such as upwelling, that did not vary spatially during the August-September period.

Similar small, stationary zones of low microwave radiance, which we again interpret as being primarily due to low ice concentration, also appear in the ESMR-5 images in the area adjacent to the Canadian Archipelago. In August-September 1975, which as said above we believe was an anomalous ice circulation year in the Beaufort Sea, two such zones appeared at 77°N and 135°W and at 80°N and 105°W. They were approximately the same size as the ones near the Siberian coast discussed above and were also near a continental shelf. The one at 77°N and 135°W persisted until late October 1975, while the one at 80°N and 105°W persisted well into the winter with gradually increasing ice concentration. Indeed, in the area where this latter one formed there were reduced microwave radiances well into spring 1976, but signatures at that time fall with the range of MY signatures so there is some ambiguity in interpretation after Fall 1975. As with the small, stationary zones near the Siberian coast, we suggest that the ones near the Canadian Archipelago could have been formed by a local ocean effect associated with the continental shelf.

In Figure 19 are shown a series of four ESMR-5 images of the Southern Beaufort and Chuckchi Seas for the period of September to 29 November 1974 along with the appropriate nomogram. Note that on 9 September a large shore lead or polynya extended from the McKenzie Delta across the coast of Alaska and into the Chuckchi Sea. By 17 November this polynya can be seen to be full of FY ice. By November all melt ponds are refrozen, therefore the radiance variations in the Chukchi Sea that occurred in the period 14 - 20 November can only be due to variations in ice concentration. Note that some areas show variations in ice concentration of about 20-30% occurring at time scales as short as several days.

4.3. Comparison of ESMR-5 and DMSP images

Initially, we believed that one way to verify what caused the low radiances in the ESMR-5 images would be to compare them to visible images of the area acquired at nearly the same time. There is a severe restraint on using visible images in this manner, because it is not possible to distinguish leads and polynyas which are full of nilas and grease ice from those which are ice free. Indeed, this ambiguity also shows up in the interpretation of radar images of the ice (Campbell et al, 1977). Also open leads and polynyas smaller than about 1 km are not resolved in the visible images, but do contribute to lower microwave radiances. The task of finding near-simultaneous visible and ESMR-5 images of the transient low radiance areas proved far more difficult than imagined. It appears that stratus clouds are almost always associated with the low radiance areas that occur in August-September. We are indebted to Dr. R. Crane for finding a visible Defense Meteorological Satellite Program (DMSP) image of part of the Arctic Ocean in the region of Svalbard that shows two polynyas and part of the ice edge in the East Greenland Seas. In Figure 20 this image, acquired on 8 April 1973, is compared to an ESMR5 image of the same area acquired on 5-7 April 1973. Three arrows connect corresponding features: 1) the edge of the sea ice in the East Greenland Sea (A), 2) a polynya north of Svalbard, and 3) a polynya north of Novaya Zemyla. The polynyas appear to be ice-free in the DMSP image, and the ESMR-5 image also shows that they are ice-free.

5. OBSERVATIONS OF SPRING AND SUMMER ICE

The variation of the microwave radiances during three successive annual cycles was discussed above (see Figure 3.). It is interesting to examine the spatial distribution of these radiances during the Spring and Summer in some greater detail. In Figures 21 and 22, radiances from the central Arctic basin are shown in fourteen ESMR-5 images beginning on 19 April and running to 14 August 1974. Based on this sequence alone, it might be concluded that a roughly linear warm front proceeds on the ice surface from Siberia to the Canadian coastline (see for instance Campbell et al. 1980). In fact, this surface warming pattern is repeated in the 1973, 1975, and 1976 data sets (not shown). This notion is further supported in Figure 23 by a sequence of Arctic Ocean Data Buoy (AODB) temperatures, albeit for 1979 (Thorndyke and Colony 1980), which show a similar pattern of ice surface warming starting at the Siberian Coast and proceeding in a roughly linear front to the Canadian coast.

A unique interpretation of the summer microwave radiance pattern changes for the Arctic Basin is not possible. To illustrate the complexity of the summer season radiances, we have selected three areas surrounding three of the AODBs to study in detail, recognizing, of course, that these data were acquired five years after the ESMR-5 data. Because of the difference in time of the two data sets, a detailed comparison is not appropriate. Still another caveat is that the AODBs move, whereas a fixed area was selected to study the ESMR radiances. Nevertheless, we assume that the average annual cycle of physical temperatures in a given Arctic basin location approximately repeats itself from one year to the next, supported in part by the repetition of the apparent linear warm front phenomenon in the 1973, 1974, 1975, and 1976 ESMR summer data sets. We also assume that the spatial variation in the temporal behavior of the physical temperatures can be neglected over the path of the buoy traverse. The three areas selected are: (1) a predominantly first-year sea ice area in the East Siberian Sea (near 76N, 166E); (2) a mixed multiyear/first-year sea ice area in the central Arctic (purple area in the 21 June image in Figure 21, near 87N, 135-180W); and (3) a heavy multiyear sea ice region north of the Queen Elizabeth Islands (near 83N 120W).

The AODB temperature data for 1979 and the nearby ESMR radiances for these three areas for 1974 are plotted in Figure 24. The first noteworthy feature of these plots is the difference in amplitudes in the temperature curves for the three AODBs. All three curves rise to about 5°C in July (consistent with the melting point of the sea ice and the offset in the buoy temperatures to be expected on the basis of solar warming of the buoys), but the February-March minimum is about -30°C for the East Siberian Sea case, -35°C for the Central Arctic, and -40°C for the area near the Queen Elizabeth Islands. Within the errors of these smoothed curves of buoy temperatures, there is certainly no obvious phase shift in the three cycles as might be expected for occurrence of an advancing linear warm front phenomenon. On the other hand, such a possibilty is not excluded and in fact is probably present as indicated by the differences in the buoy temperature minima.

Another remarkable phenomenon is observed when comparing ESMR-5 radiances and OADB temperatures for these three locations over the 1974/1979 annual cycles. Starting with the area north of the Queen Elizabeth Islands (Figure 24a), the ratio of the amplitudes of the ESMR-5 radiance and AODB physical temperature in the cycles (which are in phase even though they are five years apart) is about 0.78. This ratio of amplitude of brightness temperature to the amplitude of physical temperature implies an emissivity of 0.78 which is in reasonable agreement with the approximate 0.84 value for the emissivity of multiyear ice at the ESMR-5 wavelength of 1.55 cm. Thus, it can be seen that at this particular location an emissivity change in the snow cover on the sea ice is not evident as the physical temperature approaches the melting point in the summer (Edgerton et al., 1971, Stiles and Ulaby, 1980). In contrast, the area near the North Pole (Figure 24b), where the two cycles are again in phase, yields a ratio of about 1.1, clearly indicating a significant change in sea ice emissivity, which is assumed to be due to the presence of a liquid phase. Also consistent with such an emissivity change is the increase in the slope of the ESMR-5 radiance curve late in May (Figure 24b). The ESMR-5 radiances in the East Siberian Sea (Figure 24c) do not follow the AODB temperatures as closely as in the other two areas. The situation in this area appears to be more complex. The out-of-phase characteristic in March is likely to be related to differences in the annual cycle of the physical temperatures in 1974 and 1979. The lack of agreement in the Fall curves may also result from sea ice divergence or the formation of and subsequent freezing of melt ponds on the sea ice.

The freezing and thawing of the surfaces of melt ponds is controlled by meteorological events, such as the passage of cyclones and the presence of stratus cloud patches, that are frequent during the Arctic summer. The synoptic weather conditions associated with the pattern of brightness temperatures and snow melt in 1974 is discussed in Crane et al. (1982). Since such events move over over broad areas, it could be assumed that the response of melt ponds to them would also occur randomly over broad areas rather than in a relatively narrow band along long-term ice circulation features.

The sequential ESMR-5 images for July 1974 (Figure 22) show microwave radiance variations of sea ice occurring over broad, amorphous areas. Considering the magnitude and range of the radiance variations observed, they could be due to freezing/thawing of melt ponds. These are not nearly as low in radiance as those zones (Figure 5) which we attribute to low ice concentration. Note that there are two dominant signatures in these images: a brown/orange one which has the signature of multiyear ice with approximately 30% ice-free melt ponds; and pink areas which have the signature approximately 20 K higher and which indicate that the melt ponds are refrozen. These large, amorphous zones migrate with speeds similar to those of summer cloud systems. Comparing the July 1974 images to those for 1973, 1975, and 1976 (not shown), we find no repeating pattern in the distribution of these amorphous zones. Thus, these freeze/thaw zones in July behave in an entirely different way than the well-defined August/September lower radiance zones which are relatively narrow and appear repeatedly in the area of the transpolar drift stream.

6. CONCLUSIONS

The ESMR on the Nimbus-5 satellite provided data for the first synoptic images of the polar sea ice covers and ice sheets. It did so on an every three-day basis with occasional gaps during the first four years, and has provided data of lower quality until the present time. The ESMR data set affords a unique look at the behavior of ice on our planet. This study has focused on the interpretation of the Arctic sea ice data. Zwally et al. (1983) have performed a detailed analysis of the Antarctic sea ice data. These and other studies using ESMR-5 data have shown that the polar sea ice cover undergoes large spatial variations of ice extent, ice concentration, and ice type throughout the range of time scales from several day to interannual. ESMR-5 was a single-frequency, single polarization sensor, which measured key sea ice parameters to accuracies useful for a wide variety of studies: position of the sea ice boundaries to 25 km, ice concentration to 15%, and qualitative information on multiyear ice fraction.

In 1978 NASA launched the 10-channel Scanning Multichannel Microwave Radiometer (SMMR) on both the Nimbus-7 and Seasat satellites. These two instruments, essentially copies of each other with five frequencies each with horizontal and vertical polarization (Gloersen and Barath, 1977), represent a considerable jump in complexity over ESMR-5. The Seasat SMMR operated for the lifetime of the satellite, only 110 days, but the Nimbus-7 SMMR is still operating. Thus, the length of the high-quality SMMR data set now exceeds that of ESMR-5. The multifrequency, dual-polarized microwave data from SMMR has yielded a dramatic increase in the precision in the determination of ice parameters compared to ESMR-5. The Nimbus-7/SMMR Experiment Team has developed a sea ice algorithm that determines both total sea ice concentration and multiyear sea ice fraction making use of the polarization and multispectral information (Cavalieri et al., 1983; Gloersen et al., 1983). Other SMMR algorithms include those of Swift et al. (1983) and Svendsen et al. (1983).

The combined ESMR-5 and SMMR data sets has provided an almost decadal record of the behavior of the polar sea ice packs at a wide range of space and time scales. This record will continue to be a valuable source of information for many air-ice-sea interaction, weather, and climate studies.

8. REFERENCES

1. Campbell, W. J., Gloersen, P., Nordberg, W., and Wilheit, T. T., Dynamics and morphology of Beaufort Sea ice determined from satellites, aircraft, and drifting stations, Proceedings of the COSPAR Symposium on Approaches to Earth Survey Problems through use of Space Techniques, Constanz, F.R.G., May 1973; pp. 311-327, 1974.
2. Campbell, W. J., Ramseier, R. O., Weeks, W. F. and Gloersen, P., An integrated approach to the remote sensing of floating ice, Proceedings of the Third Canadian Symposium on Remote Sensing, Edmonton, Alberta, 22-24 September 1975; pp. 39-72, 1975a.
3. Campbell, W. J., Weeks, W. F., Ramseier, R. O., and Gloersen, P. Geophysical studies of floating ice by remote sensing, J. Glaciology, vol. 15, pp. 305-328, 1975b.
4. Campbell, W. J., Gloersen, P., and Ramseier, R. O., Synoptic ice dynamics and atmospheric circulation during the Bering Sea Experiment in USSR/US Bering Sea Experiment: Proceedings of the Final Symposium on the Results of the Joint Soviet American Expedition, Lenigard, 12-17 May 1974, K. Ya. Kondratyev, Yu, I. Rabinovich, and William Nordberg, eds., pp. 164-185, 1975c.
5. Campbell, W. J., Gloersen, P., Webster, W. J., Wilheit, T. T., and Ramseier, R. O., Beaufort Sea ice zones as delineated by microwave imagery, J. Geophys. Res., vol. 82, pp. 1103-1110, 1976.
6. Campbell, W. J., Gloersen, P., Zwally, H. J., Ramseier, R. O. and Elachi, C., Simultaneous Passive and active Microwave Observations of Nearshore Beaufort Sea Ice, Proceedings of the 9th Annual Offshore Technology Conference, 2-5 May 1977, Houston; pp. 287-294, 1977.
7. Campbell, W. J., Wayneberg, J., Ramseyer, J. B., Ramseier, R. O., Vant, M. R., Weaver, R., Redmond, A., Arsenault, L., Gloersen, P., Zwally, H. J., Wilheit, T. T., Chang, C., Hall, D., Gray, L., Meeks, D. C., Bryan, M. L., Barath, F. T., Elachi, C., Leberl, F., and Farr, T., Microwave Remote Sensing of Sea Ice in the AIDJEX Main Experiment, Boundary-Layer Meteorol., vol. 13, pp. 309-337, 1978.
8. Campbell, W. J., Ramseier, R. O., Zwally, H. J., Gloersen, P., Arctic Sea-Ice Variations from Time-Lapse Passive Microwave Imagery, Boundary Layer Meteorol., vol. 18, pp. 99-106, 1980.
9. Campbell, W. J., Ramseier, R. O., Zwally, H. J., and Gloersen, P., Structure and Variability of Bering, Okhotsk Sea Ice Covers by Satellite Microwave Imagery, in Energy Resources of the Pacific Region, ed. Michel T. Halbouty; Library of Congress Catalog Card No: 80-69782-ISBN (0-8918-061-1), 1981.
10. Carsey, F. D., Microwave Observation of the Weddell Polynya, Monthly Weather Review, vol. 18, 1980.
11. Cavalieri, D. J., Gloersen, P., and Campbell, W. J., Determination of sea ice parameters with the Nimbus-7 SMMR, submitted to J. of Geophysical Research, 1983.
12. Comiso, J. C. and Zwally, H. J., Antarctic Sea Ice Concentrations Inferred from Nimbus 5 ESMR and Landsat Imagery, J. of Geophysical Research, vol. 87, pp. 5830-5844, 1982.
13. Edgerton, A. T., Stogryn, A., and Poe, G.: 1971, Microwave Radkiometric Investigations of Snow Pack; Final Report No. 1285R-4 for USGS Contract No. 1A-08-001-11828, Aerojet General Corporation Microwave Divison, El Monte, Calif.
14. Gloersen, P., Nordberg, W., Schmugge, T. J., Wilheit, T. T., and Campbell, W. J., Microwave Signatures of First-Year and Multiyear Sea Ice, J. Geophys. Res., vol. 78, pp. 3564-3572, 1973.
15. Gloersen, P., Wilheit, T. T. Chang, T. C., Nordberg, W., and Campbell, W. J., Microwave Maps of the Polar Ice of the Earth, Bull. Amer. Meteorol. Soc., vol. 55, pp. 1442-1448, 1974a.
16. Gloersen, P., Chang, T. C., Wilheit, T. T., and Campbell, W. J., Polar Sea Ice Observations by Means of Microwave Radiometry, Proceedings of the Symposium on Advanced Concepts and Techniques on the Study of Snow and ice Resources, National Academy of Sciences, H.S. Santeford, ed., pp. 541-550, 1974b.
17. Gloersen, P., and Salomonson, V. V., Satellites-New Global Observing Techniques for Ice and Snow, J. Glaciol., vol. 15, pp. 373-389, 1975a.
18. Gloersen, P., Ramseier, R. O., Campbell, W. J., Chang, T. C., and Wilheit, T. T., Variation of Ice Morphology of Selected Mesoscale Test Areas During the Bering Sea Experiment, in USSR/US Bering Sea Experiment: Proceedings of the Final Symposium on the Results of the Joint Soviet- American Expedition, Leningrad, 12-17 May 1974, K. Ya. Kondratyev, Yu. I. Rabinovich, and W. Nordberg, eds., pp. 196-218, 1975b.
19. Gloersen, P., Ramseier, R. O. Campbell, W. J., Kuhn, P. M., and Webster, W. J., Ice Thickness Distribution as Inferrred from Infrared and Microwave Remote Sensing During the Bering Sea Experiment, in USSR/US Bering Sea Experiment: Proceedings of the Final Symposium on the Results of the Joint Soviet-American Expedition, Leningrad, 12-17 May 1974, K. Ya. Kondratyev, Yu. I. Rabinovich, and W. Nordberg, eds., pp. 282-293, 1975c.
20. Gloersen, P., Zwally, H. J., Chang, A. T. C., Hall, D. K. Campbell, W. J. and Ramseier, R. O., Time-Dependence of Sea Ice Concentration and Multiyear Ice Fraction in the Arctic Basin, Boundary-Layer Meteorol., vol. 13, pp. 339-360, 1978.
21. Gloersen, P., Cavalieri, D. J., Chang, A. T. C., Wilheit, T. T., Campbell, W. J., Johannessen, O. M., Katsaros, K. B., Kunzi, K. F., Ross, D. B.,

Staelin, D., Windsor, E. P. L., Barath, F. T., Gudmandsen, P., Langham, E., and Ramseier, R. O., A summary of results from the first Nimbus-7 Observations, submitted to J. of Geophysical Research, (1983).

22. Meeks, D. C., Ramseier, R. O., and Campbell, W. J., A Study of Microwave Emission Properties of Sea Ice - AIDJEX 1972, Proceedings of the 9th International Symposium on Remote Sensing, ERIM, Ann Arbor, Mich. pp. 307-322, 1974.
23. Norwegian Remote Sensing Experiment (NORSEX) Group, The Norwegian Remote Sensing Experiment (NORSEX) in a marginal ice zone, Geophysical Institute, University of Bergen, Bergen, Norway, in press.
24. Ramseier, R. O., and Gloersen, P., and Campbell, W. J., Variation of the Microwave Scattering and Emision from the Earth, E. Schanda, ed., pp. 87-93, 1974.
25. Ramseier, R. O., Gloersen, P., and Campbell, W. J., and Chang, T. C., Mesoscale Description for the Principal Bering Sea Experiments: Proceedings of the Final Symposium on the Results of the Joint Soviet-American Expedition, Leningrad, 12-17 May 1974, K. Ya. Kondratyev, Yu. I. Rabinovich, and W. Nordberg, eds., pp. 234-270, 1975.
26. Stiles, W.H., and F.T. Ulaby, The Active and Passive Microwave Response to Snow Parameters, 1. Wetness, J. Geophys. Res. 85, 1037-1044 (1980).
27. Svendsen, E., Kloster, K., Farrelly, B., Johannessen, O. M., Johannessen, J. A., Campbell, W. J., Gloersen, P., Cavalieri, D., and Mätzler, C., Norwegian Remote Sensing Experiment: Evaluation of the Nimbus-7 Scanning Multichannel Microwave Radiometer for Sea Ice Research, J. Geophys. Res., vol. 88, 1983.
28. Swift, C.T., W.J. Campbell, D.J. Cavalieri, L.S. Fedor, P. Gloersen, N.M. Mognard, S. Peterherych, and H.J. Zwally, Cryospheric Observations from Satellites, Chapter 11 in The Seasat-Nimbus-7 Experience, to be published in Advances in Geophysics (1983).
29. Thorndike, A. S. and Colony, R., Arctic Ocean Buoy Program Data Report, 1 January 1979 - 31 December 1979, University of Washington Polar Science Center, Contract Report, NOAA Grant NA 80-AA-D-00015.
30. Vant, M. R., A Combined Empirical and Theoretical Study of the Dielectric Properties of Sea Ice in the Frequency Range 100 MHz to 40 GHz, Environment Canada, Ocean and Aquatic Sciences/ Carl University, Ottawa, Technical Report Ph.D. Thesis, Carleton University, Ottawa.
31. Wilheit, T. T., Nordberg, W. Blinn, J., Campbell, W. J., and Edgerton, A., Aircraft Measurements of Microwave Emissions from Arctic Sea Ice Remote Sensing Environ., vol. 2, 129-139, 1972.
26. Yakovlev, G. N., Ice Routes of the Arctic, Translation by Department of National Defense, Ottawa, Canada, 342 pp., 1977.
32. Zwally, H. J., and Gloersen, P., Passive Microwave Images of the Polar Regions and Research Applications, The Polar Record, vol. 18, pp. 431-450, 1977.
33. Zwally, H. J., C. L. Parkinson, J. C. Comiso, F. D. Carsey, W. J. Campbell, P. Gloersen, Antarctic Sea Ice Cover 1973-1976 from Satellite Passive Microwave Observations, NASA SP-459, 1983.

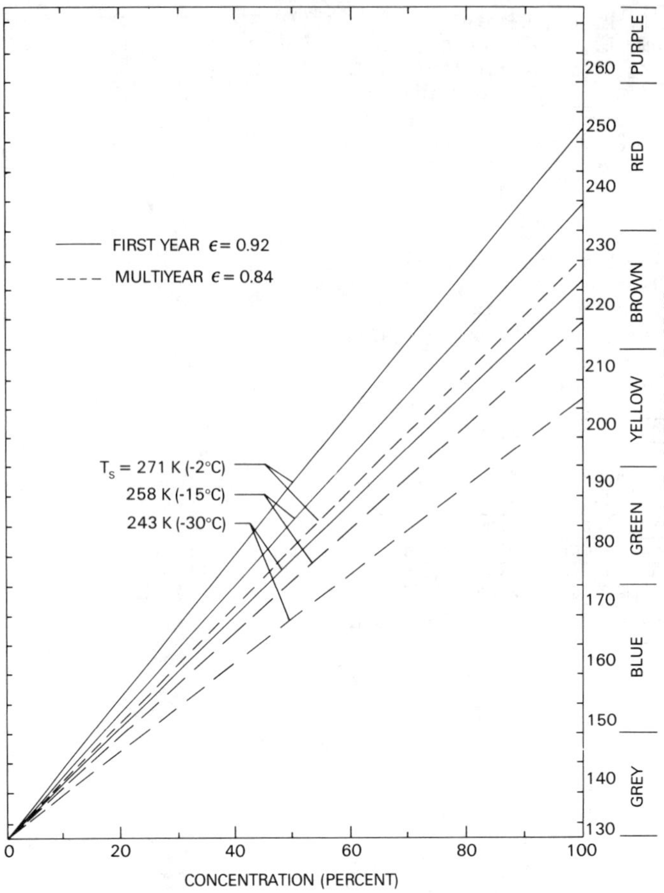

Figure 1. Sea ice concentration versus microwave radiance at a wavelength of 1.55 cm for three different radiating temperatures for first-year (FY) and multiyear (MY) sea ice.

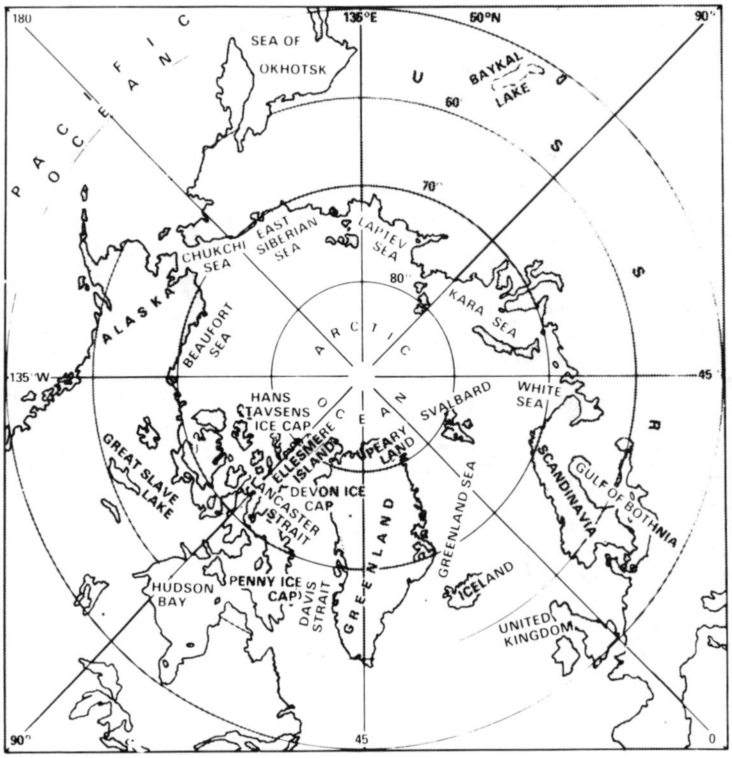

Figure 2. Arctic location map.

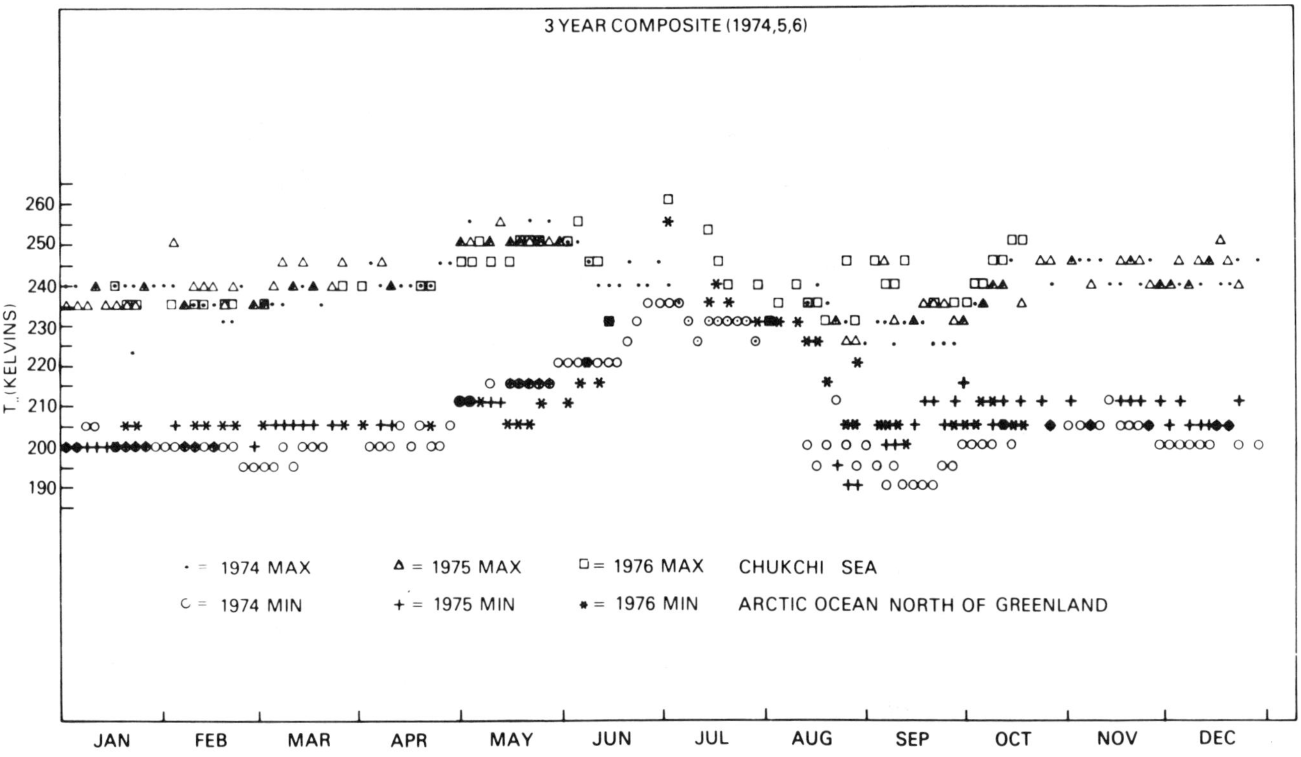

Figure 3. Three-year composite (1974-76) of seasonal variations of microwave radiance of first-year and multiyear sea ice as observed by the Nimbus-5 ESMR.

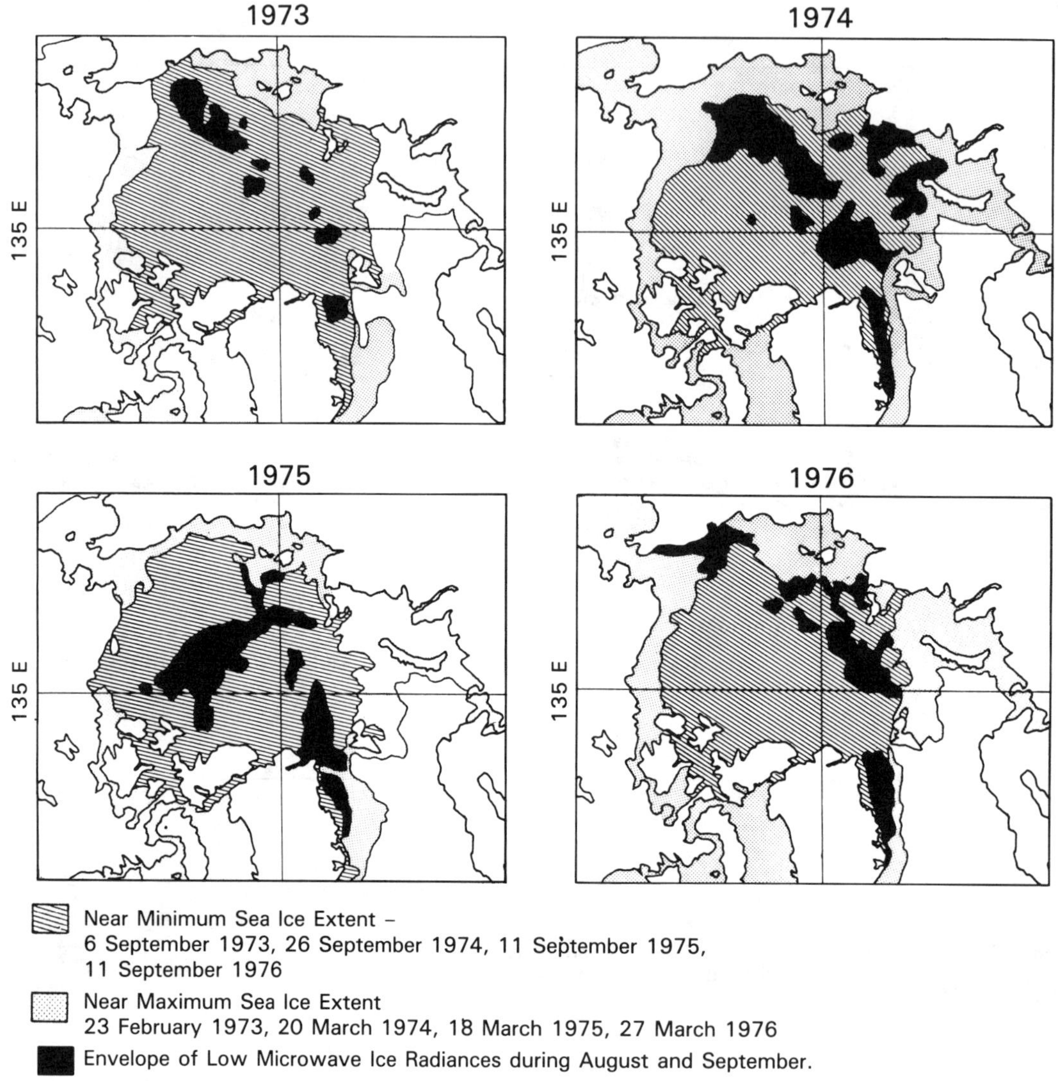

Figure 5. Near maximum and near minimum sea ice extent for the Arctic and envelopes of August-September areas of low microwave radiances (<197.5 K) derived from ESMR-5 images for 1973 through 1976.

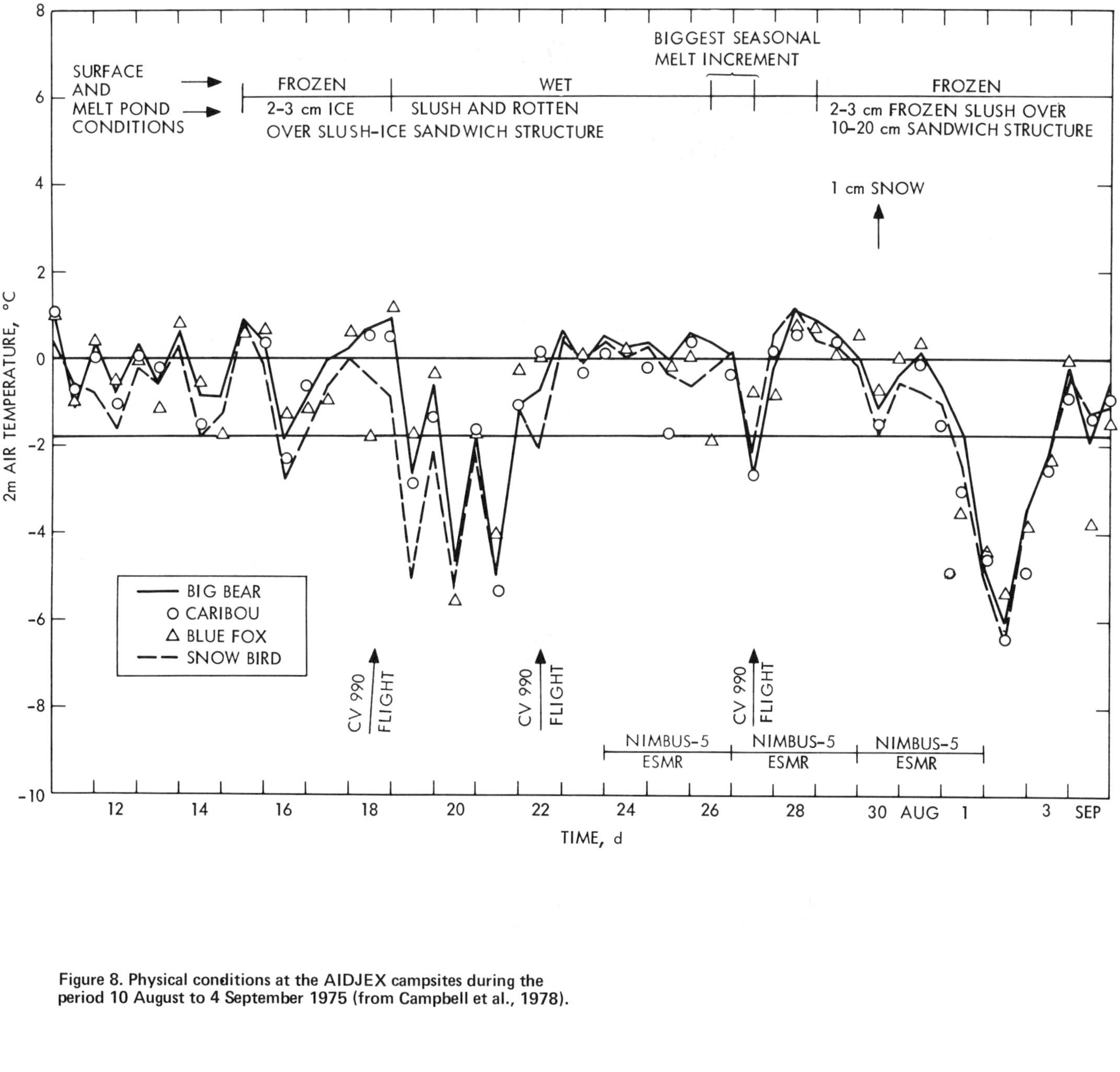

Figure 8. Physical conditions at the AIDJEX campsites during the period 10 August to 4 September 1975 (from Campbell et al., 1978).

Temporal Variations of Brightness Temperatures along a NIMBUS-5 ESMR Transect.

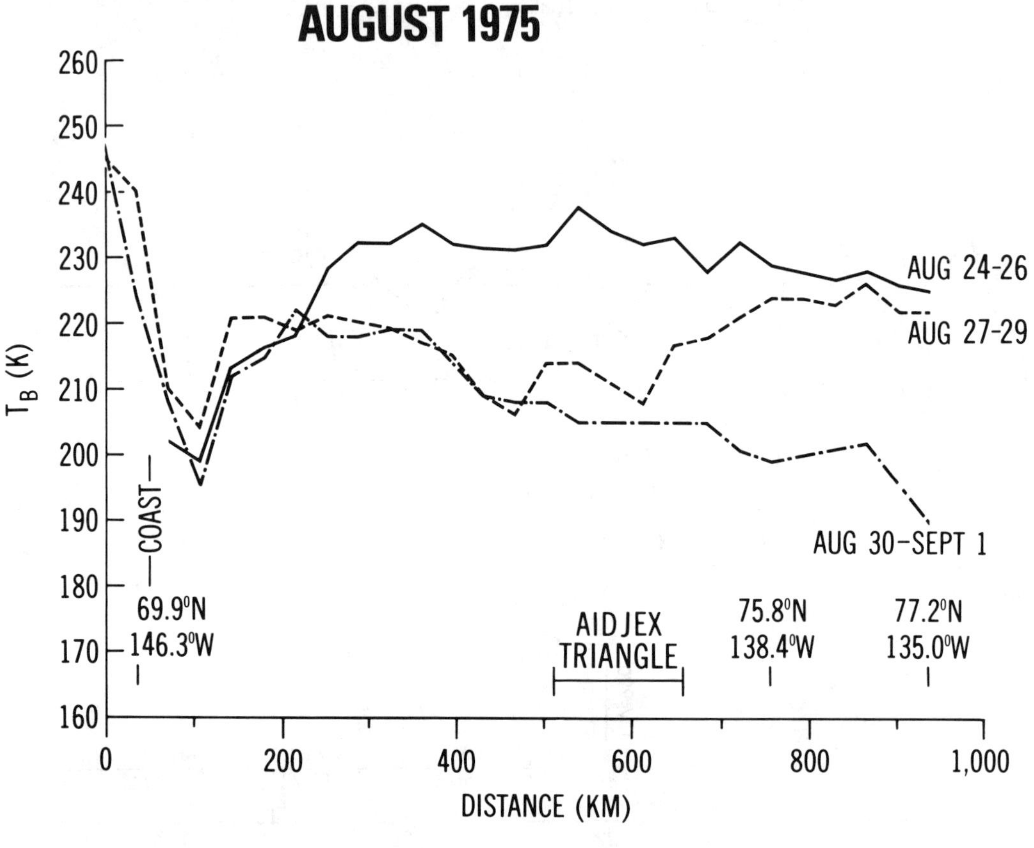

Figure 11. Illustration of the rapid fluctuations in brightness temperature with time in late August 1975 along the transect indicated in Figure 10 (from Gloersen et al., 1978).

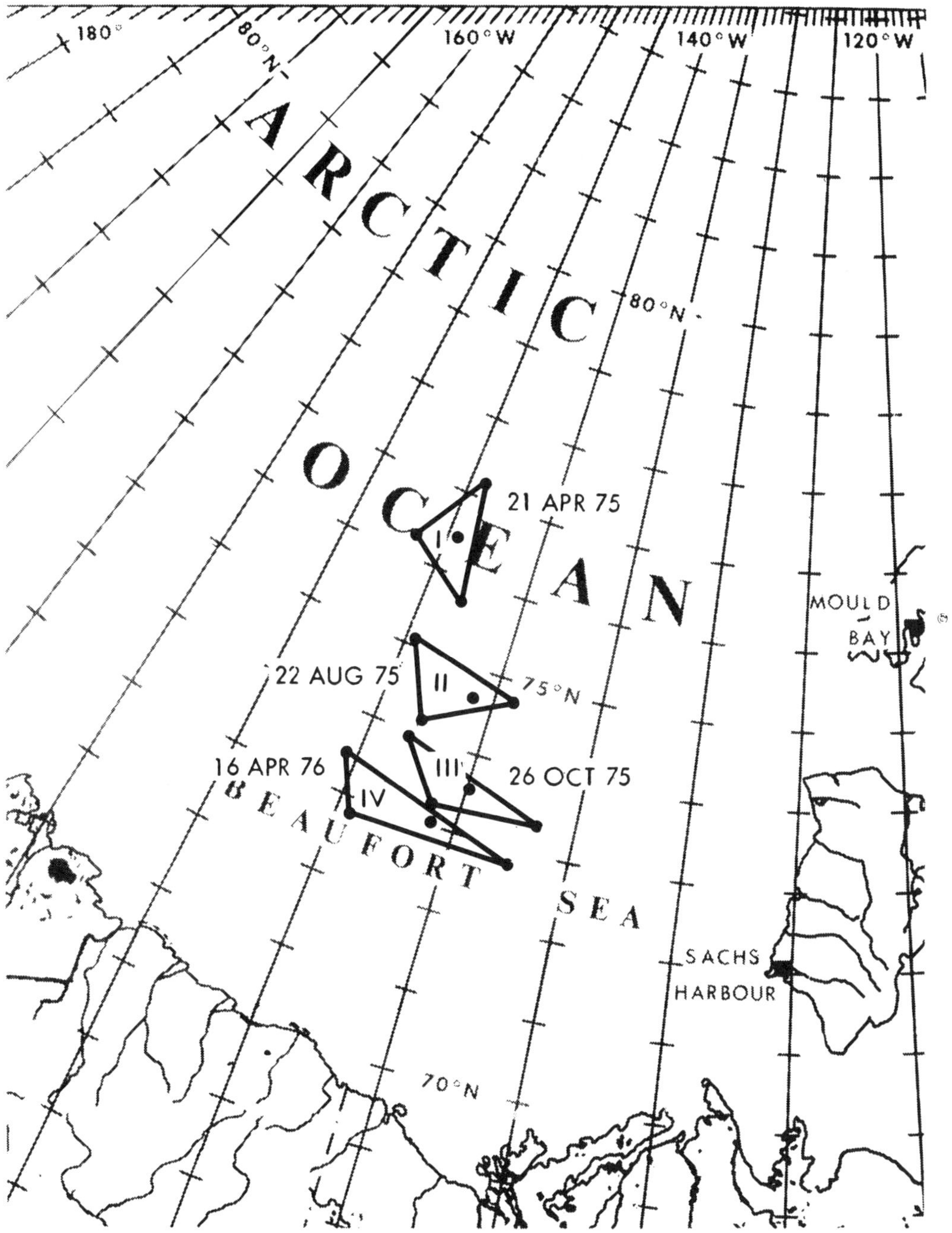

Figure 12. Map of AIDJEX area showing the locations of each drifting station during the period 21 April 1975 to 16 April 1976 when the NASA aircraft remote sensing flights were performed (from Campbell et al., 1978).

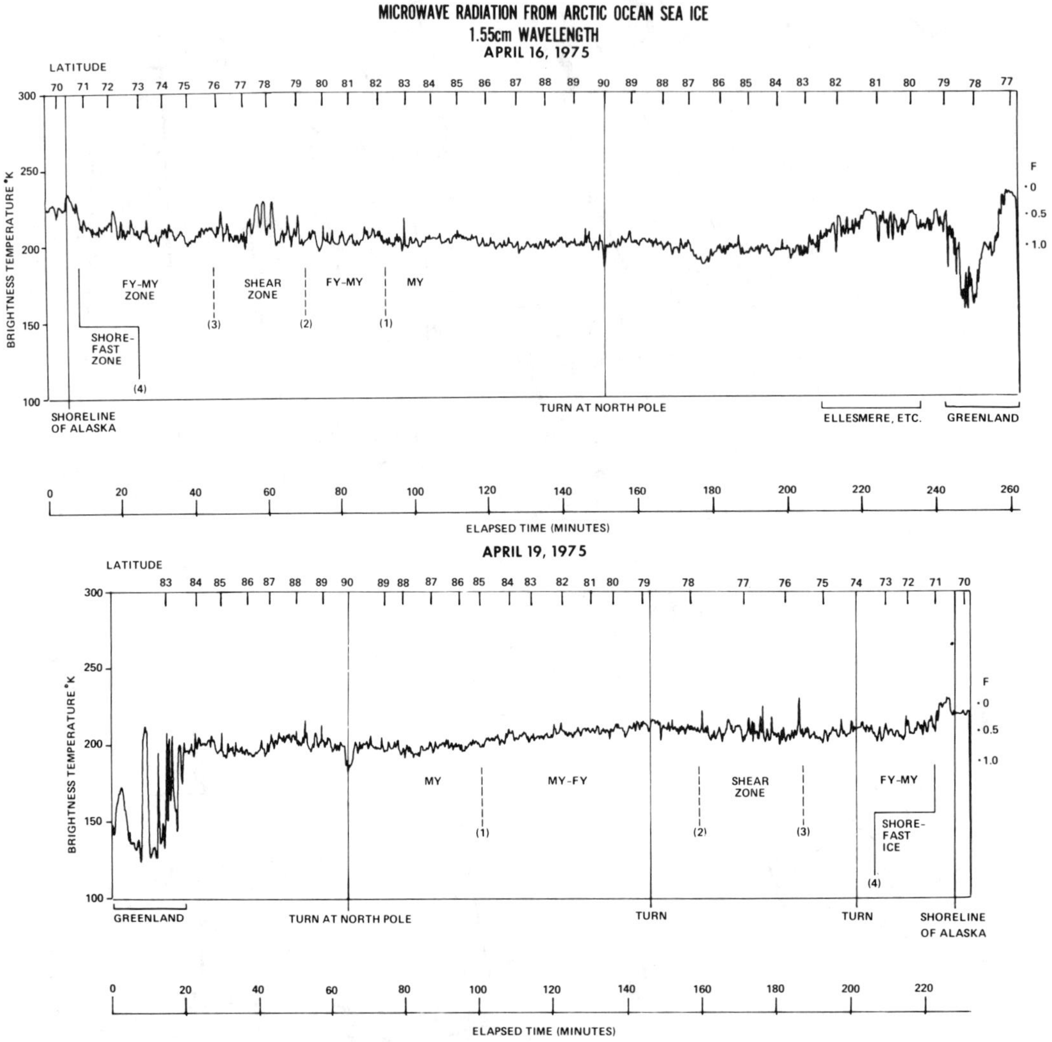

Figure 14. CV-990 aircraft ESMR profiles along the flight paths shown in Figure 13 (from Gloersen et al., 1978).

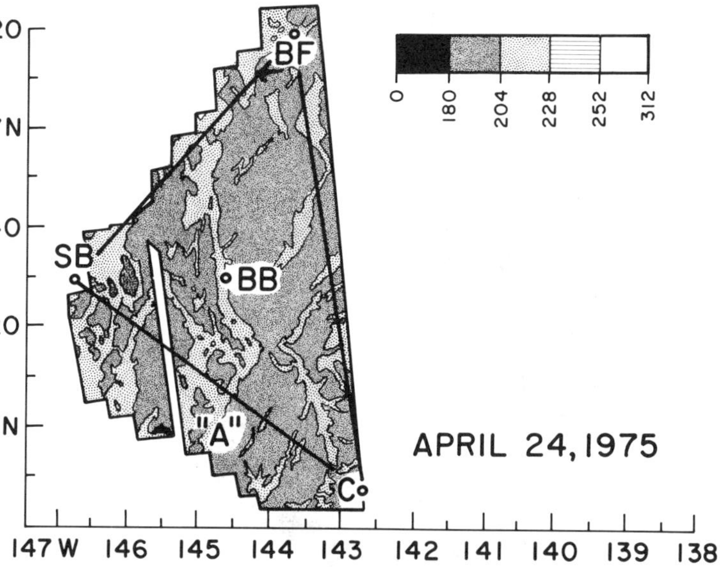

Figure 15. Aircraft ESMR image of the AIDJEX area on 24 April 1975.

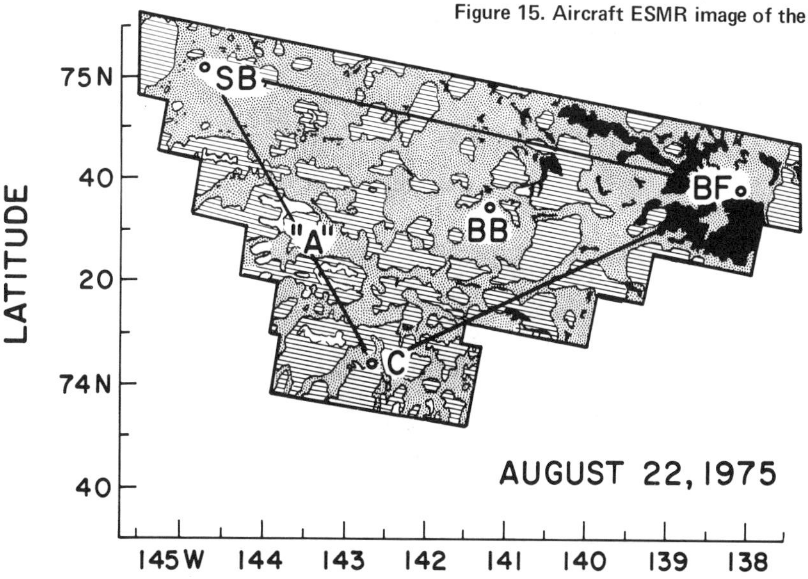

Figure 16. Aircraft ESMR image of the AIDJEX area on 22 August 1975 (from Campbell et al., 1978).

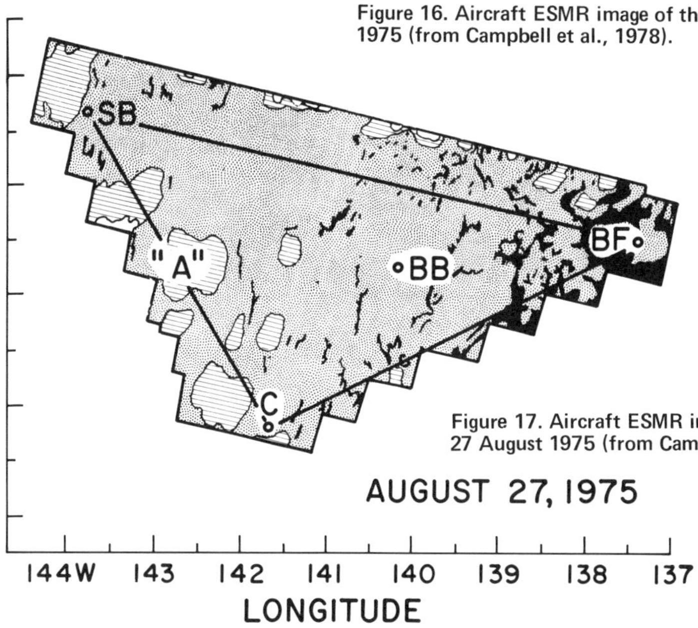

Figure 17. Aircraft ESMR image of the AIDJEX area on 27 August 1975 (from Campbell et al., 1978).

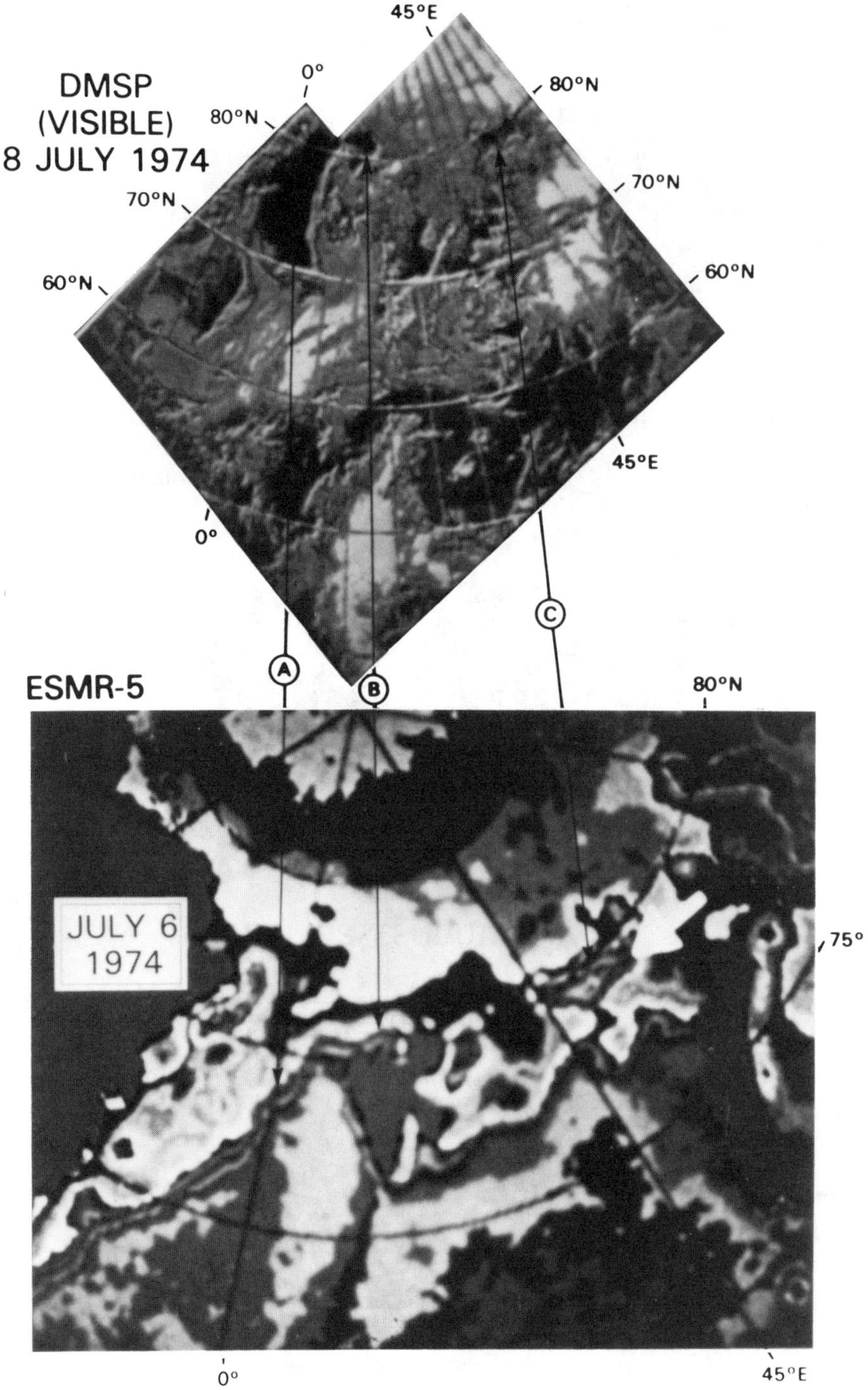

Figure 20. Nimbus-5 ESMR and DMSP images for 6 July 1973 of th eastern Arctic Ocean.

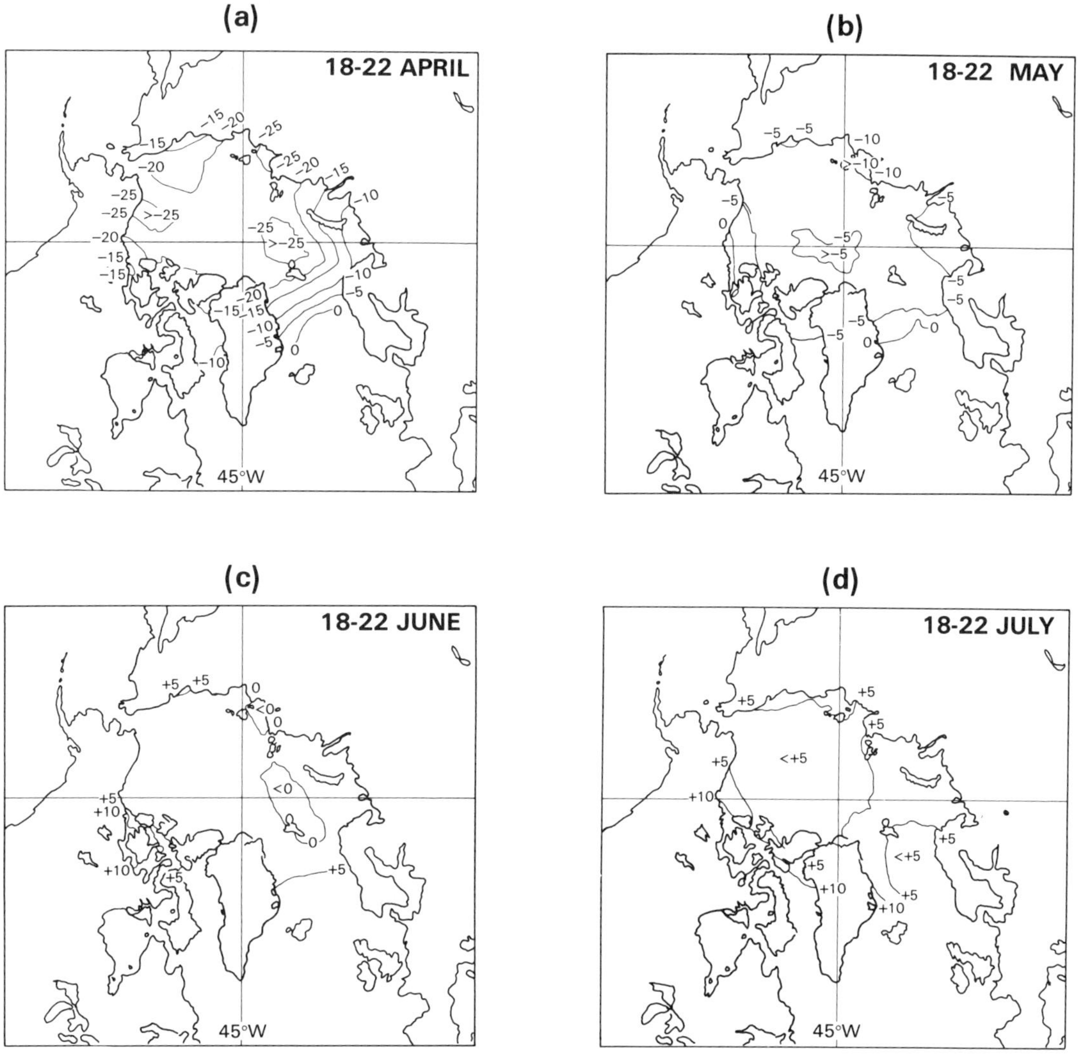

ARCTIC OCEAN BUOY SURFACE TEMPERATURE (1979)

Figure 23. Arctic Ocean Buoy temperatures fields for mid-April, mid-May, mid-June, and mid-July 1979 (data from Thorndike and Colony, 1980). In each case, the buoy data shown were averaged over a five-day interval.

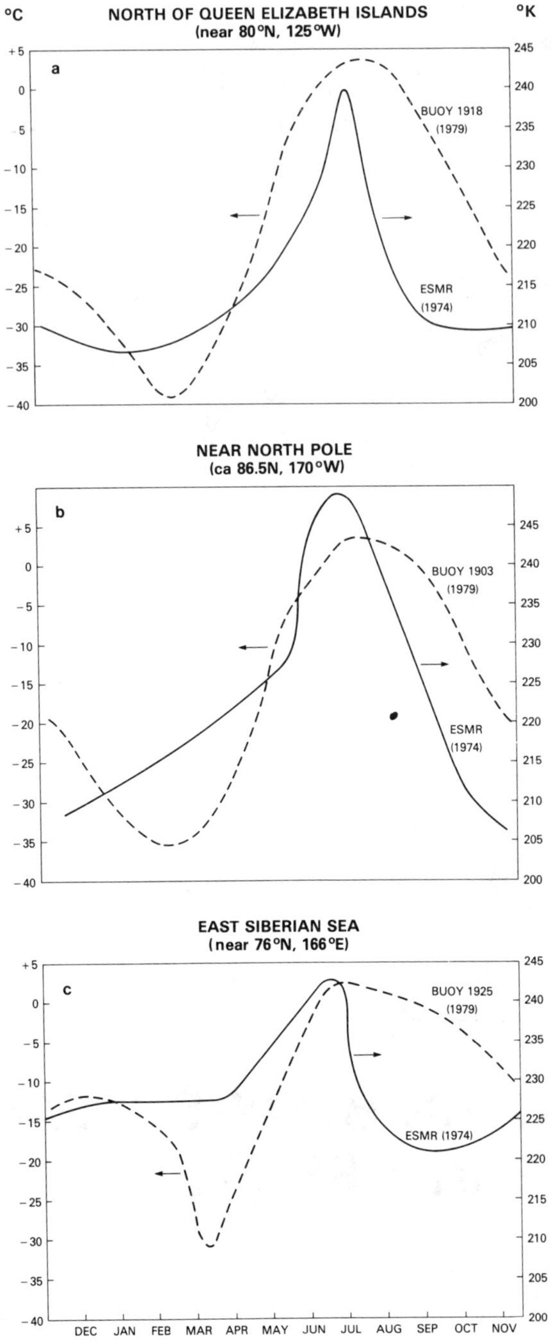

Figure 24. Temporal variation of buoy temperatures during 1979 (from Thorndike and Colony, 1980) and ESMR brightness temperatures during 1974 in the vicinity of various Arctic Ocean Buoys.

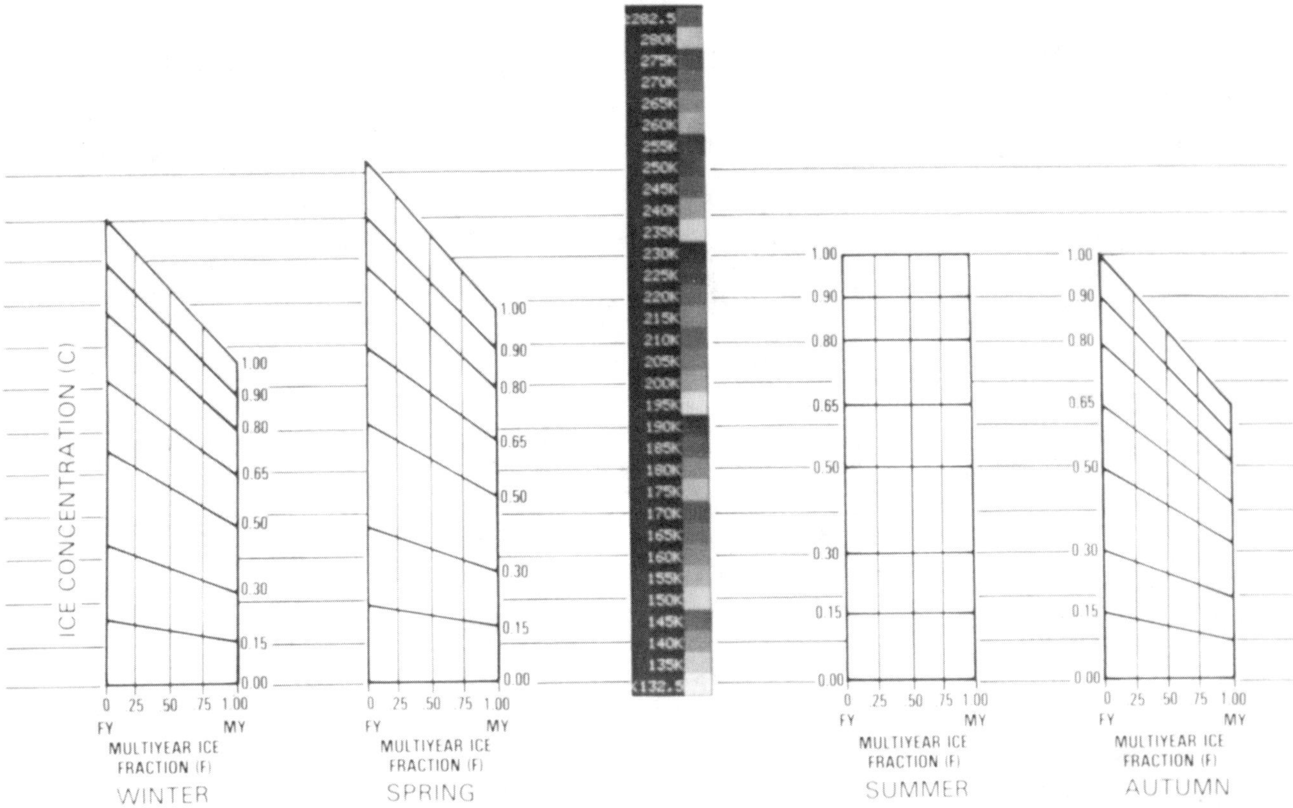

Figure 4. Seasonal nomograms of ice concentration as functions of microwave radiance and sea ice type mixtures.

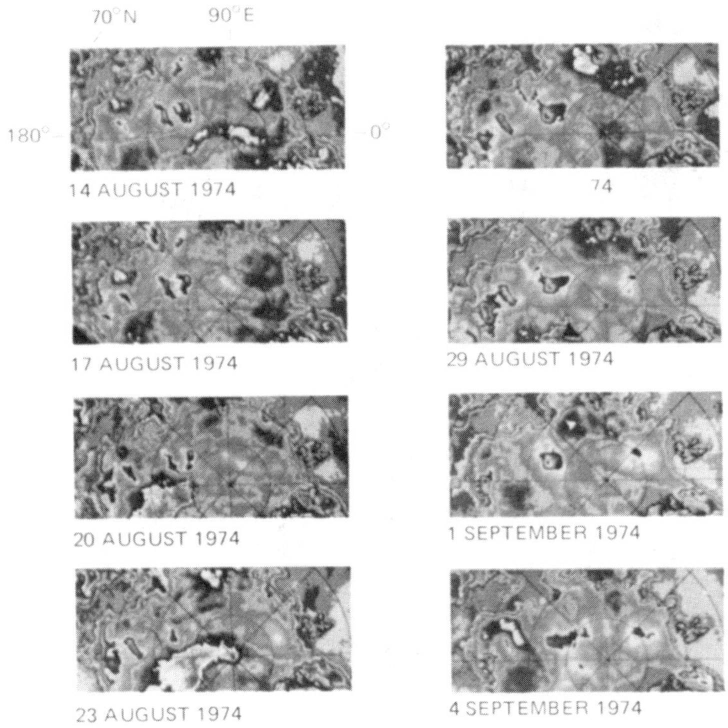

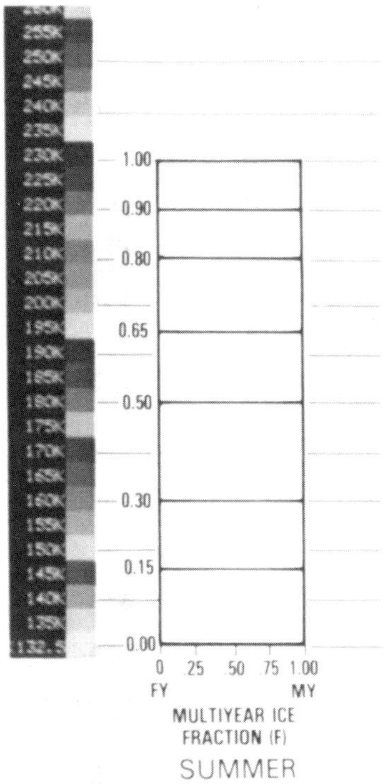

Figure 6. Sequential Nimbus-5 ESMR images of the central Arctic—14 August to 4 September 1974.

Figure 7. Sequential Nimbus-5 ESMR images of the central Arctic—7 September to 28 September 1974.

Figure 9. Arctic ESMR-5 images for 24, 27, and 30 August 1975. Transect shown in image for 27 August was used to obtain the ESMR-5 profile shown in Figure 11. The position of the AIDJEX triangle is shown in the images for 24 and 30 August (from Gloersen et al., 1978).

Figure 10. ESMR-5 images of Arctic for 2 December 1973, 21 December 1974, and 19 December 1975.

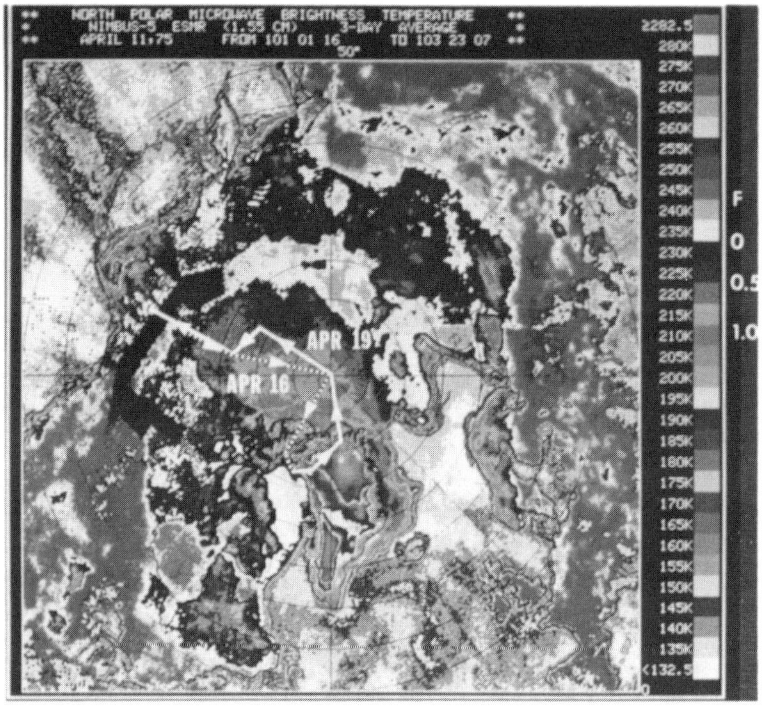

Figure 13. ESMR-5 image of the Arctic for 11 April 1975. The multiyear ice fraction (F) associated with each coded color is given in the nomogram in Figure 5. The transects on the image indicate the flight paths of the NASA CV-990 airborne laboratory on 16 and 19 April (from Gloersen et al., 1978).

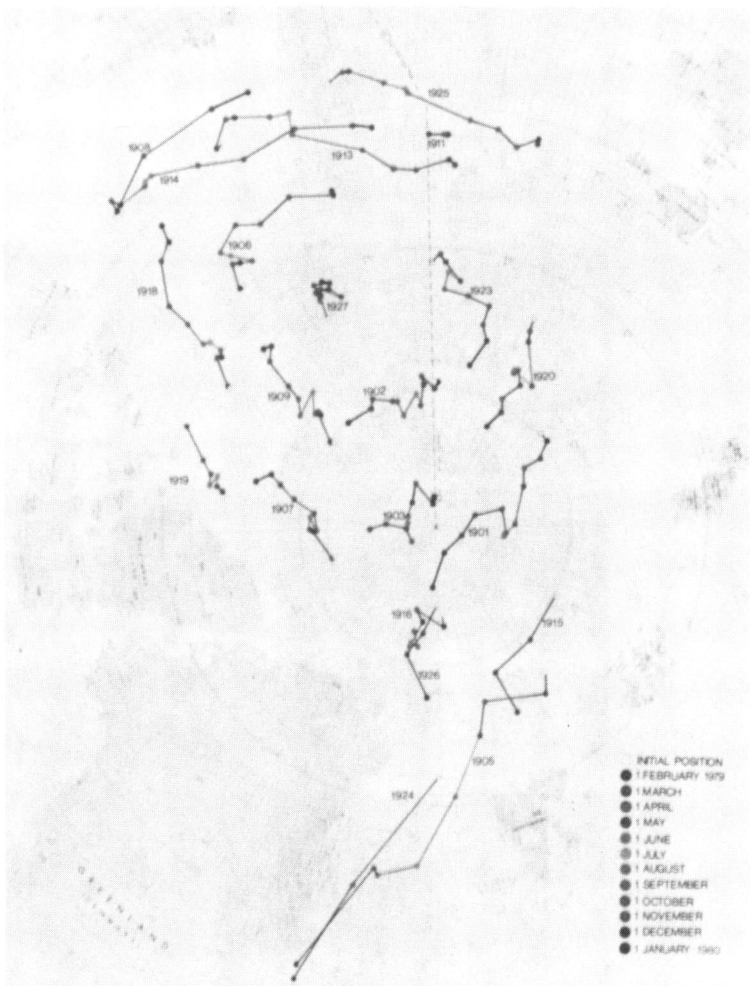

Figure 18. Drift of Arctic Ocean Buoy Network for 1979 (from Thorndike and Colony, 1980).

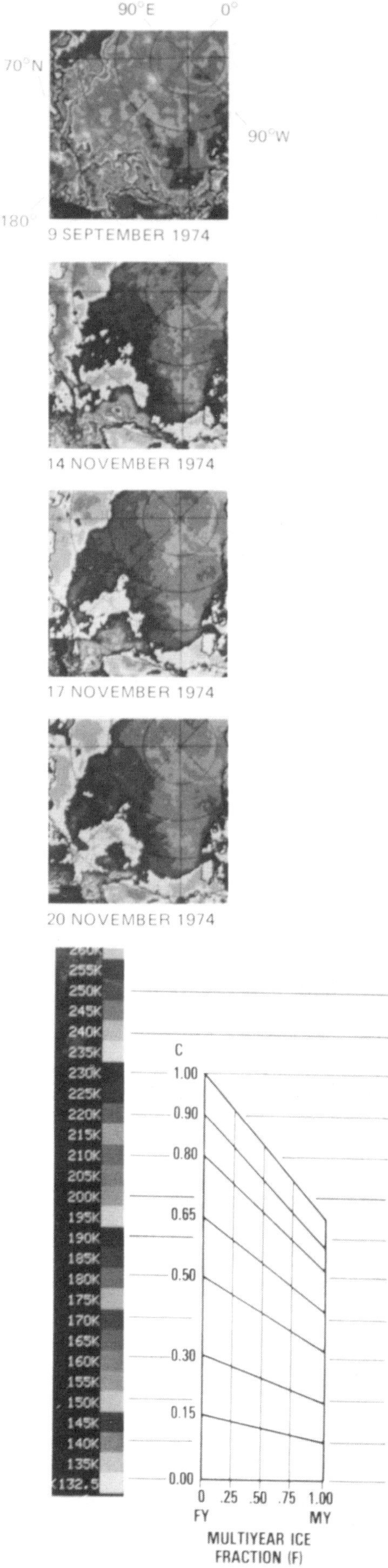

Figure 19. Sequential Nimbus-5 ESMR images of the Beaufort-Chukchi Seas area—9 September to 20 November, 1973.

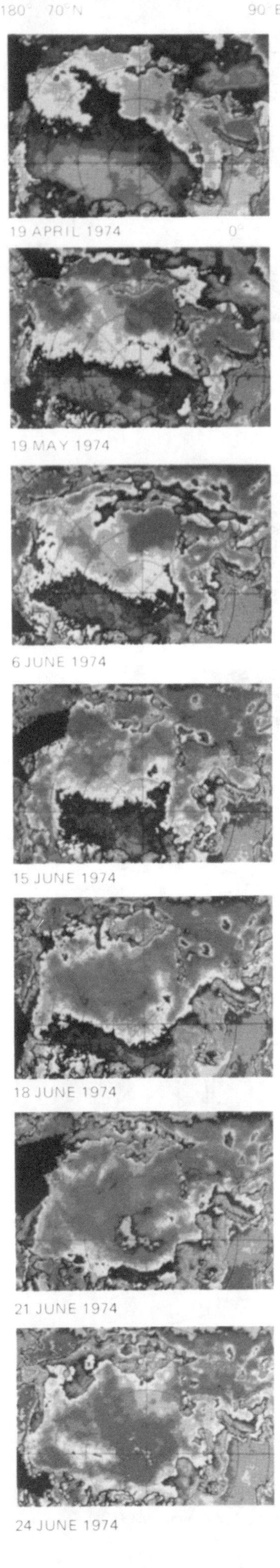

Figure 21. Sequential Nimbus-5 ESMR images of the central Arctic—19 April to 24 June 1974. Either the spring or the summer nomograms (see Figure 4) applies to Figures 21 and 22.

Figure 22. Sequential Nimbus-5 ESMR images of the central Arctic—12 July to 14 August 1974.

ARCTIC SCIENCE AND ENGINEERING III

INTRODUCTION

W. David Kingery

Department of Materials Science and Engineering
Massachusetts Institute of Technology
Cambridge, Massachusetts

Studies of the characteristics of ice and its deformation properties were first carried to elucidate the behavior of glaciers, and it has only been in the last few decades that the properties of sea ice have been seriously studied. During the early years of World War II the range of British aircraft was inadequate for protection of the North Atlantic sea lanes and several suggestions were made for using natural ice as temporary landing fields. However, icebergs are notoriously unstable, and the prospect of a landing field flipping over during an approach was not very comfortable. Sea ice platforms are also inadequate. As a result a substantial effort was devoted to developing an ice structure that would be more useful. A new ice alloy was developed (called "pykrete") and preliminary designs carried out. However, by the time it got that far the exigencies of the war situation had changed and the project was not carried to completion. This story seems to be symptomatic of our up-and-down approach to arctic endeavors. That was certainly the case when in 1961 we carried out successful experimental landings of the entire Air Force inventory of aircraft at Thale, Greenland, including a constructed parking area suitable for the heaviest aircraft. Soon thereafter the Air Force lost interest. Let us hope that our present arctic policy will include a sustained effort at understanding the properties and behaviors of arctic ice. The papers presented this afternoon will make it clear that there is still much to learn.

In this session we shall look to the behavior of arctic sea ice from two quite different points of view. First, Dr. Pritchard will take a macroscopic view of sea ice behavior on the scale of hundreds and tens of kilometers, modeling its behavior in terms of the imposed forces and characteristic ice properties. The problem is a difficult one since the forces acting on sea ice are complex and come from different origins which can be evaluated and described qualitatively, but their quantitative estimation is still most difficult. Similarly, the reaction of sea ice to the opposed forces is a complex one and depends on the source of the ice and time, temperature, and structural considerations that are most difficult to simplify in a form for which macroscopic computations are useful. Nevertheless, a lot of progress has been made the last several years, and Dr. Pritchard will inform us about the present state of affairs.

Sea ice behavior can also be approached from the point of view of materials properties in which the characteristics of the ice are related to its crystalline structure, its composition, its microstructure and the influence of various impurities which may be present. This is also a complex subject since ice forms in various ways. We can not expect that the properties of sea ice which contains a good deal of entrapped brine will be similar to the higher purity, larger grain size glacier ice carved from the shores of Greenland and Ellesmere Island. Since its mechanical properties are of interest in a temperature range quite close to its melting point, temperature and strain rate are also important variables. Since several variables influence the resulting properties, a certain complexity must be addressed and this is the task which Dr. Weeks and Dr. Mellor have undertaken.

MODELING OF ARCTIC SEA ICE FIELDS

Robert S. Pritchard
Flow Industries, Inc.
Research and Technology Division
Kent, Washington 98031 USA

The capability to model sea ice dynamics behavior has increased substantially during the last decade. The literature describing macroscale and mesoscale modeling in the Arctic Basin, especially the Beaufort Sea, the Chukchi Sea and Bering Strait region, and the Baltic, Bering and Greenland Seas is reviewed. Basic contributions to model development are identified with the intent of determining the essential similarities in approach. Features needed to describe ice behavior in a physically acceptable way are discussed. Although models have come into more common use, many of these models have not incorporated the results of recent studies of small-scale processes. The sea ice modeling community is challenged to take this better understanding of individual physical processes and to incorporate it into the models.

1. INTRODUCTION

During the last decade, the understanding of sea ice behavior has increased substantially. As in other engineering and scientific fields, modeling of this behavior has been able to keep pace because of the great advances in computer hardware and software and, to a lesser extent, because of advances in numerical solution techniques for solving complicated problems. The techniques needed to simulate sea ice behavior are some of the most sophisticated in use anywhere today.

The purpose of this review is to survey the models currently in use and to examine and compare their basic components. In this way, the differences and, even more, the similarities in the various approaches can be identified. It is hoped that this analysis will encourage and help other investigators to focus on improving our understanding and description of sea ice processes and thereby adapt and extend present models.

Models are used in a diverse range of applications, extending from the need for more fundamental scientific knowledge to a variety of specific operational needs. These applications include understanding climate dynamics, determining oil spill trajectories, estimating loads that might be applied to a fixed structure operating within the ice cover, estimating noise generated by the ice cover, or warning of possible future ice invasions during petroleum drilling operations in open-water conditions. Time and space scales in each of these problems differ significantly.

On macroscales of hundreds of kilometers, sea ice behavior affects climate dynamics. The effects of sea ice motions and deformations are realized primarily through the formation of open water, which increases heat transfer from the ocean to the atmosphere. The ice stress is not of direct physical concern. On mesoscales of tens to hundreds of kilometers, both ice motions and stresses are important. Ice motions would influence an oil spill by transporting it. Also, the large-scale ice stress could be transmitted shoreward to generate forces on offshore drilling structures. The geophysical scale forces from winds and currents over wide areas of the ice pack cause its motion. These effects can be realized at large distances when internal ice stresses are high. These large-scale forces and stresses cannot be measured directly. To compare with observed behavior, our attention must be focused on the motions, the only directly observable variable. On small scales of a hundred meters and less, these forces affect the design and use of offshore drilling structures. Artificial islands, docks, breakwaters, and conical platforms must be designed to withstand the forces that sea ice can apply. For scales between a hundred meters and ten kilometers, sea ice behavior is difficult to categorize. There is not a clear separation of scales and several problems fall within the mesoscale and small-scale bounds.

In current sea ice models, the physical behavior of sea ice is described by accounting for mass, momentum and energy balance and the constitutive laws relating deformation to stress and thickness redistribution and thermal growth. In this review, the numerical schemes used to solve the mathematical models are discussed briefly. Computational considerations include the choice of a Lagrangian or Eulerian description, a finite element or finite difference approximation, and an implicit or explicit time integration.

This review is not complete in the sense that not all ice modeling publications are discussed. Attention is focused on mesoscale and smaller modeling in which stress, velocity and ice conditions are equally important. Past reviews by Rothrock [1,2] and Hibler [3,4] on macroscale modeling are updated, but models developed primarily for studying climate dynamics are only included as they apply to mesoscale problems. Also, small-scale engineering studies are ignored. These engineering studies will, in the future, provide input to modeling individual processes, but, at this time, have not been included in larger scale models. Many valuable engineering studies are to be found in proceedings of the POAC and OTC conferences as well as the numerous recent ASME and ASCE conferences with Arctic sessions. The NTIS has also prepared a published search on sea ice. The works referenced in this paper can provide detailed information on specific topics and references to further relevant studies.

The remainder of this paper is divided into four sections. A lengthy review of literature on sea ice models appears in Section 2. Then, computational considerations needed to solve problems with these models are presented in Section 3. In Section 4, an overview of the fundamental components of a model needed to completely describe ice dynamics behavior is given. Recommendations for future extensions are also made. Finally, in Section 5, a brief conclusion evaluates the current status of ice dynamics modeling.

2. LITERATURE REVIEW

The literature reviewed in this section is extensive. It includes macroscale and mesoscale models of almost all arctic and subarctic ice-covered oceans. Since different length and time scales, geographic regions and specific applications are involved, it is difficult to separate the presentation into logical subsections. In general, the discussion is broken down by geographic region. Separate subsections are presented on models developed for use in the Arctic Basin and Beaufort Sea region, the Chukchi Sea and Bering Strait area, the Baltic Sea, and the Bering and Greenland Seas. The first subsection is an exception, however, in that it is a general discussion of free drift. This seems reasonable because free-drift models, in neglecting ice stress, do not show strong regional differences.

2.1 Free Drift

Free drift of sea ice is defined as the condition where all forces act on the ice cover except ice stress divergence. In most present models these other forces include inertial and Coriolis acceleration, air and water drag, sea surface tilt and, at an ice edge, wave radiation drag. Many authors have discussed free drift, with early references to Nansen [5] and Zubov [6]. Recent articles typically trace the momentum balance in free drift to Campbell [7]. Rothrock [1,2] and Leppäranta [8] have presented recent reviews.

McPhee [9] has used AIDJEX data from the summer of 1975 to analyze the relationship between winds and ice motions. This time period was chosen because it appears that ice motions then were unaffected by internal ice stress. The study had as a primary goal the determination of an oceanic drag law. This work has continued, and a comprehensive report has been published comparing various drag laws [10]. Overland and Pease [11] and McPhee [12] have developed similar models of the oceanic mixed layer for shallow seas. Pease et al. [13] have used a free-drift model with quadratic drag laws to determine drag coefficients near the ice edge in the Bering Sea. Another ocean model that is useful to note is the multilevel ocean model of Liu and Leendertse [14], which treats ice as a linear viscous layer but includes most ocean transport processes in great detail.

All of these water drag models may be incorporated into more complex and complete ice dynamics models, but have been exercised to date only in free-drift models (except Thomas [15] has used McPhee's shallow ocean model in his nearshore Beaufort Sea studies).

Colony and Thorndike [16] and Thorndike and Colony [17] have analyzed ice motion data and geostrophic wind data from Arctic Basin buoy deployments in 1975-76, 1979-80, 1980-81 and 1981-82 to determine how well a linear model relates daily average motions. They found that at distances further than 400 km from shore, free drift describes about 85 percent of the variance in ice velocity in winter and about 95 percent in summer. Velocity vector errors between observed and modeled velocities have a standard deviation of about 4.0 cm/s. Nearer to shore where ice is often in contact with the coast, about 50 percent of the variance is described. Mean vector errors over monthly periods are nearly constant and may be attributed to long-term ocean currents. This may be the best available estimate of the ocean currents because of the sparcity of direct measurements.

Thomas and Pritchard [18] and Pritchard [19] have used a free-drift model with quadratic drag laws to analyze AIDJEX data for the Beaufort Sea in 1975-76. They found that modeled velocity vectors have monthly mean errors less than 1.0 cm/s and standard deviations of about 3.6 cm/s in summer and up to 10 cm/s in winter. This model is also used with 25 years of barometric pressure data for the Beaufort Sea to estimate the range of ice motions that may be expected for this region. The study updates a similar estimate by Sater et al. [20].

Dimensionless variables have been introduced by Pritchard and Thomas [21] and Leppäranta [8]. The choice of reference variables differs and dimensionless parameters appear in different combinations, but similar results are obtained.

McPhee [22] has used free drift to study inertial oscillations in the Beaufort Sea. When ice stress is negligible, inertial and Coriolis forces respond to wind drag and produce inertial oscillations. In the subarctic waters of the Baltic and Bering Seas, the thinner ice means that inertial and Coriolis accelerations are smaller and, therefore, produce smaller inertial oscillations (Leppäranta [8]). Pease et al. [13] have shown that air and water drag are nearly in balance in the Bering Sea because the ice is thinner.

2.2 Arctic Basin and Beaufort Sea

The strain that may occur in an ice floe before failure is small compared to the deformation due to failure by opening, rafting and ridging. A model may describe behavior on a scale small enough to resolve variations from floe to floe or may average over a number of floes and describe a set of average properties. This set of properties must be complete enough so that the desired output variables, such as velocity and stress, are determined in an average sense at this length scale. Then, even though there are known variations on smaller scales, it is not necessary to resolve them explicitly.

The velocities of individual particles of ice on a floe differ, especially when floes are large compared to the scale of resolution. These velocity variations cannot be described well as a continuum velocity field because deformations are localized at floe boundaries rather than being distributed as continuous straining of the floes. Nye and Thomas [23] have shown how ice velocities vary over an area several hundred kilometers long by using satellite images. On smaller scales, Hibler et al. [24] have presented similar results. This work has been reviewed by Rothrock [2]. More recently, Thorndike and Colony [25] and Colony and Thorndike [16] have analyzed ice motion data from the Beaufort Sea and the central Arctic Basin to determine the portion of observed motion that can be explained by a continuum velocity field and the portion that appears as subgrid-scale velocity inhomogeneities. Their work focuses on large-scale behavior, 100 km and larger in resolution. Velocity variations as large as 0.4 cm/s in winter and 1.1 cm/s in summer cannot be described by a continuum velocity field. Strains calculated from buoys about 100 km apart have an uncertainty of 0.4 to $1.1 \times 10^{-7}\ s^{-1}$. An uncertainty of this magnitude makes it difficult to determine whether an area is diverging or converging. An uncertainty in shear and rotation is less severe because these quantities are larger than dilatation. The impact of this uncertainty is that the stress can be in substantial error because it is determined from the strain through a constitutive law and limits the accuracy of a continuum model for a selected length scale.

Sea ice dynamics modeling in the Beaufort Sea advanced at a rapid pace with the concentrated efforts during AIDJEX, as discussed by Untersteiner [26]. Coon [27] has summarized the model developed during that program. One of the major contributions of the AIDJEX model is the attempt to describe large-scale ice behavior by describing individual processes such as lead formation, rafting and ridging.

Ice conditions, which are expected to control the relative importance of individual processes, are described in terms of the thickness distribution by Thorndike et al. [28]. The large-scale heat flux between ocean and atmosphere that affects climate dynamics throughout the world is also affected directly by ice conditions [29,30]. Thermal growth and ablation of sea ice has been described by the climatological average rates for each thickness throughout the year. This approach was used for short-term simulations [31]. Thermal growth has also been accounted for by simulating the heat budget directly. This approach was used by Hibler [32] to simulate year-long behavior, and it is the approach needed for climate dynamics simulations.

Large-scale ice stress is related to deformations by an elastic-plastic constitutive law [33,34]. Ice strength has been determined by Rothrock [35] by equating the energy dissipated in plastic deformations to two small-scale energy sinks. These two sinks, which have been identified by Parmerter and Coon [36] in their study of the mechanisms prevalent in ridging, are increases in gravitational potential energy as ice conditions change and frictional dissipation as ice blocks slide through the rubble of a ridge. This AIDJEX model provides the framework within which all recent modifications and improvements are discussed.

Simulations of sea ice behavior using the AIDJEX model have indicated that strength values less than 1. x 10^4 N/m are unrealistically low [37]. Bugden [38] modified the redistribution function to allow thin ice to be ridged into a range of thicker ice categories, instead of assuming that all ice ridges into ice 5 times its original thickness as assumed by Coon et al. [31] and Thorndike et al. [28]. Rothrock [2] reviews this aspect of the model development. Hibler [32] has introduced an alternative formulation of the redistribution function in a simulation of Arctic Basin ice drift for 1962-63. This redistribution also accounts for ice being ridged into a range of thicker categories, and it more accurately reflects the morphology of the ridged ice cover and the square-root thickness scaling found by Parmerter and Coon [36]. This assumption increases the strength to more acceptable levels for accurately simulating ice trajectories. Pritchard [39] summarized the framework of the AIDJEX model and generalized the constitutive law by adding an energy sink to account for shearing deformations. He assumed that shearing occurs along narrow bands rather than over wide areas and, as a result, does not redistribute ice. These assumptions have allowed more accurate simulations to be made, while still retaining the physically realistic model developed initially during AIDJEX. Accurate simulations of nearshore ice dynamics have been performed with this model and with an ideal plastic model [40].

Rothrock [2] has called for a more thorough validation of the performance of these models. Rothrock et al. [41] presented a method for evaluating model performance by calculating the stress divergence and all other forces in the momentum balance using ice motion, wind and current observations directly. They concluded that an ideal plastic model with a teardrop yield surface did not satisfy momentum balance for the three times it was tested during the AIDJEX experiment. Kolle and Pritchard [42] have determined velocity vector errors for a series of simulations of Beaufort Sea ice dynamics. Errors in daily average ice velocities are shown to have a mean of less than 1.0 cm/s and a standard deviation of 3.6 cm/s, including both nearshore and far offshore sites. These errors appear to hold for simulations throughout the AIDJEX field experiment in 1975-76.

Hibler [3,43] has developed a viscous-plastic constitutive law to describe the mechanical response. This change from elastic to viscous behavior at low stress levels is a valuable extension because it allows different numerical techniques to be used to solve the model. No definitive study has yet been made to determine the range of conditions for which viscous or elastic behavior is superior in numerical simulations. However, the U.S. Navy has adapted the viscous model for simulations and predictions of ice behavior over the whole Arctic Basin and the elastic model for regional ice behavior analysis.

Both the elastic-plastic and the viscous-plastic constitutive laws were developed to describe sea ice that is nearly rigid at low stress levels but that succumbs to large deformations at a limit stress. These properties are consistent with the ridging mechanism model of Parmerter and Coon [36], who found that deformations are rate independent and that as a limit height is reached the ridge extends horizontally. This behavior could be described reasonably well by a rigid-plastic constitutive law in a model seeking to describe sea ice motions, but such a constitutive law poses serious numerical difficulties and the assumption of rigidity prohibits determination of stress in the ice cover. The viscous-plastic model of Hibler [3,43] also has difficulties in determining stress because of the presence of a static pressure term that treats pressure in the ice similarly to pressure in a gas. This term induces a pressure in the ice equal to half the isotropic yield strength whenever deformations are zero, a result that is not acceptable near an ice edge, for example. McKenna [44] has circumvented this shortcoming by eliminating the static pressure term.

McKenna has studied the effects of constitutive law on ice behavior for short time periods [44]. A finite element formulation was used, and both elastic-plastic and viscous-plastic laws were studied. He found little difference in the resultant ice behavior. Although the finite element formulation is limited to allow only triangular elements, the consideration of arbitrary yield surface shapes and both elastic and viscous behavior is a valuable step toward developing one computer program that can solve problems using any of the models already developed.

In addition to mass and momentum balance, which have been included in the ice models previously described, the mechanical energy balance has been described by Coon and Pritchard [45]. The mechanical energy budget has been evaluated by Pritchard et al. [46] for the Beaufort Sea in 1975-76. During two strong storm events, the ice dynamics model represented each term in this mechanical energy budget accurately. Further work has shown that one of the terms in the budget, the energy dissipated by ridging, explains about half of the variance in low-frequency background noise observed during the winter of 1975-76 [47].

In the Canadian Beaufort Sea, drilling activities offshore of the MacKenzie River delta have led to the need for a smaller scale model to predict ice motions. Leavitt et al. [48] have adapted the viscous-plastic model by introducing a four-component characterization of the ice thickness distribution, first presented in [49], to characterize this primarily first-year ice.

In the Alaskan Beaufort Sea, offshore tracts in the Diapir field from Demarcation Point to Barrow have been leased. The environmental impact statement requires that the probability of impact in case of an oil spill be determined for the Alaskan shoreline. To determine ice motions that affect this probability, Thomas [15] has adapted the AIDJEX model presented by Pritchard [39] to estimate the range of ice trajectories that can be expected. Daily ice motions were simulated. The computational grid size was typically about 40 km, but was as small as 5 km in some cases. This small-scale resolution allowed the effects of the shoreline to be included. Even at moderate ice strengths of 5. x 10^4 N/m, the smallest resolution was found to be unnecessary for determining the range of ice trajectories. Most simulations were performed with the larger grid. Some 2250 ice trajectories were calculated for 30 starting sites. The range of motions shows how ice behaves under different ice, wind and current conditions.

2.3 Chukchi Sea and Bering Strait

When a shorter fetch is available for winds and currents to act on the ice, the internal stress divergence becomes more important than the other driving forces. In the Chukchi Sea, the smaller scales are caused by surrounding land masses. Thus, the Chukchi Sea differs from the Beaufort Sea because it is bounded by land on all sides but its northern edge. Moreover, at its southern edge the funneling effect near the Bering Strait reduces the ice cover to a mere 80 km in width. The behavior of sea ice here is dominated by the internal ice stress. The dimensionless parameter that reflects the ratio of ice stress divergence to water drag provides a measure of the relative importance of the ice stress [21].

The motion of the Chukchi Sea ice cover is generally toward the northwest. However, along the Alaskan shore the ice is extremely mobile, often moving alongshore in one direction or the other (see Shapiro and Burns, [50]). Sodhi [51], Pritchard et al. [52] and Kovacs et al. [53] have used plasticity models of the ice to describe the formation of the structural arch that forms across the Bering Strait. They also studied breakouts of ice from the Chukchi into the Bering Sea when the arch collapses. Reimer et al. [54] have related breakouts to the intensity of winds and currents and have shown that currents provide the primary driving force that controls breakouts. Reimer et al. [55] give accurate simulations of ice motions in this region, including the high-speed ice motions near shore.

The strong influence of internal stress in the vicinity of the Bering Strait has allowed some study of ice strength values. Reimer et al. [54] have estimated the unconfined compressive strength to be on the order of 1.5 x 10^5 N/m. Kovacs et al. [53] have pursued the determination of yield strength and estimated the ice cover on this scale to have an internal friction of 30 degrees by considering the wedge-shaped protrusions of ice around Fairway Rock, which is southeast of the Diomede Islands in the Bering Strait. This approach is attractive because it allows a direct evaluation of one parameter, the angle of internal friction, in terms of a directly measurable quantity, the location of a slip line in the plasticity solution. Pritchard and Reimer [56] have determined the mathematical characteristics associated with the two-dimensional plasticity models; these give a means for interpreting lead directions in terms of slip lines as suggested by Marco and Thomson [57,58]. Sodhi [51] and Kovacs et al. [53] use results of a soil model adapted for sea ice, which provided consistent results. Studies by R. T. Hall (personal communication) have attempted a similar approach using large-scale lead patterns observed in the Beaufort Sea, but poor correlations between characteristic directions and lead directions were found.

2.4 Baltic Sea

Year-round shipping in the Baltic Sea has led to an active effort to develop a model to predict the effect of an ice cover on icebreaker performance. Field experiments carried out by both Finnish and Swedish investigators as early as 1974 have provided a comprehensive set of data on ice conditions, ice motions, winds and currents (see, for example, Blomquist et al. [59]; Udin and Omstedt, [60]; Omstedt and Sahlberg [61]). Mathematical models of ice motion in the Bay of Bothnia have been developed and used in Finland (Valli and Leppäranta [62]; Leppäranta [8]) and in Sweden (Udin and Ullerstig [63]; Udin and Omstedt [60]; Omstedt [64]). Research to develop the models has continued. An outstanding description of ice mechanics and modeling of sea ice behavior in the Bay of Bothnia is given by Leppäranta [8]. This comprehensive study discusses both observed ice behavior and models used to predict short-term ice motions and conditions.

Leppäranta [8] uses the thickness distribution concept to characterize ice conditions. He separates the thickness distribution into flat ice and a distribution of ridged ice. This allows the shape of a ridge sail and keel to be introduced simply and ridging intensity to be used as a means of observing the result of ridging. Nondimensional variables are defined to estimate the importance of each term in momentum balance, inertial and Coriolis acceleration, air and water drag, sea surface tilt and ice stress divergence. The important dimensionless groups are expressed as ratios of each force to the inertial force, a common technique in fluid mechanics. The stress-deformation law is assumed as a viscous law, an unfortunate shortcoming in light of recent success with plasticity models. The mechanical energy budget in the Bay of Bothnia has also been derived and compared with observed values. However, no attempt appears to have been made to use this relationship to estimate mesoscale ice strength or other properties. Hibler et al. [65,66] have presented a similar model that is under development for use by the Swedish Meteorological and Hydrological Institute for short-term ice forecasts. This is an application of the macroscale model of Hibler that uses the viscous-plastic constitutive law [3,4,32,43].

Leppäranta [67] has questioned captains of the icebreakers that operate in the Baltic during winter to evaluate the accuracy and usefulness of a model for predicting icebreaker performance in this area. These individuals gave a somewhat mixed response, but agreed that the model was useful. While this evaluation of model performance is only qualitative, it is nevertheless a critical test of model performance, i.e., whether or not it helps those who must operate in the sea ice cover.

2.5 Bering and Greenland Seas

Discussions of modeling efforts for these two regions are combined because the ice dynamics processes are relatively similar. The essential similarities are that an ice edge is present, tracking its position is necessary in most applications, and ice is usually thin enough that internal ice stress divergence is not the dominant force in the momentum balance. However, ice stress must be included in any model used to predict ice edge motion because onshore ice motions compact the ice and, thus, make ice stress an important factor. The set of processes that affects ice motion and deformation differs under the compact and dispersed ice conditions that accompany onshore and offshore motions. The plasticity models developed for Beaufort and Chukchi Sea ice behavior are applicable for describing compact ice behavior here. The plasticity models automatically produce zero stress levels in dispersed ice if a thickness distribution is included to describe the ice conditions. Therefore, they are immediately adaptable to the thinner ice conditions occurring in these subarctic waters, and they provide the best available simulations of ice behavior. However, these models lack a tuning of the constitutive laws and a description of the more complex oceanographic features and their effects on the ice.

The essential processes controlling ice behavior in the Bering Sea have been categorized by Pease [68] from data taken in March 1979. During that time, ice advected southward in a "conveyer belt," melting at the southern edge and growing in polynyas adjacent to the northern land masses by freezing. Thomas and Pritchard [69] deployed buoys on the ice during the winter of 1980-81 and found that, while the "conveyer belt" is active at times, northward motions also occur frequently. The northward motions persisted during this winter so that ice was advected northward through the Bering Strait into the northern Chukchi Sea.

A comprehensive field program is sheduled for the 1982-83 ice season as part of MIZEX [70]. The data will include ice motions and deformations, winds, currents, and ice conditions, as well as other meteorological and oceanographic observations near the ice edge. Some ice motion observations are planned within the ice pack to allow comparison with motions observed near the ice edge. A conceptual model of forces applied to the ice has been presented by Muench et al. [71] to describe forces applied by wind and current drag, wave radiation and internal wave drag. However, little effort has been focused on modeling the physical processes governing ice behavior or developing the constitutive laws relating deformations to ice stress and ice conditions. Several different mechanisms have been offered to explain the bands, streaks and patches of ice that occur during off-ice winds. Wadhams [72] and Martin et al. [73] have shown how forces due to wave radiation separate ice floes that already are isolated from the pack and cause them to accelerate away from the pack. McPhee [10] describes how the melting of ice at the ice edge reduces the ocean drag coefficient and allows higher floe velocities. Together these two mechanisms could explain how bands and streaks occur.

Tucker and Hibler [74] and Tucker [75] have simulated ice behavior east of Greenland for October and November 1979 using the viscous-plastic model discussed earlier [32,76]. The thickness distribution included two categories of ice, open water and thick ice; therefore, compactness and average ice thickness were simulated. Data provided by two buoys that had drifted through this region during the simulation period allowed simulated ice velocities to be compared with actual velocities. The average velocity error vector was about 19 cm/s and the standard deviation of daily velocity errors was 18 cm/s. Correlation coefficients for each velocity component for the two buoys ranged from 0.48 to 0.57. The ice thickness and ice edge motion were reasonable, but excessive. Inaccurate thermal growth rates were felt to be at fault. Since this was a first, rough calculation to determine which processes are most important, results were judged acceptable.

3. COMPUTATIONAL CONSIDERATIONS

Either a Lagrangian or an Eulerian description is used in sea ice dynamics modeling, depending upon the application. Both approaches provide the ability to describe ice velocity and stress fields.

The Lagrangian description refers each grid location to an initial particle position. It is most frequently used to describe the behavior of solid materials that are strained up to perhaps 50 to 100 percent, but no more. An advantage is that individual ice floes may be tracked simply. A disadvantage is that grids can become too distorted over long periods of time. The Lagrangian approach is most useful for simulations of fairly short duration, say 5 to 10 days, where total strains do not exceed the usual 50 to 100 percent. When observed ice motions are prescribed as boundary conditions, the Lagrangian description directly accepts these data as boundary conditions (e.g., Pritchard [39]).

The Eulerian description refers each grid location to a fixed point in space. It is most frequently used to describe the behavior of fluids. An advantage with this approach is that, since the grid does not distort, arbitrarily large strains may be simulated. A disadvantage is that numerical dispersion occurs as the ice cover moves through the grid. The Eulerian description is most useful when long-term simulations are needed. An example is the Arctic Basin simulation of Hibler [32] in which yearlong ice motions are calculated.

Since mathematical relationships relate time derivatives in either description, the two approaches are essentially the same. The advection through the fixed Eulerian grid adds terms to each of the equations for conservation of mass, momentum, and energy. However, for ice dynamics, the magnitude of these advective terms is negligible in almost all cases (Hibler et al. [66]; Leppäranta [8]). Therefore, they may be neglected for determining instantaneous solutions and only need to be considered when velocities are accumulated to determine long-term displacements.

Finite difference schemes were developed for early ice dynamics modeling efforts (Pritchard and Colony [77]; Hibler [78]). More recently, finite element schemes have been developed (Leavitt et al. [48]; McKenna [44]). Both methods have potential benefits. The most valuable feature of the finite element formulation is its flexibility. This method allows arbitrary connectivity of elements, simple specification of boundary tractions, and a variety of time integrators. The use of finite element formulations does, however, increase computer costs, which sometimes makes finite difference formulations more desirable.

Implicit or explicit time-stepping procedures may be used with either finite element or finite difference schemes. The time step is restricted for explicit schemes, with the size depending on either the elastic moduli or viscosities. Reasonable estimates of the time step limits are:

$$\text{Elastic} \qquad \Delta t < \frac{1}{2} \frac{\Delta x}{(M/\rho h)^{1/2}}$$

$$\text{Viscous} \qquad \Delta t < \frac{1}{2} \frac{\Delta x^2}{(\nu/\rho h)}$$

where

- Δt is time step (s),
- Δx is roughly the smallest grid spacing (m),
- ρh is mass density times thickness (kg/m^2),
- M is elastic modulus, which must exceed 200 times the ice strength if elastic strain is limited to 0.5 percent, and
- ν is kinematic viscosity, which must exceed 2×10^7 s times the ice strength if viscous strain is limited to 0.5×10^{-7} s^{-1} (about 0.5 percent per day).

The explicit numerical scheme of Pritchard and Colony [77] requires time steps as small as a few minutes for winter simulations in the Beaufort Sea. This scheme is cost-effective, though, because each time step requires little computer time. Solutions at each node are found independently of surrounding nodes at each step, so no large systems of nonlinear equations must be solved. Under similar conditions, the explicit viscous law requires time steps of only a few seconds, and in just about all realistic situations, the viscous time step is smaller than the elastic time step. Therefore, explicit numerical schemes, while effective for elastic-plastic models, do not appear to be useful for viscous-plastic models.

For useful implicit schemes, the time step is not restricted, except that it must be small enough to resolve the physical processes. This is often about 1 day. However, since solutions at all nodes are determined simultaneously, large systems of simultaneous nonlinear equations must be solved. The number of computer operations needed to determine the solution at one time step, therefore, is far greater for an implicit scheme than for an explicit scheme. The viscous-plastic constitutive law of Hibler [78] is attractive because it allows the easy use of an implicit finite difference scheme. The treatment of advection requires that the time step be limited by a Courant condition

$$\Delta t < \Delta x / |2\tilde{v}| ,$$

where $\tilde{v}$ is the maximum ice velocity. Time-step limitations are about 0.25 day for the same parameters used to estimate time-step limitations for explicit schemes.

In applying the finite element method, most codes use matrix notation and store all solutions in arrays. It is natural, then, to formulate a set of simultaneous equations whether an explicit or implicit time integrator is used. Therefore, most finite element codes offer both explicit and implicit time integrators.

4. MODEL COMPONENTS AND FUTURE EXTENSIONS

The physical behavior of sea ice is described by accounting for mass, momentum and energy balance. In addition, the constitutive laws defining thermal growth and relating stress to deformation and to the redistribution of ice must be given. The equations describing a specific model are not given here; general formulations can be found in Rothrock [2], Hibler [3,4], Leppäranta [8] and Pritchard [39]. Different levels of detail are presented in these works for the various components of the models. However, in total they provide a rather comprehensive view of conservation laws and constitutive laws, as well as descriptions of the important physical processes described by the models. In this paper, brief verbal descriptions are given for each component of a hypothetical model that reflects the basic physical processes that occur in sea ice. The model is generic in the sense that the different levels of importance of various parameters in different regions are ignored.

Ice conditions can be described by the thickness distribution or by a simpler function that describes only a few of the thickness categories. It is necessary, however, to include open water and thin ice in this function because these fractions strongly influence ice strength and vertical heat flux. Mass conservation is satisfied by describing the evolution of the thickness distribution, i.e., by accounting for mechanical redistribution of ice between categories as the ice cover opens, rafts and ridges and by accounting for thermal growth and ablation. Mechanical redistribution is assumed to depend linearly on the magnitude of plastic stretching so that the evolution of the thickness distribution is rate independent. The fraction of open water is specified for all stretching states, including both dilatation and shear. The ridging function that describes the distribution of categories into which ice is rafted and ridged is assumed. The effects of thermal growth and ablation appear in the thermal growth rates for each ice category. Thermal growth and ablation also redistribute ice as its thickness changes. Either a fixed growth rate table or an active model of the heat budget may be used.

Although the thickness distribution provides an excellent framework for describing ice conditions, our knowledge of ice redistribution is inadequate. Data are needed to estimate the fraction of open water formed in shearing, the fraction of ice participating in redistribution and the range of thicknesses into which ice is ridged. These, in fact, are all of the constitutive relationships defining redistribution, and all are inadequately understood.

The effects of velocity inhomogeneities should also be included in the equation describing the evolution of the thickness distribution. A relationship that determines part of the thickness distribution in terms of a statistical measure of the velocity inhomogeneity should be reasonable, and it should describe better the effects of this subgrid-scale process than do present models. Although these velocity inhomogeneities cannot be resolved in a deterministic way, their effects can be taken into account statistically. They generate leads, the leads freeze and thin ice results. This process alters the thickness distribution as do the rafting and ridging processes. Primarily, it is the thin ice category in the thickness distribution that is altered. Any change in the fraction of thin ice affects the ice strength, which in turn affects the large-scale stresses.

Thermal growth rates are fairly well understood, but only because thermal growth is less important in the short-term problems usually addressed in mesoscale modeling. However, at an ice edge, where freezing can cause the ice cover to extend hundreds of kilometers in a few days, better models are required. Here, the heat budget of the upper ocean will likely be necessary.

The forces exerted on the ice cover are from winds, currents, waves, sea surface tilt, and internal ice stress. Acceleration must account for both inertial and Coriolis components. The relative importance of each force and the accelerations differ with time scale and location. The conservation of horizontal momentum provides the equation relating forces and accelerations. The force from wind is described as a quadratic law. The force from currents is also included as a quadratic function of the currents relative to the ice, though it can be a more complicated function of this relative velocity, especially in shallow waters where bottom effects are important. Wave radiation forces and internal wave drag also influence ice floe motions at an ice edge and the motion of the edge itself. Incoming waves break the ice floes in addition to adding to the water drag. Furthermore, when off-ice winds form bands, streaks and patches, wave radiation in the leads accelerates the ice floes and disperses the ice. Sea surface tilt may be prescribed from the known dynamic topography or from the geostrophic current if appropriate. Tidal currents can in some locations add to the short-term motions. They can be added to the currents or can be included in an active dynamical ocean model. The inertia of the ice is negligible when averaged over a day. However, on shorter time scales, significant inertial oscillations can be generated by the balance of inertial and Coriolis accelerations. On these shorter time scales, however, the inertia of the water column becomes just as important and should be included along with the inertia of the ice.

Although the forces acting on sea ice have been described, we are not yet able to estimate them accurately. For example, our knowledge of current drag is improving, but there remains a need for extending the model to treat large-scale currents in a shallow sea. Also, even though wave radiation forces can be estimated, the effect on ice dispersion must be determined. It is likely that this effect will be described as a subgrid-scale process and not calculated directly from the divergence of the ice velocity field. The problem of sea ice behavior near an edge will rightly receive extensive study soon.

Mechanical energy is typically input by the wind and transferred into the ocean by the current drag accompanied by ice motion. In addition, energy is input by wave radiation and sea surface tilt. The kinetic energy of the ice cover is modified as ice velocity changes under the influence of these forces. The internal ice stress dissipates part of this energy through deformation. In addition, the stress flux divergence transfers energy horizontally through the ice. Effects of ice stress divergence are therefore felt long distances away from locations at which energy is input by other forces acting on the ice.

In ice dynamics models, the strength of the ice cover may be determined by assuming that the work done by the ice stress during deformations is equal to the energy dissipated by small-scale processes. But only two small scale sinks have been included in the mechanical energy budget to date: gravitational potential energy and frictional sliding. A shear sink has been shown to improve model performance, but the mechanism has not yet been analyzed. Also, since energy dissipated by ridging correlates well with observed background noise, other energy sinks such as fracture of the ice sheet should be reexamined. While too small to affect ice motion, energy released during fracture might explain more of the noise. Finally, other energy sinks, especially dissipation by velocity inhomogeneities, should be estimated and included in the mechanical energy budget.

By studying the processes that control ice behavior, investigators have generally concluded that plasticity offers the best mathematical model for describing macroscale and mesoscale sea ice behavior. This agreement has not been quick in coming, but its universal acceptance now seems apparent. The constitutive law relating stress to deformation is assumed to be elastic-plastic or viscous-plastic. Both are capable of describing the wide range of strain rates observed in the ice cover without changing material parameters in a nonphysical way. A variety of yield surface shapes are available. Yield surfaces are usually isotropic, allow no tensile stress (or only negligible values) and have a maximum compressive strength determined by failure in rafting and ridging during convergence. An associated flow rule is assumed in all models. A nonzero value of uniaxial compressive strength is essential to describe arching across narrow openings between land masses such as straits.

The evidence for selecting a yield surface and flow rule come from theoretical reasoning and from evaluating model performance in ice dynamics simulations. In a few special cases, individual parameters, namely, the isotropic and uniaxial compressive strengths, have been inferred under conditions where winds and currents were known and the ice was stationary. Many more special cases are needed to confirm these data and to estimate other parameters. What would be ideal is a set of cases that allows isolation of individual parameters rather than complicated simulations. Mathematical characteristics of the model offer some potential here.

5. CONCLUSION

The past decade has seen major advances in sea ice dynamics modeling capabilities, largely because of the concentrated efforts of the government-sponsored AIDJEX program. However, the past few years have shown far less coordination in funding of sea ice modeling research by the U.S. government. On the other hand, Canadian, Finnish and Swedish governments and the U.S. petroleum industry have increased their interest and funding. During this time several new groups have developed computer programs to solve a range of specific problems. While our understanding of the basic physics involved in individual ice processes has increased, this information has not yet found its way into the ice dynamics models. A few attempts have been made to validate the models by quantitative comparison with observations, but many more quantitative comparisons are needed. It is good that more knowledgeable scientists and engineers have become involved in arctic research, and there is an increasing need for both basic scientific knowledge and practical application of this knowledge. Hopefully, then, if these past few years are to prove fruitful, we are poised for some new ideas and some basic progress. If this paper helps to give perspective to our present capabilities and identifies some of the effort needed to make this progress, then it will have served its purpose.

REFERENCES

1. Rothrock, D. A., The Mechanical Behavior of Pack Ice, in Annual Review of Earth and Planetary Sciences, vol. 3, pp. 317-342, 1975.
2. Rothrock, D. A., Modeling Sea-ice Features and Processes, J. Glac., vol. 24, no. 90, pp. 359-375, 1979.
3. Hibler, W. D. III, Sea Ice Growth, Drift, and Decay, in Dynamics of Snow and Ice Masses, ed. S. C. Colbeck, pp. 141-209, Academic Press, New York, 1980.
4. Hibler, W. D. III, Numerical Modeling of Sea Ice in the Seasonal Sea Ice Zone, Cold Regions Sci. and Technol., vol. 2, pp. 299-321, 1980.
5. Nansen, F., The Oceanography of the North Polar Basin, The Norwegian North Polar Expedition 1893-1896, Scientific Results, vol. 3, pp. 1-427, 1902.
6. Zubov, N. N., L'dy Arktiki (Arctic Ice), Izdatel'stvo Glavsevmorputi, Moscow, 1945 (English translation by U. S. Naval Oceanographic Office and American Meteorological Society, San Diego, 1963).
7. Campbell, W. J., The Wind-Driven Circulation of Ice and Water in a Polar Ocean, J. Geophys. Res., vol. 70, no. 14, pp. 3279-3301, 1965.
8. Leppäranta, M., On the Structure and Mechanics of Pack Ice in the Bothnian Bay, Finnish Marine Research No. 248, Helsinki, pp. 3-86, 1981.
9. McPhee, M. G., An Analysis of Pack Ice Drift in Summer, in Sea Ice Processes and Models, ed. R. S. Pritchard, pp. 62-75, University of Washington Press, Seattle, 1980.
10. McPhee, M. G., Sea Ice Drag Laws and Simple Boundary Layer Concepts, Including Application to Rapid Melting, Cold Regions Research and Engineering Laboratory Report No. 82-4, Hanover, NH, 1982.

11. Overland, J. E., and Pease, C. H., Wind-Driven Ice Dynamics in a Shallow Sea, J. Geophys. Res., vol. 88 (in press), 1983.
12. McPhee, M. G., The Turbulent Boundary Layer in Shallow Water, Appendic C in Annual Report on R.U.567 to Outer Continental Shelf Environmental Assessment Program by D. R. Thomas and R. S. Pritchard, Flow Industries, Inc., Kent, Wash., 1982.
13. Pease, C. H., Salo, S. A., and Overland, J. E., Drag Measurements for First-Year Sea Ice Over a Shallow Sea, J. Geophys. Res., vol. 88 (in press), 1983.
14. Liu, S. K., and Leendertse, J. J., A 3-D Oil Spill Model With and Without Ice Cover, Proc. Int. Symp. on Mechanics of Oil Slicks, pp. 249-265, International Association for Hydraulic Research, 1981.
15. Thomas, D. R., Oiled Ice Trajectories in the Beaufort Sea, RTD Report No. 252, Flow Industries, Inc., Kent, Wash., 1983.
16. Colony, R., and Thorndike, A. S., The Horizontal Coherency of the Motion of Summer Arctic Sea Ice, J. Phys. Oceanog., vol. 10, pp. 1281-1289, 1980.
17. Thorndike, A. S., and Colony, R., Sea Ice Motion in Response to Geostrophic Winds, J. Geophys. Res., vol. 87, no. C7, pp. 5845-5852, 1982.
18. Thomas, D. R., and Pritchard, R. S., Beaufort and Chukchi Sea Ice Motion, Part 1. Pack Ice Trajectories, Flow Research Report No. 133, Flow Industries, Inc., Kent, Wash., 1979.
19. Pritchard, R. S., Beaufort Sea Ice Motion, in The Alaskan Beaufort Sea, eds. P. Barnes, E. Reimnitz and D. Schell, to appear, Academic Press, New York, 1983.
20. Sater, J. E., Walsh, J. E., and Wittmann, W. I., Impingement of Sea Ice on the North Coast of Alaska, in The Coast and Shelf of the Beaufort Sea, eds. J. C. Reed and J. E. Sater, pp. 85-105, Arctic Institute of North America, Arlington, VA, 1974.
21. Pritchard, R. S., and Thomas, D. R., The Range of Influence of Boundary Parameters in the AIDJEX Model, in Sea Ice Processes and Models, ed. R. S. Pritchard, pp. 113-123, University of Washington Press, Seattle, 1980.
22. McPhee, M. G., A Study of Oceanic Boundary-Layer Characteristics Including Inertial Oscillations at Three Drifting Stations in the Arctic Ocean, J. Phys. Oceanog., vol. 10, pp. 870-884, 1980.
23. Nye, J. F., and Thomas, D. R., The Use of Satellite Photographs to Measure the Movement and Deformation of Sea Ice, AIDJEX Bulletin No. 27, pp. 1-22, University of Washington, Seattle, 1974.
24. Hibler, W. D., Weeks, W. F., Kovacs, A., and Ackley, S. F., Differential Sea-Ice Drift, I. Spatial and Temporal Variations in Sea-Ice Deformation, J. Glac., vol. 13, no. 69, pp. 437-455, 1974.
25. Thorndike, A. S., and Colony, R., Estimating the Deformation of Sea Ice, in POAC77, vol. 1, pp. 506-517, Memorial University of Newfoundland, St. Johns, 1977.
26. Untersteiner, N., AIDJEX Review, in Sea Ice Processes and Models, ed. R. S. Pritchard, pp. 3-11, University of Washington Press, Seattle, 1980.
27. Coon, M. D., A Review of AIDJEX Modeling, in Sea Ice Processes and Models, ed. R. S. Pritchard, pp. 12-27, University of Washington Press, Seattle, 1980.
28. Thorndike, A. S., Rothrock, D. A., Maykut, G. A., and Colony, R., The Thickness Distribution of Sea Ice, J. Geophys. Res., vol. 80, no. 33, pp. 4501-4513, 1975.
29. Maykut, G. A., Energy Exchange Over Young Sea Ice in the Central Arctic, J. Geophys. Res., vol. 83, no. C7, pp. 3646-3658, 1978.
30. Maykut, G. A., Large-Scale Heat Exchange and Ice Production in the Central Arctic, J. Geophys. Res., vol. 87, no. C10, pp. 7971-7984, 1982.
31. Coon, M. D., Maykut, G. A., Pritchard, R. S., Rothrock, D. A., and Thorndike, A. S., Modeling the Pack Ice as an Elastic-Plastic Material, AIDJEX Bulletin No. 24, University of Washington, Seattle, pp. 1-105, 1974.
32. Hibler, W. D. III, Modeling a Variable Thickness Ice Cover, Mon. Wea. Rev., vol. 108, no. 12, pp. 1943-1973, 1980.
33. Coon, M. D., Mechanical Behavior of Compacted Ice Floes, J. Petroleum Tech. Trans., vol. 257, pp. 466-470, 1974.
34. Pritchard, R. S., An Elastic-Plastic Constitutive Law for Sea Ice, J. Appl. Mech., vol. 42, series E, no. 2, pp. 379-384, 1975.
35. Rothrock, D. A., The Energetics of the Plastic Deformation of Pack Ice by Ridging, J. Geophys. Res., vol. 80, no. 33, pp. 4514-4519, 1975.
36. Parmerter, R. R., and Coon, M. D., Model of Pressure Ridge Formation in Sea Ice, J. Geophys. Res., vol. 77, no. 33, pp. 6565-6575, 1972.
37. Pritchard, R. S., The Effect of Strength on Simulations of Sea Ice Dynamics, in POAC77, vol. I, pp. 494-505, Memorial University of Newfoundland, St. Johns, 1977.
38. Bugden, G. L., The Deformation of Pack Ice by Ridging, J. Geophys. Res., vol. 84, no. C4, pp. 1793-1795, 1979.
39. Pritchard, R. S., Mechanical Behavior of Pack Ice, in Mechanics of Structured Media, part A, ed. A. P. S. Selvadurai, pp. 371-405, Elsevier, Amsterdam, 1981.
40. Pritchard, R. S., A Simulation of Nearshore Winter Ice Dynamics in the Beaufort Sea, in Sea Ice Processes and Models, ed. R. S. Pritchard, pp. 49-61, University of Washington Press, Seattle, 1980.
41. Rothrock, D. A., Colony, R., and Thorndike, A. S., Testing Pack Ice Constitutive Laws with Stress Divergence Measurements, in Sea Ice Processes and Models, ed. R. S. Pritchard, pp. 102-112, University of Washington Press, Seattle, 1980.
42. Kolle, J. J., and Pritchard, R. S., A Comparison of Two Sea Ice Trajectory Models with AIDJEX Observations, Proc. 6th Annual Energy-Sources Tech. Conf. and Exhibition (in press), Houston, 1983.
43. Hibler, W. D. III, Modeling Pack Ice as a Viscous-Plastic Continuum, Some Preliminary Results, in Sea Ice Processes and Models, ed. R. S. Pritchard, pp. 163-176, University of Washington Press, Seattle, 1980.
44. McKenna, R. F., The Influence of Changes in Constitutive Law on the Dynamics of Sea Ice, masters thesis, University of Waterloo, Waterloo, Ontario, 1982.
45. Coon, M. D., and Pritchard, R. S., Mechanical Energy Considerations in Sea-Ice Dynamics, J. Glac., vol. 24, no. 90, pp. 377-389, 1979.
46. Pritchard, R. S., Coon, M. D., and Thomas, D. R., Modeling the Mechanical Energy Budget of the Beaufort Sea Ice Cover, Report No. 137R, Flow Industries, Inc., Kent, Wash., 1979.
47. Pritchard, R. S., Arctic Ocean Background Noise Caused by Ridging of Sea Ice, RTD Report No. 254, Flow Industries, Inc., Kent, Wash., 1983.
48. Leavitt, E., Sykes, J., and Wong, T. T., A Sea Ice Model Developed for Use in a Real Time Forecast System, in POAC81, vol. III,
49. Pritchard, R. S., and Coon, M. D., Canadian Beaufort Sea Ice Characterization, in POAC81, vol. III, pp. 609-618, Université Laval, Quebec, 1981.
50. Shapiro, L. H., and Burns, J. J., Satellite Observations of Sea Ice Movement in the Bering Strait

Region, in Climate of the Arctic, eds. G. Weller and S. A. Bowling, pp. 379-386, Geophysical Institute, University of Alaska, Fairbanks, 1975.

51. Sodhi, D. S., Ice Arching and the Drift of Pack Ice Through Restricted Channels, Cold Regions Research and Engineering Laboratory Report No. 77-18, Hanover, NH, 1977.

52. Pritchard, R. S., Reimer, R. W., and Coon, M. D., Ice Flow Through Straits, in POAC79, vol. 3, pp. 61-74, Norwegian Institute of Technology, Trondheim, 1979.

53. Kovacs, A., Sodhi, D. S., and Cox, G. F. N., Bering Strait Sea Ice and the Fairway Rock Ice-foot, Cold Regions Research and Engineering Laboratory Report No. 82-31, Hanover, NH, 1982.

54. Reimer, R. W., Pritchard, R. S., and Coon, M. D., Beaufort and Chukchi Sea Ice Motion, Part 2. Onset of Large Scale Chukchi Sea Ice Breakout, Report No. 133, Flow Industries, Inc., Kent, Wash., 1979.

55. Reimer, R. W., Schedvin, J. C., and Pritchard, R. S., Chukchi Sea Ice Motion, in POAC81, vol. II, pp. 1038-1046, Université Laval, Quebec, 1981.

56. Pritchard, R. S., and Reimer, R. W., Mathematical Characteristics of a Plastic Model of Sea Ice Dynamics, AIDJEX Bulletin No. 40, pp. 109-151, University of Washington, Seattle, 1978.

57. Marco, J. R. and Thomson, R. E., Spatially Periodic Lead Patterns in the Canada Basin Sea Ice: A Possible Relationship to Planetary Waves, Geophys. Res. Letters, vol. 2, no. 10, pp. 431-434, 1975.

58. Marco, J. R., and Thomson, R. E., Rectilinear Leads and Internal Motions in the Pack Ice of the Western Arctic Ocean, J. Geophys. Res., vol. 82, no. 6, pp. 979-987, 1977.

59. Blomquist, A., Pilo, C. and Thompson, T., Sea Ice-75 Summary Report, Research Report No. 16.9, Winter Navigation Research Board, Swedish Administration of Shipping and Navigation and Finnish Board of Navigation, Stockholm/Norrkoping, 1976.

60. Udin, I. and Omstedt, A., Sea Ice-75 Dynamical Report, Research Report No. 16.8, Winter Navigation Research Board, Swedish Administration of Shipping and Navigation, Finnish Board of Navigation, Stockholm/Norrkoping, 1976.

61. Omstedt, A., and Sahlberg, J., Some Results from a Joint Swedish-Finnish Sea Ice Experiment, March 1977, Swedish Meteorological and Hydrological Institute, Report No. 10, Norrköping, 1978.

62. Valli, A., and Leppäranta, M., Calculation of Ice Drift in the Bothnian Bay and the Quark, Report No. 13, Winter Navigation Research Board, Helsinki, 1975.

63. Udin, I., and Ullerstig, A., A Numerical Model for Forecasting the Ice Motion in the Bay and Sea of Bothnia, Swedish Administration of Shipping and Navigation and Finnish Board of Navigation, Report No. 18, Norrkoping, 1976.

64. Omstedt, A., A Sensitivity Analysis of Steady, Free Floating Ice, Swedish Administration of Shipping and Navigation and Finnish Board of Navigation, Report No. 19, Norrkoping, 1980.

65. Hibler, W. D. III, Udin, I., and Ullerstig, A., On Modeling Mesoscale Ice Dynamics Using a Viscous Plastic Constitutive Law, in POAC81, vol. III, pp. 1317-1329, Université Laval, Quebec, 1981

66. Hibler, W. D. III, Udin, I. and Ullerstig, A., On Forecasting Mesoscale Ice Dynamics and Buildup, Annals of Glaciology, vol. 4 (in press), International Glaciological Society, Cambridge, England, 1983.

67. Leppäranta, M., An Ice Drift Model for the Baltic Sea, Tellus, vol. 33, no. 6, pp. 583-596, 1981.

68. Pease, C. H., Eastern Bering Sea Processes, Mon. Wea. Rev., vol. 108, no. 12, 1980.

69. Thomas, D. R., and Pritchard, R. S., Norton Sound and Bering Sea Ice Motion: 1981, Report No. 209, Flow Industries, Inc., Kent, Wash., 1981.

70. Wadhams, P., Martin, S., Johannesen, O. M., Hibler, W. D. III, and Campbell, W. J. (eds.) MIZEX. A Program for Mesoscale Air-Ice Ocean Interaction Experiments in Arctic Marginal Ice Zones, Government Printing Office, 701-831/144, 1981.

71. Muench, R. D., Stegen, G. R., Hachmeister, L. E., and Martin, S., Prediction of Ice Distribution and Movement in the Outer Marginal Ice Zone, to appear in POAC83, Technical Research Centre of Finland, Helsinki, 1983.

72. Wadhams, P., A Mechanism for the Formation of Ice Edge Bands, J. Geophys. Res., vol. 88 (in press), 1983.

73. Martin, S., Kauffman, P., and Parkinson, C., The Movement and Decay of Ice Edge Bands in the Winter Bering Sea, J. Geophys. Res., vol. 88 (in press), 1983.

74. Tucker, W. B. III, and Hibler, W. D. III, Preliminary Results of Ice Modeling in the Greenland Area, in POAC81, vol. III, pp. 867-878, Université Laval, Quebec, 1981.

75. Tucker, W. B. III, Application of a Numerical Sea Ice Model to the East Greenland Area, Cold Regions Research and Engineering Laboratory Report No. 82-16, Hanover, NH, 1982.

76. Hibler, W. D. III, A Dynamic Thermodynamic Sea Ice Model, J. Phys. Oceanog., vol. 9, pp. 815-846, 1979.

77. Pritchard, R. S., and Colony, R., A Difference Scheme for the AIDJEX Sea Ice Model, in Numerical Methods in Geomechanics, vol. II, ed. C. S. Desai, pp. 1194-1209, American Society of Civil Engineers, New York, 1976.

78. Hibler, W. D. III, Documentation for a Two-Level Dynamic Thermodynamic Sea Ice Model, Special Report No. 80-8, Cold Regions Research and Engineering Laboratory, Hanover, NH, 1980.

MECHANICAL PROPERTIES OF ICE IN THE ARCTIC SEAS

W.F. Weeks and M. Mellor
U.S. Army Cold Regions Research and Engineering Laboratory
Hanover, New Hampshire 03755 USA

SUMMARY

The mechanical properties are reviewed for the main types of ice in arctic seas [glacial (icebergs), shelf (ice islands), sea ice] and representative values are given. Each ice type possesses a characteristic range of structures and compositions that differentiate it from other varieties of ice and to a considerable extent, these produce large variations in mechanical properties. Factors affecting mechanical properties (temperature, brine and gas volume, crystal orientation and size, strain rate) are discussed, as are gaps, contradictions, and inadequacies in available data.

1. INTRODUCTION

Even pure ice displays complicated mechanical properties, largely because it exists in nature at high homologous temperatures, commonly above 0.95 and almost always above 0.90. When deformed at high strain rates or loaded for brief periods, it behaves elastically. By contrast, when strained slowly or when subjected to sustained loadings ice is ductile, and it can creep to large strains without breaking. At any given stage in such a creep process, the relation betwen strain-rate and stress is strongly nonlinear, i.e. ice is visco-plastic rather than linearly viscous. Because ice properties are highly sensitive to strain rate and temperature, strength can vary greatly. Furthermore, the general effects of multiaxial stress states, as represented by failure criteria, also change drastically with changes in temperature and strain-rate.

It is not our purpose to discuss the idealized properties of bubble-free, fine-grained, randomly oriented, pure ice in this paper. Instead we shall discuss the more complicated ice that occurs naturally in the sea. The sources of these materials are highly varied, ranging from ice sheets and valley glaciers (icebergs), ice shelves (ice islands), rivers and lakes (freshwater ice), and from the freezing of the sea itself (sea ice). This last material is the predominant ice type in the seas of the Arctic and it comes in a variety of types, each with its own characteristic association of grain sizes, crystal orientations, and gas and brine inclusions. We shall discuss briefly how each type develops, its internal structure and its associated mechanical properties. To do this fully is a task far beyond the present limitations of time and space. Here we simply attempt to provide a balanced, general feel for the current state of this subject and a listing of some of the more useful references. In doing this we draw heavily on more exhaustive reviews that have already been published [1-7].

Current interest in the properties of ice in the sea is neither the result of such ice being an ideal material for study, nor of the desire of materials engineers to spend their spare moments visiting the arctic ice pack, with its mobile scenery and delightful climate. Ice in the sea is the primary obstacle to effective and safe removal of the presumed large oil and gas reserves of the continental shelves of the Arctic, as well as a barrier to development of new sea routes across the Pole that would result in great changes in current patterns of marine commerce. To overcome this barrier, it is essential that engineers understand both the behavior of ice in the sea as well as its pertinent properties.

2. STRUCTURE AND COMPOSITION OF ICE IN THE SEA

To understand the mechanical behavior of the varieties of ice that occur in arctic seas, one should first understand something about the structure and composition of these materials and also of pure ice. We now briefly review this subject, starting with pure ice, then glacier (iceberg) ice, then sea ice, and finally shelf (ice island) ice thereby moving from simple to complex structures and compositions. Most attention will be given to sea ice, in that it is the most important ice type in the majority of arctic marine areas. River and lake ice will not be discussed as they are not of major importance in most marine areas.

2.1 Ice Ih

Although there are several polymorphs of ice [8], ice Ih (so-called ordinary ice) is the only one of these that exists in significant quantities under the physical conditions encountered at the earth's surface. In fact, ice Ih (which will be referred to simply as ice) is the stable polymorph even at the bottom of the world's thickest ice sheets. Ice is unusual in comparison to most materials in that the solid phase is less dense than the liquid phase. Therefore, ice floats, forming a cover over the seas, lakes, and rivers of the Arctic, causing a variety of engineering and operational problems that have largely inspired this meeting.

The general atomic structure of ice is well understood as ice Ih was one of the first substances to have its structure determined [9]. Each oxygen atom is located at the center of a tetrahedron with four other oxygen atoms located at each of the apices

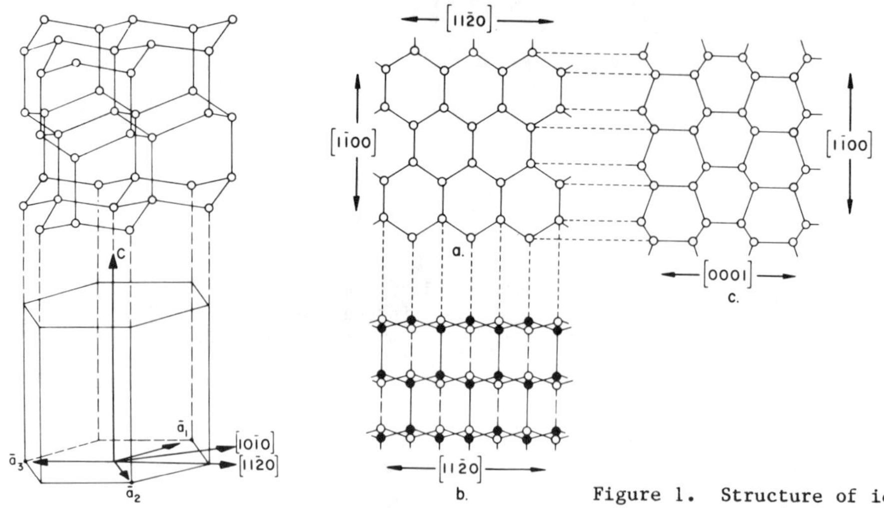

Figure 1. Structure of ice I.

(Figure 1). The tetrahedral coordination of the oxygen atoms produces an open, low density crystal structure with hexagonal symmetry. One important feature of this structure is that the oxygen atoms are concentrated close to a series of parallel planes, referred to as the basal planes. The direction perpendicular to these planes is the principal hexagonal axis, or c-axis. The arrangement is such that, in any unit cell which contains four oxygen atoms, fracture along the basal or (0001) plane involves the rupture of only two bonds, while fracture along any plane normal to this plane requires the rupture of at least four bonds. Therefore the observation that ice glides and cleaves readily on the basal plane can easily be explained in terms of its atomic arrangement. When the positions of the oxygen atoms are projected parallel to the c-axis, the resulting hexagonal array (Figure 1a) can be seen to be composed of three close-packed rows of atoms with each row paralleling the <11$\bar{2}$0> or a-axis directions. These directions, which are all equivalent, correspond to the directions of the arms of a) snow flakes growing from the vapor, b) dendritic sea ice crystals growing from the melt and c) internal melt features (Tyndall figures) that form inside ice crystals as the result of absorbed solar radiation. All are macroscopic manifestations of the atomic structure.

The structure of ice provides reasons for its characteristically low impurity content. For an impurity atom to occupy lattice sites in the atomic structure of an ice crystal, the impurity atom must be of a similar size and charge, and must form a similar type of chemical bond as the atom it is replacing. Impurities meeting these requirements for substitution into ice are rare. Possibilities are F^-, HF, NH_4^+ and NH_3, NH_4OH, NH_4F and the hydrohalogen acids and, in fact, all of these substances do substitute in the ice structure in very small amounts (mole fractions of 1 in 5000 or less). However, such materials are not present in significant amounts in natural water bodies. The amount of other more common solutes that go into solid-solution in ice crystals is so small that ice formed by freezing even concentrated solutions can be considered pure. Therefore, the phase diagrams that govern the freezing of aqueous solutions are invariably of the simple eutectic type where the pure ice that forms initially is at equilibrium with increasingly concentrated solutions of brine as the temperature decreases.

2.2 Icebergs

In the Arctic, the primary source of icebergs is the Greenland Ice Sheet, which contains 2.4×10^6 km^3 of water, has a maximum thickness of 3300 m, and annually calves about 240 km^3 of ice into the surrounding seas. If all the other permanent ice fields located in the Canadian and Soviet Arctic and in Svalbard are taken together, they contain eight times times less ice than Greenland. The iceberg production from these latter regions, although not well known, is certainly small and primarily of local importance (good examples are the small icebergs produced by the glaciers on Svalbard and the Soviet arctic islands [10]). Because of the distribution of iceberg sources, icebergs are not much of a problem in the North Pacific, the Bering Sea or most parts of the Arctic Ocean.

Estimates of the total number of icebergs spawned annually by the Greenland Ice Sheet vary from 20 to 34 thousand, with most being produced by west coast glaciers. The iceberg drift pattern is such that icebergs formed along the east coast usually drift around the southern tip of Greenland and then move north, joining the drift of the icebergs produced by the large outlet glaciers located along the west coast. This northern drift continues up to Baffin Bay, where the icebergs swing around and start moving south along the coasts of Baffin Island, Labrador and Newfoundland. They finally reach the Grand Banks and ultimately melt in the North Atlantic. Although the southern limit of iceberg drift is in general defined by the northern edge of the warm (>12°C) North Atlantic Current, icebergs have been known to transit this current in cold water eddies and have been sighted as far south as Bermuda and as far east as the Azores [11]. At present, concern about the drift and properties of icebergs is focused on the regions off the Baffin and Labrador coasts, on the Grand Banks, and to a lesser extent off the coast of West Greenland, especially in areas where exploration for offshore oil and gas is currently underway.

Few studies have been made of the characteristics of the ice in actual icebergs [11]. In the Antarctic this is not a major problem as the properties of most of the icebergs can, with some confidence, be inferred from the properties of the parent ice shelves which have been well studied [4]. Similar inferences cannot be made as readily in the case of Greenland icebergs, even though the Greenland Ice Sheet itself has been reasonably well studied. The reason for this difference is that in Greenland the inland ice invariably passes through the coastal mountains in outlet glaciers before forming icebergs. This usually produces strong deformation with resulting recrystallization changing both the crystal orientation and the grain size. The ice is then

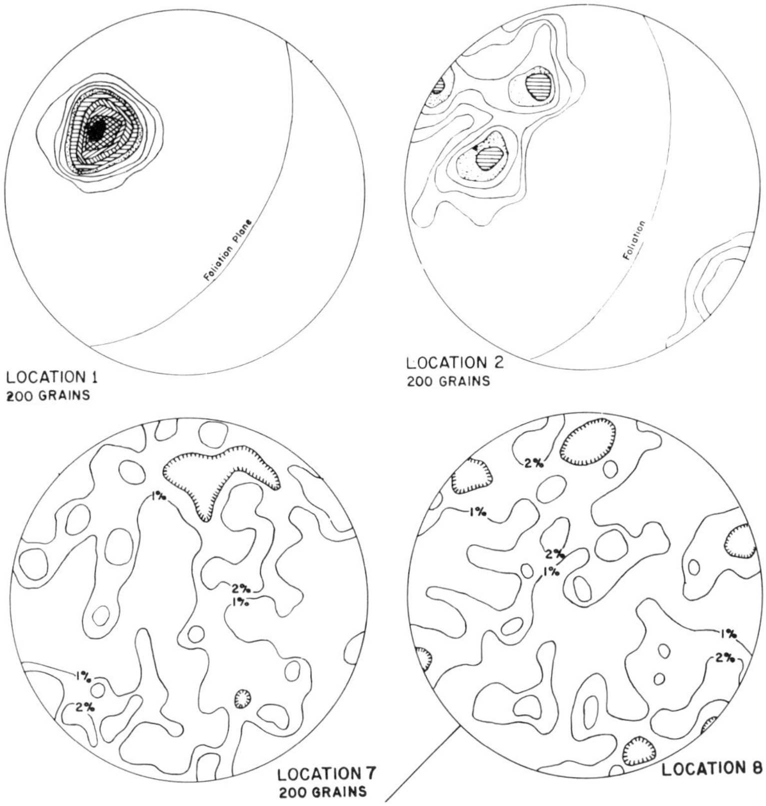

Figure 2. Four fabric diagrams of c-axis orientations in glacier ice from sites along the margin of the Greenland Ice Sheet [12].

comparable to a metamorphic rock. Figure 2 shows four fabric diagrams for ice from the margins of the Greenland Ice Sheet [12]. Two diagrams (locations 7 and 8) show a random pattern, indicating the absence of recrystallization in a strongly anisotropic stress field. Such fabrics are commonly observed in ice from the upper portions of large ice sheets, where representative crystal cross-sectional areas are 2 to 5 mm^2. The third diagram (location 1) shows an extremely strong c-axis alignment normal to the foliation (i.e. normal to the plane of implied shear). Such strong single-pole alignments are usually found in fine-grained ice that is undergoing rapid shear deformation in either temperate or polar glaciers. The fourth diagram (location 2) is of the multimaximum type, in which the individual maxima are invariably within 45° of the center of fabric symmetry. Such multimaximum fabrics are believed to develop by recrystallization in strongly deformed ice that is at or near the pressure melting point [13]. Such fabrics would be expected to be common in Greenland icebergs, as the majority of Greenland outlet glaciers are believed to be temperate. For instance, fabrics from the Moltke Glacier, a major iceberg producer in NW Greenland, are of this type [12]. Associated with this recrystallization there is characteristically a pronounced increase in grain size with cross-sectional areas ranging between 100 and 1000 mm^2. In contrast to the roughly equidimensional crystals commonly associated with the first two fabric types, the crystals showing the multimaximum fabric type are not only large, but show extremely complex interlocking shapes that makes their characterization by simple thin section analysis both difficult and time-consuming [14].

Changes in crystal alignment are usually associated with changes in characteristics of the air bubbles present in the icebergs. Ice with random c-axis orientations generally has rounded bubbles up to about 2 mm in diameter. In anisotropic ice, air bubbles tend to be tubular, with diameters between 0.02 and 0.18 mm and lengths up to 4 mm [15]. As gas pressures in bubbles in glacier ice are commonly equal to the hydrostatic pressure at a specific depth, the gas pressures in bubbles in icebergs would be expected to vary from 20 bars (roughly equal to the maximum tensile strength of ice) to some lower value, depending upon ice relaxation and gas leakage. In fact, gas pressures ranging from 2 to 20 bars have been observed [11].

As the firn limit (snowline) in Greenland is at roughly 1400 m, little or no snow or permeable ice is found in Greenland icebergs, and the ice density is presumably reasonably uniform in the range of 880-910 Mg/m^3 [16]. This means that between 86 and 89% of these icebergs are submerged (as compared to about 83% submergence for Antarctic shelf icebergs which contain snow in their upper levels).

2.3 Sea Ice

Sea ice, formed by the freezing of sea water, is different from glacier ice in both structure and composition. In contrast to glacier ice, where chemical impurities are commonly at concentrations of parts per million or lower, sea ice salt concentrations (salinities) are invariably in the parts per thousand range. The ice structure is also quite different, exhibiting a characteristic defect structure within each sea ice crystal associated with the entrapment of impurities, and also strong, distinctive crystal alignments caused by directional growth. In the Arctic, there are pronounced changes

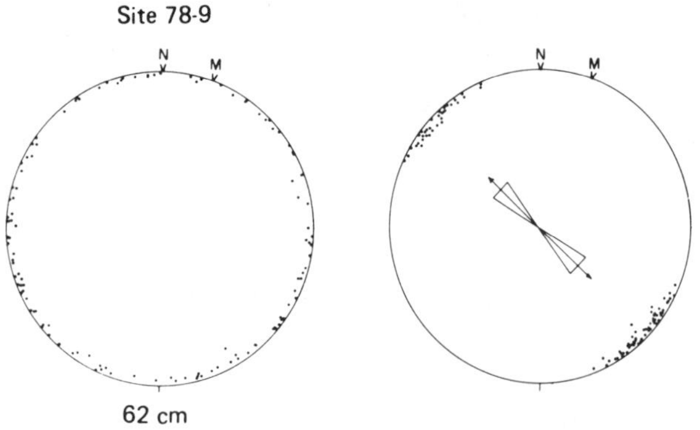

Figure 3. Representative fabric diagrams for sea ice collected along the coastline of arctic Alaska: a) random c-axis distribution in the horizontal plane, Cape Thompson, b) preferred c-axis alignment in the horizontal plane, Kotzebue Sound [20].

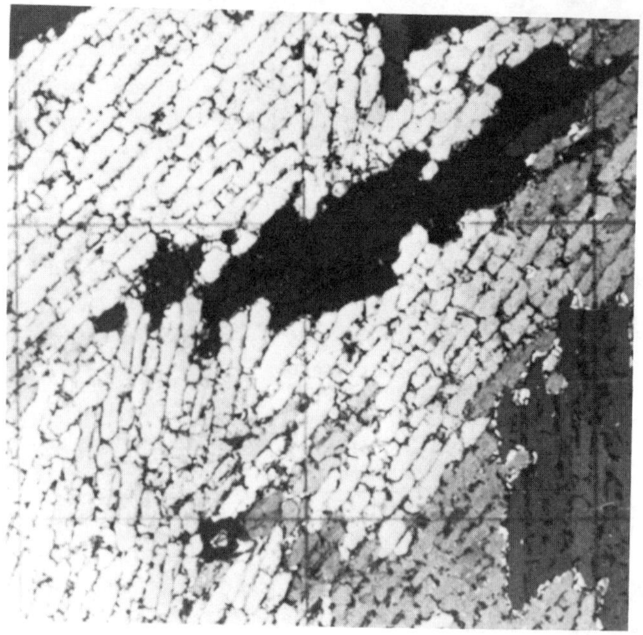

Figure 4. Photomicrograph of a thin section of sea ice showing its characteristic substructure. Grid spacing equals 1 cm.

in the properties of the sea ice that has survived one or more summer melt seasons (so-called multi-year ice). We therefore discuss the structure and composition of several different types of first year ice, of multi-year ice and of the highly deformed ice that composes pressure ridges.

Structurally, first-year sea ice is similar to a cast ingot. There is an initial skim, then a transition zone where rapid changes in crystal orientation occur, and a columnar zone formed of long crystals oriented vertically (parallel to the direction of the heat flow). Although the structure of the initial skim and the transition zone are interesting from the point of view of crystal growth, these layers are quite thin (the base of the transition layer is usually less than 30 cm below the upper surface of the ice sheet). For present purposes we only consider the properties of ice in the columnar zone. As a matter of fact, there have been no specific studies made of the mechanical properties of the ice above the columnar zone.

The structure of ice in the columnar zone is fairly uniform, with essentially all the crystals having pronounced elongation in the direction of growth. The crystal orientation is invariably c-axis horizontal, as crystals in this orientation have a growth advantage over crystals oreinted in other directions (their direction of maximum thermal conductivity is oriented parallel to the direction of heat flow [1, 17]). For years it was believed that the c-axis orientations in the columnar zone were always random in the horizontal plane [6] as a number of such fabrics had been observed. Such a material would be transversely isotropic; it would show property variations in the vertical direction associated with changes in grain size, crystal substructure, and salt content, but at any given level all directions in the horizontal plane would be identical. However, recent studies [18-21] have shown that most of the fast ice occurring over the continental shelves of the Arctic shows strong c-axis alignments within the horizontal plane. Theory, field observations and experiment [19, 22] suggest that these alignment directions are controlled by the direction of the current at the ice-sea water interface. Figure 3 shows two representative fabric diagrams for sea ice, the first showing a random c-axis orientation in the horizontal plane, the second a strong c-axis alignment. In a recent study of c-axis orientations along the Alaskan coast, over 95% of the sites sampled showed strong crystal alignments [20]. At first sight it might appear that such alignments would only develop in fast ice areas. However, observations suggest that strong alignments can develop in the pack [18, 23] if there is little rotation of the floes relative to the current direction. Such conditions do exist well offshore in the Arctic Ocean. Sea ice with such alignments is orthotropic, showing property differences along three orthogonal axes.

Associated with selective grain growth in the upper portion of the columnar zone is a marked increase in grain size with depth [1, 24]. Limited data suggest that mean grain diameter is proportional to depth in sea ice less than 60 cm thick. Mean diameters range from 0.5 to over 2 cm [6]. In thicker ice the linear increase in grain diameter with depth becomes less clear, and some decreases with depth have been observed [25]. In ice that has developed a strong c-axis alignment, it becomes difficult to distinguish one crystal from another when orientation differences are less than 5 degrees.

The most distinctive feature of sea ice, in addition to its high salt content, is the substructure within the ice crystals. In the columnar zone each sea ice crystal is composed of a number of ice platelets that are joined together to produce a quasi-hexagonal network in the horizontal plane. This substructure, shown in Figure 4, results from crystal growth with a non-planar solid-liquid interface. Similar substructures are commonly produced during the solidification of impure melts. In fact, it is the entrapment of brine between the ice plates at the non-planar interface that causes sea ice to be salty. The spacing (measured parallel to the c-axis) between the brine pocket arrays (a_o) is commonly referred to either as the brine layer spacing or as the plate spacing, and it varies inversely with growth velocity [1, 26]. Typical a_o variations range from 0.4 mm near the upper surface of the ice sheet to 1.0 mm at the base of the 2 m ice sheet. The best available study of these variations [27] was made recently at Eclipse Sound in the Northwest Territories. The results are shown in Figure 5. The inverse relation between a_o and growth velocity is clear. In thick multi-year sea ice, which presumably grows very slowly, a_o values of 1.5 mm have been observed. In the sea ice forming on the bottom of the Ross Ice Shelf at a location where the shelf is

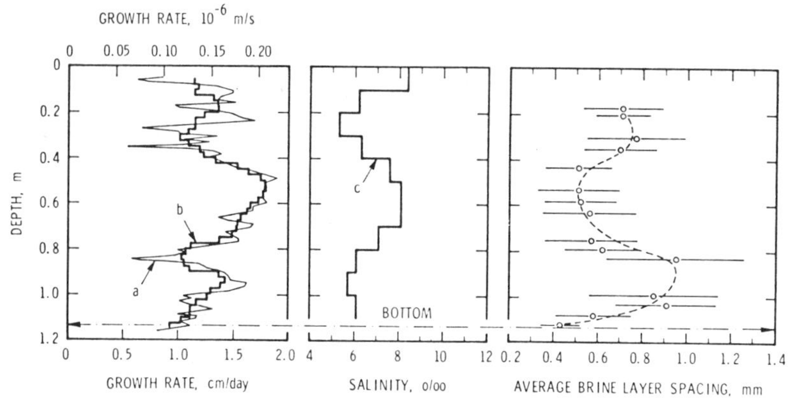

Figure 5. Profiles of growth rate, salinity and brine layer spacing. Curve b represents the mean of the calculated growth rate, curve a, for an interval of ± 50 mm for every 25 mm [27].

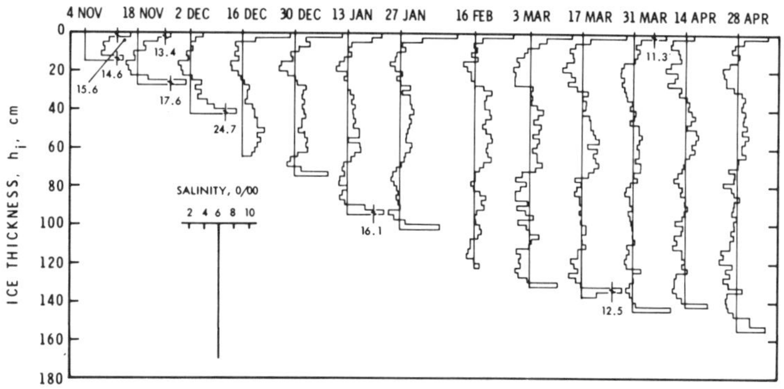

Figure 6. Salinity profiles of ice of Eclipse Sound made at two week intervals during the winter of 1977-78. Scale for salinity is shown in insert. Vertical solid lines represent a value of 6 $^{o}/oo$ and are given as reference [31].

416 m thick and the ice only grows about 2 cm/year, a_o values of 5 mm have been noted [28]. Brine layer spacing is believed to affect the strength of sea ice [29].

As mentioned earlier, the salt in the sea ice is not the result of solid solution, but is caused by the entrapment of brine between the platelets of pure ice that compose individual crystals of sea ice. The amount of salt entrapped is not constant, but varies systematically with the salinity of the water being frozen and with the ice growth velocity. Very slow growth results in near-total rejection of salt from the ice, while very rapid freezing causes near total entrapment [1, 30]. The effect of changes in growth rate on ice salinity can also be seen clearly in Figure 5. A series of representative salinity profiles for first year sea ice is shown in Figure 6 [31]. Note that the upper and lower portions of the ice characteristically have higher salinities than the ice in between, and there is a gradual decrease in the mean salinity of the ice with time.

The drainage of brine from saline ice appears to be a complicated process and several different mechanisms are believed to be involved [1]. In the present context, the most important results of brine drainage are changes in the porosity and the development of brine drainage channels. These structural features, one of which is shown schematically in Figure 7, can be considered as tubular "river"

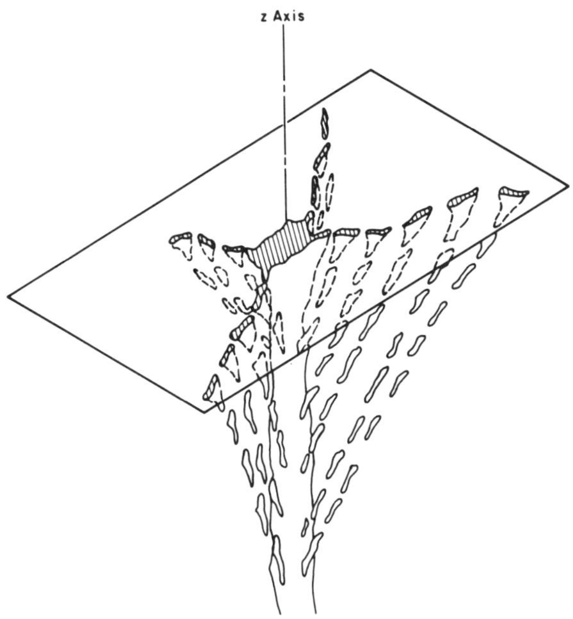

Figure 7. Schematic drawing of a cut through a brine drainage channel [32].

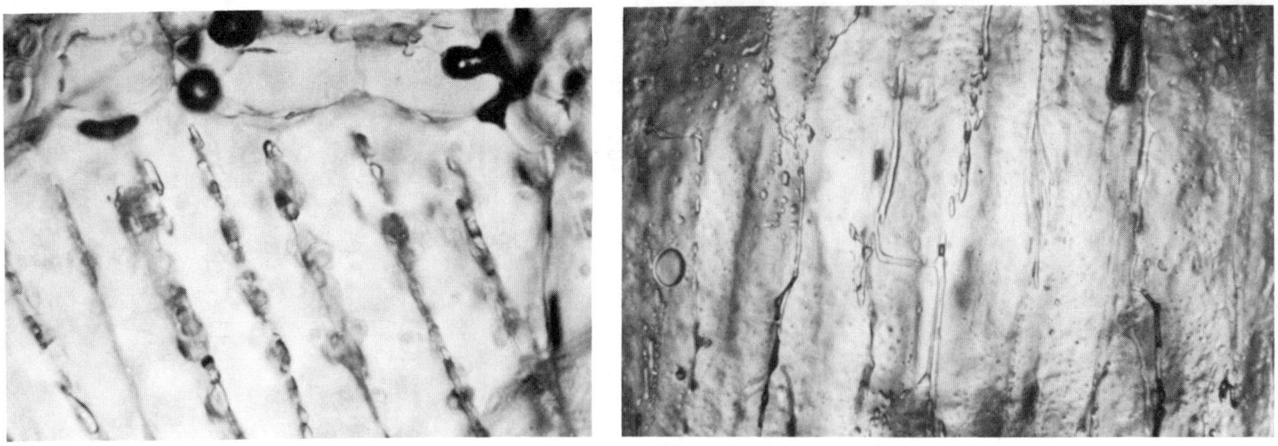

Figure 8. Shapes of brine pockets: a) horizontal view (brine layer spacing is approximately 0.5 mm), b) vertical view. Ice is from Thule, Greenland.

Figure 9. Scanning electron micrograph of a vertical section of a brine pocket at -30°C [35].

systems in which the tributaries are arranged with cylindrical symmetry around each main channel [32-33]. A representative channel diameter at the bottom of a 1.55-m-thick ice sheet is 0.4 cm and there is, on the average, one channel every 180 cm^2. Channel diameters as large as 10 cm have been noted, though most diameters range between 0.1 and 1 cm. Although these large "flaws" presumably have an effect on the mechanical properties of sea ice, no studies have been made of the matter.

Given a sample of sea ice with a specified salt content, the amount of liquid brine (the brine volume) present in the ice is a function of temperature only, because at each temperature the composition of brine in equilibrium with the ice is specified by the phase diagram [34]. Changes in the volume of brine in the sea ice are most pronounced near the melting point, where small changes in temperature cause large changes in brine volume. As most first-year sea ice has salinities in the range of 4 to 12 $^o/oo$ and temperatures between -2 and -30°C, the brine volume υ can be expected to vary between 30 and 300 $^o/oo$. Figure 8 shows detailed shapes of a series of brine pockets; they commonly are rather complex. The dark circles are gas bubbles. At lower temperatures there also are several different solid salts that precipitate in the ice (-8.7, $Na_2SO_4 \cdot 10H_2O$; -22.9, $NaCl \cdot 2H_2O$; -36.8 KCl; etc). Figure 9 is a scanning electron micrograph of a vertical section of a brine pocket at -30°C showing the solid salt crystals [35]. The effect of these solid salts on the mechanical properties of sea ice has been studied surprisingly little.

In addition to columnar sea ice, there is one other type of first-year undeformed sea ice that should be mentioned. This is frazil ice, produced by accumulation of individual discs and spicules of ice that form in the water. It has commonly been thought that, although frazil ice is frequently observed during the formation of the initial ice cover, once this cover stabilized frazil ice generation would greatly decrease. Exceptions to this would be areas near the ice edge or in large polynyas where substantial regions of open water are found. However, recent work in the Weddell Sea to the east of the Antarctic Penisula has indicated that, at least in that region, frazil ice generation is a very important ice producing mechanism [1, 36]. For instance, of the ice sampled, over 50% was frazil and the thicker the floe, the higher the percentage of frazil ice it contained. Whether such large amounts of frazil also occur in the Arctic is not known but there is no strong evidence against such a possibility. If major quantities of frazil ice form in the Arctic, there are interesting implications. First, because frazil ice forms by a completely different mechanism than columnar ice, present estimates of the amount of ice being generated in the Arctic might have to be revised. Secondly, frazil ice has a completely different crystal structure than does columnar ice. Structurally, frazil ice is commonly fine-grained, with crystal sizes of 1 mm or less, it has a crystal orientation which is presumed to be random, and there are brine pockets located mainly between the ice crystals, as opposed to within crystals. Although there have not yet been any systematic studies of the physical properties of frazil ice, they clearly would be expected to be different from those of columnar sea ice.

When sea ice goes through a summer melt period it undergoes a pronounced change in salinity produced by the percolation of relatively fresh surface meltwater down through the ice. The result is an ice sheet with very low salinities (<1 $^o/oo$) in the portion above water level and salinities of between 2 and 3.5 $^o/oo$ in the portion below water level.

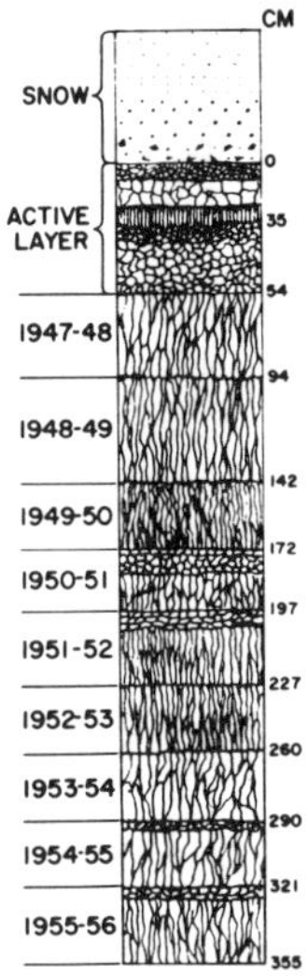

Figure 10. Cross-section of a multi-year floe [37].

Figure 11. Thin section of ice from a multi-year ridge in the Beaufort Sea showing a block of columnar ice "cemented" by fine grained granular ice. Core is 10.5 cm in diameter [43].

Once the brine has drained from the upper portion, the ice is quite porous and it may recrystallize. Ice that has survived several summers ultimately becomes a layercake of the annual layers formed during successive winter periods of growth [37-38]. A cross-section of such a floe is shown in Figure 10. In fact, much multi-year ice was probably deformed at some time in its past and would show a much more complex cross-section. Confident statements concerning the relative percentages of multi-year ice that are undeformed, deformed, columnar or frazil will have to await more adequate sampling. In general, undeformed multi-year ice in the Arctic Basin is believed to reach a steady-state thickness of 3 to 5 m, at which time the thickness ablated during the summer equals the thickness grown during the winter [39]. Although deformed sea ice can grow to greater thicknesses, rather atypical conditions are required and such ice, although known, would appear to be rare [40].

Ice thicker than 5 m is, however, rather common in the Arctic Basin. For instance, in a recent study of submarine sonar profiles of the underside of sea ice, over 40% was thicker than 5 m [41]. This thicker ice is generally believed to be pressure ridges and rubble fields that are produced by the deformation of thinner ice. Pressure ridges and deformed ice in general are common in all areas of pack ice, and are particularly common in the land-locked Arctic Ocean. Although data are limited, it currently appears that the most highly deformed, and also the thickest, ice in the Arctic occurs in a broad band starting off the NE corner of Greenland and stretching to the West, north of Ellesmere Land and then veering toward the SW down the coast of the Archipelago to the coast of Northern Alaska. The largest free-floating ridges that have been observed have sails up to 13 m high and keels up to 50 m deep. In near-coastal areas where pressure ridge keels can ground, ridge sails can be particularly high (heights in excess of 30 m have been noted.

Considering the importance of ridges, there has been surprisingly little work done on them. As first-year ridges are composed of blocks, it would be interesting to have quantitative information on block sizes and orientations from a variety of different locations and on the degree of bonding between the blocks. A limited amount of information is available for the first of these [42] but there is nothing on the latter. However it should be possible to obtain useful information on this subject from ramming tests with an icebreaker. In contrast to first-year ridges, multi-year ridges are commonly composed of massive ice, in that all the voids present in newly formed ridges have now been filled with ice. Figure 11 shows thin sections of ice from a multi-year ridge. The ice is quite complex, showing fragments of the initial ice cover that was crushed to form the ridge, plus a large amount of fine grained ice (presumably similar to frazil) that formed in the voids between the blocks [43].

2.4 Ice Islands

So called ice islands are, in fact, tabular icebergs from a relict Pleistocene ice shelf that still exists along the north coast of Ellesmere Island, the northern-most island in the Canadian Archipelago. Strictly speaking, ice islands are just a specific type of shelf iceberg, but we will discuss them in a separate category as they are unique to the Arctic Ocean and are composed of a rather complex mix of ice types. They are a particular hazard along the coasts of Northern Greenland, the Canadian Archipelago, and off the North Slope of Alaska. Ice islands can have long lifetimes. For instance, the best known ice island T-3 has been drifting around the Beaufort Gyre (the large clockwise circulation in the Beaufort Sea) for over 30 years. If current predictions of its trajectory are correct, T-3 may "die" within the next year by leaving the Arctic Ocean via the East Greenland Drift Stream and melting in the North Atlantic. Ice islands also have been known to leave the Arctic Ocean via Robeson channel (between Greenland and Ellesmere Island) and also through the Canadian Archipelago into Viscount Melville Sound. It is only after they leave the Arctic Ocean that they drift through regions where ordinary icebergs produced by glaciers are common.

As their origin is an ice shelf, ice islands are tabular with thicknesses of several tens of meters (T-3 had an initial thickness of approximately 70 m). Lateral dimensions are highly variable ranging from more than 10 kilometers to a few tens of meters for ice island fragments. There is no adequate census of the number of ice islands currently drifting in the Arctic Ocean. The numbers "sighted" at specific locations are highly variable. For instance in 1972, 433 ice islands were sighted along the Beaufort Coast. Of these 117 had lateral dimensions greater than 80 m and 1 had dimensions of over 1600 m. In 1975 there were no sightings along the same stretch of coast. It is generally believed that the large number of fragments seen in 1972 was the result of the breakup of a very large ice island

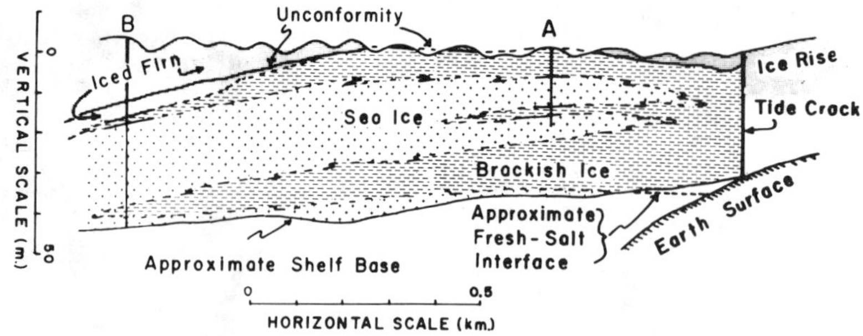

Figure 12. Interpretation of stratigraphy of part of Ward Hunt Ice Shelf, based on drill-core and laboratory studies [50].

that grounded north of Barter Island. The current interest in ice islands results from the threat they pose to offshore structures in the deeper waters of the Beaufort Shelf. The ice island problem is similar to that posed by large hurricanes in the Gulf of Mexico in that although the probability of a given structure being impacted by an ice island is small, the probability of the structure sustaining damage if a collision does occur is high.

Although the ice in ice islands has not been extensively studied, enough is known to be able to give a general description of the several different ice types involved [44-50]. The complex structures encountered in some ice islands can best be appreciated by referring to Figure 10 in Smith's study of Arlis II [47] and to Figure 12 which is presented here and which summarizes the Lyons et al. [50] picture of the structure of the ice in part of the Ward Hunt Ice Shelf. There are at least four different types of ice present. These are as follows:

Lake ice - The lake ice is the result of the freezing of elongated bodies of water that form on the surface of ice islands during the melt season. The ice can be easily recognized by its medium to very coarse grained texture, the long columnar crystals with straight grain boundaries and the long, linear, well oriented bubbles. Typical grain diameters are larger than 3 cm.

Snow ice - This ice type, which composes much of the upper part of the Ellesmere Shelf, is produced by the densification and recrystallization of snow. The crystals are equant and euhedral showing a typical mosaic texture. Crystal orientations are random and grain sizes characteristically range from 0.5 to 1.5 cm.

Sea ice - Sea ice usually occurs in the lower portion of the ice shelf. This material has the characteristics of multi-year sea ice, although the salinities are somewhat lower (< 2 °/oo). The crystals show the characteristic elongation of columnar zone sea ice and the substructure within each crystal is still evident. Some fine grained frazil ice has also been reported. In two cases [47, 49] strong preferred c-axis alignments have been noted.

Brackish ice - This ice shows a well developed stratification, which is a reflection of cyclic variations in the amount of entrapped gas and the average grain size. Typical layer thicknesses are between 20 and 25 cm, and the strata may locally be warped into a series of folds. Very large crystals are common, with some as large as 120 cm. C-axis orientations commonly are vertical and salinities are usually very low. Although several different explanations have been advanced to account for such unusual ice, the most likely explanation is that this material developed as a series of annual growth layers that formed from the freezing of a layer of brackish melt water that is known to be present beneath the ice shelf at some locations [50]. This explanation can account for both the complex interstratification of this ice with normal sea ice, and for the measured oxygen isotope ratios.

3. MECHANICAL PROPERTIES

3.1 General behaviour

Systematic knowledge of the mechanical properties of ice derives mainly from studies of non-saline polycrystalline ice which is not strongly anisotropic. The principal motivations for study have been glaciology, where the chief concern is with flow under small deviatoric stresses (< 0.2 MPa), and engineering, where the emphasis is on strength at relatively high strain rates ($> 10^{-6}\ s^{-1}$). It is now possible to unify the findings from these areas of study [51], especially since it has been demonstrated that the favoured test of glaciology, the constant load creep test, can give essentially the same information as the favoured test of ice engineering, the constant strain rate strength test [52,53].

In simple terms, ice has the following characteristics.

1. Under moderate hydrostatic pressure and moderately low temperature, ice compresses elastically with a bulk modulus of about 9 GPa. Any bubbles in the ice compress so as to equilibrate in accordance with the gas laws, and they may eventually disappear to form a clathrate. Under sufficiently high pressure, ice Ih transforms into high density polymorphs (including water), as described by the phase diagram for isothermal compression. Under intense adiabatic compression (e.g. explosive loading), discrete phase transitions are not detected, but the Rankine-Hugoniot characteristic gives a pressure-volume relation that is not much different from a "smeared-out" version of the Bridgman isotherm.

2. Under deviatoric stress, ice deforms as a non-linear viscous solid, changing its fabric and structure in the process. Under constant stress, a complete creep curve shows deceleration followed by acceleration to a final rate in the usual way, although there are possible complications. Under constant strain rate, a complete stress/strain curve shows stress rising to a peak before falling and tending asymptotically to a limit, again with a possible complication in the form of an initial yield point. For any given stage of deformation, the stress/strain-rate relation is non-linear. It is usually given as a simple power relation, although the exponent changes over the complete range of

stresses and strain rates. Below about -10°C the effect of temperature can be described by an Arrhenius relation with an activation energy of about 70 kJ/mole. However, closer to the melting point, temperature sensitivity is greater than such an equation would predict.

3. High sensitivity to strain rate and temperature causes ice to display a broad range of rheological properties. With high rates and low temperature, elastic behaviour dominates, and deviatoric straining culminates in brittle fracture. With low rates and high temperatures, ductility is predominant, and large creep deformations can occur. Very often both elasticity and non-linear viscosity make significant contributions to deformation and rupture processes.

4. In multiaxial stress states, compressive bulk stress (isotropic component of stress tensor) has little effect on the deviatoric stress/strain-rate relation when stress deviators are very low and temperature is well below 0°C. By contrast, moderate pressure suppresses internal microcracks at high strain rates, and it increases deformation resistance and "strength." Extreme pressure at typical temperatures pushes ice towards the phase transformation to water, and consequently deformation resistance and "strength" decrease with increasing pressure, almost irrespective of the magnitude of deviatoric stress or strain rate.

Strength and deformation resistance are influenced by strain rate, temperature, porosity and grain size in the following way.

Strain rate. Deformation resistance and "strength" increase with increase of imposed strain rate. Conversely, strain rate $\dot{\varepsilon}$ increases with increase of imposed stress σ. In either case, $\dot{\varepsilon} \propto \sigma^n$, where n might range from 2 at very low stress (< 0.01 MPa) to 4 at high strain rates (> $10^{-3}s^{-1}$).

Temperature. When ice is truly "solid" (below -10°C for non-saline ice, or below the eutectics of dissolved impurities), strain rate $\dot{\varepsilon}$ for a given stress σ can be described by a relation of the form $\dot{\varepsilon} \propto \exp(-Q/RT)$, where T is absolute temperature, R is the gas constant, and Q is an activation energy for the controlling deformation process, say 70 kJ/mole for diffusional creep. The corresponding relation for σ is obtained from the $\dot{\varepsilon}$-σ relation (the power law). At temperatures close to 0°C, all polycrystalline ice has liquid, or liquid-like transitional layers, at the grain boundaries. Thus there are additional thermally activated processes influencing strength and deformation, and the simple Arrhenius relation does not apply.

Porosity. In simple terms, strength and deformation resistance decrease with increasing porosity. For the complete range of material properties from dense, impermeable ice to highly compressible, permeable snow, there is a continuous decrease in strength and in deformation resistance. In saline ice, porosity is created by both air bubbles and brine cells. The effect of variations in the size and shape of pores has not received much experimental attention, although it is a tempting topic for theoretical speculation.

Grain size. Where elastic deformation and brittle fracture are significant contributors, strength σ decreases with increasing grain size d, and there is support for the idea that $\sigma \propto d^{-1/2}$. There are indications that coarse-grained ice may fracture and yield at strains smaller than the yield strains for fine-grained ice. In the range where ductility dominates, there is not yet any convincing evidence that strength and deformation are much affected by grain size.

Another factor to be considered is the size of the stressed volume, since non-metallic brittle solids typically get weaker as volume increases (increasing the probability of encountering bigger flaws). Published data for ice on this topic cover only a narrow range of volumes, but it is to be expected that fracture strength will decrease with increasing volume at high strain rates (where cracks and similar flaws control the failure). The deformation resistance is not expected to be much affected by size at very low strain rates (where the controlling flaws are thought to be dislocations).

Perhaps the most difficult variables to deal with are anisotropy and inhomogeneity. Studies of anisotropy are not very far advanced, so it is dangerous to venture generalizations. However, there is not much doubt that ice with preferred crystal orientation flows most easily when the resolved stress is parallel to the basal planes of the crystals. With high strain rates and multiaxial stress, the "strength" of columnar ice varies depending on whether the stress field is tending to push the columns together or apart.

Ice testing and experimental data. Under the best of circumstances, most mechanical tests are much more complicated than their textbook idealizations. When typical tests are applied to ice, the problems are magnified by thermal instability of the material (melting, evaporation, brine drainage, vapour and surface diffusion) and by high sensitivity to rate and temperature (changing the balance of elasticity and plasticity).

An international group has been trying for the last decade to bring some order to ice testing, but the standards of experimental work are still highly variable. For the present we have to use results from some test programs that are obviously flawed, and it is necessary to be aware of common sources of error and misunderstanding. The following points might be kept in mind when considering the data given in the remainder of this paper.

1. For tests near 0°C, lax temperature control can introduce large errors.

2. For high rate/low temperature tests (elastic/brittle), great care and very precise technique are needed to avoid errors.

3. For tests to large strains, special procedures are needed to produce representative results.

4. Most of the indirect tests, in which some material properties must be assumed, cannot be used to investigate rate and temperature effects. For example, the assumption of elasticity in beam flexure or disc compression is hard to justify when rates are low or temperatures high.

5. Blind application of test procedures from other technical fields can give misleading results (e.g. use of quasi-static tests to measure Young's modulus).

6. Large-scale field tests (e.g. on floating beams or cantilevers) can involve material which is inhomogeneous, anisotropic, and subject to appreciable temperature gradients.

There have been no truly comprehensive test programs covering all the variables in a systematic way, and so a certain amount of intelligent guesswork is needed in order to extrapolate and interpolate from the existing data. However, the broad picture is now starting to emerge, and any given set of data can be checked for consistency and plausibility against a variety of independent data sets.

3.2 Iceberg Ice

There are virtually no published data on mech-

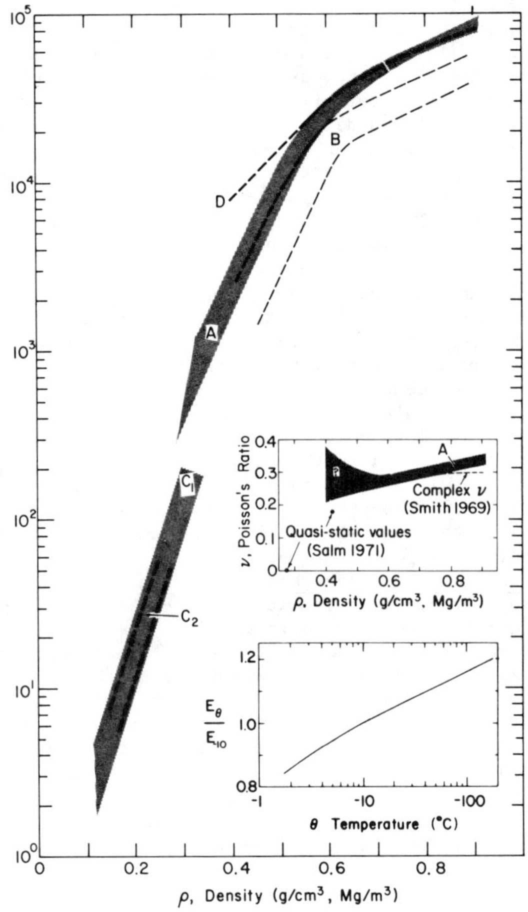

Figure 13. Summary of Young's modulus data for non-saline ice and snow (for data sources see [63].

anical properties of ice that has been collected from icebergs, but there is plenty of information about glacier ice, which is what icebergs consist of. Over the interior areas of Greenland and Antarctica, glacier ice forms by a sedimentation process involving visco-plastic compaction of dry snow. The resulting material is fine-grained and almost isotropic, with included air which forms closed bubbles when the bulk density reaches about 0.8 Mg/m^3. Only in the layers very close to the glacier bed is there significant shearing, with consequent development of preferred crystal orientation, but when ice from the interior is funneled out to the sea through ice streams and valley glaciers there is more general shearing, and some inclusion of rock debris. Therefore, in lieu of data on ice from actual icebergs we have to be content with a summary of the properties of glacier ice, of artificial ice which simulates isotropic polar glacier ice, and of some other types of non-saline ice.

Elastic moduli. For polycrystalline ice of low porosity (density $\rho \rightarrow 0.917$ Mg/m^3), high frequency dynamic measurements of Young's modulus E give values of approximately 9.0 to 9.5 GPa in the temperature range -5 to -10°C. Careful measurements of the initial tangent modulus for quasi-static uniaxial compression tests give quite similar values [54-57]. As temperature decreases, E increases nonlinearly (Fig. 13), but the effect is small for "true" Young's modulus (as opposed to "effective" values of E which include creep effects). Porosity n, which can be expressed alternatively as bulk density ρ, has a significant influence on E (Fig. 13), and it is interesting to note that E drops sharply below the density which represents close-packing of equant grains ($\rho \approx 0.55$ Mg/m^3).

Poisson's ratio ν, as measured by dynamic tests, has values close to 0.3 for non-saline ice of low porosity (Fig. 13), and there is not much variation with porosity over the range where the material is regarded as "ice" rather than "snow" ($\rho > 0.8$ Mg/m^3). Bulk modulus K is

$$K = E/3(1-2\nu) \approx E$$

and the shear modulus G is

$$G = E/2(1+\nu) \approx 0.38E .$$

In ice engineering it is frequently necessary to apply elastic analyses in situations where the ice deformation is not purely elastic. In such cases, it may be appropriate to use "effective" moduli derived from relatively slow quasi-static tests. Because these effective moduli (E') represent the combined effects of elasticity, recoverable "delayed elasticity," and irrecoverable creep, they are appreciably more sensitive to temperature, strain rate and vibrational frequency than is Young's modulus E. At low temperatures and/or high strain rates, $E' \approx E$, but at low strain rates ($\sim 10^{-7}$ s^{-1}) or relatively high temperatures ($\approx$ -10°C), E' may be as low as 25% to 30% of E. When low strain rates are combined with temperatures approaching 0°C, E' can have very low values, and the elastic approximation may cease to be useful. In comparison with the effects of temperature and strain rate on E', porosity variations over the typical range are not very significant, but there is a slight decrease of E' with increase of n (decrease of ρ).

"Effective" values of Young's modulus E' should be paired with "effective" values of Poisson's ratio, ν'. Although ν' does not receive explicit treatment in the literature, some deductions can be made [2]. As ductility increases it is reasonable to expect $\nu' \rightarrow 1/2$, representing incompressible flow, with $E'/K \rightarrow 0$.For ice which has low porosity (or water-filled pores) the bulk modulus K should not vary much with porosity, temperature, or strain rate, and for a first approximation it can be assumed equal to the true Young's modulus for zero porosity, E_o. Thus ν' can be expressed as

$$\nu' = \frac{1}{2} - \frac{1}{6}\frac{E'}{E_o}$$

which gives a systematic variation between the limits 1/3 and 1/2.

Strength and deformation resistance. Strength for any specified state of stress can be defined as the maximum stress, or deformation resistance, for a given strain rate. For ductile yielding of fine-grained ice, constant strain rate strength tests give essentially the same information as constant stress creep tests, so that "strength" can be obtained from either the peak of a conventional stress/strain curve or the inflection point of a conventional creep curve [52, 53].

The most common test is uniaxial compression. Uniaxial compressive strength σ_c for non-saline ice at -5° to -10°C varies by three orders of magnitude (0.01 to 10 MPa) as strain rate varies from about 10^{-11} to 10^{-2} s^{-1}. At high strain rates, σ_c is not highly sensitive to temperature, and at very high rates the temperature effect is expected to be comparable to that for Young's modulus (Fig. 14). At very low strain rates, the variation of σ_c with temperature (Fig. 14) can be deduced from the dependence of minimum creep rate on temperature (Fig. 15). Because the stress/strain rate relation is the same for constant strain rate and constant stress (Fig. 16), the stress/strain-rate relations developed by

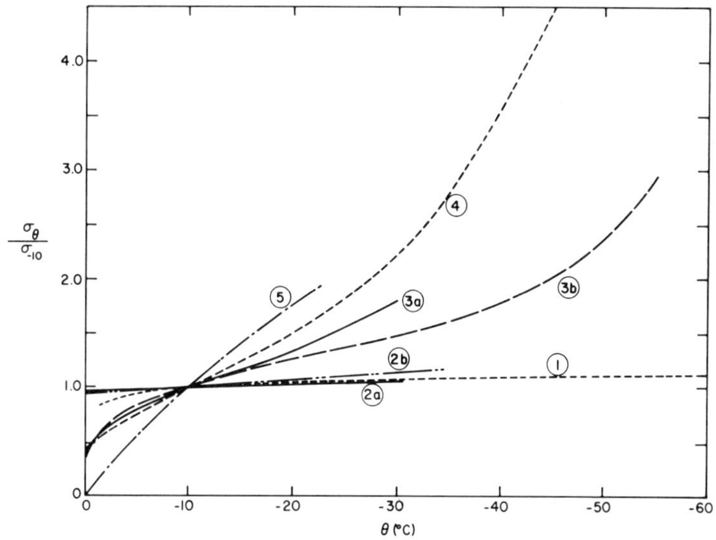

Figure 14. Compilation of temperature relationships [2]. All stress values are normalized with respect to the value for −10°C. 1) Variation of Young's modulus with temperature. 2a) Uniaxial tensile strength - data from [58]. 2b) Uniaxial tensile strength - data from [59]. 3a) Uniaxial compressive strength - data from [58]. 4) Ductile yield stress - data from [60]. 5) Pressure for phase transition from Ice Ih to water under isothermal hydrostatic compression.

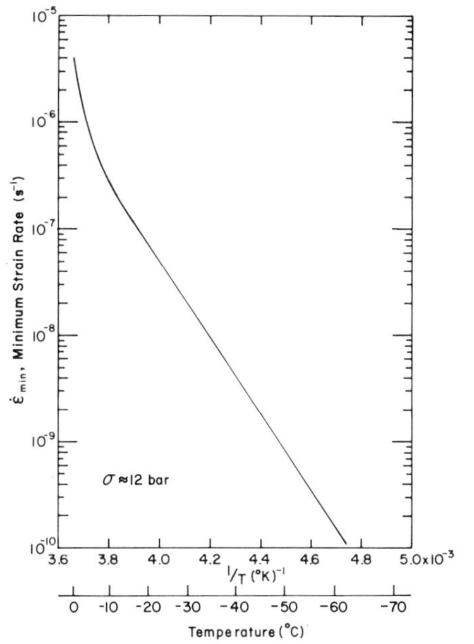

Figure 15. Empirical relation between minimum strain rate and temperature for high-stress creep [60].

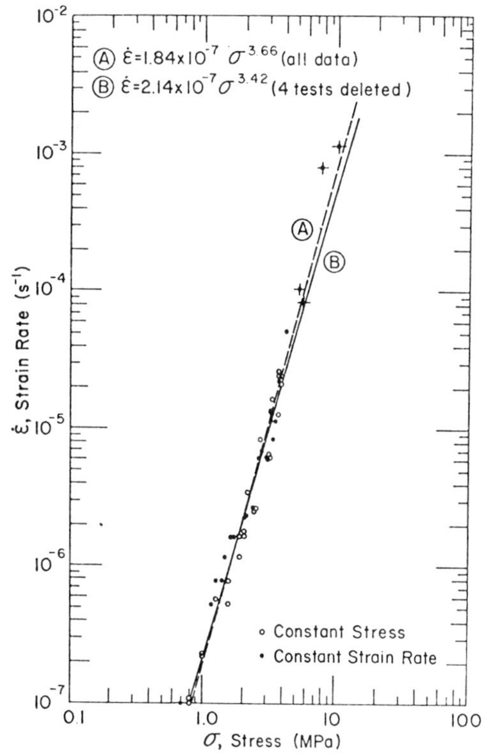

Figure 16. Data from tests under constant load and constant displacement rate. Lines A and B are regression lines, as indicated on the figure [53].

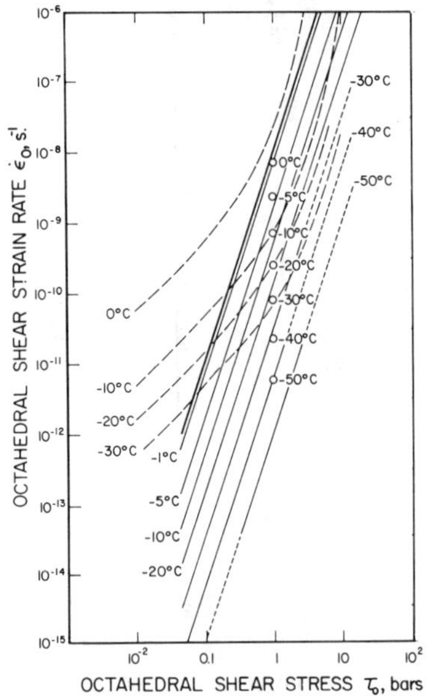

Figure 17. Stress/strain rate relations for creep of glacier ice [61]. Broken lines indicate relations derived from earlier studies [62].

glaciologists [61] for minimum creep rate (Fig. 17) can be interpreted as strength/strain-rate relations for low-rate ductile yield. It might be noted that glaciologists commonly represent axial stress σ_1 in terms of octahedral shear stress τ_{oct} ($\tau_{oct} = \sqrt{2}\ \sigma_1/3$) and axial strain rate $\dot{\varepsilon}_1$ in terms of octahedral strain rate $\dot{\varepsilon}_{oct}$ ($\dot{\varepsilon}_{oct} = \dot{\varepsilon}_1/\sqrt{2}$).

It is sometimes useful to know the "time-to-failure" t_f, defined as the time taken to reach the peak of a stress/strain curve or the inflection point of a creep curve. Figure 18 shows how t_f for fine-grained ice is inversely proportional to both strain rate and a power of stress.

At high strain rates, σ_c decreases as porosity increases (ρ decreases), as shown in Figure 19. For low strain rates, the effect of porosity on the σ_c-$\dot{\varepsilon}$ relation has to be deduced indirectly (Fig. 20).

Grain size d does not appear to affect σ_c systematically at low strain rates ($< 10^{-4}\ s^{-1}$). At high strain rates, σ_c is expected to decrease as d increases, perhaps with $\sigma_c \propto d^{-1/2}$, but adequate data are not yet available.

In cases where ice undergoes ductile yield without fracture or rupture, it is sometimes useful to know the "residual strength" at relatively large strains. The relation between residual strength and strain rate is much the same as the relation between creep rate and applied stress for large strains, and relevant data for both constant strain rate and constant stress are given in Figure 21.

Defining uniaxial tensile strength σ_T in the same way that σ_c was defined, there is little difference between σ_T and σ_c for isotropic ice at low strain rates ($< 10^{-7}\ s^{-1}$). In both cases the ice yields by shearing, and the difference of normal stress does not seem to have much effect. Above some critical strain rate ($10^{-6}\ s^{-1}$ at -7°C -- see Fig. 22) there is a bifurcation in the stress/strain-rate curves for tension and compression, presumably because internal microcracks can form and influence the failure at high rates. At high rates ($> 10^{-5}\ s^{-1}$), σ_T tends to a limiting value, while σ_c continues to increase. For fine-grained ice at -7°C, the high

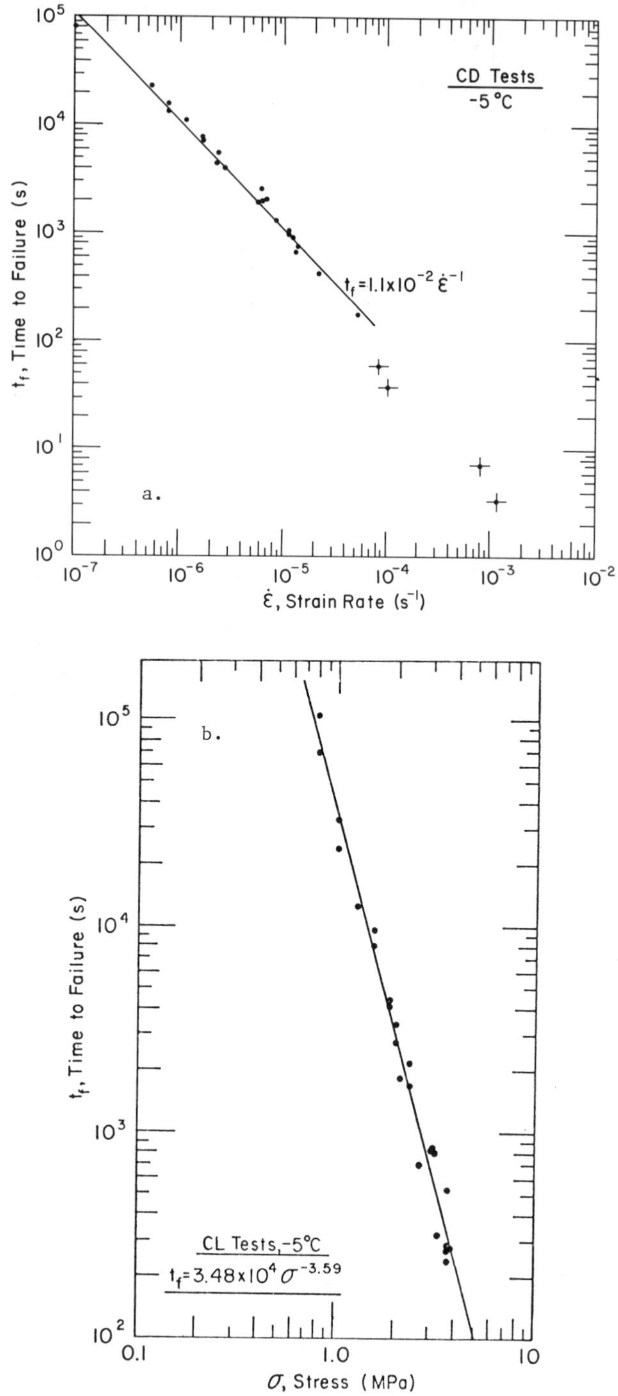

Figure 18. Time-to-failure t_f for fine grained non-saline ice. t_f is given as a function of strain rate (Fig. 18a), and as a function of stress (Fig. 18b).

rate limit of σ_T is about 2 MPa, with σ_c around 10 MPa. This gives a ratio of σ_c/σ_T well below the theoretical values of 8 or more that are predicted by Griffith theory and its derivatives (another reason why diametral compression of a disc or cylinder cannot be used to measure σ_T in ice).

The effect of temperature on σ_T is the same as its effect on σ_c at low strain rates. By contrast, the lack of sensitivity to strain rate at very high rates leads to the expectation that there will be a corresponding insensitivity to temperature in that

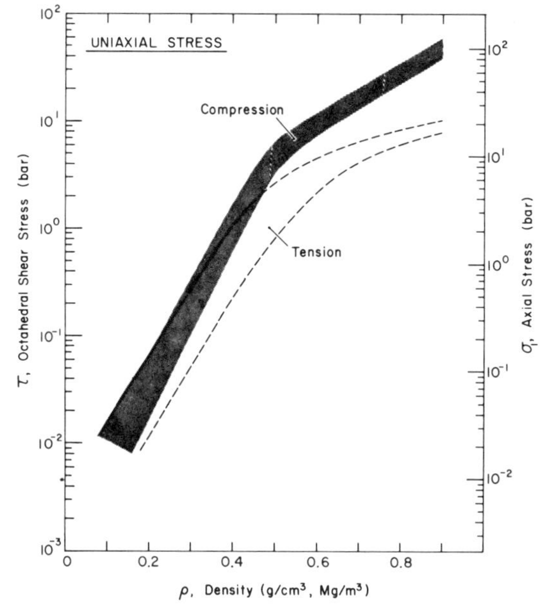

Figure 19. Summary of data on the strength of non-saline ice and snow (for details on data sources see [63]).

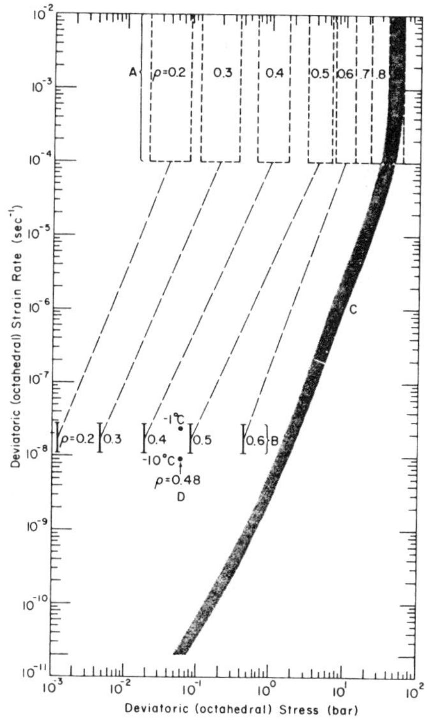

Figure 20. Deviatoric stress/strain-rate relations deduced indirectly from various data sources [63]. (A) Snow, ≈ -10°C. (B) Snow, ≈ 0° to -7°C. (C) Ice, ≈ 0.9 Mg/m^3, -2°C to -10°C. (D) Snow, 0.49 Mg/m^3, -1°C to -10°C.

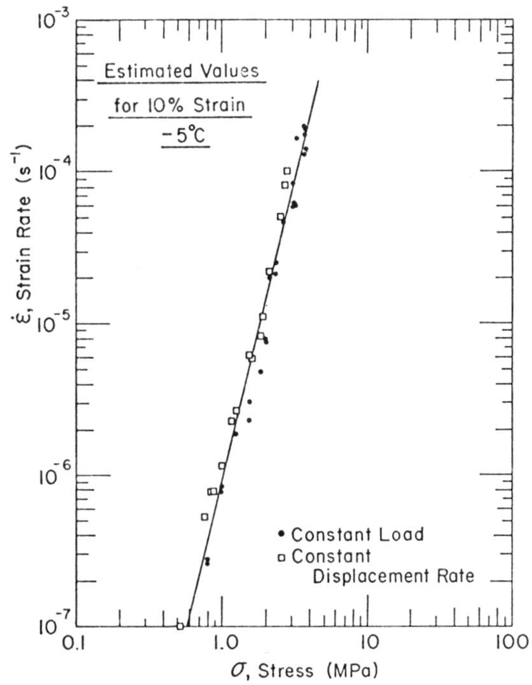

Figure 21. Stress/strain-rate data for 10% axial strain, obtained from extrapolation of constant load and constant displacement rate tests at -5°C. The regression line represents the combined data set [53].

range. The limited experimental data support this idea (Figs. 14 and 23).

At high strain rates, σ_T decreases with increasing porosity (decreasing density), as indicated in Figure 19. At low strain rates, the trend is expected to follow that for σ_c (Fig. 20).

The grain size d has a considerable influence on σ_T at moderate strain rates (10^{-6} s^{-1}). The effect can be described by the Hall-Petch relation

$$\sigma_T = a + b\, d^{-1/2}$$

where a and b are constants [64]. This type of behaviour is expected to prevail at strain rates higher than 10^{-6} s^{-1}, but at very low strain rates d may not have much effect. For practical purposes the value $\sigma_T \approx 2$ MPa for fine-grained ice can be regarded as an upper limit. More coarse-grained non-saline material that is encountered in glacier ice, lake ice, and old sea ice will usually have tensile strength much lower than 2 MPa (typically 1 MPa or less).

Failure strains and yield strains. Traditionally there has been rather little interest in absolute value of the strains at which fracture and ductile yielding occur in ice, and yield criteria have always been formulated in terms of stress. However, while yield stresses for ice vary by orders of magnitude, the strains for fracture and ductile yield stay within much more limited ranges.

When fine-grained ice is strained in uniaxial compression at rates less than about 10^{-4} s^{-1}, under either constant strain-rate or constant stress (Fig. 24), there is a well-defined ductile yield at axial strains of approximately 1% [52]. This shows up either as a peak stress on a stress/strain curve, or

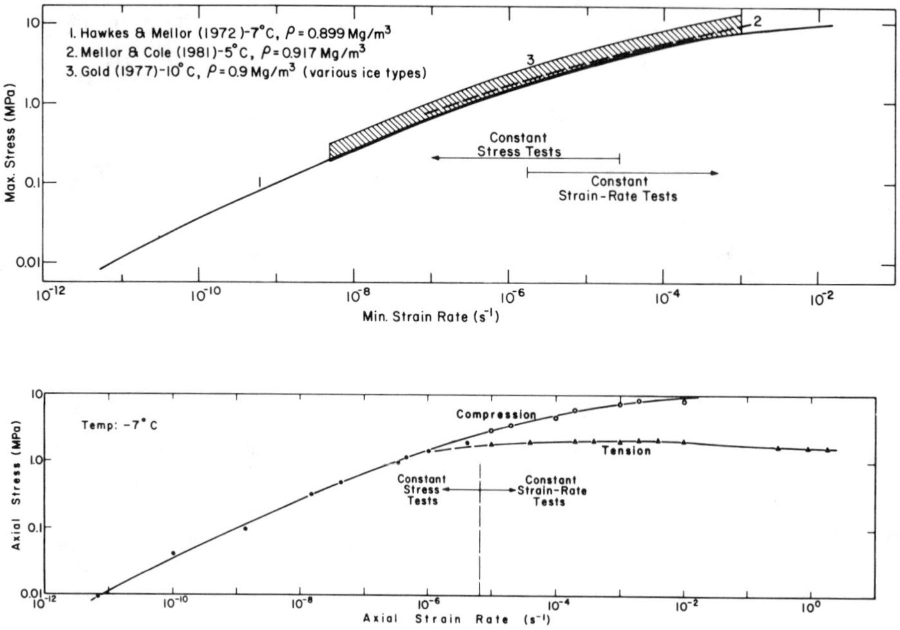

Figure 22. Effect of strain rate on σ_C and σ_T for non-saline ice. Figure 22a indicates σ_C values from various sources and for various ice types. Figure 22b compares σ_C and σ_T for fine-grained non-saline ice [55].

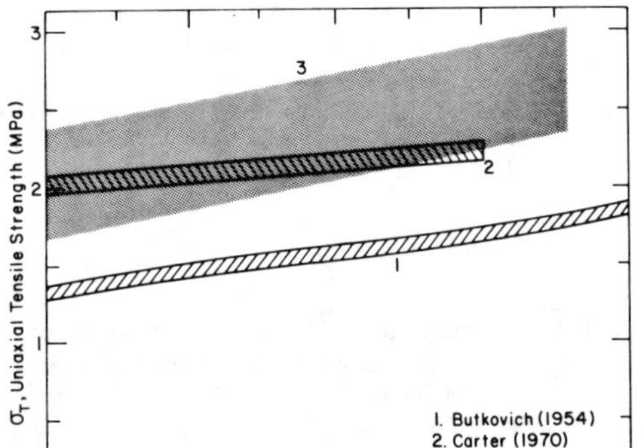

Figure 23. Variation of σ_T with temperature for non-saline ice. Data from [58, 59, 71].

as a strain-rate minimum on a creep curve. Even when the compressive stress is cycled at varying frequencies (Fig. 25), the mean creep curve still shows minimum creep rate at about 1% axial strain [65]. With strain rates in the range 10^{-7} to 10^{-4} s^{-1}, the same ice has an initial yield point which occurs at smaller strains (0.03 to 0.5%). This initial yield is associated with the onset of internal cracking. At very high strain rates (> 10^{-4} s^{-1}) the initial yield becomes the sole yield, i.e. onset of internal cracking is followed immediately by fracture of the entire specimen.

Comparable data for coarse-grained ice are not yet available, but current work on coarse-grained old sea ice of low salinity suggests that there may be only one identifiable yield point, with strains at that point always well below 1%.

Multiaxial stress states. For multiaxial stress states, strength is best specified by a formal failure criterion, such as an equation or graph describing the failure envelope in principal stress space. A general criterion is hard to formulate even for isotropic ice (none of the classical criteria are broadly applicable), and for anisotropic ice there are very great difficulties. Consequently, engineers often have to get by with the most primitive of assumptions, e.g. failure occurring when the major principal stress reaches σ_T or σ_C, depending on the nature of the problem.

There have been speculations about the qualitative forms of failure criteria for isotropic ice [66], drawing on the observed facts that: (i) hydrostatic pressure has little effect on shearing at very low creep rates, (ii) moderate pressure increases strength and deformation resistance at high strain rates, (iii) high pressure lowers the deformation resistance at all rates, (iv) the envelope intersects the principal stress axes at σ_T and σ_C, and intersects the hydrostat at the pressure for the phase transition from Ice Ih to water.

In addition to data for σ_T and σ_C, and for phase transition pressures, there are a few data sets from triaxial tests ($\sigma_2 = \sigma_3$) in the compression-compression quadrants [67], and in the tension-tension quadrants [68]. There are also some data from biaxial tests (σ_3=0) [69]. However, when the number and range of potential variables are considered, these results are too fragmentary to provide a clear picture for this review.

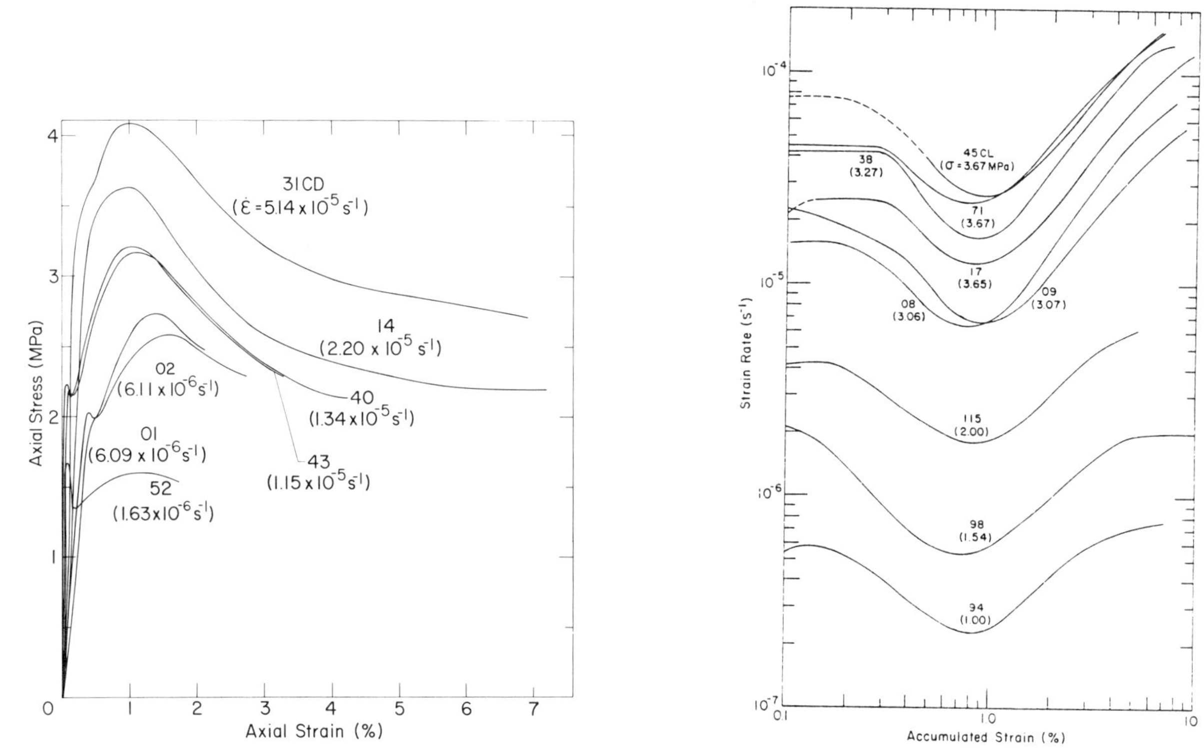

Figure 24. Example of (a) stress/strain curves and (b) creep curves for fine-grained non-saline ice at -5°C [52].

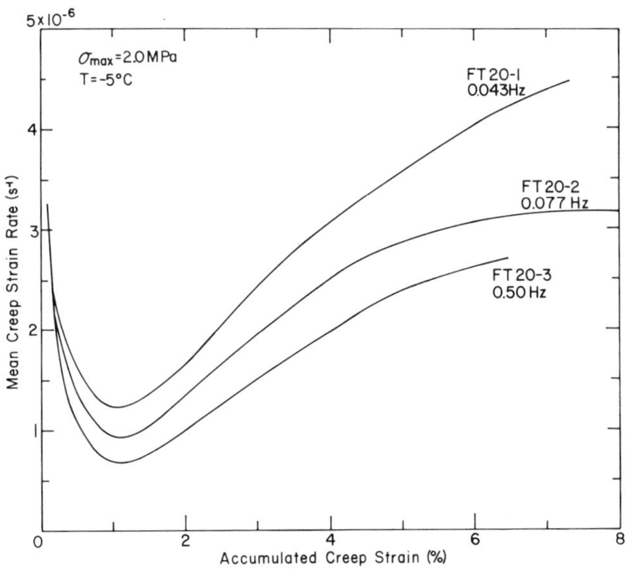

Figure 25. Examples of mean creep curves for tests on fine-grained ice in which compressive stress cycles between 0 and 2 MPa at the frequencies indicated [65].

Fracture toughness. The fracture toughness of ice has attracted considerable interest in recent years, but published data have to be approached with caution. The general concept is clearly applicable to ice under elastic/brittle conditions , but it is irrelevant when rates and temperatures are such that ice yields by flow and recrystallization rather than by the nucleation and propagation of cracks. The flexural tests typically used to measure K_{Ic}, the critical stress intensity factor for "Mode I" crack extension, can be interpreted by elastic theory at high rates and low temperature, but the elastic assumptions became progressively worse as rates and temperatures produce greater ductility.

Where ice is elastic and brittle, K_{Ic} should be predictable from Young's modulus E ($\approx$ 10 GPa) and the specific surface energy γ ($\approx$ 0.1 J/m^2). For plane stress,

$$K_{Ic} \rightarrow (2E\gamma)^{1/2} \approx 45 \text{ kN-m}^{-3/2}$$

Measured values of K_{Ic} do appear to have a lower limit close to this value. More typical measured values are around 100 kN-m$^{-3/2}$, implying that γ_p, the specific energy for plastic working of Orowan/Irwin theory, is about 5 γ. With this prima facie evidence that simple Griffith theory might be applicable for the elastic/brittle condition, one is tempted to calculate σ_T from E, γ and the controlling crack length 2c. For plane stress,

$$\sigma_T = \left(\frac{2}{\pi}\right)^{1/2} \left(\frac{E\gamma}{c}\right)^{1/2} \approx \frac{2.52 \times 10^4}{\sqrt{c}} \text{ Pa}$$

where c is in metres. If we make a guess that 2c is equal to the grain size of the ice, calculated values of σ_T are as shown in Figure 26, which also gives measured values of σ_T.

When conditions are such that ice has some ductility, it might be expected that "toughness" would increase with increasing temperature and decreasing strain rate. With the exception of one data set, the

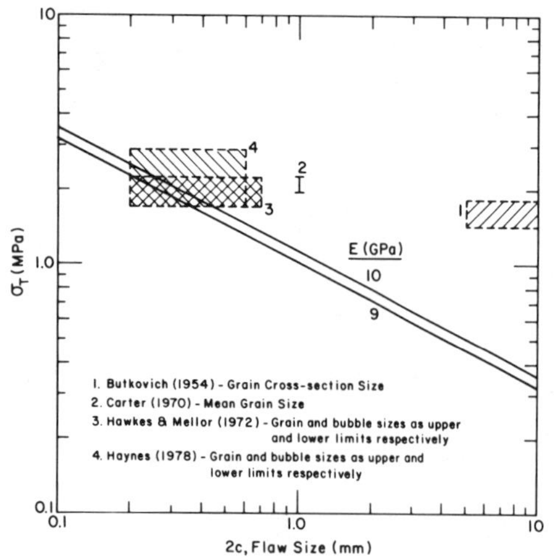

Figure 26. Comparison of theoretical tensile strength with measured values, assuming that flaw size equals grain size [70].

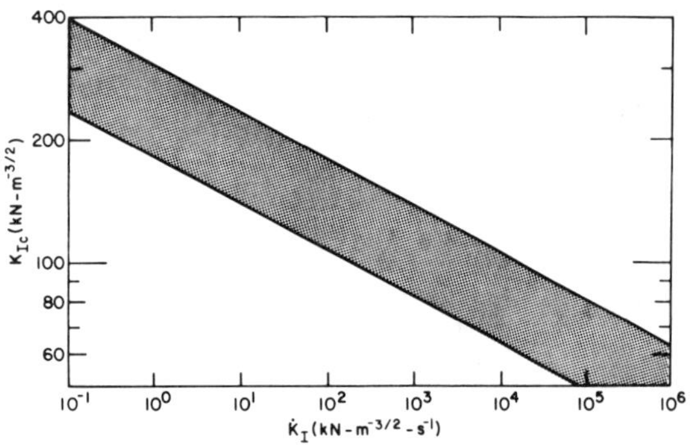

Figure 27. Effect of loading rate on K_{Ic} for non-saline ice [70, 72].

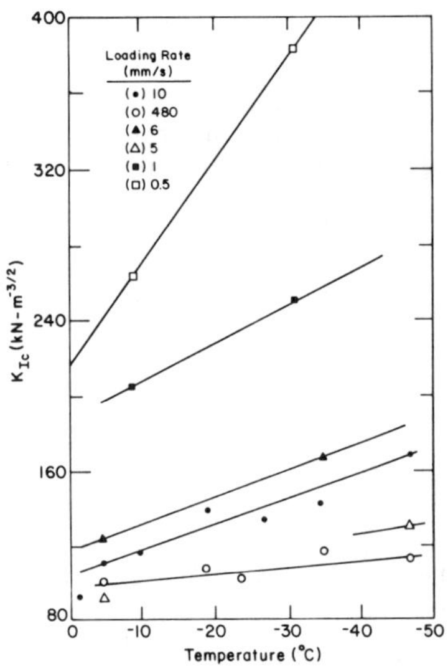

Figure 28. Variation of K_{Ic} with temperature and loading rate [73].

experimental evidence supports this expectation for rate effects (Fig. 27). However, test data show that K_{Ic} increases as temperature decreases (Fig. 28), contradicting the simple expectation. Looking at things another way, K_{Ic} can be expressed as

$$K_{Ic} = \sigma_T (\pi_c)^{1/2}$$

and this suggests that K_{Ic} might vary with rate and temperature in the same way as σ_T. The expectation would then be for K_{Ic} to be insensitive to rate, and for K_{Ic} to increase slowly with decreasing temperature. The latter appears to be borne out when testing rates are high (Fig. 28). However, another possibility is that the existence of liquid films or liquid-like layers in the grain boundaries might lead to a Rehbinder, or Joffe, effect [2]. Since the liquid/solid value of γ is about 30% of the vapour/solid value, K_{Ic} could be halved by intrusion of a liquid film into a growing crack. Such intrusion would probably not occur where rates are very high, irrespective of temperature, but it could occur with a combination of high temperature and low test rate.

A more complete discussion of the fracture toughness of ice, and the underlying theory, is given elsewhere [2].

3.3 Sea Ice

Precise measurements of the basic mechanical properties of sea ice are not plentiful, partly because the material is difficult to work with (brine mobility, complex structure), and partly because of practical demands for relatively crude field data. Consequently we have to draw upon experimental results for non-saline ice for a background picture of how sea ice might behave in a general sense. In assessing the sea ice data, it may be helpful to regard salinity as a major new variable, with freshwater ice of zero salinity representing a reference state.

Salinity has a direct influence on porosity, since salts rejected by the ice crystals during freezing form concentrated brine, which is distributed through the ice mass in pores. At any given temperature, the volume of brine-filled pores ("brine porosity") increases with increase of overall salinity. However, the brine-filled pores are not the only pores in sea ice; there are also gas bubbles, and the total porosity is the "gas porosity" plus the "brine porosity." In the past, the "gas porosity" was usually unknown, and brine porosity was substituted for total porosity. There is now a simple method for overcoming this problem [74]. As temperature decreases in ice of a given salinity, brine volume decreases, since equilibrium concentration has to be maintained. This means that temperature has a dual effect on the mechanical properties of saline ice -- it affects the ice matrix, much as it does in non-saline ice, but it also changes the porosity. Because increasing temperature and increasing porosity both tend to lower the stiffness, the deformation resistance and the strength of ice, it might be expected that temperature effects in sea ice would be stronger than those in non-saline ice.

Mechanical properties of sea ice are often plotted against porosity and against temperature. When examining such plots, it should be understood that these two variables are not normally independent of each other. Temperature is usually an implicit variable in porosity effects (porosity is varied by changing the temperature in ice of given salinity). Similarly, porosity is often an implicit variable in temperature effects.

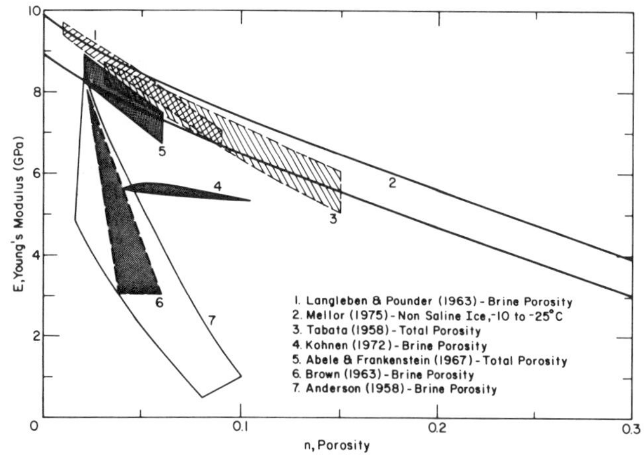

Figure 29. Young's modulus as a function of porosity [2, 63, 75, 76, 77, 78, 79, 80].

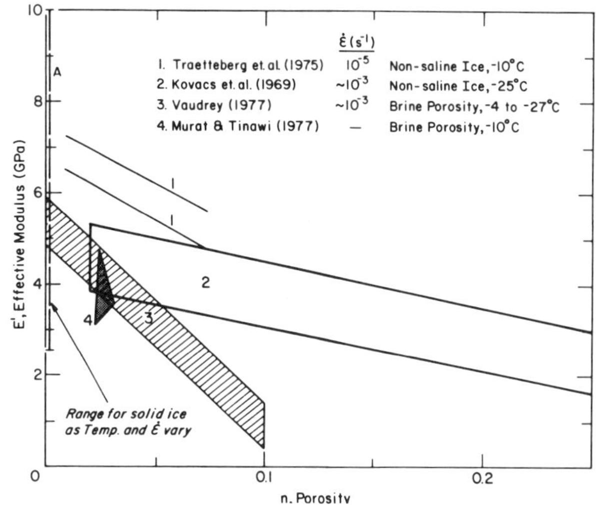

Figure 30. Summary of data for effective modulus E' plotted against porosity [2, 81, 82, 83, 84].

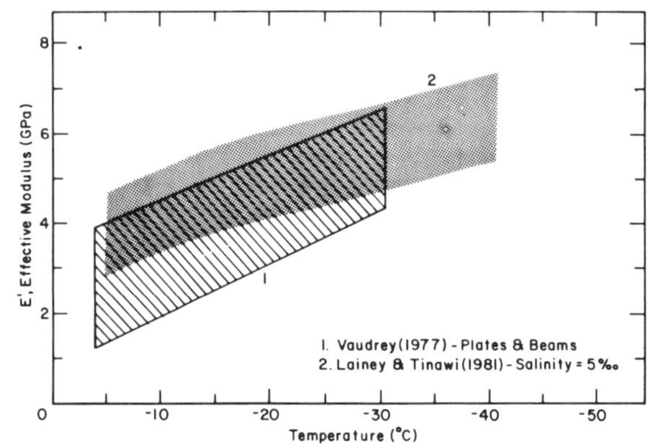

Figure 31. Summary of values for effective modulus E' plotted against temperature [2, 83, 85].

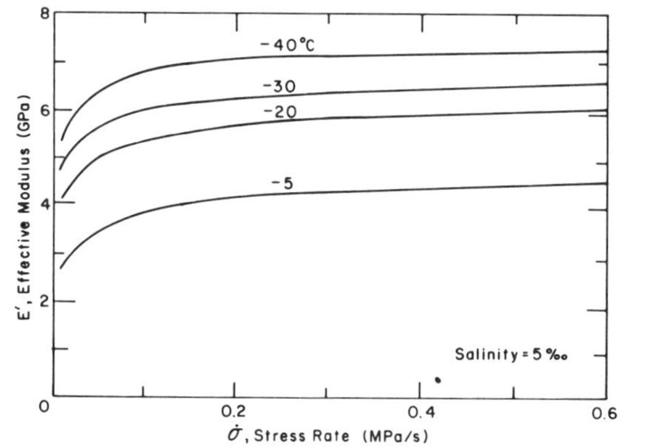

Figure 32. Variation of effective modulus E' with stress rate and temperature [85].

Elastic moduli. Measurements of Young's modulus E have been made by methods depending on high frequency waves, pulses or vibrations. Effective moduli E' have been measured by quasi-static tests (typically beam flexure), both in the laboratory and in the field.

Figure 29 gives a general impression of E values for sea ice over a range of porosity. It also provides a comparison with non-saline ice. The results for small-scale samples (data bands 1, 3, 5) show good agreement with non-saline ice (data band 2). The values obtained from seismic tests and flexural waves over wide areas are mostly much lower than E for non-saline ice, but this is not too surprising when all of the complications of the ice sheet are considered.

Values of E' are indicated in Figure 30, and are compared with E' values for non-saline ice. There is close agreement between sea ice and non-saline ice at low porosity, and the low values for sea ice at higher porosities can probably be attributed to the combined effects of porosity and temperature. Data band 3 for E' in Figure 30 agrees well with data bands 6 and 7 for E in Figure 29.

The effect of temperature on E' is indicated in Figure 31. The slightly steeper trend in data band 1 is probably due to the fact that all tests were made at a single value of salinity. Figure 32 shows how E' varies with stress rate and temperature at a fixed value of salinity. At very low temperature and high stress (or strain) rate, E' approaches E.

Effective values of Poisson's ratio ν' for sea ice were measured in beam flexure experiments by Murat and Lainey [86]. For very low stress rates, ν' tended to the expected limit value of 0.5 (see page 244). For the highest test rates (0.6 MPa/s, and up to $1.6 \times 10^{-4}\ s^{-1}$), ν' had values between 0.35 and 0.4. The mean value for high rate decreased with decreasing temperature, from about 0.40 at -5°C, to 0.37 at -30° and -40°C. This observed temperature trend supports a speculation made much earlier by Weeks and Assur [6] on the basis of Soviet seismic data. Such a trend is opposite to what would be expected with air-filled pores, but it can be explained by expressing ν in terms of E and the bulk modulus K [2]:

$$\nu = (3K-E)/6K = (3-E/K)/6$$

Since K is about the same for ice and water, a small increase in the volume of water-filled pores should have little effect on the overall value of K. Thus the variation of ν with porosity will be controlled largely by variations in E, which decreases as porosity increases. The equation therefore predicts an increase of ν with increase of porosity. Because porosity increases with increasing temperature in saline ice, ν should increase with increasing temperature.

In columnar sea ice, anisotropy may have a

greater influence on ν than do temperature or porosity. Wang [88] found that sea ice was much stiffer in a direction parallel to the long axes of columns (vertical) than in the perpendicular direction (horizontal), giving ν' in the range 0 to 0.2 vertically and 0.8 to 1.2 horizontally.

Bulk modulus K and shear modulus G are not commonly measured as such in sea ice, but given pairs of values for E and ν they are easily calculated (see page 244).

Strength and deformation resistance. Uniaxial stress tests provide clear and unambiguous data if they are done well. Uniaxial compression tests have been applied to sea ice by many investigators, but uniaxial tension has rarely been attempted. The most common strength tests have involved flexure of beams or cantilevers. For laboratory experiments small beams are cut from an ice sheet, or saline ice is produced artificially. For large scale field tests, beams or cantilevers are sawn in the ice sheet, with the "fixed" ends still attached to the sheet (limited flexure at the beam root is still possible). For the reasons mentioned earlier (page243), beam tests can give misleading results, and beam data for non-saline ice have been ignored in this review. However, in the absence of adequate uniaxial data for sea ice, we have to make use of beam data, which do have special value when beam tests are regarded as analogue tests for plate flexure. To distinguish the "modulus of rupture" or "flexural strength" from uniaxial tensile strength σ_T, we use the symbol σ_{FT}. We have completely disregarded results from once-popular Brazil tests and ring tests (diametral compression of discs and annuli), since these test have proved unsuitable for ice [89].

The uniaxial compressive strength of sea ice is expected to vary with strain rate, temperature and porosity in a manner qualitatively similar to non-saline ice. Figure 33 represents some data by Wang [90] which conform to a power relation between strain rate and stress, with an exponent of about 4. Some other data, selected from results by Schwarz [91], are shown against strain rate in Figure 34. In Figure 35, the effect of temperature on σ_c is shown for sea ice and for some roughly comparable fresh-water ice (lake and river ice).

The effect of porosity on σ_c and other mechanical properties has traditionally been displayed by plotting the property against the square root of brine volume. For reasons discussed elsewhere [2], this practice is not followed here; brine volume is represented simply as brine porosity. Variation of σ_c with brine porosity at high strain rates ($\approx 10^{-3}\ s^{-1}$) is shown in Figure 36, which also brings out the well established fact that σ_c is strongly dependent on the direction of loading in columnar ice. Further evidence is given

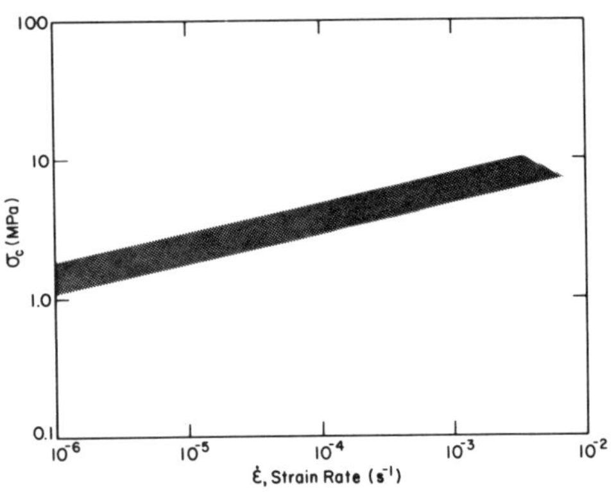

Figure 33. Uniaxial compressive strength of sea ice as a function of strain rate (data selected from [88]).

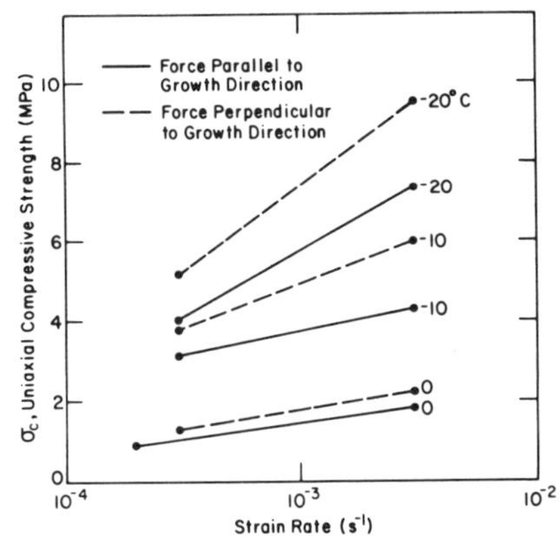

Figure 34. Variation of σ_c with strain rate, temperature and grain orientation (data selected from [91]).

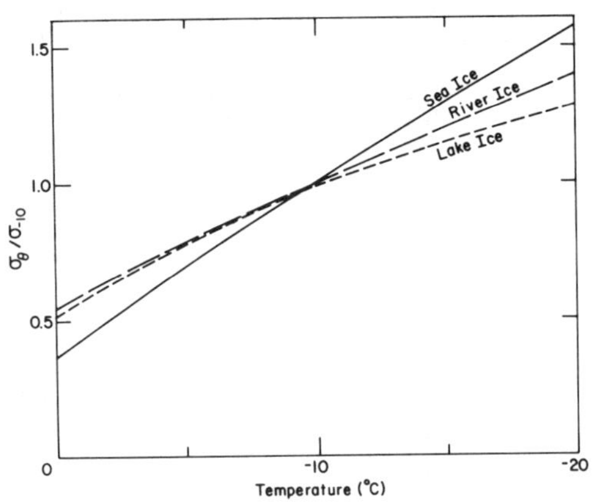

Figure 35. Variation of σ_c with temperature for three ice types (data from [91]).

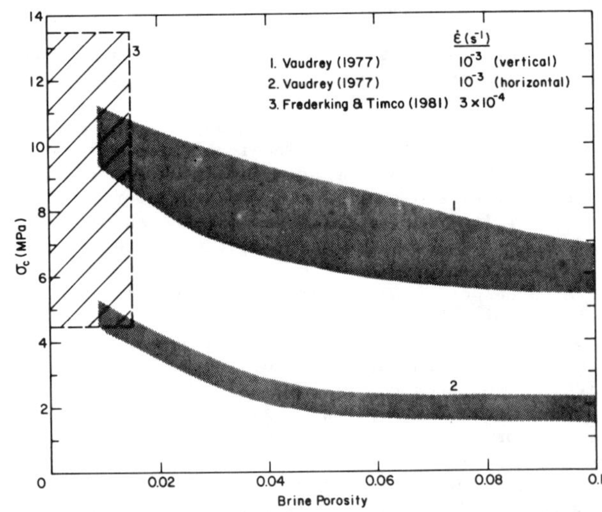

Figure 36. Summary of data for σ_c as a function of brine porosity [83, 92].

by Figure 37, which shows the ice to be weakest when major principal stress is at 45° to the direction of the c-axes (i.e., at 45° to the horizontal plane).

Figure 38 provides confirmation that σ_c decreases with increase of grain size at fairly high strain rates ($\geq 10^{-5}\ s^{-1}$).

Uniaxial tensile tests on sea ice have been rare because of the difficulty in maintaining perfect specimen axiality and in avoiding perturbations of the stress field. These difficulties are now being overcome, but new data have not yet appeared. The only usable published data seem to be those represented in Figure 39, where σ_T is plotted against temperature for two salinities and two loading directions. As might be expected, σ_T increases with decreasing temperature and decreasing salinity. It is greater for vertical specimens of columnar ice (Fig. 39a) than for horizontal specimens (Fig. 39b). The temperature effect implies a porosity effect, which is illustrated by a re-plot of mean values from Figure 39 in Figure 40.

Flexural strength σ_{FT} has been measured in many test programs involving both laboratory work and field work. The variability of results is somewhat daunting, as can be seen from the summary in Figure 41, where σ_{FT} is shown against brine porosity. Figure 42 summarizes some data on the variation of σ_{FT} with temperature, but the results should be

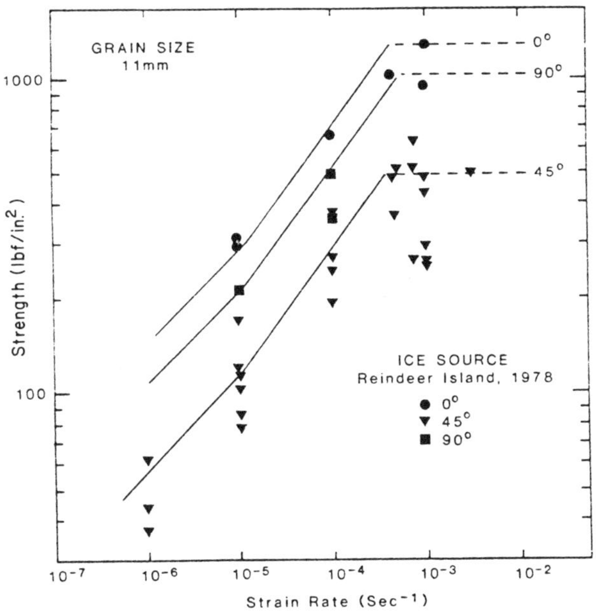

Figure 37. Variation of σ_c with strain rate in columnar sea ice at -10°C. The three data sets show the effect of crystal orientation [93].

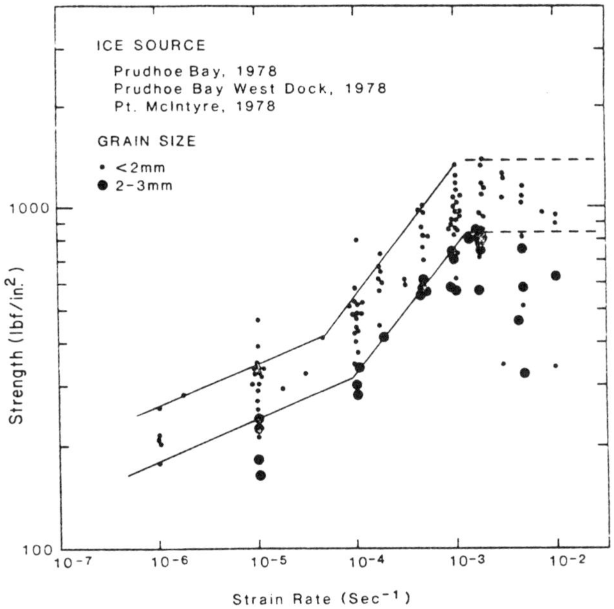

Figure 38. Variation of σ_c with strain rate in granular sea ice at -10°C. The effect of grain size is also indicated [93].

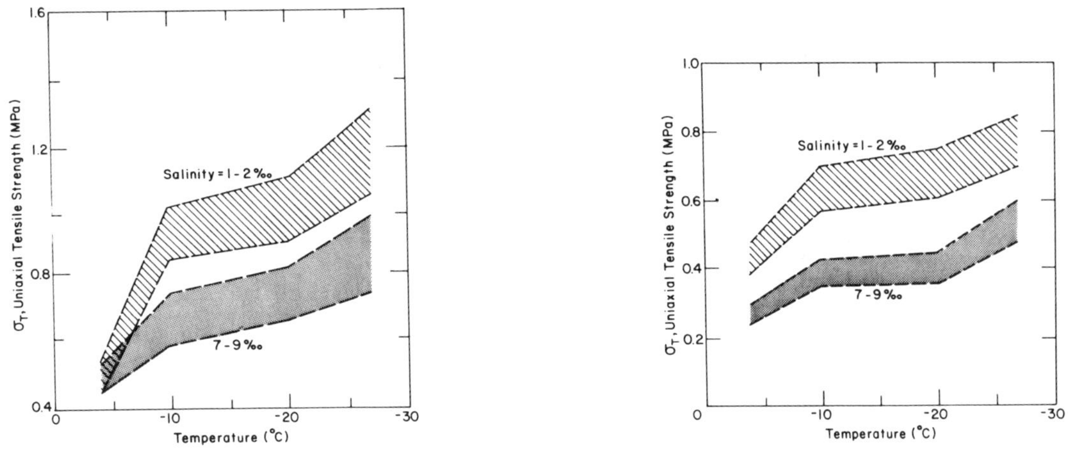

Figure 39. Variation of uniaxial tensile strength with temperature and salinity for sea ice: (a) vertical specimens, (b) horizontal specimens [94].

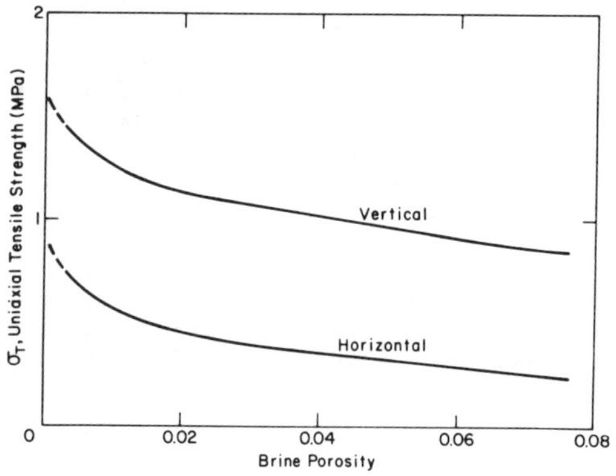

Figure 40. σ_T as a function of brine porosity [94].

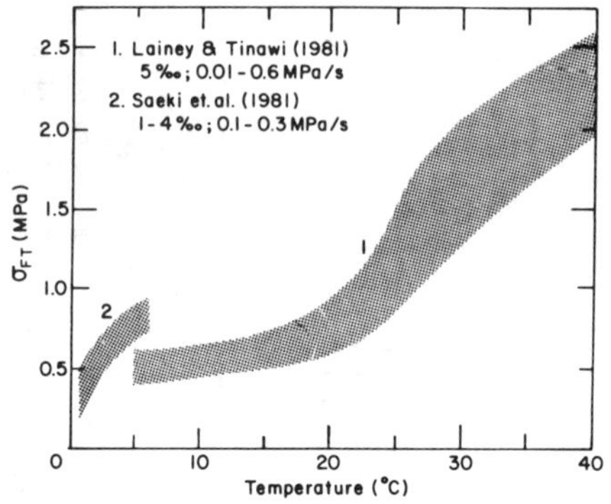

Figure 42. Variation of flexural strength with temperature [85, 98].

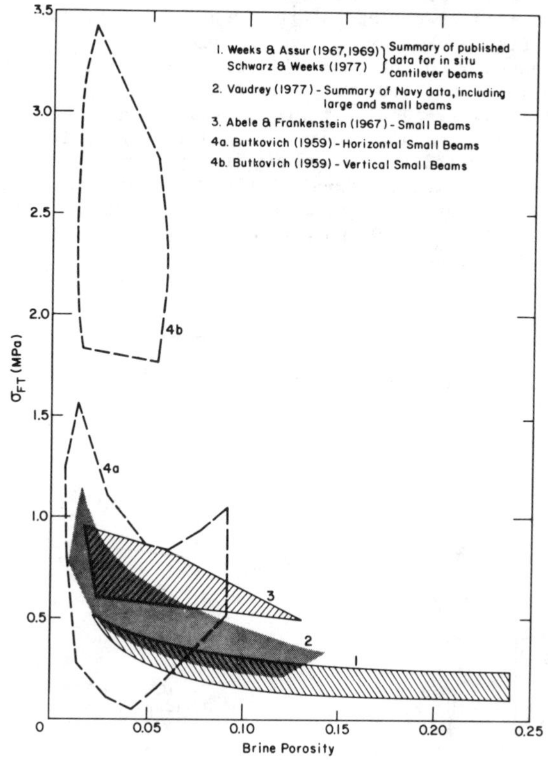

Figure 41. Summary of data for σ_{FT} as a function of brine porosity [2, 6, 7, 78, 83, 96].

treated with caution because of probable departure from ideal elasticity at high temperatures. A discussion of rate effects, temperature effects and other complications is given elsewhere [2], but it might be worth mentioning here that in situ beams are subject to large variations in grain structure, steep temperature gradients, beam-root stress concentrations, scale effects and, at high rates, inertial effects in the underlying water.

Fracture toughness measurements on sea ice have been reported from several studies [72, 83, 99, 100, 102). The effect of loading rate on K_{Ic} is shown in Figures 43 and 44, and it can be seen that, for high loading rates, K_{Ic} tends to values that are close to the theoretical "Griffith" value for non-saline ice (see page 249). In Figure 43, temperature seems to have very little effect on K_{Ic}, in contrast to the trend shown earlier in Figure 28 for comparable deflection rates. One study [102] has given some evidence of a decrease in K_{Ic} with increase of brine porosity; another study [83] purports to show the same trend, but the data points have no significant correlation. In both Figure 43 and Figure 44, it seems that K_{Ic} increases with an increase in grain size.

Some conventional triaxial tests ($\sigma_1 \neq \sigma_2 = \sigma_3$) have been made on sea ice at moderately high rates. Soviet tests [103] on artificial and natural sea ice show the major principal stress σ_1 increasing with σ_2, σ_3, and the maximum shear stress, $(\sigma_1 - \sigma_2)/2$, increasing nonlinearly with the normal stress on the plane of maximum shear. The failure value of σ_1 was an order of magnitude higher than σ_c with σ_2 around 4 MPa. Under confining pressure, the failure stress decreased with increasing salinity and increasing temperature, just as it does in the uniaxial stress

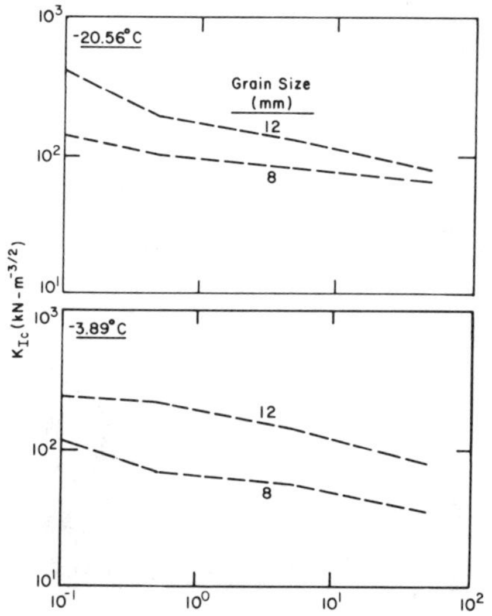

Figure 43. Variation of K_{Ic} with loading rate, grain size and temperature [109] for columnar fresh water ice.

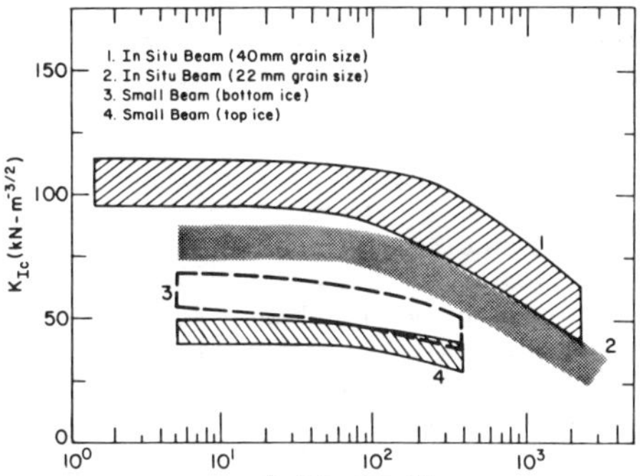

Figure 44. Effect of loading rate on K_{Ic} for sea ice (data from [72]).

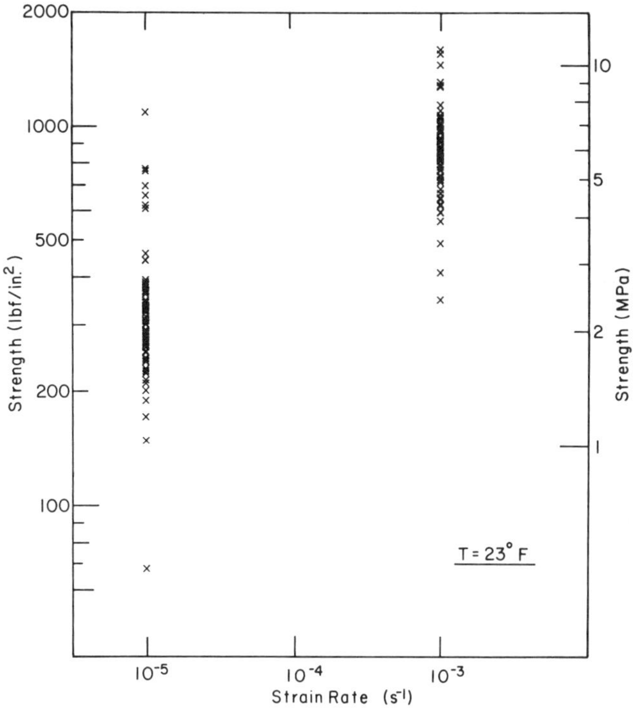

Figure 45. Uniaxial compressive strength of multiyear sea ice at -5°C and strain rates of 10^{-5} and 10^{-3} s^{-1} [43].

state. Isotropic fine-grained ice behaved differently from anisotropic ice, and strength varied with loading direction in anisotropic ice. Values of ϕ for the Mohr-Coulomb criterion were mostly in the range 30° to 50°, with extreme values of 14° and 55°.

"True" triaxial tests ($\sigma_1 \neq \sigma_2 \neq \sigma_3$) have been made on anisotropic saline ice [104], but so far the results are too complicated to be summarized concisely.

One potential problem in triaxial testing is that failure could be influenced by differences of loading rates for the principal stresses, or by variations in the ratio σ_1/σ_2. Because a complex stress field in natural ice is likely to fluctuate with the ratio of principal stresses staying fairly constant, a new triaxial test device has been developed to keep the ratio σ_1/σ_2 constant throughout a test [105].

Very little work has been done on the mechanical properties of multi-year sea ice, and it is still hard to generalize about differences between "new" and "old" ice. Some studies suggest that multiyear ice is weaker than first-year ice [106], but recent detailed studies [43] do not support this idea. Actually, it is not easy to make comparisons because of the great variability of strength in multiyear ice, which contains many different types of ice. Figure 45 gives some values of σ_c at strain rates of 10^{-5} and 10^{-3} s^{-1}, with a temperature of -5°C. If Figure 45 is compared with Figures 37 and 38, it can be seen that the strength range for multiyear ice is similar to that for granular first-year ice at -10°C. At 10^{-3} s^{-1}, the multiyear ice is about the same strength as columnar first-year ice, but at 10^{-5} s^{-1} the columnar ice is weaker than multiyear ice except when tested in the "hard fail" direction.

Uniaxial tension tests on multiyear ice are being made as part of a current program at CRREL, and the indications so far are that σ_T is always less than 1 MPa.

3.4 Ice Island Ice

The earlier notes on ice islands indicate a rather complicated structure, with four different types of ice distinguishable. Since mechanical tests have not been made on ice islands to any significant extent, it is useful to simplify the earlier picture so as to draw some conclusions about probable mechanical behaviour.

Of the listed constituents for ice islands, two are non-saline ice: lake ice, and snow ice (which is just glacier ice). The properties of lake ice as such have not been reviewed here, but they can be inferred from the general properties of granular and columnar non-saline ice. A more specific review of lake ice properties can be found elsewhere [7]. Snow ice is the material that makes up glaciers, ice caps and ice shelves. Unless it has been metamorphosed by strong shearing, it is typically isotropic, with grain size and porosity varying. For polar glacier ice, grain size varies from around 1 mm in "snow," where the bulk density is less than 0.8 Mg/m^3, to around 5 mm in very dense ice (porosity $\rightarrow$ 0) from the deep layers. The properties of porous glacier ice have already been summarized, and the properties of "snow" and "snow ice" for lower ranges of bulk density are dealt with elsewhere [63, 107].

The remaining two constituents of ice islands are saline ice types (sea ice, brackish ice), but salinity is likely to be very low. The sea ice component is very old sea ice, and therefore perhaps comparable to some of the multiyear ice that has just been discussed. Not much is known about ice formed from brackish water, but a first guess might be that it would be similar to lake ice.

Actually, the small ice island fragments that drift into shallow coastal waters have suffered ablation from both top and bottom surfaces, and they may not include all the ice types that have been listed. From impressions gained during visits to a number of small ice islands, and from drilling and blasting work on ice islands, one of us is inclined to regard them simply as small icebergs.

3.5 Fragmented Ice

There is a tendency to assume that the problems created by floating ice are solved once the ice is broken. In fact, accumulations of fragmented ice can, in some circumstances, resist ship movement and load structures much more severely than an unbroken ice sheet.

Fragmented ice covers a broad spectrum, from fine-grained mush ice, through blocky brash ice, up to floe ice, rafted ice, and ridged ice. As far as ships and structures are concerned, the accumulations of fragmented ice that are of most concern are mush, brash, and first-year pressure ridges.

Mush ice is something like waterlogged snow, with fluid properties while it is floating freely, and high cohesion when it is compacted or drained. Although it can cause real problems for ships and marine structures, systematic study of mush ice is only just starting and, apart from a small amount of information in the snow mechanics literature, data are not yet available.

Brash ice has more or less equant particles in the size range 0.02 to 2.0 m. The tendency has been to treat it analytically as a granular "$c-\phi$" material that conforms to a Mohr-Coulomb failure criterion, but for horizontal penetration of a uniform layer the stress-free upper and lower boundaries appear to permit yielding in conformance to a criterion of the von Mises type [108]. However, measurements of ϕ have been made in a number of studies.

Large ice blocks pushed together into pressure ridges also form a "granular" material which initially has cohesion c and internal friction (shear resistance) ϕ. The scale of this material is too big for conventional measurements of its bulk properties, but some deductions can be made from analysis of natural processes.

The properties of fragmented ice are summarized and discussed elsewhere [2].

4. CONCLUSION

From this brief survey of the different varieties of ice encountered in arctic waters, it is evident that there are large gaps in the available data. There are virtually no data derived from direct studies on icebergs and ice islands. Although major studies on these ice masses are perhaps unnecessary, some exploratory studies would be useful in order to confirm that icebergs and ice islands are, in fact, similar to their parent ice bodies. Studies on first-year sea ice have been made almost entirely on ice which has formed near the coast. This kind of sea ice is believed to have strong c-axis alignments in the horizontal plane, but only in a few cases has the orientation of the stress field relative to c-axis orientation been taken into consideration. The most significant shortcoming at the present time is lack of information on multiyear ice, which is the most common material in the central arctic pack, and probably the most threatening material for offshore structures. Finally, little is known about the properties of marine frazil ice. It may be more common than was previously thought to be the case, and the fine-grained coherent ice formed from frazil may be stronger than first year congelation ice.

As far as the acquisition and interpretation of test data is concerned, the field of sea ice research is moving into a new phase. Some of the older test methods, and the data generated by them, have had to be discarded, a process that has forced both of the present authors to junk some of their older work. Current activity is based on test techniques that are more refined and more carefully selected, and there are moves towards standardization of test techniques worldwide. Tests can be designed, conducted and interpreted with better appreciation for the relevant constitutive relations and failure criteria, and with less slavish conformance to methodologies borrowed from other technical fields. Some experimental areas are still deficient, and more emphasis needs to be given to multiaxial stress states and to loadings of long duration.

The things we call basic mechanical properties are, of course, only meaningful within the framework of underlying theory, and they have to be applied to engineering design through the appropriate theory. At present there is only loose coordination of ice engineering research in the areas of theoretical mechanics, experimental determination of properties, and solution of practical boundary value problems. Consequently, there is a danger that efforts in these various areas might be mismatched. For example, theoreticians might be calling for highly complex data from polyaxial tests on rate-dependent anisotropic material at a time when design engineers are struggling to progress beyond simple elastic analysis and maximum principal stress failure criteria. Some measure of coordination is needed in order to satisfy the legitimate demands of both basic research and practical engineering.

In spite of all these difficulties, the general situation is encouraging. Ice mechanics has made considerable progress in recent years and may, in fact, be making some new contributions to applied mechanics and materials science. There is a good collection of basic data, and many practical problems can be tackled with confidence. Perhaps the greatest problem at the present time is that the accelerated leasing schedule for the Beaufort, Chukchi and Bering Seas will create a heavy demand for high quality data, while the number of experienced research people, the physical facilities for ice research, and the levels of funding support are all severely limited. To overcome this problem, the very least that is needed is an expansion and intensification of collaboration between academia, industry and government.

REFERENCES

1. Weeks, W.F. and Ackley, S.F., The growth structure and properties of sea ice. USACRREL Monograph 82-1, 130 pp., 1982.
2. Mellor, M., Mechanical behavior of sea ice. USACRREL Monograph 83-1, 92 pp., 1983.
3. Coon, M.D., Brown, C.B., Cox, G.F.N., Ralston, T.D., Shapiro, L. and Weeks, W.F., Research in Sea Ice Mechanics, Marine Board, Assembly of Engineering, National Res. Council, National Acad. Press, 80 pp., 1981.
4. Weeks, W.F. and Mellor, M., Some elements of iceberg technology, in Iceberg Technology, ed. A.A. Husseiny, pp. 45-98, Pergamon Press, 1978.
5. Schwarz, J. and Weeks, W.F., Engineering properties of sea ice. J. Glaciology, vol. 19, no. 81, pp. 499-530, 1977.
6. Weeks, W.F. and Assur, A., The mechanical properties of sea ice. USACRREL Monograph II - C3, 80 pp., 1967.
7. Weeks, W.F. and Assur, A., Fracture of lake and sea ice. USACRREL Research Report 269, 77 pp., 1969.
8. Hobbs, P., Ice Physics, Oxford Univ. Press, 837 pp., 1974.

9. Bragg, W.H., The crystal structure of ice. Proc. Phys. Soc., vol. 34, pp. 98-103, 1922.
10. Lepparänta, M., Observations of icebergs in the Barents Sea in July 1980. Iceberg Research, No. 2, pp. 3-4, 1982.
11. Robe, R.Q., Iceberg drift and deterioration, in Dynamics of Snow and Ice Masses, ed. S.C. Colbeck, pp. 211-59, Academic Press, New York, 1980.
12. Rigsby, G.P., Study of ice fabrics, Thule area, Greenland, Snow, Ice and Permafrost Research Establishment Report 26, 6 pp., 1955.
13. Gow, A.J. and Williamson, T., Rheological implications of the internal structure and crystal fabrics of the West Antarctic ice sheet as revealed by deep core drilling at Byrd Station. Geol. Soc. Amer. Bull., vol. 87, pp. 1665-77, 1976.
14. Rigsby, G.P., The complexities of the three-dimensional shape of individual crystals of glacier ice. J. Glaciology, vol. 7, no. 50, pp. 233-52, 1968.
15. Scholander, P.F. and Nutt, D.C., Bubble pressure in Greenland icebergs, J. Glaciology, vol. 3, no. 28, pp. 671-678, 1960.
16. Smith, E.H., Arctic ice, with special reference to its distribution in the North Atlantic Ocean, U.S. Coast Guard Bull. No. 19: I-X, 221 pp., 1931.
17. Weeks, W.F., The structure of sea ice: a progress report, in Arctic Sea Ice, U.S. Nat. Acad. Sciences - Nat. Res. Council Pub. 598, pp. 96-98, 1958.
18. Cherepanov, N.V., Spatial arrangement of sea ice crystal structure. Prob. Arkt. i Antarkt., vol. 38, pp. 176-181, 1971.
19. Weeks, W.F. and Gow, A.J., Preferred crystal orientations along the margins of the Arctic Ocean. J. Geophys. Res., vol. 84, no. C10, pp. 5105-5121, 1978.
20. Weeks, W.F. and Gow, A.J., Crystal alignments in the fast ice of arctic Alaska, J. Geophys. Res., vol. 85, no. C2, pp. 1137-1146.
21. Kovacs, A. and Morey, R., Radar anisotropy of sea ice due to preferred azimuthal orientation of the horizontal c-axes of ice crystals, J. Geophys. Res., vol. 83, no. C12, pp. 6037-6046, 1978.
22. Langhorne, P., Laboratory experiments on crystal orientation in NaCl ice. Annals of Glaciology, vol. 4, 1983.
23. Kovacs, A. and Morey, R.M., Investigations of sea ice anisotropy, electromagnetic properties, strength, and under-ice current orientation, CRREL Report 80-20, 18 pp., 1980.
24. Tabata, T. and Ono, N., On the crystallographic study of several kinds of ice, Low Temp. Sci., vol. A20, pp. 199-214, 1962.
25. Gow, A.J. and Weeks, W.F., The internal structure of fast ice near Narwhal Island, Beaufort Sea, Alaska, CRREL Report 77-29, 8 pp., 1977.
26. Lofgren, G. and Weeks, W.F., Effect of growth parameters on the substructure spacing in NaCl ice crystals, J. Glaciology, vol. 8, no. 52, pp. 153-164, 1969.
27. Nakawo, M. and Sinha, N.K., Brine layer spacing of first-year sea ice. Draft manuscript, 1982.
28. Zotikov, I.A., Zagorodnov, V.S. and Raikovski, J.V., Core drilling through the Ross Ice Shelf (Antarctica) confirmed basal freezing. Science, vol. 207, no. 4438, pp. 1463-1465, 1980.
29. Weeks, W.F. and Assur, A., Structural control of the vertical variation of the strength of sea and salt ice, in Ice and Snow - Processes, Properties and Applications, ed. W.D. Kingery, pp. 258-276, MIT Press, 1963.
30. Weeks, W.F. and Lofgren, G., The effective solute distribution coefficient during the freezing of NaCl solutions, in Physics of Snow and Ice, ed. H. Oura, Inst. Low Temp. Sci., Hokkaido, vol. I, no. 1, pp. 579-597, 1967.
31. Nakawo, M. and Sinha, N.K., Growth rate and salinity profile of first-year sea ice in the high Arctic, J. Glaciology, vol. 27, no. 96, pp. 315-330, 1981.
32. Lake, R.A. and Lewis, E.L., Salt rejection by sea ice during growth, J. Geophys. Res., vol. 75, no. 3, pp. 583-597, 1970.
33. Niedrauer, T.M. and Martin, S., An experimental study of brine drainage and convection in young sea ice, J. Geophys. Res., vol. 84, no. C3, pp. 1176-1186, 1979.
34. Assur, A., Composition of sea ice and its tensile strength, in Arctic Sea Ice, U.S. Nat. Acad. Sciences - Nat. Res. Council Pub. 598, pp. 106-138, 1958.
35. Sinha, N.K., Technique for studying structure of sea ice, J. Glaciology, vol. 18, no. 79, pp. 315-323, 1977.
36. Gow, A.J., Ackley, S.F., Weeks, W.F. and Govoni, J.W., Physical and structural characteristics of Antarctic sea ice. Annals of Glaciology, vol. 3, pp. 113-117, 1982.
37. Cherepanov, N.V., Using the methods of crystal optics for determining the age of drift ice. Problemy Arktiki, vol. 2, pp. 179-184, 1957.
38. Schwarzacher, W., Pack-ice studies in the Arctic Ocean, J. Geophys. Res., vol. 64, pp. 2357-2367, 1959.
39. Maykut, G. and Untersteiner, N., Some results from a time-dependent thermodynamic model of sea ice, J. Geophys. Res., vol. 67, pp. 1550-1567, 1971.
40. Walker, E.R. and Wadhams, P., Thick sea ice floes, Arctic, vol. 32, no. 2, pp. 140-147, 1979.
41. Wadhams, P., Sea ice topography of the Arctic Ocean in the region 70° W to 25°E. Phil. Trans. Roy. Soc. London, vol. A302, no. 1464. pp. 45-85.
42. Tucker, W.B., Sodhi, D. and Govoni, J., Structure of first-year pressure ridge sails in the Prudhoe Bay region, in The Beaufort Sea - Physical and Biological Environment, eds. P. Barnes, D. Schell and E. Reimnitz, Academic Press, 1983.
43. Cox, G.F.N., Richter, J., Weeks, W.F., Mellor, M. and Bosworth, H., The mechanical properties of multi-year sea ice, Phase I: Test results. USACRREL Report 83- , 1983.
44. Crary, A.P., Arctic ice island and ice shelf studies, in Scientific Studies at Fletcher's Ice Island, T-3, 1952-1955, Vol. III, ed. V. Bushnell, pp. 1-37, Air Force Cambridge Research Center Geophysical Research Paper No. 63 (AFCRC-TR-232(3)), 1960.
45. Marshall, W.F., Structure and stratigraphy of T-3 and the Ellesmere Ice Shelf, in Scientific Studies at Fletcher's Ice Island, T-3, 1952-1955, Vol. III, ed. V. Bushnell, pp. 45-57, Air Force Cambridge Research Center Geophysical Research Paper No. 63 (AFCRC-TR-232(3)), 1960.
46. Ragle, R.H., Blair, R.G. and Persson, L.E., Ice core studies of Ward Hunt Ice Shelf, 1960. J. Glaciology, vol. 5, no. 37, pp. 39-59, 1964.
47. Smith, D.D., Ice lithologies and structure of ice island Arlis II. J. Glaciology, vol. 5, no. 37, pp. 17-38, 1964.
48. Nakaya, U., Muguruma, J. and Higuchi, K., Glaciological studies on Fletcher's Ice Island (T-3). Arctic Institute of North America Research Paper No. 21, 1962.
49. Cherepanov, N.V., Structure of sea ice of great

thickness. Trudy Arkt. i Antarkt. N.I. Institut, vol. 367, pp. 13-18, 1964.
50. Lyons, J.B., Savin, S.M. and Tamburi, A.J., Basement ice, Ward Hunt Ice Shelf, Ellesmere Island, Canada. J. Glaciology, vol. 10, no. 58, pp. 93-100, 1971.
51. Mellor, M., Glacier mechanics, IAHR Internat. Sympos. on Ice, Quebec, Canada, Vol. II, pp. 455-473, 1981.
52. Mellor, M. and Cole, D., Deformation and failure of ice under constant stress or constant strain-rate, Cold Regions Sci. Tech., vol. 5, pp. 201-219, 1982.
53. Mellor, M. and Cole, D.M., Stress/strain/time relations for ice under uniaxial compression, Cold Regions Sci. Tech., 1983.
54. Gold, L.W., Some observations of the dependence of strain on stress for ice, Canad. J. Phys., vol. 36, no. 10, pp. 1265-1275, 1958.
55. Hawkes, I. and Mellor, M., Deformation and fracture of ice under uniaxial stress. J. Glaciology, vol. 11, no. 61, pp. 103-131, 1972.
56. Sinha, N.K., Short-term rheology of polycrystalline ice, J. Glaciology, vol. 21, no. 85, pp. 457-473, 1978.
57. Sinha, N.K., Rheology of columnar-grained ice, Experimental Mechanics, vol. 18, no. 12, pp. 464-470, 1978.
58. Carter, D., Brittle fracture of snow ice, Proceed. IAHR Sympos. Reykajvik, Iceland, Paper 5.2, 1970.
59. Haynes, F.D., Effect of temperature on the strength of snow-ice, USACRREL Report 78-27, 18 pp., 1978.
60. Mellor, M. and Testa, R., Effect of temperature on the creep of ice, J. Glaciology, vol. 8, no. 52, pp. 131-145, 1969.
61. Paterson, W.S.B. and Budd, W.F., Flow parameters for ice sheet modeling, Cold Regions Sci. Tech., vol. 6, no. 2, pp. 175-177, 1982.
62. Budd, W.F., The dynamics of ice masses. Austrailian Nat. Antarctic Research Expedition, Scientific Report 108, 1969.
63. Mellor, M., A review of basic snow mechanics. Snow Mechanics Sympos., Grindelwald, IAHS Pub. No. 114, pp. 251-291, 1974.
64. Currier, J.H. and Schulson, E.M., The tensile strength of ice as a function of grain size. Acta Metallurgica, vol. 30, pp. 1511-1514, 1982.
65. Mellor, M. and Cole, D., Cyclic loading and fatigue in ice, Cold Regions Sci. Tech., vol. 4, pp. 41-53, 1981.
66. Mellor, M., Mechanical properties of polycrystalline ice. Physics and Mechanics of Ice, IUTAM Symposium, Copenhagen, Springer-Verlag, pp. 217-245, 1979.
67. Jones, S.J., Triaxial testing of polycrystalline ice. Third Internat. Conf. on Permafrost, Edmonton, pp. 670-674, 1978.
68. Haynes, F.D., Tensile strength of ice under triaxial stresses. USACRREL Research Report 312, 1973.
69. Frederking, R.M.W., Plane-strain compressive strength of columnar-grained and granular snow ice. J. Glaciology, vol. 18, no. 80, pp. 505-516, 1977.
70. Deleted.
71. Butkovich, T.R., Ultimate strength of ice, USA SIPRE Research Report 11, 1954.
72. Urabe, N. and Yoshitake, A., Strain rate dependent fracture toughness (K_{Ic}) of pure ice and sea ice. Internat. Sympos. on Ice, IAHR, Quebec, pp. 410-420, 1981.
73. Miller, K.J., The application of fracture mechanics to ice problems, in Physics and Mechanics of Ice, ed. P. Tryde, Springer-Verlag, Berlin, pp. 265-277, 1979.
74. Cox, G.F.N. and Weeks, W.F., Equations for determining the gas and brine volume in sea ice samples. CRREL Report 82-30, 7 pp., 1982.
75. Langleben, M.P. and Pounder, E.R., Elastic parameters of sea ice, in Ice and Snow - Properties Processes and Applications, ed. W.D. Kingery, MIT Press, pp. 69-78, 1963.
76. Tabata, T., Studies of the visco-elastic properties of sea ice, in Arctic Sea Ice, U.S. Nat. Acad. Sci. - Nat. Res. Council Pub. 598, pp. 139-147, 1958.
77. Kohnen, H., Seismic and ultrasonic measurements on the sea ice of Eclipse Sound near Pond Inlet, N.W.T. in northern Baffin Island. Polarforschung, Jahrg 4, no. 2, pp. 66-74, 1972.
78. Abele, G. and Frankenstein, G., Snow and ice properties as related to runways in Antarctica. USACRREL Technical Report 176, 37 pp., 1967.
79. Brown, J.H., Elasticity and strength of sea ice, in Ice and Snow: Properties, Processes and Applications, ed. W.D. Kingery, pp. 79-106, MIT Press, 1963.
80. Anderson, D.L., Preliminary results and review of sea ice elasticity and related studies, Trans. Engineering Inst. Canada, vol. 2, no. 3, pp. 116-122, 1958.
81. Traetteberg, A., Gold, L.W. and R. Frederking, The strain rate and temperature dependence on Young's modulus of ice, International Symposium on Ice Problems, IAHR, Hanover, NH, USA, pp. 479-486, 1975.
82. Kovacs, A., Weeks, W.F. and F. Michitti, Variation of some mechanical properties of polar snow, Camp Century, Greenland, CRREL Research Report 276, 33 pp., 1969.
83. Vaudrey, K.D., Ice engineering - study of related properties of floating sea-ice sheets and summary of elastic and viscoelastic analyses, U.S. Naval Civil Engineering Lab. Tech. Rept. R860, 1977.
84. Murat, J-R. and R. Tinawi, Sea ice testing in flexure, Proceedings of POAC 77, St. John's, Newfoundland, vol. 2, pp. 638-653, 1977.
85. Lainey, L. and Tinawi, R., Parametric studies of sea ice beams under short and long term loading, IAHR Internat. Sympos. on Ice, Quebec, Canada, pp. 440-452, 1981.
86. Murat, J.R. and Lainey, L.M., Some experimental observations on the Poisson's ratio of sea ice, Cold Regions Science and Technology, vol. 6, no. 2, pp. 105-113, 1982.
87. Deleted.
88. Wang, Y.S., Uniaxial comparison testing of arctic sea ice, POAC, Quebec, Canada, vol. 1, pp. 346-355, 1981.
89. Mellor, M. and I. Hawkes, Measurement of tensile strength by diametral compression of discs and annuli, Engineering Geology, vol. 5, pp. 173-225, 1971.
90. Wang, Y.S., Crystallographic studies and strength tests of field ice in the Alaskan Beaufort Sea, Proceedings of POAC 79, Trondheim, vol. 1, pp. 651-665, 1979.
91. Schwarz, J., The pressure of floating ice-fields on piles. (In International Association of Hydraulic Research. IAHR Symposium: ice and its actions on hydraulic structures, Reykjavik, Iceland, 8-10 September 1970. Delft, International Association of Hydraulic Research, paper 6.3., 1971.
92. Frederking, R.M.W. and Timco, G.W., NRC ice property measurements during the CANMAR KIGORIAK trials in the Beaufort Sea, winter 1969-80, Nat. Res. Council Canada, Div. Building Res., DRB Paper No. 947, 1981.
93. Wang, Y.S., Sea ice properties, in Technical Seminar on Alaskan Beaufort Gravel Island Design, Exxon, U.S.A., 1979.

94. Dykins, J.E., Ice engineering - tensile properties of sea ice grown in a confined system, U.S. Naval Civil Engineering Lab. Tech. Rept. R689, 1970.
95. Deleted.
96. Butkovich, T.R., On the mechanical properties of sea ice, Thule, Greenland, 1957. U.S. Snow, Ice and Permafrost Research Establishment. Research Report 54, 11 pp. and appendices, 1959.
97. Deleted.
98. Saeki, H. and Ozaki, A., Experimental study on flexural strength and elastic modulus of sea ice. POAC, Quebec, vol. 1, pp. 536-547, 1981.
99. Urabe, N., Iwasaki, T. and Yoshitake, A., Fracture toughness of sea ice. Cold Regions Sci. Tech., vol. 3, no. 1, pp. 29-37, 1980.
100. Urabe, N. and Yoshitake, A., Fracture toughness of sea ice - in situ measurement and its application. POAC, Quebec, Canada, vol. 1, pp. 356-365, 1981.
101. Deleted.
102. Shapiro, L.H., Metzner, R.C. and Johnson, J.B., Fracture toughness of sea ice. Report to Shell Development Company.
103. Panov, V.V. and N.V. Fokeev, Compression strength of sea ice specimens under complex loading, Problemy Arktiki i Antarktiki, vol. 49, pp. 81-86, (English translation pp. 97-104), 1977.
104. Häusler, F.V., Multiaxial compressive strength tests or saline ice with brush-type loading plates, International Symposium on Ice, Quebec, IAHR, pp. 526-539, 1981.
105. Mellor, M., Cox, G.F.N. and H.W. Bosworth, The mechanical properties of multi-year sea ice, Phase I: Techniques for measuring the mechanical properties of ice, USACRREL Report for Shell Development Co., (to be published by CRREL), 1983.
106. Timco, G.W. and Frederking, R.M.W., Compressive strength of multi-year ridge ice, in Workshop on Sea Ice Ridges, Calgary, Alberta, Nat. Res. Council of Canada, 1980.
107. Mellor, M., Engineering properties of snow, Journal of Glaciology, vol. 19, no. 81, pp. 15-66, 1977.
108. Mellor, M., Ship resistance in thick brash ice, Cold Regions Science and Technology, vol. 3, no. 4, pp. 305-321, 1980.
109. Hamza, H. and D.B. Muggeridge, "Plain Strain Fracture Toughness (K_{Ic}) of Fresh Water Ice," POAC 79, vol. 1, pp. 697-707, 1979.

CLOSING REMARKS: PROLOGUE and EPILOGUE

Ned A. Ostenso
National Oceanic and Atmospheric Administation
National Sea Grant College Program

The opportunity to provide the closing remarks for this series of distinguished lectures and the seminar on Arctic Technology and Policy is a triple treat for me. One is because of my long, albeit late, association with Arctic research. Another is to be identified with the Sea Grant program at MIT. And finally to be here among old friends.

But having said that I am faced with capping off an intense session of presentations and discussions spanning the intellectual gamut from "Legal Regime of the Arctic" to "Concrete Structures for the Arctic" to "Tectonics of the Arctic Ocean." In ranging over this seemingly disparate spectrum for a focus to concluding remarks, I am struck by a common thread (other than the obvious Arctic connection) and that is the pioneering nature of all of these fields of endeavor. This, indeed is the nature of oceanography itself.

Most disciplines of human endeavor can be measured by milestones of significant new understanding. In virtually every case, these milestones stretch back in history for centuries, if not millenia; from Aristarchus to Copernicus to Kepler to Herschel in Astronomy; from Hero of Alexandria to Galileo to Newton to Maxwell to Einstein in Physics; from Hippocrates to Harvey to Hooke to Pasteur to Crick and Watson in Medicine; from Hammurabi to Justinian to Marshall in Law.

Oceanography has no such long history, which is its thrill and challenge. To make my point, I will argue that there are seven salient milestones in our understanding of the oceans:

First, is that the oceans in fact exist--that is as deep terrestrial basins versus relative thin laminae of water over an otherwise homogeneous (excepting surficial geology) Earth. Geologists refer to these as the first order features--continents and ocean basins. Even this most primitive understanding is relatively recent. In 1854, Matthew Fontaine Murray published the first bathymetric map of any ocean based upon accumulated records of deep sea soundings in the North Atlantic. Acoustic ranging of the seafloor was developed between the World Wars and the recording bathymeter did not come into common usage until after World War II. Leadtime soundings were tedious and imprecise. Few were made beyond depths to assure safety of navigation. Illustrative of this paucity of data is the first modern chart of the Arctic Ocean Basin published by K. O. Emery in 1949, This chart was based upon only 152 soundings deeper than 915m, or an average of one for each 12,000 square miles. His bathymetric map portrayed a single oceanic depression varying little in detail from that constructed by Nansen in 1904 based upon sounding data from the 1893-96 drift of the FRAM. In 1962, I published what I believe to be the first bathymetric map of the Arctic Ocean that indicated the presence of three trans-basin ridges. In preparing this chart, I had available, either published or by reconstruction from overlaying Soviet drift and high latitude airlifted expedition stations locations over their bathymetric charts, a scant few thousand soundings in the entire ice-covered Arctic other than my own data.

Second, is the awareness that the waters of the ocean have a certain ordered motion, or dynamics. The first described in publication was that of the Gulf Stream by Benjamin Franklin in 1770. The Cromwell Current which is 3,500 miles long, almost as swift and carrying nearly half the flow volume of the Gulf Stream was not discovered until 1952.

Third, is the realization that the ocean basins are not featureless depressions but, rather, that they are geologically complex like their terrestrial counterparts. The first clue to "structure" in an ocean basin came in 1873 when a rise in the middle of the Atlantic was discovered by the British research ship CHALLENGER. However, it was not until 1956 that Maurice Ewing and Bruce Heeze predicted an interconnected global-encircling oceanic ridge system. It required another decade to describe this largest structural feature on the surface of the Earth (being over 40,000 miles in length) in sufficient detail to show its major features of block faulting, central grabens, systematic offsets, etc.

The fourth and fifth intellectual milestones in oceanography were nearly contemporaneous in their inceptions in the mid-1960's. One was the realization that not only did the ocean contain structural geologic features but that the controlling tectonic processes were going on at rates many orders of magnitude faster than the "Uniformitarianism" of geologic orthodoxy could readily comprehend. This grand discovery of seafloor spreading and global tectonism showed the Earth to be constructed of some two-dozen structurally competent plates that are driven by forces, still poorly understood, in differential motion relative to each across the surface of the Earth. Plate tectonics, as this process is now commonly referred to, is a scientific concept of power and beauty rivaling those of Hutton, Darwin and Kepler; yet the simple descriptive phase of this discovery is barely completed and the processes can only be imagined.

The other was the realization that oceanic currents are but a small fraction of the internal dynamics of the ocean, indeed only about 2 percent. In addition, there are great eddies and gyres in the ocean that are analogous to the high and low pressure cells of the atmosphere. Put somewhat poetically, we learned that the ocean like the atmosphere has both climatic and meteorological scales of mass, energy and momentum transfer. The description of these mesoscale eddys in the ocean is incomplete and their physics poorly understood today.

Sixth is the effort to achieve a new sense of legal order and international orderliness through the powerful concept of the oceans being a global commons. The ocean has been the subject of codification since a 1493 Papal Bull divided the Atlantic between Spain and Portugal. This tidy arrangement was upset in 1588 when England and the Netherlands defeated the Spanish armada and "rights" based upon the power to exercise them were replaced by the concept of freedom of the seas as articulated by Hugo Grotius in his 1609 tract Mare Liberum. A subsequent large body of explicit and customary law has developed to define and protect individuals rights on, in, and under the ocean. In 1973, a series of conferences on the Law of the Sea began with the express purpose of codifying not an individual's but rather the world's rights to the ocean under the concept of it being the common heritage of mankind. This is at once one of the most noble and difficult efforts at international legislation since the Antarctic Treaty. Although the complexity of the issues and the diversity and numbers of the participants augers ill for success, the very process of the conventions has set a new tone for human relations. Surely the prospects of new international accords must be viewed as a major resource derived from the ocean.

Seventh, and finally, the recently discovered biological, geochemical, geological, and physical processes at seafloor spreading loci is one of the salient discoveries of this decade, if not the century.

A whole new form of life has been discovered that is completely decoupled from the photosynthetic process. The only comparable life hitherto known to exist on Earth rely on a photosynthetic substrate for their existence. These purely chemosynthetic animals are a biological marvel as alien to our prior understanding as creatures from another world. What promise might they hold for exotic new drugs and pharmaceuticals? Temperature differentials as great as 400°C occur over a span of just a few meters. Through our Ocean Thermal Conversion program, we are spending tens of millions of dollars to capture the energy of a temperature differentiation of only a few degrees over a kilometer! A megawatt of energy is produced from a small fumarole not much larger than this podium. Polymetallic sulfides are being deposited at rates that may exceed human consumption. Are metals a renewable resource!?

The oceans have been viewed as a passive catchbasin accumulating all the wastes, natural and anthropogenic, that wash off the continents. Other than depositional and some bioalteration processes, the ocean's accumulation of such "contaminants" was ever increasing. We now comprehend that along the locus of seafloor spreading massive quantities of material, probably every element in the periodic table, are being injected into the ocean's waters. We must learn to understand the oceans as the great chemical processing plan that it surely must be. The assimilative capacity of the oceans to digest man's waste products in the seafloor, water column and even the atmosphere must be completely reassessed. To date, our uses of the ocean has been largely limited to waging war, a medium of transportation, and a source of food and fiber from a primitive hunting and gathering process. On this intellectually bankrupt foundation, we have tried repeatedly and failed successively to develop a national ocean policy.

Now, with these new discoveries and the application of other scientific advances, such as genetic engineering, microbiology and salt water fermentation the possibility is real that the ocean can be a virtually unlimited cornucopia of metals, energy, protein, pharmaceuticals, industrial feedstocks and other resources.

The point of this effort to summarize your seminar is to emphasize the intellectual adolescence, if not youth, of the ocean physical, biological and social sciences and engineering. In our nascent field, we must not be lulled into the mental tidiness of an Aristotle but rather, like Galileo, to look with wonder and openmindedness for what we see that is new. The diversity of topics discussed here will surely be exceeded by the unanticipated and exciting pathways these basic research, applied technologies and policy considerations will lead mankind. I am honored to be a part of this collegium and am grateful for your participation.

ABOUT THE AUTHORS

Arthur B. Baggeroer is an MIT professor with a joint appointment in the Department of Ocean Engineering and Department of Electrical Engineering and Computer Science. He is also a Guest Investigator at Woods Hole Oceanographic Institution. Professor Baggeroer has undertaken research in adaptive array processing studies, multichannel seismic array design and testing on Georges Bank and dereverberation of deep-water multiples in the East Atlantic Continental Margin Program. He has been part of FRAM II and FRAM IV, two Arctic programs sponsored by the U.S. Office of Naval Research. He has also conducted studies of long-range, low- frequency reverberation in the Canada Basin. At present he is also working on an MIT Sea Grant project investigating acoustic telemetry for untethered, underwater vehicles and sensors. Professor Baggeroer completed his undergraduate studies at Purdue University and received his S.M. and Sc.D. from MIT.

P. Douglas Bruchet is Assistant Project Manager for Petro-Canada's Arctic Pilot Project. He directs the development of policies, programs and strategies for selecting pipeline and terminal sites for liquid natural gas (LNG) and for assessing the environmental, socio-economic and safety effects. He is also responsible for directing the project throughout the various federal and provincial regulatory processes. He serves on the Environmental Advisory Committee on Arctic Marine Transportation which reports to Canada's Ministry of Environment; the committee is currently formulating guidelines for marine transportation throughout the Northwest Passage.

William J. Campbell completed his undergraduate work in physics at the University of Alaska and received his M.S. and Ph.D. in atmospheric physics and oceanography from the University of Washington. He was a Fulbright Scholar at Cambridge University, England where he worked with the Scott Polar Institute. During the past twenty-five years he has done extensive field work in the Arctic and Antarctic. In 1964 Dr. Campbell joined the U.S. Geological Survey and in 1969 became the Chief of its Ice Dynamics Project at the University of Puget Sound, where he also serves as a Research Professor in the Physics Department.

Chryssostomos Chryssostomidis is a Professor in MIT's Department of Ocean Engineering and the Director of the MIT Sea Grant College Program. He is recognized internationally for his research and teaching of ship design and offshore marine operations. Currently, Dr. Chryssostomidis is studying the dynamics of marine risers and is pursuing research in computer aided design and executive systems that have multidisciplinary applications. He received a B.Sc. in naval architecture from the University of Durham at the University of Newcastle upon Tyne, England. He received from MIT an S.M. in naval architecture and marine engineering, the engineers degree in naval architecture, and a Ph.D. in ship systems analysis.

Gregory L. Duckworth is a Ph.D. candidate in a joint MIT and Woods Hole Oceanographic Institution program (expected completion March 1983). His thesis is on the use of data adaptive signal processing algorithms for the analysis of Arctic refraction and long-range propagation data. As a staff member and consultant at MIT's Lincoln Laboratory in 1979 and 1980, he helped to develop and evaluate hardware and programming for passive acoustic location systems. In 1980 and 1982 he undertook experimental work in ocean and seismic acoustics in the Arctic Ocean as a participant in the East Arctic programs. Mr. Duckworth was graduated summa cum laude from Rice University in Houston and received his S.M. from MIT.

F.P. Dunn joined Shell Oil in 1961 and was assigned within the year as a Project Engineer in an offshore production division working on facilities and small platforms. He has been Manager of the Offshore Construction and Design Group for over 13 years. The group designs all Shell Oil platforms and consults on many problems related to offshore and Arctic construction. Mr. Dunn spent five years as an electronic countermeasures officer and navigator in the U.S. Air Force after being graduated from Ohio State University with a B.S. and M.S. in civil engineering.

Ira Dyer, a Professor in MIT's Department of Ocean Engineering, is well known for his underwater acoustics research. Most recently, he has been studying the acoustic properties of the Arctic Ocean, including ambient noise and continental margin backscattering. He leads the Acoustics Sub Panel of the Science Group for the Marginal Ice Zone Experiment (MIZEX). With Professor Arthur Baggeroer of MIT, he has led acoustic studies on the FRAM and CANBARX projects. Before joining the MIT faculty in 1971, Dr. Dyer was Vice President for Bolt Beranek and Newman, Inc. A graduate of the Institute with his S.B., S.M., and Ph.D. in physics, he served from 1973-1975 as Director of the MIT Sea Grant Program and from 1971-1981 as Head of the Ocean Engineering Department. He is a member of the National Academy of Engineering, and has received awards from the Acoustical Society of America, the U.S. Coast Guard, and the IEEE Oceanic Engineering Society.

Ben C. Gerwick, Jr. has been a Professor of Civil Engineering at the University of California, Berkeley since 1971, focusing on construction engineering and management. From 1974-1978, he served as President of Federation Internationale de la Precontrainte (FIP) and from 1979-1980 was Chairman of the Marine Board of the National Research Council. Professor Gerwick has been a consultant on offshore platforms including Ekofisk, Beryl A., Ninian Central, Statfjord platforms A, B, and C, and the Tarsiut Caission-Retained Island. He has worked on conceptual designs for concrete offshore structures in the Beaufort and Bering seas and offshore eastern Canada. He is a graduate of the University of California, Berkeley with a degree in civil engineering.

John G. Gissberg has handled fisheries and natural resource affairs and litigation for the Alaska Attorney General's Office since 1976. He taught at the University of Singapore in 1972 and was in private practice in Anchorage and Tokyo before joining the Alaska Department of Law. A graduate of the College of Fisheries at the University of Washington (B.Sc., 1965), he also has degrees from the University of Michigan (J.D., 1970; Ph.D. in fisheries, 1971). Born and raised in Juneau, Dr. Gissberg is a fourth generation Alaskan resident.

Per Gloersen received his M.S. and Ph.D. in physics from Johns Hopkins University, Baltimore, Maryland after doing undergraduate work there. His thesis and dissertation were on production and analysis of infrared spectra of hydrogen bromide and the molecules of hydrogen and its isotopes. From 1956-1970 he carried out research in plasma physics at the General Electric Space Sciences Laboratory. When he joined the NASA Goddard Space Flight Center in 1970 he directed his research toward obtaining and interpreting microwave spectra of the earth and its atmosphere from sensors located on aircraft and spacecraft. This work has contributed to an improved microwave instrument, the Scanning Multichannel Microwave Radiometer (SMMR), which is currently operating on board the Nimbus-7 satellite. At present he is currently a Senior Scientist at the Goddard Laboratory.

Chris A. Graham received his Bachelor of Science and Master of Engineering degrees in civil engineering and has practiced geotechnical engineering for ten years. Prior to joining Gulf-Canada Resources, Inc. in May 1972, he worked with geothermal and geotechnical industry consultants and aided in the evaluation and design of pipelines and structures on permafrost, artificial islands in the Beaufort Sea, and conventional geotechnical studies in non-permafrost regions. His main responsibilities at Gulf are for geotechnical evaluation for well completions through permafrost, offshore drilling structures, and subsea and onshore pipelines for pre-development studies.

Arthur Grantz joined the U.S. Geological Survey (USGS) in 1949 after graduating from Cornell University with an A.B. in geology. Both his M.S. and Ph.D. were taken at Stanford University where his doctoral dissertation was written on strike slip faults in Alaska. Dr. Grantz, Geologist with USGS, has had extensive experience in geologic mapping and in stratigraphic, aeromagnetic and petroleum resource studies in Alaska. Since 1969 he has been involved in geologic and petroleum resource investigations of the continental shelf and slope in the Chukchi and Beaufort seas.

Donn K. Haglund is Chairman of the University of Wisconsin, Milwaukee Geography Department and Vice President of Comite' Arctique International, the organization of Arctic specialists which sponsors conferences on polar subjects and coordinates international Arctic expeditions and research projects. His research has primarily focused on the cultural and economic geography of the Arctic and sub-arctic regions, with his field experience taking him to those areas almost every year since 1956. Professor Haglund teaches the only university-based Arctic winter field course in the western world. An active participant in a number of professional societies, he has chaired the Committee on Polar Resource Development of the U.S. National Academy of Sciences and recently completed a three-year term as Chairman of the Board of Governors of the Arctic Institute of North America. He received his B.A. in geography and romance languages from Drake University, and his M.A. and Ph.D. in geography and economics from the University of Pennsylvania.

John Lawrence Hargrove is the Director of Studies of the American Society of International Law. He is also the current Director of the Department of Legal Studies of the Johns Hopkins University School of Advanced International Studies. A graduate of Baylor University, Dr. Hargrove holds an L.L.B. from New York University and a Ph.D. from Harvard. He is now serving as a member of the U.S. government Advisory Committee on Antarctic Resources.

William D. Hibler, III has been a Research Physicist with the Cold Regions Research and Engineering Laboratory in Hanover, New Hampshire since 1970 with two interruptions when he was a Visiting Fellow in the Geophysical Fluid Dynamics Program at Princeton University. At the Cold Regions Laboratory his primary research interests have been in large-scale numerical modelling of sea ice and ice covered oceans and lakes; the application of models to ice forecasting and nearshore ice characteristics; and the relationship of sea ice to climatic change. A 1965 graduate of the University of Missouri in physics, Dr. Hibler received his Ph.D. from Cornell University.

Ola M. Johannessen is an Associate Professor at the Geophysical Institute, University of Bergen, Norway, the school from which he was graduated in 1965. From 1966 to 1970 he was on the faculty at McGill University in Montreal working with ice drift problems in the Gulf of St. Lawrence. After four years at SACLANT ASW Research Center in LaSpezia, Italy, he returned to the University of Bergen to work on marginal ice zone problems. In 1981-1982 he held the Office of Naval Research Arctic Chair at the Naval Post Graduate School in Monterey, California where he coordinated the formation of the International Marginal Ice Zone Program (MIZEX). At present Dr. Johannessen is serving as the Chief Scientist for MIZEX.

G. Leonard Johnson has been the Director of the Arctic Program in the U.S. Office of Naval Research since 1980. For five years prior to that assignment he had been the Physical Science Administrator for the Arctic Program and had served in the U.S. Naval Oceanographic Office as Deputy Director of Ocean Floor Analysis Division and Oceanographer. He has been a Lecturer at the University of South Carolina in Columbia and Catholic University in Washington D.C. Dr. Johnson received a B.A. from Williams College and a M.S. from New York University in geology. His drPhil was completed in marine geology at the University of Copenhagen, Denmark.

Ian J. Jordaan as Head of Research and Development for Det norske Veritas (Canada) Ltd. oversees the organization's research in cold climate technology, offshore structures and risk analysis. Before joining this organization in 1982, he was a Professor of civil engineering at the University of Calgary, Alberta. He managed numerous research projects and delivered lectures at the university on statics, dynamics, mechanics, probability theory and materials. As a Research Assistant, Kings College and consultant at Ove Arup & Partners, London, Dr. Jordaan specialized in the design of concrete structures. He received his B.S. and M.S. in engineering from the University of Witwatersrand, Johannesburg, South Africa and his Ph.D. from the University of London, Kings College, England.

Nancy J. Kellner is a Research Associate and Assistant to the Director of Studies of the American Society of International Law. She was graduated from Amherst College and holds an M.A. from the Johns Hopkins School of Advanced International Studies.

Judith Tegger Kildow, Associate Professor of Ocean Policy at MIT, has specialized in analyzing the responses of complex political systems to technological change; her work, which emphasizes the multidisciplinary aspects of marine management, has focused on Law of the Sea, coastal zone and ocean resource management issues, and on the problems of international technology transfer. Professor Kildow, who received her M.A., M.A.L.D. and Ph.D. from The Fletcher School, Tufts University, Boston, joined the MIT faculty in 1973. From 1976-1978 she was a Visiting Associate Professor at the University of California at San Diego in the Department of Political Science and a Researcher at the Scripps Institution of Oceanography.

W. David Kingery joined the MIT faculty in 1951 and became a full Professor in the Department of Materials Science and Engineering in 1962. The graduate education program he established at the Institute has had an influence on ceramic curricula throughout the world. His ceramic science contributions are considered to be seminal in the fields of diffusion, thermal stresses, thermal conductivity, microstructure effects, sintering and liquid-phase sintering, refractory corrosion and grain-boundary phenomena in ceramics. Professor Kingery is the recipient of the Ross Coffin Purdy Award of the American Ceramic Society, the John Jeppson Award, and the Robert Sosman Memorial Lecture Award. He was elected to the National Academy of Engineering in 1975.

Lawrence A. Lawver after receiving a B.S. from Stanford University and a Ph.D. from the University of California, San Diego completed his post-graduate work as a Research Geophysicist at Scripps Institution of Oceanography in 1977. Before coming to MIT in 1980 as a Research Associate, he was a geophysicist with the USGS in their Geothermal Studies Project for two years. At present Dr. Lawver's research interests are in heat flow and tectonics of the South Atlantic and Scotia Sea, as well as the Canada Basin. He is a member of the American Geophysical Union and the Royal Astronomical Society. Dr. Lawver is currently supervising the Tectonics of Southern Oceans Program for Dr. John G. Sclater at MIT.

Samuel V. Maxwell is the Manager of Frontier Development Engineering for Gulf Canada Resources Inc. His responsiblities touch upon all the developmental projects in the frontier regions north and east of Canada. A management expert, Dr. Maxwell has helped plan large- and small-scale projects in areas around the world from the Gulf of Suez to China. As Vice President of Parsons Brinckerhoff, Inc., he had major responsibilities in managing a complete program to design the U.S. Department of Energy Strategic Petroleum Reserve. He holds a B.S. in petroleum engineering from the University of Oklahoma, an M.S. in engineering from the University of Maracaibo, Venezuela, and a Ph.D. in engineering from the University of Salvador.

David McGonigal received his Bachelor of Engineering degree in mechanical engineering in 1975. During six years with Gulf Canada's Frontier Development Division, he has been responsible for field and analytical research projects in the Beaufort Sea on ice characteristics, scour, ice/structure interaction, ice management, and oceanography, and for the application of this information to the design and operation of offshore systems for oil exploration and production.

Linda R. Meinke is a Research Specialist in MIT's Department of Earth and Planetary Sciences. Ms. Meinke received a B.A. in mathematics from the University of California, San Diego and focused her graduate studies on computer science at the University of Washington, Seattle.

Malcolm Mellor is an Engineer and Scientist at the Cold Regions Research and Engineering Laboratory (CRREL) in Hanover, New Hampshire; he is also a private engineering research consultant. At CRREL he serves as a technical advisor for engineering research on snow, ice and frozen ground. His personal research covers material properties, machine design for drilling, excavating, cutting, trenching and tunneling, explosions technology, avalanche protection, and iceberg technology. Dr. Mellor's field work has taken him to Antarctica, Greenland, Alaska, the Canadian arctic, Siberia, Korea, and the Yukon. He received his B.Sc. in civil engineering from Nottingham University, England and his M.Sc. and D.Sc. in applied science from Melbourne University in Australia. His Ph.D. in engineering is from Sheffield University, England.

Ned A. Ostenso is the Acting Assistant Administrator for Research and Development with the National Oceanic and Atmospheric Administration (NOAA) and the Director of the National Sea Grant College Program which is under the aegis of NOAA. An alumnus of the University of Wisconsin, Dr. Ostenso completed his B.S. and M.S. in geology and his Ph.D. in geology and geophysics. He helped to found the Geophysical and Polar Research Center at the University. Dr. Ostenso's extensive research experience in solid-earth and marine geophysics has been honored with his name designating a major

Antarctica mountain and an Arctic Ocean seamount. He has led the U.S. delegations to the United Nations and Buenos Aires and was the Chief U.S. delegate for the Bilateral Agreement between the U.S. and the People's Republic of China in the field of cooperation in marine and fishery scientific technology and with the Soviet Union Bilateral Agreement on World Ocean Studies.

Robert S. Pritchard is an Associate of Flow Industries, Inc. in Kent, Washington. Formerly, he served the firm as Senior Research Scientist and Director of the Geo/Solid Mechanics Department. Prior to that, he had been a Principal Scientist in the Arctic Ice Dynamics Joint Experiment at the University of Washington and Research Engineer at the Eric H. Wang Civil Engineering Research Facility in Albuquerque, New Mexico. He specializes in solid mechanics, applied mathematics, plasticity, numerical analysis and the transfer of technology from mechanics to applied problems in geomechanics. For the past 10 years he has focused on modelling sea ice dynamics. Dr. Pritchard has authored or co-authored over 60 publications and reports, primarily in the areas of ice, soil and rock mechanics. He received his B.S. from Lehigh University and his M.S. and Ph.D. from the University of New Mexico.

Michael R. Robertson is the Manager of Regulatory and Environmental Affairs with Petro-Canada's Arctic Pilot Project (APP) and has been closely involved with the National Energy Board and other hearings that have been held in conjunction with this project. Before joining APP, he was Director of the Canadian Wildlife Service, responsible for western Canada and the Northwest Territories. He also has considerable experience in the governmental side of the regulatory process. His graduate studies were done at the University of Alberta, Canada in limnology and he is a member of several professional biological societies.

R. Tucker Scully is the Director of the Office of Oceans and Polar Affairs in the U.S. Department of State. He joined the Department of State in 1965 and served abroad both in Lebanon and Greece. Since returning from Greece, eight years ago, he has been involved in oceans policy matters. He has served on U.S. delegations to the Law of the Sea Conference and represented the U.S. in a variety of other negotiations relating to ocean and polar matters. Most recently he has acted as U.S. spokesman in negotiations relating to Antarctica and Antarctic resources. He attended the University of Virginia, graduating in 1962 with a B.A. in foreign affairs. He is a member of the Raven Society and Phi Beta Kappa. He undertook graduate work at the University of Virginia in political science and at the University of Rhode Island from which he received a Masters of Marine Affairs Degree in 1975.

Robert E. Smith completed his Ph.D. at the University of Texas with a concentration on structures and soil mechanics. Dr. Smith joined Atlantic Richfield in 1964 in the Research and Development Department where he worked on horizontal fracturing, underwater technology, and offshore technology. In 1966 he moved to the General Engineering staff in Dallas as a structural and soil engineer. Since that time he has been active in offshore design, soil mechanics, and Arctic engineering and environmental studies. Dr. Smith, currently the Director of Civil Engineering for ARCO Oil and Gas Company-Dallas, has been involved in several major Alaskan projects and studies including the Cook Inlet Development, Prudhoe Bay and Kuparuk Fields, Trans-Alaska Pipeline and the Alaska Natural Gas Transmission System.

George P. Vance is the Arctic Technology Development Group Leader for Mobil Research and Development Corporation in Farmers Branch, Texas. He joined Mobil in 1980 following three years as a Research Engineer at the U.S. Army Cold Regions Research and Engineering Laboratory in Hanover, New Hampshire and twenty years with the U.S. Coast Guard, (ten as a Professor of Marine and Ocean Engineering at the U.S. Coast Guard Academy). Dr. Vance is an expert in modeling systems for vessels in ice and has published numerous papers on the subject. A graduate of the Coast Guard Academy, he received a M.S. in naval architecture and marine engineering and nuclear engineering from the University of Michigan. His Ph.D. in ocean engineering is from the University of Rhode Island.

Norbert Untersteiner, Professor of Atmospheric Sciences and Geophysics, Director of the Polar Science Center, Applied Physics Laboratory at the University of Washington, Seattle, is a participant in many national and international cold region efforts. In 1981 he organized and directed a NATO Advanced Study Institute on Air-Sea-Ice Interaction. Prior to that, he developed and directed a large interdisciplinary project, AIDJEX, in the Arctic Ocean. Dr. Untersteiner has served on numerous national and international committees. At present he is on the National Academy of Science Committee on the Polar Regions and Climate Changes and serves as Chairman of the Science Working Group on Passive Microwave Data for Sea Ice Research for NASA. Dr. Untersteiner was graduated from the Realgymnasium, Salzburg, Austria and completed his Ph.D. at the University of Innsbruck, Austria.

Peter Wadhams is Assistant Director of Research, Scott Polar Research Institute, University of Cambridge, England. Educated at Churchill College, Cambridge (B.A. in physics, 1969), he was an assistant to the Senior Scientist in "Hudson-70" expedition of Bedford Institute (first circumnavigation of Americas) from 1969-1970. During 1970-1974 he undertook graduate research on wave-ice interaction at SPRI (Ph.D. 1974). He worked with the Frozen Sea Research Group, Victoria B.C., working on the Beaufort Sea Project, a study of Beaufort Sea ice morphology in 1974 and 1975. And in 1976 he became leader of Sea Ice Group, SPRI, carrying out research on Arctic sea ice thickness profiling, wave-ice interaction, ice edge dynamics and similar sea ice problems. From 1980-1981 he was Visiting Professor of Arctic Marine Science at the Naval Postgraduate School, Monterey, California.

Wilford F. Weeks joined the U.S. Army Cold Regions Research and Engineering Laboratory in Hanover, New Hampshire in 1962 after two years with the Air Force Cambridge Research Center and five years with the Department of Earth Sciences at Washington University. He is currently a Research Glaciologist in the Snow and Ice Branch where his field experience has extended geographically from the Arctic Ocean to Antarctica. He is also an Adjunct Professor at Dartmouth College. Dr. Weeks has been an active participant on many scientific and engineering committees including the Steering Committee on NASA Initiatives Concerning Long-Term

Changes Affecting the Habitability of the Globe, the National Academy of Science Task Force on Arctic Science Policy and the National Petroleum Council's Production Engineering Task Group on Arctic Oil and Gas Resources. He received his B.S. and M.S. at the University of Illinois, and Ph.D. from the University of Chicago in geochemistry.

Paul C. Xirouchakis, an Assistant Professor of Ocean Engineering at MIT, has focused his research on the non-linear response of structures under load, including buckling, creep, and visco-elastic deformations. At present he is working on a Sea Grant project on rational selection of strengthening criteria for navigation in ice. In 1982, Professor Xirouchakis was a guest researcher at Det norske Veritas in Oslo, Norway. He is an associate member of the Society of Naval Architects and Marine Engineers and a member of the Panel MS-9, "Ocean Engineering in Frigid Environments." He received a Diploma in mechanical engineering, N.T.U.A., Athens, Greece, a Diplome D'Ingenieur Grande Ecole, E.N.S.T.A., Paris and a M.S. (1975) and Ph.D. (1978) from MIT.

H.J. Zwally is Head of the Ice Section, Oceans and Ice Branch in NASA's Goddard Laboratory for Atmospheric Sciences. He received his Ph.D. in physics with a minor in mathematics from the University of Maryland. Dr. Zwally has been extensively involved in the application of remote sensing to snow and ice research. He has been the Program Manager for Remote Sensing and Glaciology for the National Science Foundation's Division of Polar Programs and a NASA Study Scientist directing a definition study for an advanced satellite mission for ice, ocean, and climate research. Currently he manages and conducts research in polar oceanography and glaciology including variability of polar ice, ice/ocean/atmosphere interactions, sea ice dynamics, ice sheet and shelf dynamics, associated physical processes and remote sensing of ice and ocean parameters.

AUTHOR INDEX

SUBJECT INDEX